CALORIMETRY IN HIGH ENERGY PHYSICS

Proceedings of the Fifth International Conference on

CALORIMETRY IN HIGH ENERGY PHYSICS

Brookhaven National Laboratory

September 25–October 1, 1994

Editors

Howard A. Gordon

Brookhaven National Laboratory

Doris Rueger

Brookhaven National Laboratory

World Scientific

Singapore • New Jersey • London • Hong Kong

Published by

World Scientific Publishing Co Pte Ltd
5 Toh Tuck Link, Singapore 596224
USA office: 27 Warren Street, Suite 401-402, Hackensack, NJ 07601
UK office: 57 Shelton Street, Covent Garden, London WC2H 9HE

British Library Cataloguing-in-Publication Data
A catalogue record for this book is available from the British Library.

CALORIMETRY IN HIGH ENERGY PHYSICS

ISBN-13 978-981-02-2381-6
ISBN-10 981-02-2381-1

PREFACE

The Fifth International Conference on Calorimetry in High Energy Physics was held September 25 – October 1, 1994 at Brookhaven National Laboratory. Previous conferences were held at Fermilab in 1990, on Capri in 1991, at Corpus Christi in 1992, and on Elba in 1993.

The conference began with three invited overview talks:

- Jacques Colas (LAPP-Annecy) spoke on Advances in Liquid Ionization Calorimeters.
- A Talk on Crystal Calorimeters in Particle Physics by Harvey Newman (CalTech).
- Marvin Weber (LBL) presented Scintillator Materials for Calorimetry.

Following, there was a session of invited talks devoted to the Experience from Large Calorimeter Systems: The D0 Calorimeter by Dean Schamberger (SUNY – Stony Brook), the Performance of the H1 LAr Calorimeter by Alexei Babaev (DESY), the Performance and Understanding of the Zeus Calorimeter by James Crittenden (U. of Bonn) and the KLOE Electromagnetic Calorimeter by Juliet Lee-Franzini (Frascati).

The remainder of the conference was divided into topical sessions of contributed talks:

- Electromagnetic Calorimeters: Crystals, Liquid and Others, Scintillator and Preshower
- Hadronic Calorimetry, including emphasis on forward calorimeters, damage
- Readout and Trigger Electronics
- Optical readout, monitoring and radiation

A detailed schedule is included in this volume. It was clear from the presentations that calorimeters continue to be crucial to a wide variety of high energy and nuclear physics experiments and that much progress is still being made. The challenges of the physics of the LHC (and the defunct SSC), for example, has led to many new concepts for EM, hadron and forward calorimetry.

We were fortunate to have many industrial representatives support the conference and set up displays on Tuesday, Sept. 27, 1994. We would like to thank specifically: Bicron Corporation; EG&G Instruments; Electron Tubes, Inc.; Hamamatsu Corporation; NE Technology, Inc.; Optovac, Inc.; Phillips Photonics; Phillips Scientific; and Ziemer & Associates, Inc.

We also would like to thank the local organizing committee who planned this successful conference: E. Kistenev, R. Majka, V. Radeka, M. Shaevitz, S. White, and C. Woody. We were assisted by a very enthusiastic and efficient secretarial staff at the Conference which tended to all the myriad requests which came up. We would like to thank Carolyn Albert, Rae Bailey, Elizabeth Mogavero, and Jackie Mooney. Finally we would like to acknowledge the efficient support of the BNL Staff Services Division and the Graphics Arts Division, and Service America Corp. who also contributed to this successful meeting.

In closing, those of us who were lucky enough to take the excursion to Montauk will always remember the fine lunch at the Duck Walk Vineyard, the surf casters landing striped bass at Montauk Point, the serendipitous sushi of blue-fin tuna from Captain Michael Potts of the Blue Fin IV and the wonderful dinner and ambiance afterwords at Gosman's.

Howard J. Gordon and Doris Rueger
February 10, 1995

CONTENTS

EM Liquid and Others

Session Chair – Bernard Aubert (LAPP Annecy)

EM Scintillator

Session Chair – Sebastian White (BNL)

EM Preshower

Session Chair – Giorgio Apollinari (Rockefeller)

Hadronic Calorimeter

Session Chair – Konstantin Goulianos (Rockefeller)

Hadronic Forward

Session Chair – Richard Wigmans (Texas Tech. University)

Readout and Trigger

Session Chair – Robert Chase (Orsay)

Optical Readout, Monitoring, Radiation Damage

Session Chair – C. Wuest (LLL)

Overview

Session Chair: Antonio Ereditato (INFN Naples)

LIQUID CALORIMETRY FOR LHC

J. COLAS
LAPP
BP 110, 74941 Annecy-le-vieux, Cedex, France

ABSTRACT

This paper points out some of the main results obtained recently in the field of sampling calorimetry using noble liquids as active media. They are the results of an active R&D programme by the RD3 and GEM collaborations. This work is now being used in the design of the ATLAS detector.

1. Introduction

This paper does not review all aspects of the now very large field of liquid calorimetry. The focus is on the use of this technique at the future large hadron collider LHC. A large R&D effort was launched in Europe (RD3) and also in the US (GEM collaboration) several years ago; this has yielded nice results as shown below. A good fraction of this work is now being used in the design of the ATLAS detector.

At LHC, the new challenge for experimentalists is the very high luminosity of the machine - in excess of 10^{34} $cm^{-2}s^{-1}$ - with consequences on the level of irradiation of the detector and on the amount of pile-up. Noble liquid calorimetry is intrinsically radiation resistant; however, this was less than obvious for the readout electronics. A lot of work has been done and the required performances seems now feasible as reported by M. Citterio[1] at this conference. An other worry was the large amount of pile-up expected if the detector could not be made granular and fast enough. As shown below, the "accordion" design[2] answers successfully these two problems for electromagnetic (em) calorimetry. For hadronic calorimetry where fine segmentation is not required, the concept of electrostatic Transformer (EST) was developed to enable fast signal readout[3]. Recent test beam results are reported.

In a collider experiment, the "accordion" geometry fits well the barrel part of the detector. In the end-cap region it had to evolve into a new, similar geometry : the "Spanish fan".

Calorimeters should not only measure energy but also shower position and direction. A light Higgs particle decaying into two γ's sets very stringent requirements. This has led to the development of very sophisticated preshower detectors. Three different flavors of preshowers have been tested during the summer of '94 and preliminary results have been reported at this conference.

2. The Accordion and EST concepts

Signal formation in liquid ionization calorimeters is simple and well understood[4]. As sketched in figure 1, free charges liberated by an incoming ionizing radiation, start drifting toward the electrodes producing a current with a sharp leading edge. Electrons with their much faster velocity (~ 4.3 mm/μs @ 10kV/cm) dominate this fast signal. The gap geometry defines the longest drift distance and thus the signal duration. For instance, a uniform ionization along the gap gives the characteristic triangular shape shown in figure 1. The large number of drifting electrons (~ 5 10^6 e/GeV) ensures a reproducible signal shape with small statistical fluctuations; the initial current thus contains all the information about the calorimetric energy measurement.

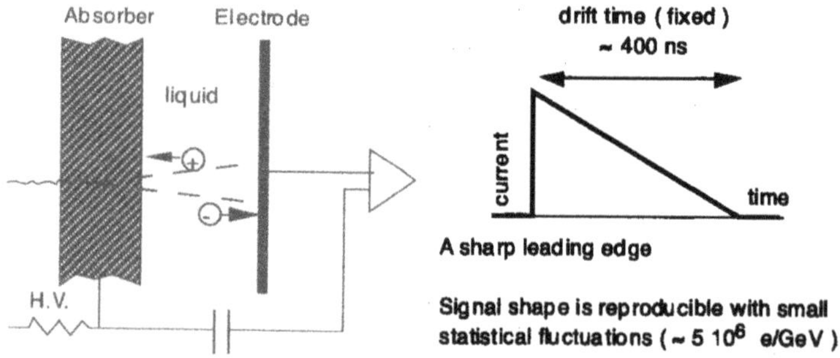

figure 1 : Sketch of the signal formation in liquid ionization calorimeters.

However, the fast rise time of the current signal is easily lost in the readout as explained in figure 2. Inductances in the leads to the preamplifier, detector capacitances, non negligible input impedance of the preamplifier, set a lower limit to the time needed to transfer the signal from the detector to the readout electronics. To achieve a short transfer time, one has to :

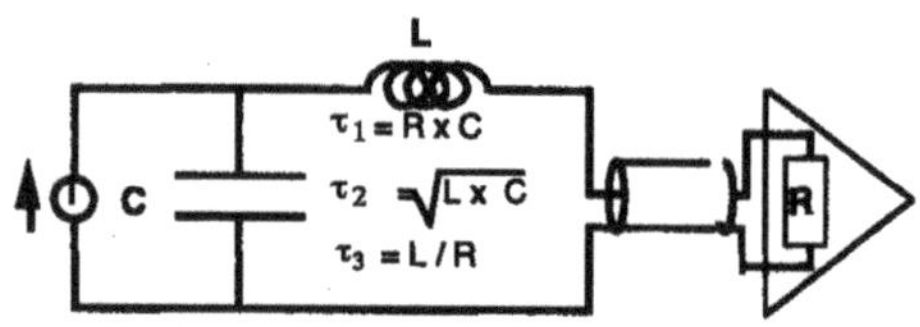

figure 2 : Schematic of the readout electronics showing contributions to the rise-time of the signal.

- reduce the detector capacitances;
- reduce the inductance of the connections;
- reduce the input impedance of the preamplifier.

All three techniques have been used. The high granularity requirement of em calorimetry at LHC implies small cells with relatively small capacitance. This fits the speed requirement. For hadronic calorimetry where only a much coarser granularity is needed, to reduce the detector capacitance, some of the ionization gaps can be connected in series rather than in parallel. This is the EST concept[3].

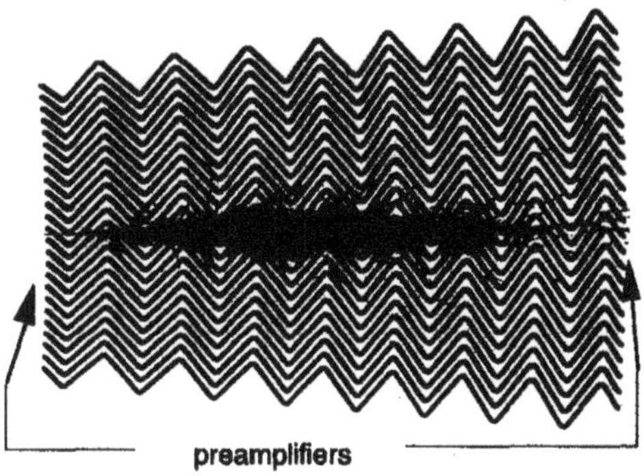

figure 3 : GEANT simulation of an em shower in an accordion geometry. Readout preamplifiers are located at the front and back faces.

The "Accordion" design introduced by D. Fournier solves the problem of connections and related inductances in an elegant way[2]. The electrodes themselves bring the signal to preamplifiers located at the front and back face of the calorimeter in the shortest possible way without introducing cracks which would prevent high granularity (see fig. 3). A good example of the high granularity possible with this geometry is given by the development of integrated preshowers described below.

The development of large FET's with high transconductance enables now preamplifiers with input impedance as low as 10 Ohms. This subject is detailed in the presentation by D. Camin at this conference[5].

3. Performances of "Accordion" EM Calorimeters

3.1 The RD3 and GEM prototypes

To further test the accordion design, the RD3 collaboration built a large scale liquid argon (LAr) em prototype with a fully pointing structure, already conceived as a sector of the ATLAS calorimeter. It extends over 2m along the z axis, which corresponds to the pseudo rapidity range $0 \leq \eta \leq 1.08$, and covers 27° in azimuth ϕ. In the collider reference system adopted here, the z axis is along the LHC beam line while ϕ and θ are the azimuthal and polar angle. Figure 4 shows the detector during assembly. The total thickness is 25 X^0 at $\eta = 0$. The 2m long absorbers are made of 1.8 mm thick lead plates which are sandwiched in-between two 0.2 mm thick stainless steel sheet. At large rapidity ($\eta > 0.7$) the lead thickness is reduced to compensate for the increase in the effective thickness seen by particles hitting at a shallow angle. The electrodes are multi-layered copper-kapton boards separated from the converter plates by a 1.9 mm LAr gap. Three consecutive kapton boards are grouped together at the readout level giving an effective granularity $\Delta\phi \approx 0.020$, while the transverse segmentation $\Delta\eta \approx 0.018$ is obtained by etching projective strips on the conductive layers of the electrodes. More details can be found in reference 6.

figure 4 : Picture of the RD3 "2m" prototype during assembly.

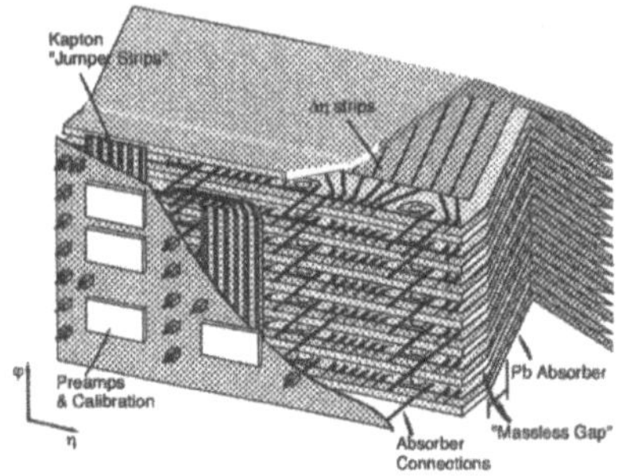

figure 5 : Sketch of the LKr GEM prototype.

The GEM collaboration built and tested during the summer of '94 an "accordion" em prototype filled with liquid krypton (LKr). The detector is rather similar to the one just described, except for its smaller size. But, since liquid krypton has a shorter radiation length ($X^0 \sim 4.8$ cm), the thickness of the lead absorber is reduced to 1 mm. A sketch of the detector is shown on figure 5. Preliminary results have been reported at this conference by D. Lissauer[7].

3.1 Speed and Noise Performances

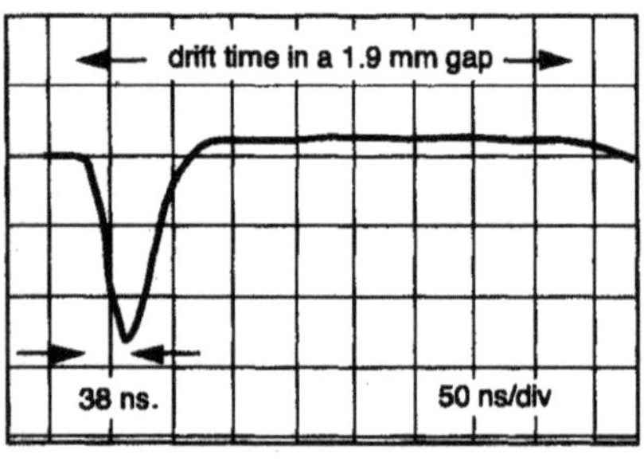

figure 6 : Scope picture of the pulse from a 60 GeV electron hitting the RD3 prototype

Figure 6 shows a scope picture of a pulse from a 60 GeV electron hitting the RD3 em calorimeter prototype. The electronics processing used in this test, differentiates the current pulse and limit the bandwidth to reduce electronics noise. The negative peak corresponds to the leading edge of the current pulse, the positive flat top to the linear decrease of the current. The pulse ends, as expected, after $\sim$ 420 ns corresponding to the maximum electron drift time in a 1.9 mm LAr gap. The pulse is fast enough that the r.m.s. of the noise coming from the pile-up of several events, from successive bunch crossings 25 ns apart, is only 25% higher than the noise from the pile-up of events from a single crossing. The electronics noise for these very fast shaping time is

~ 300 MeV summing over an area large enough to measure the energy of electromagnetic showers. This low figure allows to detect muons above the noise as shown in figure 7.

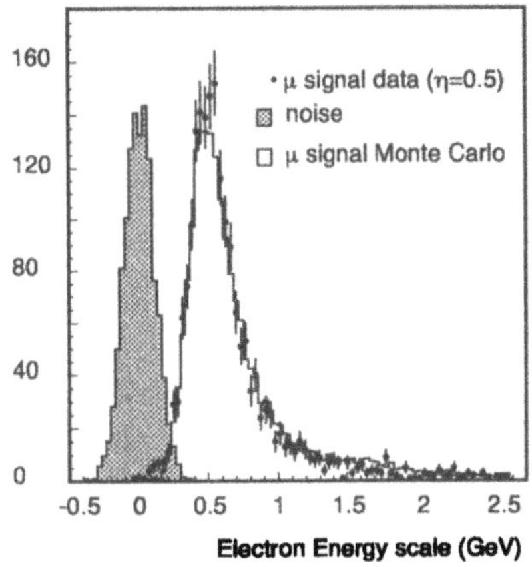

figure 7 : Signal from 300 GeV muons in the RD3 prototype superimposed with the noise.

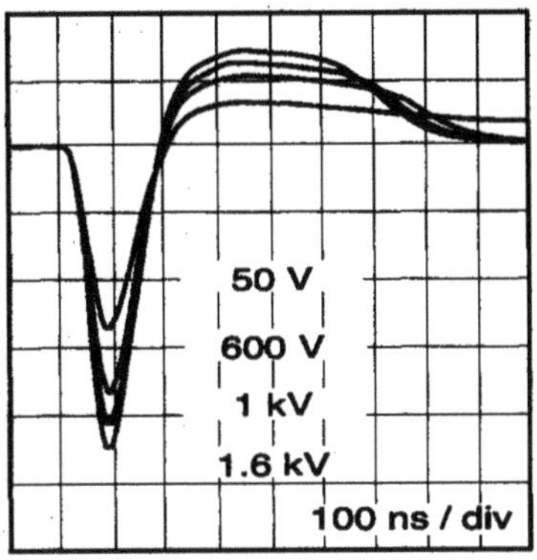

Figure 8 : Pulses from the GEM LKr prototype for various HV settings.

Figure 8 shows wave forms obtained by the GEM collaboration with their krypton prototype. The dependence of signal amplitude and drift time with electric field is moderate. Sampling the pulse with an 18 ns period, its time of arrival is measured with a precision better than 200 ps for an energy above 10 GeV[8].

Results by these two teams clearly prove that the accordion geometry enables fast readout with a noise performance adequate for experimentation at large energy.

3.2 Limits from the geometry : local constant term

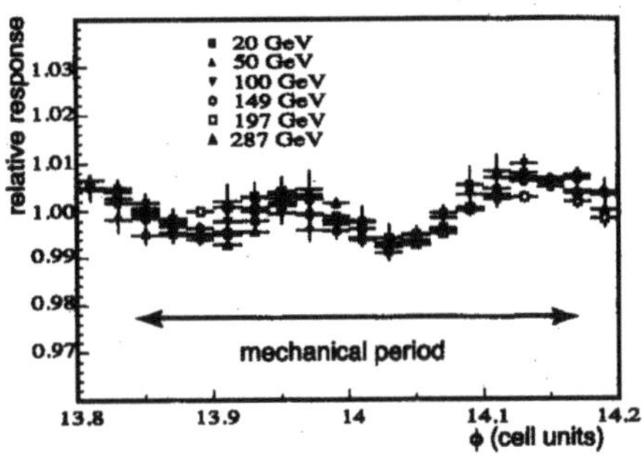

figure 9 : Residual modulation of the energy response as a function of φ, due to the accordion geometry.

Ideally, the sampling fraction of a calorimeter should be independent of the particle impact point. An accordion geometry with acute corners could satisfy this constraint; however in practice bends have a non zero radius of curvature (~ 3 mm). Optimization of the absorber geometry and of their relative position have been performed using detailed GEANT simulations. A small residual variation of the energy response with the azimuthal position was expected due to this purely geometrical effect and was indeed measured by RD3 (fig. 9). This small effect (< ± 1%) can be corrected for using information from the calorimeter alone. This has been done in figure 10 which gives the measured energy resolution as function of the electron energy. The constant term (c) which describes local imperfection is small ~ 0.3%. The "accordion" geometry does not introduce sizable imperfections. The resolution is given for two different impact points with different incidence angles (74° and 44°) corresponding to two absorber thicknesses. The sampling term (a) is similar (~10%) at the two positions. The effect of shallower incidence angle is effectively compensated by a thinner lead absorber.

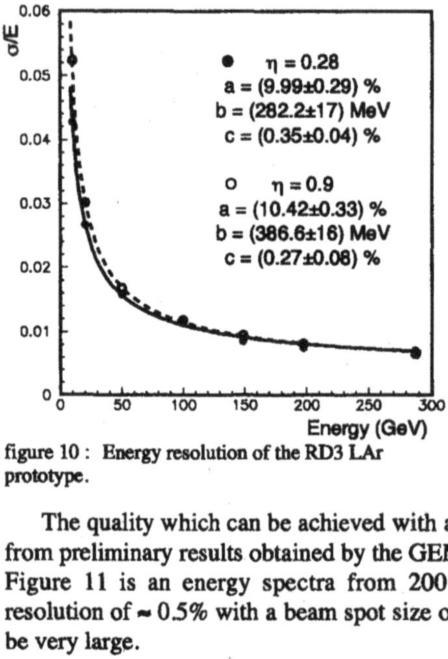

figure 10 : Energy resolution of the RD3 LAr prototype.

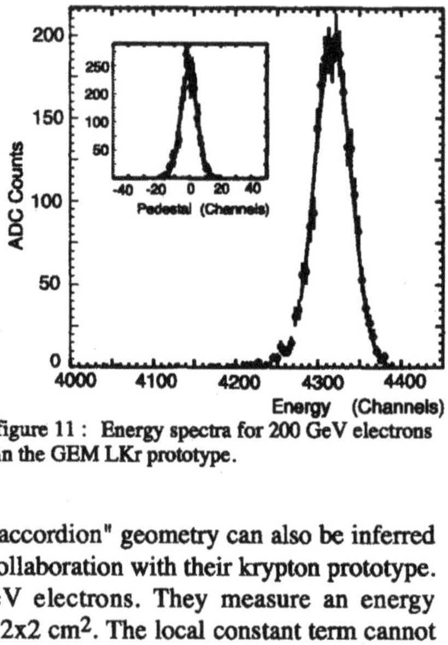

figure 11 : Energy spectra for 200 GeV electrons in the GEM LKr prototype.

The quality which can be achieved with an "accordion" geometry can also be inferred from preliminary results obtained by the GEM collaboration with their krypton prototype. Figure 11 is an energy spectra from 200 GeV electrons. They measure an energy resolution of ~ 0.5% with a beam spot size of ~ 2x2 cm^2. The local constant term cannot be very large.

3.3 Long range uniformity

The peak should not only be narrow, its position should also be independent of the impact point of the particle. Liquid ionization calorimeters with well controlled geometry should satisfy this requirement. The "2m" liquid argon RD3 prototype is already large enough to provide a sensible test. Figures 12 shows the result of a scan with 287 GeV electrons over a ~ $1m^2$ area. About 150000 events enter the plot and are distributed over 123 cells. The measured non uniformity is ~ 0.6% (r.m.s.). The residual ϕ modulation was already discussed; several others terms add to this spread. Figure 13 shows the measured contribution of a variation in the lead thickness. The thicker the absorber, the lower the energy response; yet, the loss in sampling fraction is partially compensated by an increased signal due to a higher electric field from a smaller liquid gap. The table below gives a break down of the contributions to the constant term as estimated by the RD3 collaboration for their prototype.

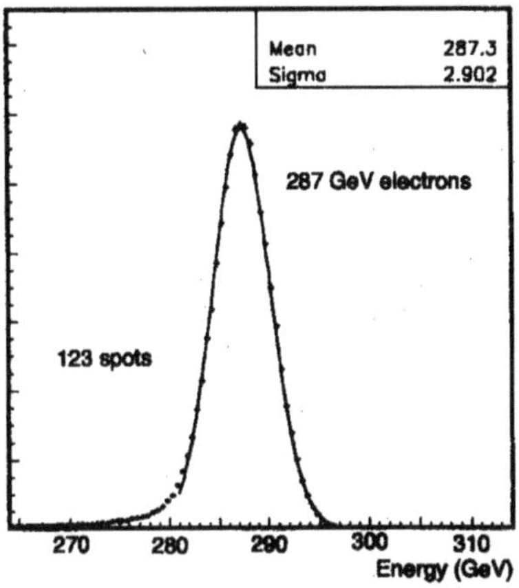

figure 12 : Overall energy spectrum for 287 GeV electrons hitting 123 cells of the RD3 prototype.

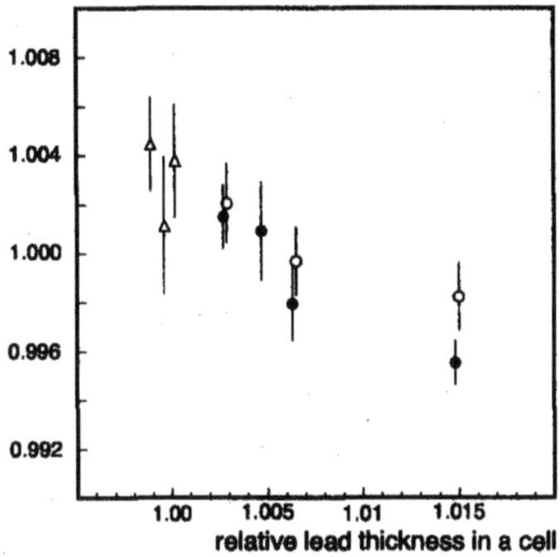

figure 13 : Variation of the energy response as function of the relative absorber thickness.

	Contribution to the constant term (%)	
Mechanics	residual ϕ modulation :	0.3
	absorber thickness :	0.3
	gap non uniformity :	0.15
Electronics	calibration pulse non uniformity :	0.3
	timing precision :	0.2
	readout calibration	0.2
	residual cross-talk	0.15
total ($\oplus$)		0.63

Given the results above, the ATLAS collaboration[9] decided to use the "accordion" design for its em barrel calorimeter. Extensive simulations have been done and are still underway to optimize all the parameters (including the choice of the liquid, argon or krypton) within the constraints imposed by the ATLAS detector (cost and geometry).

The challenge, as for any large calorimeter system, is to keep the "global" constant term at the same or below the low level obtained with prototypes.

4. EST and hadronic calorimetry

The segmentation of hadronic calorimeters is much coarser than for em detectors. With the usual ganging scheme of liquid gaps, this leads to large capacitances (tens of nanofarads) and as explained in §1 to a slow response. In the EST concept, as shown in figure 14, a fraction of the ionization gaps are connected in series reducing the capacitance seen by the preamplifier. This is achieved with a suitable orientation of the electric field in the gaps. This concept can be merged with the "accordion" design and the RD3 collaboration built and tested such a prototype. Results obtained in test beams are discussed more fully by J. Schwindling at this conference[10].

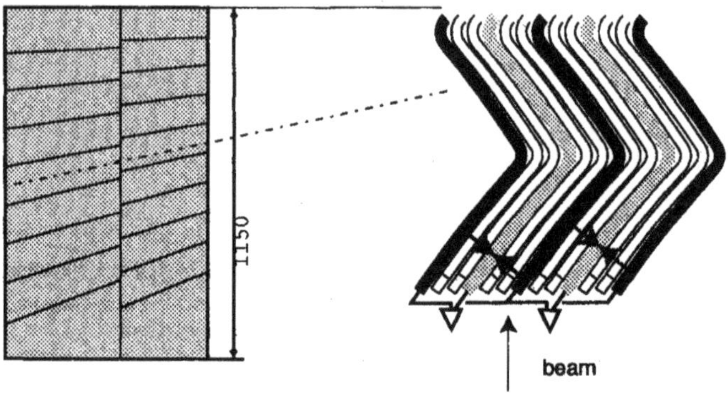

figure 14 : Schematic representation of the RD3 hadronic calorimeter prototype using the "accordion" and EST concept.

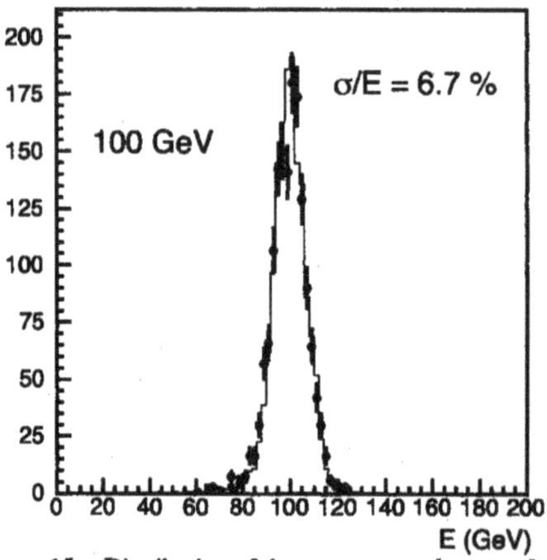

figure 15: Distribution of the reconstructed energy for 100 GeV pions. The histogram represents the data, dots are the simulation results.

One potential problem with this technique is an increase cross-talk between cells adjacent in rapidity. Indeed most of the absorbers are cut into tiles along η to form calorimeter cells (see fig. 14) and cross-talk capacitances are formed between the (1 cm) thick absorber edges. To reduce this coupling, tiles are spaced by a 5 mm gap. This cross-talk induces some charge delocalisation. Nevertheless, according to simulation, these effects are small compared to intrinsic shower fluctuations. This has been confirmed experimentally and, as shown in figure 15, satisfactory energy resolution is obtained after applying a weighting method as pioneered by the H1 collaboration[11]. Note the good agreement between data and simulation. The simulation program is the GEANT-GHEISHA package with all cross-talk effects added.

This work has confirmed the validity of the EST concept. The price to pay is an increase in the cost of the mechanics : profiling and cutting the absorber in tiles. As for experimentation at LHC the requirements on hadronic calorimetry are less stringent, the ATLAS experiment, decided to limit the use of LAr hadronic calorimetry to the end-cap region where radiation resistance is a must. The detector concept is sketched in figure 16. The calorimeter has four readout segments in depth and is about 10 λ thick. The granularity is about $\Delta\eta \times \Delta\varphi = 0.1 \times 0.1$. To reduce the cost of the mechanics, absorbers are flat 25 mm thick copper plates, not cut in tiles. The large (8mm) argon gap is readout by thin tiles forming an EST[9]. Several preamps with a summed output are used to readout a single calorimeter cell.

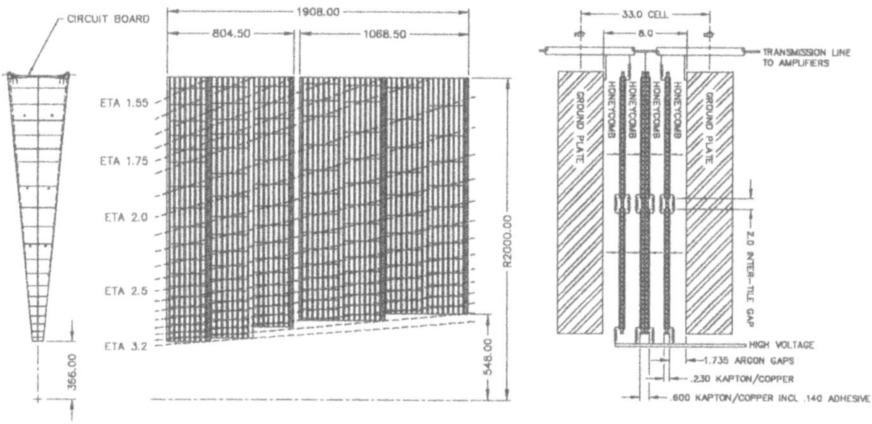

figure 16 : R-ϕ view and R-Z view of one module of the ATLAS LAr hadronic end-cap. On the right, detail of the electrode structure.

5. From "Accordion" to "Spanish-Fan"

The "accordion" geometry fits well the barrel calorimeter of a collider experiment. Azimuthal symmetry and signal readout at the front and back faces of the detector are the main characteristics one would like to keep in an end-cap geometry. This lead to the "Spanish fan" design illustrated in figure 17.

As for the "accordion", bent absorber plates zig-zag along the line of flight of the incident particles. To match an end-cap geometry, the wave amplitude and folding angle

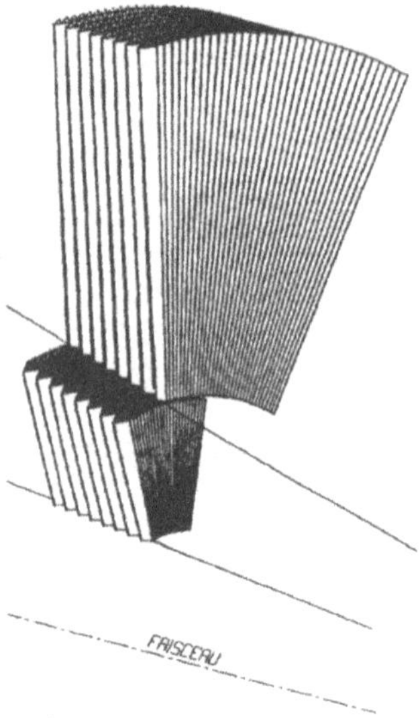

figure 17 : Artistic view of a "Spanish fan" end-cap em detector.

increase with the distance to the beam line (R). To stay within practical limits, the large rapidity span of the end-cap ($1.4 < \eta < 3.2$) is covered by two concentric donuts ($1.4 < \eta < 2.4$ and $2.4 < \eta < 3.2$). The absorber thickness and the liquid argon gap vary with R in order to keep the sampling fraction and calorimeter density constant over the whole area of the detector. The electric field in the gap is kept constant increasing high voltage as a function of the radius.

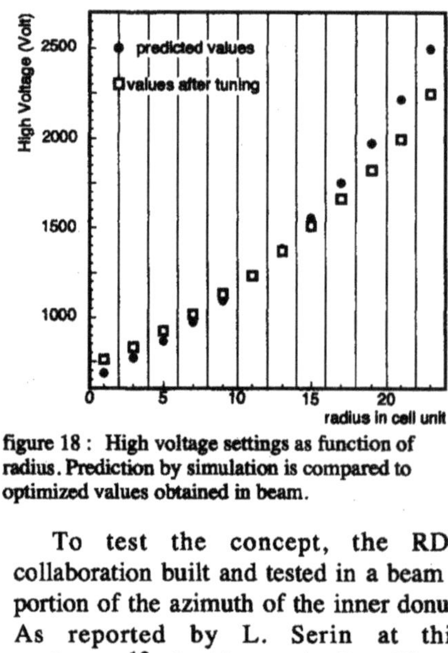

figure 18 : High voltage settings as function of radius. Prediction by simulation is compared to optimized values obtained in beam.

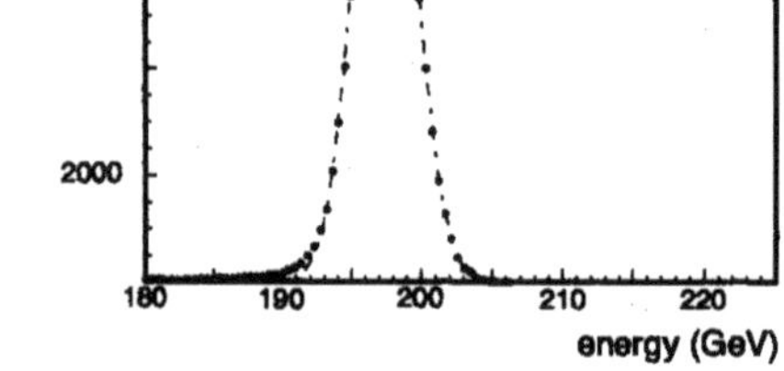

figure 19 : Energy spectrum for 197 GeV electrons hitting 42 cells of the "Spanish fan" prototype.

To test the concept, the RD3 collaboration built and tested in a beam a portion of the azimuth of the inner donut. As reported by L. Serin at this conference[12], the idea works fine. Figure 18 compares the high voltage settings expected from simulation and the values tuned in the beam. The small difference can be entirely attributed to our limited knowledge of the drift velocity as a function of the applied electric field. After correction for the response as a function of rapidity, a first notion of the uniformity of the detector is given in figure 19 which shows the result of a scan of 3 lines at constant ϕ with 197 GeV electrons. The results are quite promising.

6. Preshowers

The capability to detect a possible light Higgs particle decaying into two photons puts severe constraints on the required calorimeter performances :

- the calorimeter energy resolution must be excellent over a large acceptance;
- photon directions must be measured to better than 40 mrad/$\sqrt{E}$;
- photons from π^0 decays must be rejected. A γ/jet rejection factor of 10^4 is required.

In addition, a good measurement of the average ϕ for an em shower has been shown to be very useful in reconstructing the electron momentum, degraded by bremsstrahlung, by combining the inner tracker information with the shower measurement.

A particular concern for noble liquid calorimetry is the amount of dead material upstream of the active part of the detector (cryostat walls,...) which becomes non negligible at large rapidity. A thin presampler layer just after the cryostat wall could be used to correct for this energy loss. For this purpose, but also to accurately measure the shower position and to be able to reject π^0's, the RD3 collaboration has developed a separate detector which sits in front of the calorimeter. This detector has two layers, highly segmented to give a good accuracy on the shower position in the two views ϕ and η. The first layer (ϕ) is after $2X^0$ of material. A 1 X^0 thick lead absorber sheet is inserted between the two preshower layers to increase the γ conversion efficiency. More information can be found in the report by J. Collot at this conference[13].

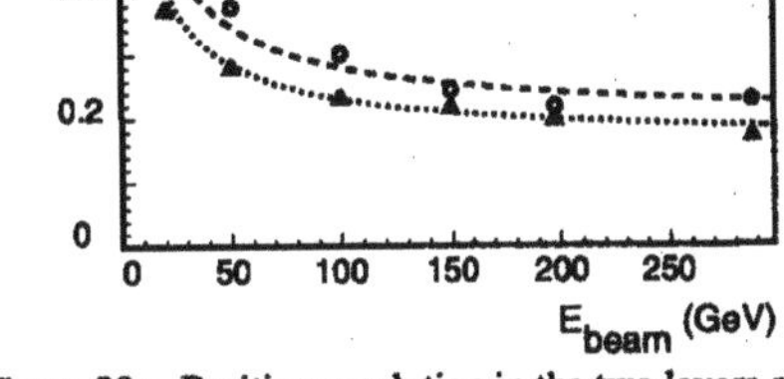

figure 20 : Position resolution in the two layers of the RD3 preshower prototype.

Preliminary test beam results obtained with this detector are reported on figure 20. After correction for a standard S-shape response and unfolding of the beam chambers resolution, a space resolution better than 300μ is measured for energies above 100 GeV. The better resolution in the η layer is understood as coming from a more developed shower deeper in the detector ($3X^0$).

An other approach was pushed forward by the GEM collaboration who proposed to highly segment the first layer of the calorimeter drawing 4.7 mm wide strips on the kapton electrodes[7]. This measures the η coordinate of the shower. To keep the number of channel under control, more electrodes are ganged together in ϕ (see fig. 5). To compensate for the energy lost upstream, lead is removed at the beginning of the absorber. This form a region (massless gap) with an increased sampling fraction. Its depth is optimized by simulation. Figure 21 shows preliminary results on the measured space resolution. The signal is mostly contained in a single strip. This is nice for π^0 rejection but somewhat limits the space resolution at high energy.

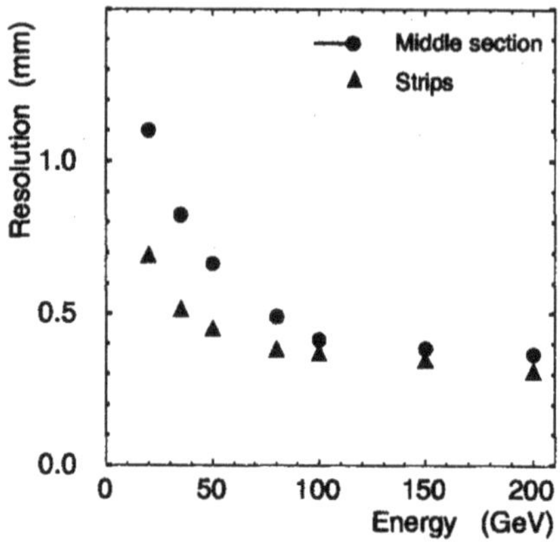

figure 21 : Space resolution obtained with the GEM LKr prototype.

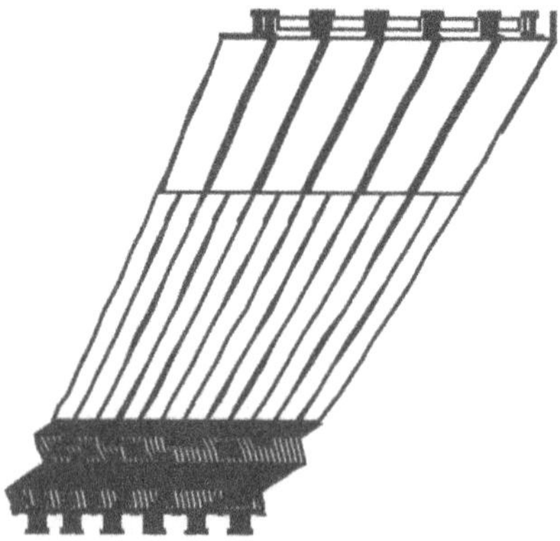

figure 22 : Kapton electrode with U strips in the first layer. Once bent in an accordion shape, the strips are at constant U.

Following similar concepts, but with a different drawing of the strips on the kapton electrode (fig. 22), the RD3 collaboration built a prototype of an integrated preshower measuring two views $u = \eta + \phi$ and $v = \eta - \phi$. The strip width is ~ 5 mm. U and V electrodes are interleaved when stacking the module. The strip layer is ~ 5.6 X^0 deep. Figure 23 gives the measured profile of a 200 GeV electron. A very narrow core sits on a broader distribution. The measured position resolution in the two directions η and ϕ is better than 300 μm above 100 GeV. Using the position measured in the strip and middle layer and the corresponding lever arm as estimated by simulation (~ 10 cm) the shower direction is measured. The angular resolution is below 6 mrad above 100 GeV (see fig. 24). Refer to the report by I. Wingerter-Seez at this conference[14] for more information.

Using these results and simulation of the actual ATLAS geometry, this collaboration has decided to use a combination of these approaches in its barrel em calorimeter. An independent detector, based on the technique developed for the RD3 preshower will provide an independent measurement of the energy lost in the dead material upstream of the active part of the calorimeter. This loss can then be corrected for optimally at each energy. The first layer of the calorimeter will be finely segmented with strips (either η or UV) to provide π^0 rejection and angular measurement.

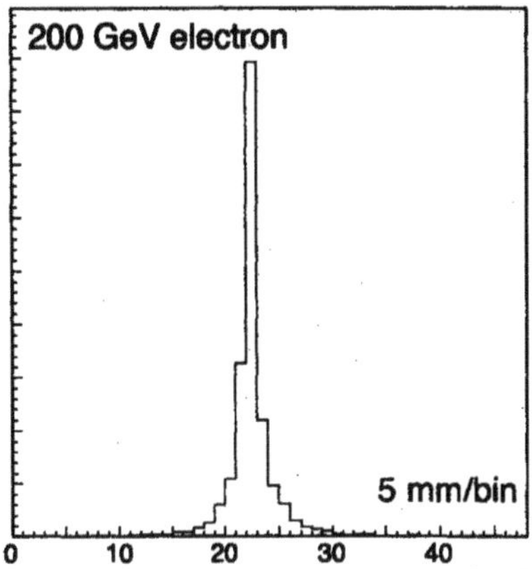

figure 23: Typical shower profile of a 200 GeV electron in the UV preshower prototype.

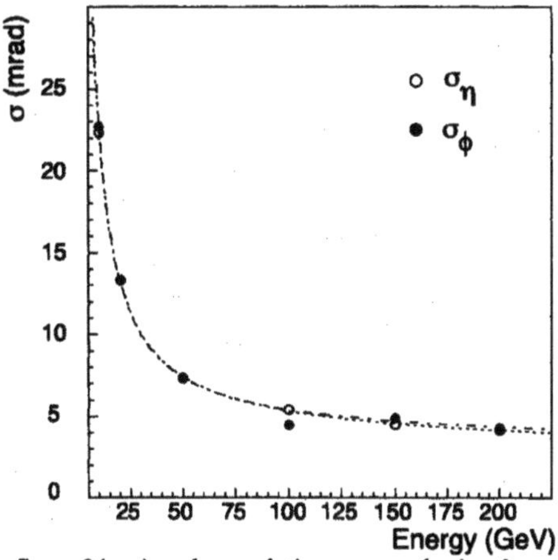

figure 24 : Angular resolution measured using the strip and middle layers of the RD3 UV prototype as a function of the electron energy.

7. Conclusions

R&D on the use of noble liquid calorimetry at the future LHC has been very successful. The readout speed and the granularity of the detector have been vastly improved without spoiling the good properties of liquid ionization calorimetry : uniformity and time stability of the response. In addition rad-hard, low noise electronics have become available. All these results are being used in the design of the calorimeters of the ATLAS experiment. Now, the challenge is to built these large and complex objects even improving over the performances obtained with prototypes.

References

1 M. Citterio, 5th Int. Conf. on Calorimetry in HEP, Brookhaven, '94

2 B.Aubert et al. (RD3 collaboration), Nucl. Inst. and Meth. **A309** (1991) 438
B.Aubert et al. (RD3 collaboration), Nucl. Inst. and Meth. **A321** (1992) 467
B.Aubert et al. (RD3 collaboration), Nucl. Inst. and Meth. **A325** (1993) 118

3 J. Colas et al., Nucl. Inst. and Meth. **A325** (1993) 118

4 V. Radeka & S. Rescia, Nucl. Inst. and Meth. **A294**(1990) 583

5 D. Damin, 5th Int. Conf. on Calorimetry in HEP, Brookhaven, '94

6 D.M. Gingrich et al. to be published in Nucl. Inst. and Meth.

7 D. Lissauer, 5th Int. Conf. on Calorimetry in HEP, Brookhaven, '94

8 O. Benary et al. report BNL-60279, to be published in Nucl. Inst. and Meth.
9 ATLAS Technical Proposal, CERN/LHCC/94-43
10 J. Schwindling, 5th Int. Conf. on Calorimetry in HEP, Brookhaven, '94
D.M. Gingrich et al. to be published in Nucl. Inst. and Meth.
11 W. Braunschweig et al. (H1 calorimeter group)
Nucl. Inst. and Meth. **A265** (1988) 419, Nucl. Inst. and Meth. **A275** (1989) 246
B. Andrieu et al.(H1 calorimeter group), report DESY 93-047
12 L.Serin, 5th Int. Conf. on Calorimetry in HEP, Brookhaven, '94
13 J. Collot, 5th Int. Conf. on Calorimetry in HEP, Brookhaven, '94
14 I. Wingerter-Seez, 5th Int. Conf. on Calorimetry in HEP, Brookhaven, '94

SCINTILLATOR MATERIALS FOR CALORIMETRY

MARVIN J. WEBER
Lawrence Berkeley Laboratory
University of California
Berkeley, CA 94720 USA

ABSTRACT

Requirements for fast, dense scintillator materials for calorimetry in high energy physics and approaches to satisfying these requirements are reviewed with respect to possible hosts and luminescent species. Special attention to given to cerium-activated crystals, core-valence luminescence, and glass scintillators. The present state of the art, limitations, and suggestions for possible new scintillator materials are presented.

1. Introduction

The past decade has witnessed a veritable renaissance in research and development of scintillator materials, prompted to a major degree by the need for scintillators for precision calorimetry in high energy physics, but also by the needs for high light output scintillators for medical imaging, geophysical exploration, and numerous other scientific and industrial applications. Improved experimental techiques for studying scintillator materials and more knowledgeable, systematic surveys have led to the development and better understanding of many promising scintillator materials for calorimetry and numerous other applications. The current state of the art and progress in the search for improved scintillator materials is well documented in the published proceedings of a recent workshop[1] and a symposium[2] devoted to scintillator materials.

Scintillators may be in the form of crystals, glasses, liquids, and gases and composed of organic and inorganic materials.[3] Here we restrict consideration to inorganic solids. A history of the discovery of important inorganic scintillator materials—important in the sense that they either became commercially available and widely used (or have the potential of becoming so in the case of recently discovered materials) or triggered further developments or new research directions—is shown in Fig. 1. The century of discovery may be divided into three phases. The first phase included the earliest scintillators: $CaWO_4$ first used in the year following Roentgen's discovery of x-rays, uranyl salts used by Becquerel in 1896 to discover radioactivity, and ZnS used by Crookes to detect and count radioactivity and by Rutherford to study alpha particle scattering. This period of visual scintillation counting ended with the development of photomultiplier tube and the discovery of scintillation in naphthalene and led in mid-century to the second phase and Hofstadter's discovery of NaI(Tl). In a burst of exploration during the next few years, the scintillation properties of most pure and activated alkali halide crystals were investigated. Lithium containing compounds to detect neutrons and the first glass scintillators were also developed in the

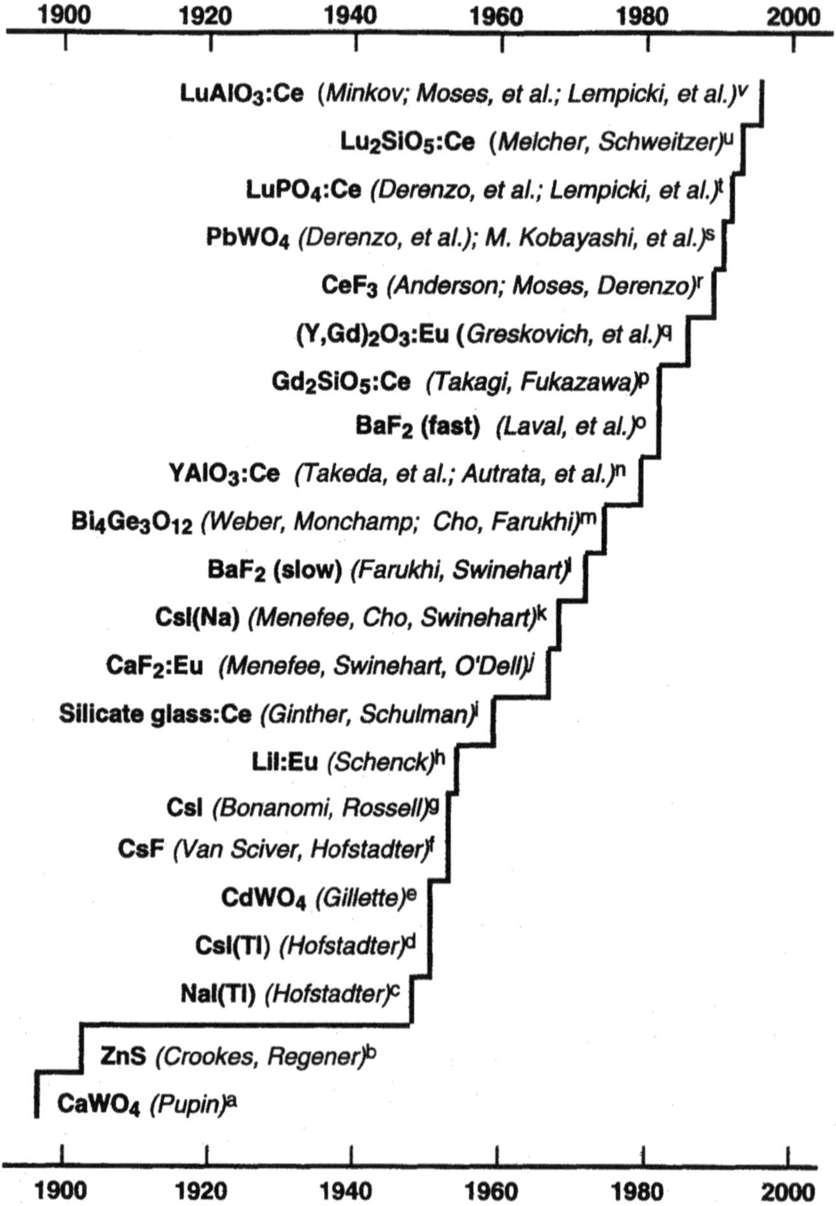

Figure 1. Discovery of important inorganic scintillator materials.

1950s. A steady precession of material discoveries followed leading to the third phase and the explosive growth of activity during the past decade. These discoveries and improvements in scintillator material benefited throughout by concurrent research on photoluminescence, cathodoluminescence, and x-ray phosphors.

Current scintillator materials of choice for calorimetry and their properties are reviewed in several articles in references 1 and 2 and in many papers in this volume and will not be repeated here. Instead we address the following questions. Are there still better scintillators to be found? What do we mean by better? How much better could scintillators be? Have we exhausted the Periodic Table? We do this by considering the requirements for calorimetry, the processes governing scintillation efficiency, and various luminescent species and host compounds that may have the potential for satisfying the requirements of high energy physics calorimetry.

2. Properties of Scintillator Materials

2.1 Requirements

Today we have a large number of well-characterized scintillator materials including crystals, glasses, plastics, and liquids. No single material is superior for all applications; in most cases improvements in one or more properties are desirable. Properties that one must keep in mind in selecting a material for a specific application are shown in Fig. 2. The material may be a crystal or glass in bulk, fiber, or sheet form. The emission wavelength and the light yield will determine the best photodetector to use. Fast signal rise and decay times are important for timing and high event rate applications or time-of-flight (TOF) modes of operation; the absence of afterglow is important in medical imaging. Stability includes several factors that must be known or controlled: environmental or chemical durability, ruggedness and mechanical shock resistant, and variation of light output with temperature and time. Radiation damage, which may be irrelevant for detectors for most imaging applications, is extremely important in high radiation environments such as in detectors for use with high luminosity colliders. High density and stopping power (i.e., large effective atomic number) are important for reducing the amount of material needed. Whereas for detecting very energetic particles light yield may not be too critical, for applications where the particle energy is smaller or fixed, e.g., in positron emission tomography, increased light yield is important for improving accuracy and spatial resolution. Energy resolution is also dependent on light yield. The price of raw materials, the method of preparing the scintillator material, and fabricating the material into the desired size and shape all enter into the final cost. These and other factors not included in Fig. 2 (we rarely have a lobotomy) differ in their relative importance in selecting materials for a particular application.

Some specific requirements for calorimetry and other applications are compared in Table 1. High density and high Z materials are of almost universal interest because increased stopping power reduces the amount of scintillator material needed. For shower containment, the radiation length and the Moliere radius, which are proportional to Z, are of special interest.[5] High energy physics (HEP) calorimetry differed from low- or intermediate energy calorimetry and most other applications in the requirements for (1) light

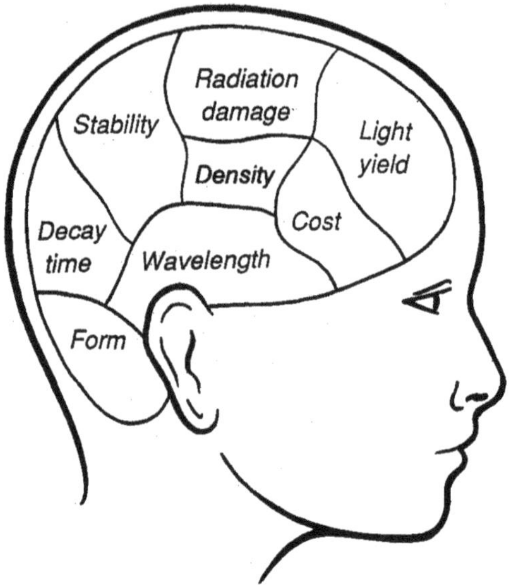

Figure 2. Properties of scintillators to be considered when selecting materials.

yield, (2) decay time, and (3), for high-luminosity, high-intensity colliders, radiation hardness. Because of the high energies of the incoming radiation, which may be 100 GeV or more, light yields in terms of luminescent photons per MeV can be very modest. In general an output of >200 photons/MeV (about 0.5% of that of NaI(Tl)) is sufficient to provide adequate accuracy and energy resolution. The light yield also affects the type of photodetector required (e.g., photomultiplier tube, photodiode, avalanche photodiode). The decay time must be short and consistent with the event rate. For precision calorimetry, uniformity of light output is essential. Therefore accumulated radiation damage, which appears as induced optical absorption bands that absorb part of the scintillation light, must be controlled.

TABLE 1 Requirements for various scintillator applications (adapted from ref. 4).

Application	Density (g/cm^3)	Z	Light yield (phot./MeV)	Decay time (ns)	Radiation hardness (Mrad)
HEP calorimetry	high	high	>200	<20	>10
Low–int. energy calor.	high	high	high	varies	+/-
Nuclear physics	high	high	high	varies	-
Astrophysics	high	high/low	high	less import.	-
Medical imaging	high	high	high	≤ 1 (TOF)	-
Industry	high	high	high	less import.	-
Neutrons	high	B,Li,Gd	high	10 – 100	-

With present and proposed detectors for high energy physics experiments having grown to gargantuan dimensions requiring thousands of crystals with quantities of materials measured in cubic meters, the cost of the scintillator materials has become a major concern.[6,7] Low- or moderate-cost crystal production methods such as Bridgman or Czochralski growth are highly desirable. Glass, because of its low-cost, large-volume production, is attractive for scintillator materials but has other limitations that are discussed later.

Below we consider possible luminescent species and host materials that may provide the high density, fast decay, and light output required for HEP calorimetry. Although radiation hardness is a prime consideration, because of space limitations it will not be discussed here (the reader is referred to papers in references 1 and 2 for details of this issue).

2.2 Scintillation Efficiency

The basic processes in scintillation may be divided into three stages:[8-11] (i) the absorption of the incident radiation or particle by the host and conversion of the energy into thermalized electrons and holes, (ii) transfer of some fraction of the electron and hole excitation to luminescence centers, and (iii) the luminescence process. The quantum efficiency of the scintillation process is given by

$$\eta = \beta \bullet S \bullet Q, \tag{1}$$

where β is the conversion efficiency for creating electron-hole pairs, S is the transfer efficiency, and Q is the radiative efficiency of the luminescence centers. Because the incident energy E_{in} of a particle will usually be very much bigger than the band gap E_g of the material, the number of electron-hole pairs and resultant scintillation photons may be very large, thus yielding huge quantum efficiencies. In terms of energy efficiency, however, the performance of scintillators is less impressive. For a scintillation photon of energy E_s, this efficiency is given by

$$\eta(E_s/E_{in}) \sim (E_{in}/\alpha E_g)\, S \bullet Q\, (E_s/E_{in}) = (E_s/\alpha E_g)\, S \bullet Q\,. \tag{2}$$

In Eq. (2) it is assumed that it takes on the average an energy α times the band gap to create a thermalized electron-hole pair. Various treatments of polaron and plasmon models and electron-phonon scattering losses have shown that α is typically about 2–3 for semiconductor and insulator materials (see Ref. 12 for an good review of this problem). Thus from Eq.(2), for a material having transfer and luminescence efficiencies S and Q of unity and a scintillation photon energy approaching that of the band gap, the energy efficiency should be ~25–30%, which is about what has been obtained for the best phosphor materials. For one of the best scintillators, NaI(Tl), E_s is equal to approximately $E_g/2$ and the reported efficiency is 12%, therefore S and Q must again be near unity. CsI(Tl) has an even higher efficiency than NaI(Tl).[13] Thus scintillator materials already exist with near the maximum achievable efficiency and only small improvements in light yield are possible. For HEP calorimetry, however, light yield is usually not the main issue. Rather, a fast decay time, good stopping power for shower containment, radiation hardness, and reasonable cost are the principal concerns for the scintillator material.

2.3 Luminescent Species

Table 2 summarizes the wide variety of luminescent species possible for inorganic scintillators. The luminescence of many of these, such as transition metal ions, f–f transitions of lanthanides ions, and filled shell ions, involve transitions of varying degrees of forbiddeness and hence are generally slow (>1 μs). Exciton decay rates may be fast but vary widely with exciton type and host. To ensure the requirement of fast decay, we consider luminescence involving only allowed electric-dipole transitions of Ce^{3+} and core-valence transitions.

TABLE 2. Inorganic luminescent species.

Type	Transition	Examples
Transition metal	nd – nd	Ti^{3+}, Cr^{3+}, . . . Cu^{2+}, Mo^{3+}
Lanthanide/actinide	nf – nf	Pr^{3+}, Nd^{3+}, . . . Yb^{3}, U^{3+}
Lanthanide/actinide	nf – (n+1)d	Ce^{3+}, Eu^{2+}
Filled shell	nd^{10} – (n+1)s	Cu^{+}, Ag^{+}
Post-transition group	ns^{2} – nsnp	Sn^{2+}, Sb^{3+}, Tl^{+}, Pb^{2+}, Bi^{3+}
Molecular complex	charge transfer	WO_4^{2-}, TaO_4^{3-}
Exciton	e – h recombination	CsI, BaF_2 (STE)
Core–valence	np(cation) – n'p(anion)	BaF_2, CsF

The dependence of scintillation on temperature is a further consideration for precision calorimetry. At room temperature (T_R) the luminescence and decay of Ce^{3+} and core-valence transitions exhibit little variation with temperature. In contrast, most excitonic, post-transition-group ion, and molecular complex luminescence processes are temperature dependent and the scintillation yield may vary by more than 1%/K, thus requiring a temperature stabilized environment for reliable accurate calibration.

The decay times and emission intensities from molecular complexes and post transition group elements depend on the degree of thermal quenching and or changing level populations with temperature and can vary greatly with the structure and chemical composition of the host material. This accounts for the large variation in scintillation yields of different tungstate, bismuth, and lead compounds. The decay time of such materials can be reduced, if necessary for calorimetry, by increasing the temperatue to increase the rate of nonradiative decay but with a concomitant loss of light output. The scintillation of $PbWO_4$ is an example of a material that naturally exhibits the features of fast decay and low light output at room temperature and is a candidate for a planned HEP detector.[7]

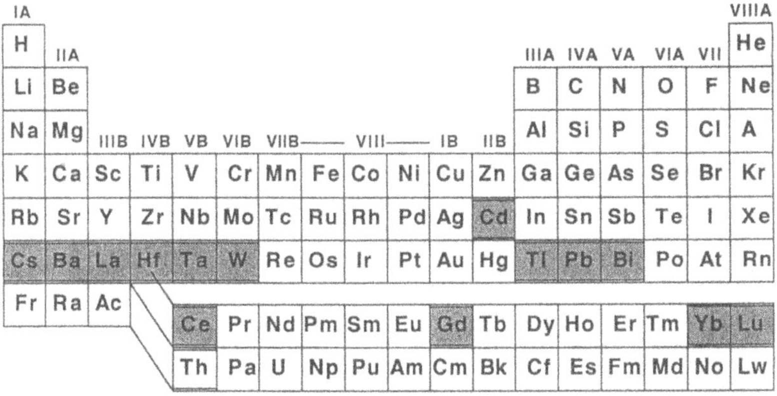

Figure 3. Cations used in dense scintillator materials.

2.4 Host Materials

Cations that have been exploited for dense scintillator materials are indicated in Fig. 3. The post transition group elements Tl, Pb, and Bi are the highest Z, nonradioactive elements in the Periodic Table which accounts for the wide spread use of $Bi_4Ge_3O_{12}$ (BGO) and the current interest in $PbWO_4$. These materials have densities of 7.1 and 8.3 g/cm^3 and radiation lengths of 1.1 and 0.9 cm., respectively. There are, however, many more dense Pb and Bi containing compounds with densities in the range 9-10 g/cm^3 (for some representative examples, see Table 3 of Ref. 14). These may either be Pb or Bi emitting materials or hosts for other activator ions or luminescent species.

As noted above, although the light yields of many compounds containing post transition group elements or molecular complexes such as tungstate or tantalate groups may be small at room temperature, they may still be adequate for high energy calorimetry. Since many of these are very dense materials, they warrant further investigation to quantify their yields and dependences on temperature.

The refractive index n of the host material is a further consideration. Since the probability of electric-dipole transitions of an activator is proportional to $n(n^2 + 2)^2$, the radiative lifetime is reduced if a high-refractive-index host is used. The refractive index also enters into the transport and coupling of the scintillation light to the photodetector.

Lutetium compounds have become favorite hosts for high-efficiency Ce^{3+}-activated scintillators because of their high density and high light yield, although they have a background count rate due to radioactive ^{176}Lu which may be detrimental for some applications. Many lutetium aluminate, silicate, and phosphate compounds have been investigated (see Section 3). Crystals of simple Ce-doped Lu_2O_3 (density 9.4 g/cm^3) have been prepared and its photoluminescence and scintillation properties studied (the results will be reported elsewhere). Another group of compounds that we are studying are the lutetium borates, i.e., $LuBO_3$ (6.9 g/cm^3) and Lu_3BO_6 (7.4 g/cm^3). Other dense compounds that are possible hosts for Ce^{3+} and have not been reported (to my knowledge include the 2:1 compound of the Lu_2O_3–Al_2O_3 system $Lu_4Al_2O_9$ (8.4 g/cm^3), the 6:1 compound of the phosphate system $Lu_{12}P_2O_{23}$ (9.6 g/cm^3), Lu_2O_2S (8.9 g/cm^3), $LuLaO_3$ (9.1 g/cm^3), $LuGaO_3$ (8.8 g/cm^3), $LuInO_3$ (8.3 g/cm^3), and Lu_3InO_6 (10 g/cm^3). The $LnTaO_4$ group of compounds are known phosphors and can be doped to emit from the ultraviolet to the red.[15] Of these, $LuTaO_4$ has a density of 9.75 g/cm^3. These are examples of a much larger group of high density compounds including lutetium tantalate, tungstate, and bismuth compounds. The ease of growing large, high-optical-quality crystals of any of the above materials is of course a paramount issue.

A host material with the density and efficiency of lutetium compounds but without the high cost of Lu would be welcomed for those applications involving relatively large quantities of scintillator materials. Although the price of lutetium materials is coming down,[16] it may always be prohibitively expensive for use in large detectors for HEP experiments.

3. Cerium Activated Scintillator Materials

Cerium-activated scintillator materials have been known and used for several decades but have received renewed interest for many applications because of the favorable spectroscopic properties of Ce^{3+} and the ability to incorporate Ce^{3+} into many different host materials.[17] Among the attractive properties of Ce^{3+} are (1) its luminescence which occurs in the visible-near ultraviolet region and is well matched to the spectral response of high efficiency photodetectors, (2) the emission which involves an allowed electric-dipole transition and thus is fast for good timing resolution, (3) the luminescence intensity and decay time which are stable and insensitive to temperature for most hosts at room temperatures, (4) there are no long-lived radioactive isotopes, and (5) cerium is the least rare of the rare earths. The density, dominant decay time, and relative light output (photons/MeV) of a number of Ce^{3+}-activated materials are given in Table 3. In many instances the optimum concentration of Ce remains to be established.

There is a large variation in the Ce^{3+} scintillator efficiencies in Table 3, ranging from values approaching that of NaI(Tl) for $LuAlO_3$ and Lu_2SiO_5 to values two orders of magnitude smaller for some glasses. All of the materials in Table 3—both crystals and glasses—exhibit intense photoluminescence (except for PbF_2) with luminescence lifetimes at room temperature characteristic of the expected probability for radiative decay, thus their radiative quantum efficiencies Q are near unity. Although the conversion efficiency β may vary somewhat for crystals and glasses of different chemical compositions, the large variation in Ce^{3+} scintillation efficiency of these materials is due predominantly to differences in the transfer efficiency S. Electrons and holes created in the conduction and valence bands may combine radiatively or nonradiatively, be trapped by various defects, form mobile or trapped excitons, or migrate to the vicinity of a Ce^{3+}

TABLE 3. Properties of cerium-activated scintillator materials.

Material	Density (g/cm^3)	Decay time (ns)	Relative light yield
Reference			
NaI(Tl)	3.7	230	100
Crystals			
$LuAlO_3$ (LuAP)	8.3	10–20	80
Lu_2SiO_5 (LSO)	7.4	~40	75
Gd_2SiO_5 (GSO)	6.7	30–60	20-25
$LuPO_4$ (LOP)	6.5	24	33
$YAlO_3$ (YAP)	5.6	28	40
CeF_3	6.2	~5, 30	5–9
BaF_2	4.9	50	6
PbF_2	8.2		~0
Glasses			
silicate (GS1)	2.5–2.7	~60*	~5–10
borate	2.4–2.5	~30	~3
phosphate	~2.6	~30	<1
fluoride (HfF_4)	~6	10,3 0	<1

* Plus slower components.

ion and excite it. The relative probabilities of these processes and the position of the Ce^{3+} electronic energy levels in the gap between the valence and conduction bands enter into the transfer efficiency and are material dependent.

The scintillation efficiencies of lutetium aluminum perovskite[18] and lutetium orthosilicate[19] are the highest reported for any Ce^{3+} activated material and nearly equal to that of NaI(Tl), thus they are of interest for many applications. Since for these materials $E_s \sim (1/2)E_g$, this efficiency suggests high values for both S and Q. The scintillation outputs of LuAP and LSO have, however, been found to vary significantly with growth and annealing conditions. Lutetium aluminate crystals have shown the presence of both the perovskite and garnet phases which alters the effective light yield.[16] Measurements of light output for a number of different LSO samples have shown a convincing anti-correlation between trap-related integrated thermoluminescence output and scintillation light output over a range of several orders of magnitude.[20] Thus defects and their effect on transfer efficiency can account for the large variation in scintillation found for different LSO crystals. The existence of very deep traps in LSO crystals is also evident from the report[21] of phosphorescence lasting for >2000 s. Therefore if the number of trapping centers in these and other materials can be reduced, slightly higher Ce^{3+} light yields may be possible.

The best cerium-activated glasses (silicates) have scintillation efficiencies of only about one-tenth that of NaI(Tl); other glasses (borates, phosphates, fluorides) are reported to have lower efficiencies.[1] This is not surprising given that glass is a disordered medium with a potentially large number of defects and traps to reduce the transfer efficiency. The small scintillation efficiency of Ce-doped glasses is due to the large number of defects that trap electrons and holes and prevent or delay the excitation of and eventual recombination at the activator. Defects in glass are defined as deviations from short-range order and may be intrinsic (for example, three-coordinated Si, oxygen vacancies, or interstitial oxygen in the case of SiO_2), broken bonds, or extrinsic intrinsic (for example, impurities). Many different thermally-induced defects are also possible. The number and type of traps vary with the glass composition, structure, and the thermal treatment. The low efficiency of heavy metal fluoride glasses is probably associated with the more ionic, weaker bonding in these glasses. These are dense glasses and although their light output is low, the yield is sufficient for HEP calorimetry.[7]

Several years ago Spowart[22] investigated Ce^{3+}-activated Li-Mg-Al silicate and found that the scintillation efficiency and thermoluminescence glow curves varied with glass composition but no detailed study was made. Recently Bliss, Craig, et al.[23] have begun a systematic investigation of the effects of composition and microstructure on the scintillation efficiency of Ce-doped silicate glasses. In one study, a large systematic variation in scintillation efficiency was found by varying the alkaline earth component in otherwise identical glasses.[24] Measurements of the absorption and emission spectra and decay curves for these glasses showed only very small changes, as expected for such small compositional variations. That the change in scintillation efficiency is related to defects and their effect on the transfer efficiency was demonstrated by the anti-correlation between integrated thermoluminescence glow curve intensity and the scintillation efficiency.[25]

The effect of defects also appears in the temperature dependence of the scintillation output. At low temperatures, $\ll 300$ K, much of the electron-hole excitation is trapped before it can reach the Ce^{3+} centers and the scintillation yield is low. With increasing temperature the excitation becomes untrapped and the scintillation light output increases. Eventually, usually at temperatures above T_R, the lowest 5d level of Ce^{3+} begins to decay nonradiatively by multiphonon processes and the light output decreases.[26] Thus the scintillation intensity exhibits a peak similar to that observed for Tl-activated crystals.[27,28] For many materials this peak occurs, fortunately, in the vicinity of room temperature (see, for example, reference 29). (CeF_3 shows a different temperature-dependent behavior[30] because of the large fraction of direct excitation of Ce^{3+}).

As illustrated above, defects play a dominant role in determining the scintillation efficiency of many activated materials. The overall scintillation process is best investigated and understood by photoluminescence, radioluminescence, and thermoluminescence measurements combined with excitation spectra recorded using synchrotron radiation. Recently the latter was used to examine the relative differences in light output resulting from direct excitation into the 5d bands of Ce^{3+} and from excitation of valence band and core electrons.[31] Large differences in excitation spectra associated with the transfer process were observed for the two extreme cases in Table 3—lutetium crystals and glasses. Theoretical calculations of electronic structure of the host and luminescence centers[32] can also be useful in understanding the behavior of new scintillator materials and in improving our ability to predict and tailor chemical compositions and structures for specific applications.

4. Core Valence Luminescence Materials

In the decade since the first identification of fast scintillation due to core valence transitions, many materials have been reported to exhibit this luminescence and the phenomenon is reasonably well understood.[33] The first and classic example of core-valence luminescence (CVL) is BaF_2 where a hole created in the Ba 5p core band is filled by an electron from the F 2p valence band. Because the energy difference between the valence band and the 5p core levels is less than the band gap of the material, Auger decay is energetically forbidden and the decay is radiative. Since the process involves an anion-cation crossover transition, it has been called cross luminescence (or Auger-free decay).

Core-valence luminescence involving np – n'p transitions have been reported for Ba, K, Rb, and Cs in numerous binary and ternary materials.[34] The emission is in the ultraviolet, fast (~1 ns), and independent of temperature. The light yield is typically about 1500–2000 photons/MeV and thus sufficient for HEP calorimetry if the CVL cation is present in greater than 10 cation percent. Examples of the scintillation properties of CVL materials are given in Table 4. The emission of Cs compounds is particularly attractive because it occurs at wavelengths that are well matched to common photodetectors. The emission of K and Ba compounds is at shorter wavelengths that require appropriate transmissive and photoemissive materials to detect the CVL.

TABLE 4. Examples of core-valence luminescence observed at 300 K (from ref. 34).

Crystal	Density (g/cm^3)	Wavelength (nm)	Decay time (ns)
KF	2.5	156	
$KLuF_4$	5.2	170–200	1.5
RbF	3.6	203, 234	1.3
$RbCaF_3$	3.6	240–300	2.8
CsF	4.1	390	2.9
CsCl	4.0	240, 270	0.9
CsBr	4.4	250	0.07
BaF_2	4.9	195, 220	0.8
$BaLiF_3$	5.2	190, 230	$\lesssim$ 1.0

While the decay rates and light yields of many CVL materials are sufficient for for HEP calorimetry, higher density and radiation hard materials are desirable. The densities of the CVL materials reported to date are low to moderate, thus the radiation lengths are relatively long and large crystals are required to achieve adequate stopping power for high energy calorimetry (for example, 50-cm long crystals were considered for the GEM BaF_2 detector[35]). Examination of the NIST crystal diffraction database[36] reveals a number of more dense, wide band gap materials containing CVL cations. Some examples include Ba_2YbF_7 (6.1 g/cm^3), $BaThF_6$ (6.7 g/cm^3) (if the radioactivity of Th can be tolerated), and $CsPbF_3$ (6.0 g/cm^3). Lutetium containing compounds are again attractive because of their high densities and large band gaps. In addition to $KLuF_4$ in Table 4, another example of a potential lutetium CVL material is $CsLu_4F_{13}$ (5.8 g/cm^3). The possibility of CVL in dense lead and bismuth compounds such as $RbBiF_4$ (6.3 g/cm^3) or $RbPb_2F_5$ (8.5 g/cm^3) will depend on the various band gaps.

The materials in Table 4 and those discussed above are all halides with large band gaps, but there are a number of oxides that also have large band gaps. For example, BaB_2O_4 and CsB_3O_5 (crystals developed for use as nonlinear optical materials) transmit down to 200 nm. We have observed fast (~1 ns) x-ray excited scintillation from these crystals and, in addition, from $Ba_3(PO_4)_2$, Cs_2CO_3, and Cs_3PO_4, Excitation and emission spectra are needed to confirm that this is cross luminescence. Some examples of dense oxide materials containing CVL cations are $CsLuO_2$ (7.8 g/cm^3), $RbYbO_2$ (7.4 g/cm^3), and $RbLuO_2$ (7.6 g/cm^3). Possible CVL from glasses is discussed in Section 5.4.

As evident above, there are a number of more dense materials that may be fast CVL scintillators suitable for HEP calorimetry. Other crossover transitions such as np - n'd are potential sources of core-valence luminescence,[33] but thus far none have been reported. In the case of crystalline materials where the CVL ions are intrinsic components of the material, band structure calculations of the positions of the conduction, valence, and core levels can be used to predict CVL. Cluster calculations have also been used to account for general spectral features and decay times of CVL.[33,37]

5. Glass Scintillators

As a host material, glass has the advantages that it can be cast in various sizes and shapes, of high optical quality and produced using relatively inexpensive production techniques. Because of its compositional versatility, many physical and spectroscopic properties can be tailored (within limits) for specific applications. However, the number of potential point defects and trapping centers in a disordered medium such as glass is large which, as noted for Ce-activated crystals, can reduce the transfer efficiency and resultant light yield. Although the low light output of glass scintillators is unattractive for many applications, an output of ~1% of NaI(Tl) satisfies the requirement for HEP calorimetry in Table 1. Another consideration for the usage of glass is radiation hardness. Glasses are usually not radiation hard. Ways of attempting to cope with this issue include (1) compositional variations and additives to reduce color center formation, and (2) optical or thermal bleaching.[38]

5.1 Cerium-activated Glasses

Of the cerium-doped glasses investigated thus far, silicate glasses have the highest reported light yield of ~1500–5000 photons/MeV.[39,40] The Ce^{3+} luminescence decay is nonexponential (as expected due to the existence of physically different sites in a disordered medium) with the principal Ce decay time of ~50 ns characteristic of radiative 5d–4f transitions, but with longer decay components. Measurements of a Li-Mg-Al silicate glass (GS1) by Angelini et al.[40] were fitted with decay components having characteristic times approaching a millisecond. They also showed that only about 60% of the light was emitting in the first microsecond following excitation. This indicated that much of the excitation in glass was stored in deep traps.

While the light output of the best Li-Mg-Al silicates is adequate for HEP calorimetry, the density is low, ~2.5 g/cm^3, and the radiation length is large, about 10 cm.[39] Other reported phosphate and borate glasses have lower light outputs and similarly low densities and long radiation lengths. Heavier silicate glasses containing lanthanides, lead, and other heavy cations are known but none of these glasses nor bismuth germanate glasses are reported to have any reasonable scintillation output at room temperature. Extremely dense (8.2 g/cm^3) PbO-Ga_2O_3–Bi_2O_3 glasses are also known,[41] but visible transmission begins at ~500 nm and, therefore, is unsuitable for Ce^{3+} emission. This glass could, however, be a host for activators emitting at longer wavelengths (e.g., Ti^{3+}).

Heavy metal fluoride glasses, especially those containing Hf, can have densities in the range ~6 g/cm^3 with radiation lengths of ~1.6 cm. The scintillation light yields are very small, but do just satisfy the criterion in Table 1. Such glasses have been melted in 100 kg batches for infrared optical windows and lenses, therefore they are attractive from the viewpoint of production. Their radiation resistance, however, is low which has thus far made them unacceptable for HEP calorimetry.[7]

5.2 Glass Ceramics

Vitroceramics are materials having a mixture of crystalline and glassy. phases If the crystals are small compared to the wavelength of light or of the same refractive index as the glass, the material may be optically transparent. If cerium can be incorporated into the crystalline phase of such materials, one could preserve the feature of high scintillation efficiency characteristic of a crystal with the potential low-cost, large-scale production of a glass ceramic. The question of radiation hardness remains.

Most known transparent glass ceramics do not have a crystalline phase with a cation site suitable for substitution of a trivalent rare earth. Recently, however, a transparent vitroceramic doped with trivalent lanthanides was reported.[42] The composition (mol.%) was $30SiO_2$-$15AlO_{1.5}$-$24PbF_2$-$20CdF_2$-$11LnF_3$, where Ln was Er and Yb; the density of these materials is about 6.5 g/cm^3. The lanthanide ions were preferentially segregated from the precursor glass and dissolved into $Pb_xCd_{1-x}F_2$ microcrystals of about 20 nm in size, presumably in the form of a LnF_3-CdF_2-PbF_2 solid solution. The photoluminescence and scintillation properties of Ce^{3+} or Eu^{2+} in these materials should be investigated to establish their value for calorimetry and other scintllator applications.

5.3 Cross Luminescence Glasses

Core-valence transitions are localized phenomena and do not require long-range order or periodicity of the host, [43] hence they can occur in glass. CVL has been reported in a K-fluoroberyllate glass.[44] Glasses based on BeF_2 as the glass former can have the large band gaps, > 10 eV,[45] therefore CVL involving K, Rb, Cs, and Ba cations in such glasses should all be possible. The light yield of the CVL will be proportional to the content of the CVL cation, which in general will be ≤ 50 cation percent. The density of these glasses are low, however, ~3-4 g/cm^3. Heavy metal fluoride glasses, as noted above, can have high densities but their absorption edge is at ~200 nm. CVL should be observed from Cs and Rb in these glasses; CVL from Ba is more questionable because of possible competing Auger transitions. Fluoride glasses would combine features of moderately high density with fast ultraviolet scintillation. Cesium and rubidium CVL may also be observable from wide band gap borate, phosphate, and silicate glasses, but these glasses will generally have relatively low densities. As with all glasses, radiation damage is an omnipresent consideration for HEP calorimetry.

5.4 Organic Scintillants in Inorganic Glasses.

Organic scintillators are composed of various fluors (aromatic hydrocarbons, organic dyes) in a plastic host.[3] These are generally fast (ns) scintillators but their densities are very low, typically ~1.0-1.1 g/cm^3, and hence of limited usefulness for HEP calorimetry. Incorporating organic molecules into more dense inorganic hosts is limited by the melting temperatures of the latter which are usually well above the decomposition temperature of the organic scintillant. Although various organic materials were introduced into boric acid glasses for luminescence and scintillation studies many years ago,[46-48] these are low density glasses with poor chemical stability.

Several years ago a new class of low-melting temperature tin fluorophosphate glasses was discovered[49] and various organic dyes were incorporated into these glasses for nonlinear optical applications.[50] The glasses are durable with moderate densities (3.5-5.5 g/cm^3), refractive indices of 1.6-1.9, and Knoop hardnesses of 90-120. Several organic fluors have been incorporated into these glasses and their scintillation properties investigated by Smith, et al.[51] All glasses exhibited intense photoluminescence. Fast (~1 ns) scintillation was observed, but the light output was low. This was due to the small concentration of fluor added and or inefficient transfer from the host. Exactly how the organic molecules are incorporated into the glass and the mechanism and efficiency of host-fluor energy transfer are unknown. The concentration of the organic fluor in the glass represents a fundamental problem for the practical application of these scintillators.

Tin fluorophosphate glasses warrant further investigation for general calorimetry where fast plastic scintillators are now used because they offer the possibility of greater stopping power with increased hardness and physically durability. Glass fibers provide good spatial precision for tracking of ionizing particles in the vicinity of a high energy interaction.[39] Since these glasses can be drawn into fibers with smaller diameters than is now possible with plastics, they could also be useful in certain tracking applications.

6. Conclusions

The scintillation light output and decay time of most Ce^{3+}-activated and core-valence luminescence materials are satisfactory for HEP calorimetry. These materials have the additional advantage that, in contrast to many other scintillator materials, their output is not very sensitive to temperature at ambient temperatures. Materials having higher densities would be beneficial and some suggested approaches to achieving this objective were presented. Further systematic investigation and theoretical treatment of materials incorporating the post transition group elements or molecular complexes are also needed to develop relationships of structure and bonding to scintillation properties and thereby guide the search for improvements in this class of materials. Radiation damage was not dealt with here but is acknowledged to be an extremely important consideration. Finally, after having discovered a promising scintillator material for HEP calorimetry, one of the most significant questions is whether large crystals of high purity can be grown in the sizes required using economically acceptable methods.

7. Acknowledgments

It is pleasure to acknowledge many informative and stimulating discussions of scintillation phenomena and scintillators materials with Stephen Derenzo, William Moses, Charles Melcher, Jeffrey Schweitzer, Mary Bliss, and Richard Craig. This work was supported in part by the Director, Office of Energy Research, Office of Health and Environmental Research, Medical Applications and Biophysical Research Division of the U.S. Department of Energy under contract No. DE-AC03-76SF000098, and in part by Public Health Service Grant R01 CA48002 awarded by the National Cancer Institute, Department of Health and Human Services.

8. References

1. *Heavy Scintillators for Scientific and Industrial Applications*, eds. F. De Notoristefani, P. Lecoq and M. Schneegans (Editions Frontieres, Gif-sur-Yvette, 1993).
2. *Scintillator and Phosphor Materials*, eds. M. J. Weber, P. Lecoq, R. C. Ruchti, C. Woody, W. M. Yen and R. Y. Zhu (Materials Research Society Symposium Proceedings Vol. **348**, 1994).
3. G. F. Knoll, *Radiation Detection and Measurement*, 2nd edition (John Wiley & Sons, New York, 1989).
4. C. W. E. van Eijk, in *Heavy Scintillators for Scientific and Industrial Applications*, eds. F. De Notoristefani, P. Lecoq and M. Schneegans (Editions Frontieres, Gif-sur-Yvette, 1993) 601.
5. H. J. Hilke, in *Heavy Scintillators for Scientific and Industrial Applications*, eds. F. De Notoristefani, P. Lecoq and M. Schneegans (Editions Frontieres, Gif-sur-Yvette, 1993) 43.
6. H. Newman, this volume.
7. P. Lecoq, this volume.

8. G. F. Garlick, in *Luminescence of Inorganic Solids*, ed. P. Goldberg (Academic Press, New York, 1966) 685.
9 G.W.Ludwig and J. D. Kingley, *J. Electrochem. Soc.* **117** (1970) 348 and 353.
10. A. Lempicki, A. J. Wojtowicz and E. Berman, *Nucl. Instr. and Meth.* **A333** (1993) 304.
11. G. Blasse, *J. Lumin.* **60 & 61** (1994) 930.
12. P. A. Rodnyi, P. Dorenbos and C. W. E. van Eijk, in *Scintillator and Phosphor Materials*, eds. M. J. Weber, P. Lecoq, R. C. Ruchti, C. Woody, W. M. Yen and R. Y. Zhu (Materials Research Society Symposium Proceedings Vol. **348**, 1994) 379 and references cited therein.
13. J. D. Valentine, D. K. Wehe, G. F. Knoll and C. E. Moss, *IEEE Trans. Nucl. Sci.* **40** (1993) 1267.
14. S.E. Derenzo, W. W. Moses, M. J. Weber and A. C. West, in *Scintillator and Phosphor Materials*, eds. M. J. Weber, P. Lecoq, R. C. Ruchti, C. Woody, W. M. Yen and R. Y. Zhu (Materials Research Society Symposium Proceedings Vol. **348**, 1994) 39.
15. L. H. Brixner, *Mater. Chem. Phys.* **16** (1987) 253.
16. W. W. Moses, S. E. Derenzo, A. Fyodorov, M. Korzhik, A. Getkin, B. Minkov and V. Aslanov, *IEEE Trans. Nucl. Sci.* (in press).
17. C. W. E. van Eijk, J. Andriessen, P. Dorenbos, and R. Vissor, *Nucl. Instr. and Meth.* **A348** (1994) 546.
18. A. Lempicki, C. Brecher and D. Wisniewski, *IEEE Trans. Nucl. Sci* . (in press).
19. C. L. Melcher and J. S. Schweitzer, *Nucl. Instr. and Meth.* **A314** (1991) 212 and *IEEE Trans. Nucl. Sci.* **NS-39** (1993) 502.
20. R. Vissor, C. L. Melcher, J. S. Schweitzer, H. Suzuki and T. A. Tombrello, *IEEE Trans. Nucl. Sci.* **41** (1994) 689.
21. P. Dorenbos, C. W. E. van Eijk, A. J. J. Bos and C. L. Melçher, *J. Lumin.* **60 & 61** (1994) 979.
22. A. R. Spowart, *J. Phys. C: Solid State Phys.* **12** (1979) 3369.
23. M. Bliss, R. A. Craig, P. L. Reeder, D. S. Sunberg and M. J. Weber, in *Scintillator and Phosphor Materials*, eds. M. J. Weber, P. Lecoq, R. C. Ruchti, C. Woody, W. M. Yen and R. Y. Zhu (Materials Research Society Symposium Proceedings Vol. **348**, 1994) 195.
24. M. J. Weber, M. Bliss, R. A. Craig, and D. S. Sunberg, *Radiation Effects and Defects in Solids* (in press).
25. R. A. Craig (private communication).
26. Li-Ji Lyu and D. S. Hamilton, *J. Lumin.* **48&49** (1991) 251.
27. J. C. Robert, J. G. Lynch and W. Jack, *Proc. Phys. Soc. (London)* **78** (1961) 1188.
28. J. D. Valentine, W. W. Moses, S. E. Derenzo, D. K. Wehe and GF. Knoll, *Nucl. Instr. and Meth.* **A325** (1993) 147.
29. P. Dorenbos, R. Vissor, C. W. E. van Eijk, R. W. Hollander and P. Shotanus, *Nucl. Instr. and Meth.* **A310** (1991) 236.
30. D. F. Anderson, *Nucl. Instr. and Meth.* **A287** (1991) 606.

31. C. Pedrini, A. N. Belsky, A. N. Vasil'ev, D. Bouttet, C. Dujardin, B. Moine, P. Martin, and M. J. Weber, in *Scintillator and Phosphor Materials*, eds. M. J. Weber, P. Lecoq, R. C. Ruchti, C. Woody, W. M. Yen and R. Y. Zhu (Materials Research Society Symposium Proceedings Vol. **348**, 1994) 225.
32. See, for example, R. Vissor, J. Andriessen, P. Dorenbos and C. W. E. van Eijk, *J. Phys.: Conden. Matter* **5** (1993) 5887 and J. Andriessen, P. Dorenbos and C. W. E. van Eijk, in *Scintillator and Phosphor Materials*, eds. M. J. Weber, P. Lecoq, R. C. Ruchti, C. Woody, W. M. Yen and R. Y. Zhu (Materials Research Society Symposium Proceedings Vol. **348** 1994) 355.
33. P. A. Rodnyi, *Sov. Solid State Phys.* **34** (1992) 1053 and references cited therein.
34. C. W. E. van Eijk, J. Lumin. **60&61** (1994) 936.
35. See H. Newman, in *Heavy Scintillators for Scientific and Industrial Applications*, eds. F. De Notoristefani, P. Lecoq and M. Schneegans (Editions Frontieres, Gif-sur-Yvette, 1993) 5.
36. NIST Crystal Data Identification File (1993), International Centre for Diffraction Data, Newtown Square, PA, U.S.A.
37. C. W. E. van Eijk, in *Heavy Scintillators*, eds. F. De Notoristefani, P. Lecoq and M. Schneegans (Editions Frontieres, Gif-sur-Yvette, 1993) 161 and references cited therein.
38. P. Pavan, G. Zanella and R. Zannoni, *Nucl. Instr. and Meth*. **B61** (1991) 487.
39. Atkinson, J. Fent, et al., *Nucl. Instr. and Meth*. **A254** (1987) 500.
40. C. Angelini, W. Beusch, et al., *Nucl. Instr. and Meth*. **A281** (1989) 50.
41. W. H. Dumbaugh, *Phys. Chem. Glasses* **27** (1986) 119.
42. Y. Wang and J. Ohwaki, *Appl. Phys. Lett*. **63** (1993) 3268.
43. P. A. Rodnyi, I. H. Munro, M. A. Macdonald, E. N. Mel'chakov, S. S. Kotel'nikov, and A. S. Voloshinovsky, *Nucl. Instr. and Meth.* **B 88** (1994) 407.
44. R.T. Williams, D.J. Nagel, P.H. Klein, and M.J. Weber, *J. Appl. Phys.* **52** (1981). 6279.
45. K.L. Bedford, R.T. Williams, W.R. Hunter, J.C. Rife, M.J. Weber, D.D. Kingman, and C.F. Cline, *Phys. Rev.* **B 27** (1983) 2446.
46. E. Tiede, *Chem. Ber*. **55** (1922) 588.
47. F. E. F. Germann, G. H. Lee, F. F. Brown and L. M. Guthrie, *J. Opt. Soc. Am*. **44** (1954) 196.
48. P. A. Egelstaff, *Nucl, Instr.* **1** (1957) 197.
49. P. A. Tick, *Phys. Chem. Glasses* **25** (1984) 149.
50. W. R. Tompkin, R. W. Boyd, D. W. Hall and P. A. Tick, *J. Opt. Soc. Am.*. **B4** (1987) 1030.
51. C. Smith, N. F. Borrelli, et al., in *Scintillator and Phosphor Materials*,, eds. M. J. Weber, P. Lecoq, R. C. Ruchti, C. Woody, W. M. Yen and R. Y. Zhu (Materials Research Society Symposium Proceedings Vol. **348**, 1994) 209.

8.1 References for Figure 1

a. M. Pupin, *Radiologic Science for Technologists* (C. V. Mosby, St. Louis, 1975).

b. E. Rutherford, J. Chadwich, and C. D. Ellis, *Radiations from Radioactive Substances* (Cambridge University Press, 1930) 54; Regener, Verh. d. D. Phys. Ges. **19** (1908) 78 and 351.

c. R. Hofstadter, "Alkali halide scintillation counters," *Phys. Rev.* **74** (1948) 100.

d. R. H. Gillette, "Calcium and cadmium tungstate as scintillation counter crystals for gamma ray detection," *Rev. Sci. Instr.* **21** (1950) 294.

e. R. Hofstadter, "Properties of scintillator materials," *Nucleonics* **6** (1950) 70; H. Grassman, E. Lorenz and H. Moser, "Properties of CsI(Tl) - renaissance of an old scintillation material,". *Nucl. Instr. and Meth.* **228** (1985) 323.

f. W. Van Sciver and R. Hofstadter, "Gamma- and alpha-produced scintillations in cesium fluoride," *Phys. Rev.* **87** (1952) 522; M. Moszynski, C. Gresset, J. Vacher, and R. Odru, "Properties of CsF, a fast inorganic scintillator in energy and time spectroscopy," *Nucl. Instr. and Meth.* **179** (1982) 271.

g. J. Bonanomi and J. Rossell, *Helv. Phys. Acta* **25** (1952) 525; B. Hahn and J. Rossell, *ibid.* **26** (1953) 271; H Knoepfel, F. Luepfe, and P. Stoll, Caesium iodide as a scintilation phosphor," *Helv. Phys. Acta.* **29** (1956) 241; S. Kubota, S. Sakuragi, S. Hashimoto, et al., "A new scintillation material: pure CsI with 10 nsec decay time," *Nucl. Instr. and Meth,* **A268**, (1988) 275.

h. J. Schenck, "Activation of lithium iodide by europium," *Nature* **171** (1953) 518; R. B. Murray, "Use of $Li^{6}I$(Eu) as a scintillator detector and spectrometer for fast neutrons, *Nucl. Instr.* **2** (1958) 237.

i. R. J. Ginther and J. H. Schulman, " Glass scintillators," *IRE Trans. Nucl. Sci.* **NS-5** (1958) 92; R. J. Ginther, "New cerium activated scintillating glasses," *ibid.* **NS-7** (1960) 28; D. G. Anderson, J. Dracass and T. P. Flanagan, "Luminescent glasses for use in the detection of nuclear radiation," *J. Electron. Control* **7** (1959) 463.

j. J. Menefee, C. F. Swinehart and E. W. O'Dell, "Calcium fluoride as an x-ray and charged particle detector," *IEEE Trans. Nucl. Sci.* **NS-13** (1966) 720.

k. J. Menefee, Y. Cho, and C. Swinehart, "Sodium activated cesium iodide as a gamma ray and charged particle detector," *IEEE Trans. Nucl. Sci.* **NS-14** (1967) 464.

l. M. R. Farukhi and C. F. Swinehart, "Barium fluoride as a gamma ray and charged particle detector," *IEEE Trans. Nucl. Sci.* **NS-18** (1971) 200.

m. M. J. Weber and R. R. Monchamp, "Luminescence of $Bi_4Ge_3O_{12}$: spectral and decay properties," *J. Appl. Phys,* **44** (1973) 5495; Z. H. Cho and M. R. Farukhi, "Bismuth germanate as a potential scintillation detector in positron cameras," *J. Nucl. Med.* **18** (1977) 840.

n. T. Takeda, T. Miyata, F. Muramatsu and T.Tomiki, *J. Electrochem. Soc.* **127** (1980) 438; E. Autrata, P. Schauer, J. Kvapil and Jos. Kvapil, "A single crystal of $YAlO_3$-Ce^{3+} as a fast scintillator in SEM," *Scanning* **5** (1983) 91; V. G. Baryshevsky, et al., *Nucl. Instr. and Meth.* **B58** (1991) 291.

o. M. Lavel, M. Moszynski, R. Allemand, E. Cormoreche, P. Guiner, R. Odru, and J. Vacher, "Barium fluoride–inorganic scintillator for subnanosecond timing," *Nucl. Instr. and Meth.* **206** (1983) 169.

p. K. Takagi and T. Fukazawa, "Cerium activated Gd_2SiO_5 single crystal scintillator," *Appl. Phys. Lett.* **42** (1983) 43.

q. C. D. Greskovich, D. A. Cusno, and F. A. DiBianco, United States Patent 4,518,546 (1986).

r. D. F. Anderson, "Properties of the high-density scintillator cerium fluoride," *IEEE Trans. Nucl. Sci,* **NS-36** (1989) 137; W. W. Moses and S. E. Derenzo, "Cerium fluoride: a new, heavy fast scintillator," *IEEE Trans. Nucl. Sci,* **NS-36** (1989) 173.

s. S. E. Derenzo, W. W. Moses, R. Perera, et al., "Prospects for new inorganic scintillators," *IEEE Trans. Nucl. Sci,* **NS-37** (1990) 203; L. Nagornaya and V. Ryzhikov, "Fast scintillator based on large "heavy" tungstate single crystals," in *Heavy Scintillators for Scientific and Industrial Applications,* eds. F. De Notaristefani, P. Lecoq, M. Schneegans (Editions Frontieres, Gif-sur-Yvette, 1993) 367: M. Kobayashi, M. Ishii, Y. Usuki, and H. Yahagi "$PbWO_4$ scintillator at room temperature," *ibid.* 375; L. V. Miassoedov, et al., "$PbWO_4$: a heavy, fast and radiation resistant scintillator for EM calorimetry," *ibid.* 437.

t. S. E. Derenzo, W. W. Moses, J. L. Cahoon, et al., "X-ray fluorescence measurements of 412 inorganic compounds," *IEEE Nuclear Science Symposium Conference Record 91 CH3100-5,* **vol. 1** (1991) 143; A. Lempicki, E. Berman, A. J. Wojtowicz, M. Balcerzyk, and L. A. Boatner, *IEEE Trans. Nucl. Sci.* **40** (1993) 384.

u. C. L. Melcher and J. S. Schweitzer, "Cerium-doped lutetium orthosilicate: a fast, efficient new scintillator," *IEEE Trans. Nucl. Sci.* **NS-39** (1992) 502; H. Suzuki, T. A. Tombrello, C. L. Melcher and J. S. Schweitzer, *Nucl. Instr. and Meth.* **320** (1992) 263.

w. B. I. Minkov, *Functional Mater.* **1** (1994) 103; W. W. Moses, S. E. Derenzo, A. Fyodorov, M. Korzhik, A. Getkin, B. Minkov and V. Aslanov. $LuAlO_3$:Ce - a high density, high speed scintillator for gamma detection," *IEEE Trans. Nucl. Sci.* (in press); A. Lempicki, M. H. Randles, D. Wisnieski, M. Bakcerzyk, C. Brecher and A. J. Wojtowicz, "$LuAlO_3$:Ce and other aluminate scintillators," *IEEE Trans. Nucl. Sci*. (in press).

Experience from Large Calorimeter Systems

Session Chair: Bruno Mansoulie (Saclay)

The DØ Calorimeter Performance and Calibration

R. Dean Schamberger
for the DØ Collaboration
Physics Department, SUNY at Stony Brook
Stony Brook, NY 11794-3800, USA

ABSTRACT

The DØ calorimeter is a large uranium–liquid argon sampling calorimeter constructed to study proton–antiproton collisions at $\sqrt{s} = 2$ TeV at the Fermilab Tevatron Collider. The DØ experiment has been using it to collect physics data for over two years. It consists of approximately fifty thousand readout channels covering all phi, $|\eta| < 4.4$ and up to nine depths. The performance of such a large system is discussed in the context of both its reliability and its stability. The calibration of the liquid argon calorimeter in the absence of a central magnetic field is also discussed, from our initial testbeam calibration to the present *in situ* calibration of both its electromagnetic and hadronic energy response.

1. Introduction

The DØ detector, located at the Fermi National Accelerator Laboritory, is a large, state-of-the-art detector that has as its primary physics focus the study of high mass states and large p_T phenomena. It consists of three major components: a central tracking system, a nearly 4π solid angle precision calorimeter, and a muon identification system. This article will discuss the calorimeter system only. For details on the overall detector, the reader is referred to Reference 1.

The calorimeter consists of three cryostats, each containing thin sheets of uranium interleaved with liquid argon to produce a stable, unity gain, hermetic, finely segmented, and radiation hard calorimeter based on detection of ionization produced in the liquid argon. These three cryostats, called CC (for Central Calorimeter), ECN (for End Calorimeter North, the direction from which protons enter the interaction region), and ECS (for End Calorimeter South) are required to hold the liquid argon and keep its temperature below the boiling point of argon. The detector is symmetric around the Tevatron beam pipe and segmented into 64 equal size segments in azimuth ϕ ($\Delta\phi \sim 0.1$ radians). It is also segmented in depth along the particle direction into eight or nine segments. The first four layers comprise the finely sampled electromagnetic (EM) section, approximately 21 X_0 in total thickness. This is followed by a more infrequently sampled hadronic section. The final outer layer, called the Coarse Hadronic Layer, is made with copper or steel to save in cost. Finally the calorimeter is segmented along the θ direction, where θ is the polar angle to the proton beam direction, in semi-projective towers of equal size in pseudorapidity (η) defined by $\eta \equiv -ln(tan(\theta/2))$. The η segmentation ($\Delta\eta$) is 0.1 for $|\eta| < 3.2$ and 0.2 for $|\eta| > 3.2$. A cross sectional view of one quarter of the detector, showing the η and depth segmentation is shown in Fig. 1. As can be seen in Fig. 1, significant

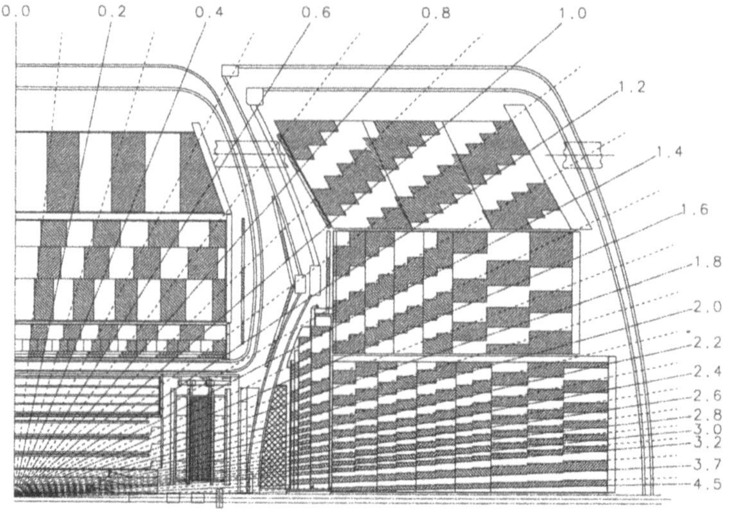

Figure 1: Cross sectional view of the DØ calorimeter and central tracking system showing the η and depth segmentation of the calorimeter. One quarter of the detector is shown with the interaction point in the lower left corner of the figure.

extra material from the cryostat walls and detector support structures exists in the region from $|\eta|$ of 0.8 to $|\eta|$ of 1.4, which we call the intercryostat region(ICD). To improve the sampling of the detector in this region, extra readout layers containing no absorber plates were added inside the cryostats. In addition, scintillators were added between the cryostats to add one extra sample in this region. Some of the other relevant parameters of the detector are listed in Table 1. The total number of nuclear interaction lengths ranges from 7 to 9 λ_0 with the minimum in the ICD region.

The noise levels in the DØ calorimeter readout are very small with respect to energies of interest, with typical pedestal RMS widths of 5 to 10 MeV per channel in the electromagnetic section and ≈50 MeV in the hadronic section. The coherent noise level is extremely low allowing one to add up 4000 channels (more than is used in the largest jet) and still not significantly increase the resolution above the incoherent contribution.

2. Performance, Stability, and Reliability

The most important parameter in the performance of the DØ calorimeter is its superior resolution for both electromagnetic and hadronic energy. This resolution can be achieved for hadronic jets because the DØ calorimeter is an almost perfectly compensating calorimeter. In general, the response of a sampling calorimeter to incident

Table 1: Parameters of the DØ calorimeter

	CC	EC
Weight(metric tons)	305	258 each
LAr Volume(Kliters)	19	12 each

	EM	Fine Hadronic	Coarse Hadronic
LAr Gap(mm)	2.3	2.3	2.3
Absorber	U^{238}	U^{238}	Cu/steel
Thickness(mm)	3 or 4	6.0	46.5
Rad Length(X_0)	21	95-122	31-65
Int Length(λ_0)	0.8	3.3-4.7	3.2-7.0

electrons and pions of the same energy need not yield the same signal ($e/\pi \neq 1$). Since hadronic jets contain some varying fraction of their energy in π^0's (which deposits energy via electromagnetic showers), the only way to obtain good hadronic jet energy resolution over a large energy range is to have the e/π ratio close enough to unity to be negligible. As measured at the testbeam this ratio is within 2% of unity[2] at 150 GeV and rises to about 9% at 10 GeV.

The energy resolution of the DØ calorimeter can be described as arising from three major sources. The first is the noise term which is due to fluctuations in the pedestal determination which is a fixed value, independent of the observed signal. The second is the sampling term which reflects statistical fluctuations in the energy deposited in the argon and therefore scales like the square root of the signal size. This is the term that is usually quoted when discussing how good a particular detector is. This is only true, however, in the limit where the other two contributions are negligible. The third is the constant term, which reflects how well the response of different parts of the detector are equalized. It therefore scales linearly with signal size, assuming the energy is distributed over approximately the same number of readout cells, independent of energy. Another source of variations that contribute to this term is response instabilities with time which have not been properly corrected for with calibrations.

We have fit the energy resolution to these three different terms for both the CC and EC detectors using the following functional form:

$$\left(\frac{\sigma_E}{E}\right)^2 = \left(\frac{N}{E}\right)^2 + \left(\frac{S}{\sqrt{E}}\right)^2 + (C)^2, \tag{1}$$

where N, S, and C are the noise, sampling, and constant terms, respectively. We fit for each of the three terms independently for electron and pion data taken in the testbeam at known energies. The data is shown in Fig. 2 with the fitted resolution function superimposed, for each of the four cases. The results of the fit are given in Table 2. Our energy resolution is clearly dominated by the sampling term, over the entire energy range of interest.

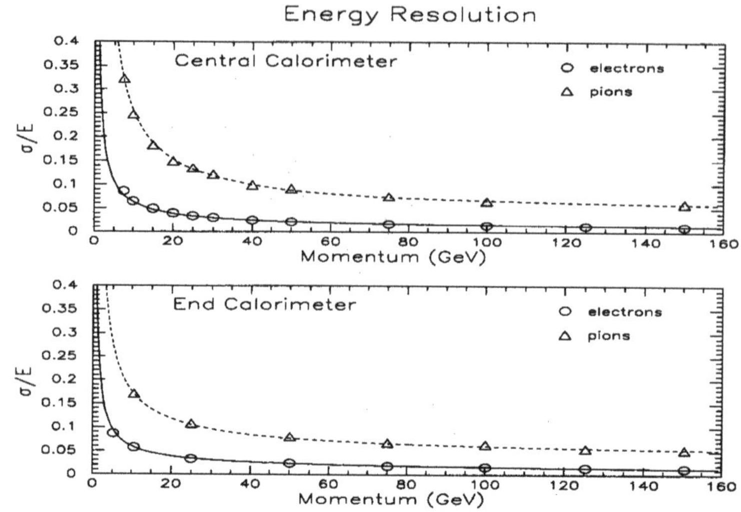

Figure 2: DØ energy resolution for pions and electrons in the CC and the ECs. The smooth curves are fits to the data using the form given by Eq. (1), and resulting in the parameters shown in Table 2.

In order to insure that the constant term does not increase as we combine data taken over longer and longer periods of time, we constantly monitor the response of the calorimeter. The response of the detector can be separated into two components: the signal induced in the readout cell due to showering particles and the electronics response to a fixed signal. We monitor the electronics response of each channel by injecting a fixed amount of charge, with approximately the same time structure as the real detector signal, through a precision resistor located at the preamplifier inputs. The only changes we have observed (at the 0.2% level) can be associated with changes in the ambient temperature which varies by a couple of degrees Celsius from

Table 2: Parameters from the fit to Eqn. 1 for each of the four cases discussed in the text. The noise term was fixed in the CC analysis to its measured value.

	Noise Term N (GeV)	Sampling Term S ($\sqrt{GeV}$)	Constant Term C
Electrons in CC	0.42 (fixed)	0.148±0.004	0.003±0.002
Electrons in EC	0.33±0.030	0.157±0.005	0.003±0.002
Pions in CC	2.00 (fixed)	0.470±0.010	0.045±0.004
Pions in EC	0.975±0.035	0.446±0.009	0.039±0.002

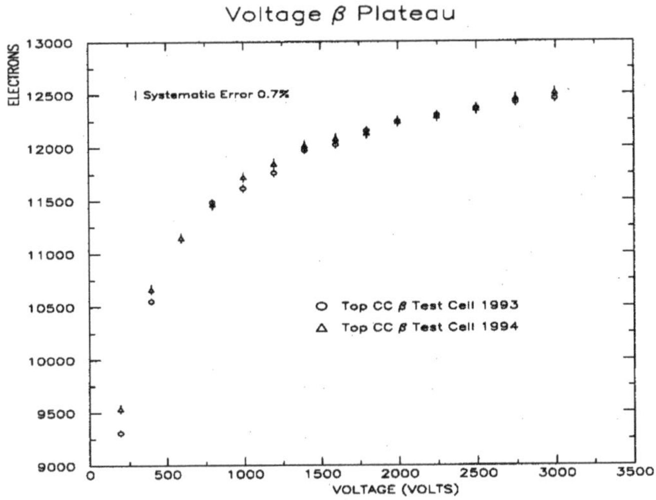

Figure 3: Response of the β cell verses high voltage applied across its 2.3 mm argon gap. The two sets of data were obtained from one of the three cryostats, taken approximately one year apart. (Note the suppressed zero) (*DØ Preliminary*)

winter to summer. The major reasons for potential changes in the detector signal are due to changes in the liquid temperature (density) or contamination of the argon. Inside each cryostat there are three special readout cells that monitor the argon by detecting the β decay of a radioactive source. The three cells are located near the top of the liquid, near the bottom of the cryostat, and approximately at the middle depth in the liquid argon. By monitoring the signal obtained from the source (and correcting for any known electronics gain shifts) we can check for time-dependent changes. We also regulate the argon vapor pressure to maintain a constant argon density, so we do not expect any changes in response unless there were an unknown source of contamination in the system. No significant change in response has been seen; the long term variations in response are less than 1% per year. We also monitor the HV plateau of the source signals, as contaminants can usually be observed as a change in the shape of the plateau curve. Fig. 3 shows that there has been no significant change in shape over the last year. The DØ calorimeter is operated with 2 KV across the 2.3 mm argon filled gap.

Another important parameter of the DØ calorimeter is the linearity of its response, particularly for electromagnetic showers, where the excellent energy resolution and fine segmentation allows us to observe resonances that decay into two electrons (or two photons) by reconstructing its invariant mass from the measured energies and directions of the two particles. To obtain the best linearity possible, without compro-

Table 3: Sampling weights determined from testbeam data in both CC and EC modules.

		CC	EC
	α(MeV)	2.956±0.005	3.742±0.008
EM1	β_1	1.308±0.004	1.47±0.03
EM2	β_2	0.852±0.005	1.00±0.01
EM3	β_3(fixed)	1.0	1.0
EM4	β_4	0.966±0.009	1.10±0.01
FH1	β_5(fixed)	1.84	1.67
	δ(MeV)	347±13	300±20

mising resolution, we use data taken in the testbeam with electrons of known energies, to obtain the relative weights for each of the readout depths (for CC and EC modules separately) by minimizing the following chi-squared like function:

$$\chi^2 = \sum_E \sum_\eta \sum_i \left(\frac{P_i - E_i}{\sigma_p} \right)^2 \tag{2}$$

where the sum over E goes from 10 to 150 GeV, η = 0.05, 0.45 and 1.05 for the CC module and η = 1.95 for the EC module, and the sum over i is over all events for each energy and η region. P_i is the testbeam measured momentum, σ_p is the calorimeter resolution at each of the average beam momentum runs, and E_i is given by the following formula:

$$E_i = \alpha \sum_{j=1}^{5} \beta_j E'_{ji} + \delta \tag{3}$$

where α, β_j and δ are the parameters we adjust to minimize the χ^2 function. α is an overall scale factor and, along with the choice of $\beta_3 \equiv 1.0$, gives the translation from ADC counts to GeV. E'_{ji} is the response, corrected for electronics gain differences and summed over all channels, in a given layer j for a given event i. δ is a constant offset needed to improve the linearity in the high energy response of the detector to electrons. This function simultaneously minimizes the resolution and any deviations from linearity over the energy range included in the sum. As long as the resolution is small with respect to the energy (true for electromagnetic showers with energies about 10 GeV), this procedure essentially minimizes the deviations of the mean from the ideal linear response. The fitted values are given in Table 3 and are similar to sampling weights calculated from the known dE/dx values for argon and uranium. The value for layer 1 includes effects due to energy lost in material in front of the detector. β_5 is held fixed at its calculated sampling weight since only a very small amount of energy actually leaks into the first layer of the fine hadronic readout section.

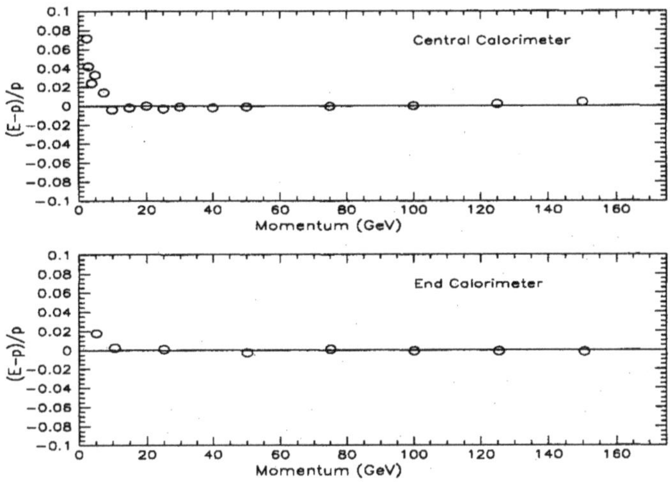

Figure 4: Residuals (measured energy minus beam momentum normalized to the beam momentum) observed in testbeam calibration runs for electrons in CC and EC modules plotted versus beam momentum. (*DØ Preliminary*)

Shown in Fig. 4 are the deviations from linearity for electron energies from 10 GeV to 150 GeV, as obtained in the testbeam for both CC and EC. Data for energies below 10 GeV are also shown, though not used in the fit.

The reliability of the calorimeter and its readout electronics has been very good. Because we welded the cryostats shut, extra care was taken to ensure the highest reliability. After the cryostats were sealed and filled with liquid argon, we found 37 out of the 47,800 channels to be dead. Most were due to cable failures which happened during the final stages of sealing and filling the cryostat. Since that time, no new internal failures have occurred. The electronics has also proven to be very reliable, with an average of 10 channels not working properly at any one time. Failures of electronics channels after the preamp can typically be fixed in a one (or more) hour accelerator access period which occurs on average every ten days. Repairing preamp related components require moving the large muon iron magnet and therefore can only happen when there is a one (or more) day accelerator access period. This occurs on average every six months. For this reason, we built duplicate power supplies for the entire preamp system, which can be remotely enabled in the case where a primary supply fails (which would have disabled 8% of the detector in a localized region). Such a failure has only happened once in over two years of operation. With typically less than 0.1% of the channels not working at any one time, and since those channels are distributed all over the detector, we typically do not bother to correct our Monte-Carlo simulations for this effect.

3. Electromagnetic Energy Scale

In order to establish the absolute energy scale for the electromagnetic part of the DØ calorimeter a number of steps must be performed. Because DØ has no central magnetic field to measure the momentum of the particles created in $p\bar{p}$ interactions we will, in the end, rely on the observation of $Z^0 \rightarrow e^+e^-$ interactions to fix the scale. Our initial energy scale was established by combining the measured ADC counts to GeV conversion obtained in the testbeam with the relative testbeam to DØ electronics response. This gives the correct normalization assuming that any differences in response of the detector average to zero. The first step in checking the normalization is to establish the relative response of each readout cell in the calorimeter. This entails two steps: establishing the relative electronics response of every channel, and then establishing the relative response of the detector. In order to calibrate the relative response of every channel in the DØ calorimeter, a precision set of pulsers was built which deliver a precise amount of charge to every channel. This system allows us to know the relative response of every electronics channel in the DØ calorimeter to better than 0.25% .

To obtain the relative detector elements response is more complicated. Both time and financial constraints did not allow us to test each and every module in a testbeam. In fact, the scheduling of the module installation into the cryostats at DØ precluded testing any of the CCEM modules actually installed in DØ (a total of 32 modules). Instead we tested "spare" modules built at the same time but never installed in DØ. There were only two monolithic ECEM modules produced (one for each end); the one that eventually was installed in the EC south cryostat was tested. The sampling weights described in the previous section are applied to all CC or both EC modules to allow combining energy from different cells in the same module. The only remaining degree of freedom is the relative response of each module. The two CCEM spare modules measured in the testbeam setup showed peak-to-peak differences of $\sim 2-3\%$ which could not be accounted for by variations in plate thickness or other assembly tolerances. We have used *in situ* data to establish a relative calibration in DØ. We exploited the ϕ symmetry of our detector to look at the rate of production of EM clusters (primarily γ's from π^0's) above a fixed threshold. We obtain the data in special runs which, at the trigger level, only require a single EM trigger tower with E_T above 7 GeV. A fixed threshold is then chosen to be high enough to avoid any potential trigger bias, but as low as possible to keep the statistics high[3]. The energy scale for each module is adjusted until the rates in each module are the same. The correction factors needed to equalize the module response, normalized to the average value, are shown in Fig. 5 (left). To check the effectiveness of these normalization factors, we split a sample of $W \rightarrow e\nu$ events into three parts; those with $f < 0.99$ (low), those with $0.99 < f < 1.01$ (med), and those with $f > 1.01$ (high). In Fig. 5 (right) is shown the difference in the W mass as determined from the three different groups, both before and after module uniformity corrections are applied. A clear improvement in detector uniformity is seen after applying these corrections.

This only leaves the three overall cryostat scale factors to be determined. For that

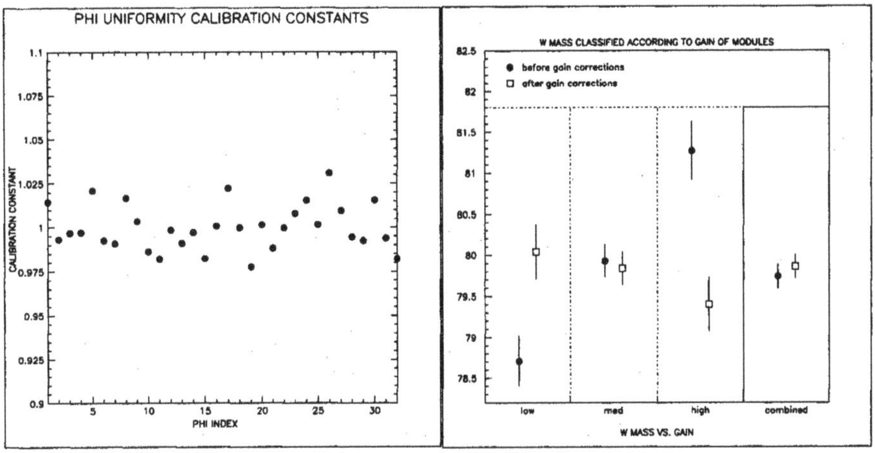

Figure 5: (left) Relative correction factors (*f*) needed to equalize the response of the CCEM modules. The rms of *f* is 1.4%. (right) Fitted W-mass binned by CC module gain correction. Y axis is in arbitrary units. (*DØ Preliminary*)

we use the reconstructed mass of the Z^0, from $Z^0 \rightarrow e^+e^-$ events. From a sample of 373 (50,38) Z^0's with both electrons in the CC (ECN,ECS) electromagnetic section, we obtain a (*DØ Preliminary*) scale factor of 1.045±0.002 (1.035±0.011,1.005±0.011) for the CC (ECN,ECS) cryostat. This scale factor relates our initial testbeam calibration to the known Z^0 mass. The extent to which these numbers are not unity, reflects the combined effects involved with transferring the testbeam calibration to DØ and differences in module responses. As we continue to collect $Z^0 \rightarrow e^+e^-$ events, we will continue to improve the precision with which we know these calibration factors. In addition we are continuing to collect and analyze more "ϕ uniformity" data in CCEM modules to allow us to further subdivide the modules to look for variations in η as well. We are also in the process of collecting and analyzing data from both ECEM modules to check their relative response.

These calibration factors implicitly assume that there is no offset in the energy response at DØ with respect to the testbeam. To place limits on any potential differences, we have studied three separate final states: $Z^0 \rightarrow e^+e^-$, $J/\psi \rightarrow e^+e^-$, and $\pi^0 \rightarrow \gamma\gamma$ (where both photons convert early enough in the tracking system that we can measure the opening angle of the two photons). The $Z^0 \rightarrow e^+e^-$ events that we used to establish the energy scale can also set a limit on any potential offset[4] because the energies of the electrons range from 25 to 80 GeV in the CC. Then if one assumes:

$$E^e(true) = \kappa E^e(meas) + \rho \tag{4}$$

where κ is the scale factor and ρ is a potential offset, one can relate the true to the measured mass by substituting Eq. (4) into the standard invariant mass formula:

$$M = \sqrt{4E_1^e E_2^e sin^2\left(\frac{\omega}{2}\right)} \tag{5}$$

where ω is the opening angle between the two electrons. Only keeping linear terms in ρ (assuming $\rho \ll E^e$) one gets:

$$M^{ee}(true) = \kappa M^{ee}(meas) + f\rho \tag{6}$$

where f is a purely kinematic factor of the decay, given by:

$$f = \left(\frac{2(E_1^e(meas) + E_2^e(meas))}{M^{ee}(meas)}\right) sin^2\left(\frac{\omega}{2}\right) \tag{7}$$

In the case of $J/\psi \rightarrow e^+e^-$, and $\pi^0 \rightarrow \gamma\gamma$ we combine their measured masses with our measured Z^0 mass to solve for κ and ρ in Eq. (6). In Fig. 6 (left) one sees the J/ψ mass distribution from which we extract a mass[5] of 3.03±0.04±0.19 GeV/c^2, where the first error is statistical and the second error systematic. The systematic error is dominated by the uncertainty in the low energy response of the DØ calorimeter. In Fig. 6 (right) one sees the π^0 symmetric mass distribution (assuming $E_1^\gamma = E_2^\gamma = E^{\pi^0}/2$) from which we extract a mass[6] of 135.4 ± 3.8±9.2 MeV/c^2. The statistical error has been dropped because it is negligible with respect to the systematic errors; the first error is the modelling error and the second error is related to the low energy response uncertainty.

The present (*DØ Preliminary*) determination of ρ from the three analyses are $\rho = -320 \pm 780$ MeV from the Z^0 analysis, $\rho = -130 \pm 410$ MeV from the J/ψ analysis, and $\rho = -245 \pm 314$ MeV from the π^0 analysis. All three are consistant with zero but are, on average, slightly negative.

4. Missing E_T and Jet Energy Scale Determination

Because of the excellent energy resolution and the nearly 4π solid angle coverage, the DØ calorimeter has excellent missing E_T ($\not{E}_T$) resolution. The $\not{E}_T$ resolution of the DØ calorimeter has been determined[7] imperically from low energy dijet data to have the form:

$$\sigma_{\not{E}_T} = a + b \cdot S_T + c \cdot S_T^2 \tag{8}$$

where a = (1.89±0.05) GeV, b = (6.7±0.7)×10^{-3}, and c = (9.9±2.1)×$10^{-6} GeV^{-1}$, and S_T is the summed transverse energy seen in the calorimeter. This parameterization is valid in the range ($70 \leq S_T \leq 330$) GeV.

Now that we have established the electromagnetic energy scale, we can use it to calibrate the jet energy scale. Because of the large fluctuations in hadronic energy depositions and ambiguities in where to assign small amounts of energy near the

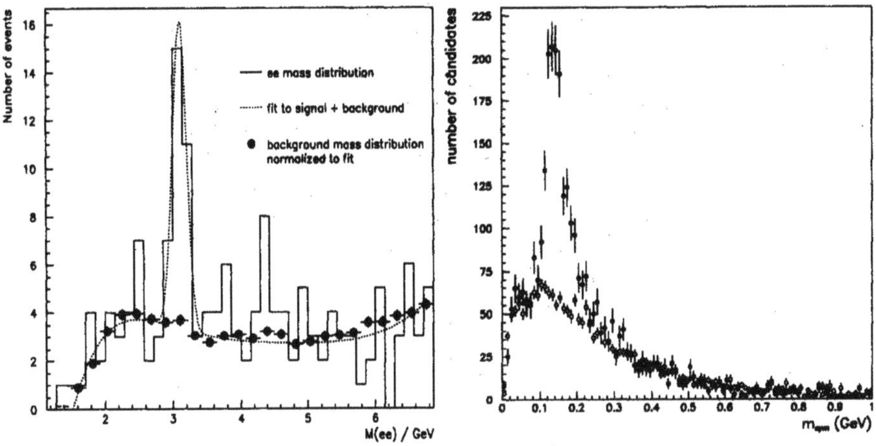

Figure 6: (left) J/ψ mass distribution used in checking the DØ energy scale and offset. The histogram is the data, the dashed curve is a fit to a single gaussian plus a fixed background shape (from a fit to the solid points) with a floating normalization only. (right) The π^0 symmetric mass distribution used in checking the DØ energy scale and offset. Solid points are the data, open circles are the background spectrum normalized outside the peak region. (*DØ Preliminary*)

edges of jets, the jet energy corrections are larger (and more uncertain) than those for electromagnetic objects. The DØ jet finding algorithm[8] defines a jet by summing up the E_T for all cells in a cone in η, ϕ space of radius R $= \sqrt{\Delta\eta^2 + \Delta\phi^2}$. We then vary the central value of η, ϕ to maximize the energy associated with the jet.

The first step in setting the jet energy scale is to use dijet events where one of the jets (called the trigger jet) is in the CC, and whose energy is all contained in the EM part of the calorimeter. These "EM jets" are either photons or jets that fragment into leading π^0's. We then use the observed $\not{E}_T$ as an indication of the hadronic energy scale error. We take the response of the calorimeter to hadrons, R_{had}, to be given by the following formula:

$$R_{had} = 1 + \frac{\vec{\not{E}}_T \cdot \hat{n}_T^\gamma}{E_T^\gamma}, \tag{9}$$

where $\hat{n}_T^\gamma$ is the unit vector along the photon direction in the plane transverse to the beam axis. This technique works well for $|\eta| \leq 1$, but is limited by statistics in the EC region. The full jet energy correction can then be expressed by three factors given in the following formula:

$$E(\text{meas}) = (1 - C)\, R_{had} E(\text{true}) + U, \tag{10}$$

where U represents the energy contributed by both the underlying event and the suppression of our asymmetric pedestals, and C represents unseen energy resulting from limitations in the algorithm used. The quantities E_T(true) and E_T(meas) denote the true and measured jet energies, respectively. All 3 of the quantities in Eq. (10) – R_{had}, U, and C – depend on E(true), η and ϕ, while C and U depend on the algorithm used as well. The quantity U is obtained from minimum bias events taken during normal running. The quantity C is obtained by combining testbeam single particle shower profiles with Monte-Carlo calculations of particle directions and momenta, passed though our jet finding algorithm. Once we have calibrated jet energies in the CC region, we use these jets to calibrate jets in the EC region. The (*DØ Preliminary*) average correction for $E_T > 20$ GeV is approximately 15-25%, with an error of about 5%.

5. Conclusions

The DØ uranium-liquid argon calorimeter has been running now for over two years with excellent stability and reliability. Its good resolution, linearity and hermeticity make it an excellent detector for the study of high P_T physics and search for high mass states. We have established the EM and jet energy scale using collider data to the precision necessary to obtain significant physics results and expect to improve them further as more data becomes available.

6. References

1. "THE DZERO DETECTOR", S. Abachi et al., *Nucl. Inst. and Methods in Phys. Research* **A338** (1994) 185.
2. "Beam Tests of the DØ Uranium/Liquid Argon End Calorimeters", S. Abachi et al., *Nucl. Inst. and Methods in Phys. Research* **A324** (1993) 53.
3. "Measurement of the W Boson Mass in Proton-Antiproton Collisions at $\sqrt{s} =$ 1.8 TeV", Q. Zhu, *PhD. Thesis*, New York University, New York, New York, (April, 1994).
4. "Energy Scale Error for W mass Measurement", U. Heintz, *DØ internal note* **DØ-1819** (July, 1993).
5. "Observation of $J/\psi \rightarrow ee$ Decays at DØ and Calibration of the Central Calorimeter Electromagnetic Energy Scale", I. Adam, *DØ internal note* **DØ-2298** (October, 1994).
6. "A Measurement of the Calorimeter Response Using π^0 Decays", U. Heintz, *DØ internal note* **DØ-2268** (August, 1994).
7. "A study of the DØ calorimeter $\not{E}_T$ Resolution Using Low E_T Jet Triggers", M. Paterno, *DØ internal note* **DØ-1782** (July, 1993).
8. "Studies of JetProduction with the DØ Detector", H. Weerts, *Proceedings of the 9th Topical Workshop on Proton-Antiproton Collider Physics*, University of Tsukuba, Ibaraki, Japan, (1993) 192.

PERFORMANCE OF THE H1 LIQUID ARGON CALORIMETER

A. BABAEV
ITEP, Moscow, Russia / DESY, Hamburg, Germany
H1 Collaboration

ABSTRACT

The report includes a description of the instrumentation, the HV system, the electronic calibration and LAr purity. The energy calibration was taken from test beam measurements. Checks of the electromagnetic and hadronic energy scales at HERA are shown. The LAr trigger performance is discribed with respect to timing and efficiency.

1. The H1 LAr Calorimeter

The storage ring HERA started its operation in 1992 and provides collisions of 27.5 GeV positrons (electrons before July 94) with 820 GeV protons. HERA works recently with 170 proton and 168 positron bunches and has several (p and e^+) pilot bunches. The time between two bunches is 96 ns. The H1 detector was designed to satisfy the HERA physical program including tests of the standard model up to high Q^2 ($\approx 10^5\ GeV^2$) and low x ($\approx 10^{-4}$) and searches for new physics.

The liquid argon calorimeter (LAr)[1] is a main component of the H1 detector. It provides a precise measurement of the particle energy and triggers for a wide class of the physical processes. The structure of the LAr calorimeter is showen in fig.1. The calorimeter consists of an electromagnetic (EM) and a hadronic (HAD) part. The absorber of the HAD calorimeter is made from stainless steel plates with a thickness of 16 mm. The liquid argon gap is 2×2.4 mm. The interaction length (λ) of the HAD calorimeter is in a range of 5-8 λ (including the EM part). The EM calorimeter has 2.4 mm lead plates as absorber and 2.35 mm liquid argon gaps. The radiation length of the EM part varies between 20 X_0 and 30 X_0 (1.3 λ).

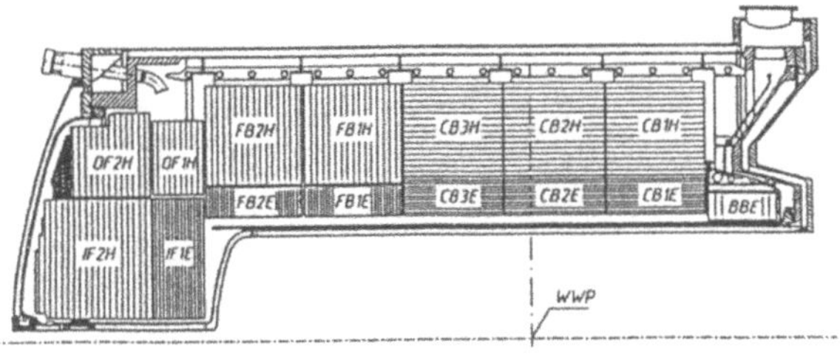

Figure 1: The structure oh the H1 LAr calorimeter. The orientation of absorber plates is indicated.

Mechanically the calorimeter is divided into eight self-supporting wheels:

inner forward (IF), outer forward (OF), forward barrel (FB), central barrel (CB) and backward barrel electromagnetic (BBE). In ϕ-direction the wheels are separated either into eight octants (FB, CB, BBE) or into two half rings (IF,OF). The high granularity of the calorimeter provides a good separation of EM and HAD showers, a good longitudinal sampling and the effective triggering of the H1 detector.

2. The HV system

CAEN HV modules are used for the LAr calorimeter HVsystem to supply 44352 calorimeter cells with 1504 HV lines. One cell is supplied by several (up to 12) HV lines to keep the homogeneity of the calorimeter response in case that a HV line has problem. An OS9 system controls the HV generators and distributors via a VME bus interface. For each HV line the current is monitored. The over current threshold is at 2 μA. The operation voltage is 1500 V which corresponds to an electric field of 625 V/mm. It is chosen to ensure reliability of the calorimeter performance especially noise rates of the calorimeter trigger due to discharging of a weak HV line. Due to HV problems inside the cryostat 59 (4%) HV lines out of 1504 are at 1000 V and 3 (0.2%) lines are disconnected. The HV CAEN modules are working sufficiently reliable after replacing the opto-coupled cards by new ones after the first 1.5 year of operation.

3. The LAr purity

The calorimeter was filled with 53 m^3 liquid argon in January 91 and up to now it is working without purifying. The purity of the liquid argon is monitored by a set of eleven small ionisation cells (4mm or 6mm gaps) which are integrated into the cryostat. The energy spectra of the radioactive decays of the β or α sources are measured by a multichannel analyser controlled by a PC. Two methods of monitoring are used:

- *The method of the amplitude* where the amplitude is measured continously with constant HV in the ionisation cell. Some corrections are applied to get an absolute value for the collected charge.
- *The method of the HV curve* where the amplitudes are measured at different HV setting in the ionisation cell or in the calorimeter stack itself. Fits to the data gives values for the impurity and the collected charge at infinite electric field.

Fig. 2a shows the relative variation of the charge collected by a cell during the leakage of a cryostat valve. Fig. 2b shows impurity as function of time. The 'natural' aging of the LAr in the cryostat is around 0.12 ppm per year, corresponding to a decrease of the calorimeter signal by a factor 0.77% per year. One can see on the picture the increase of the impurities during a bad refill of the cryostat and during the leak period. The HV curve with the LAr calorimeter itself has been measured by cosmic particles (barrel part) and by hallo muons (forward part). Taking into account the data of the LAr purity monitoring and the HV calorimeter response the charge collection efficiency is 0.879 $\pm$ 0.015 at 1500 V.

4. The Electronic chain and calibration

The front-end electronic reads out the 44352 calorimeter cells. The cell

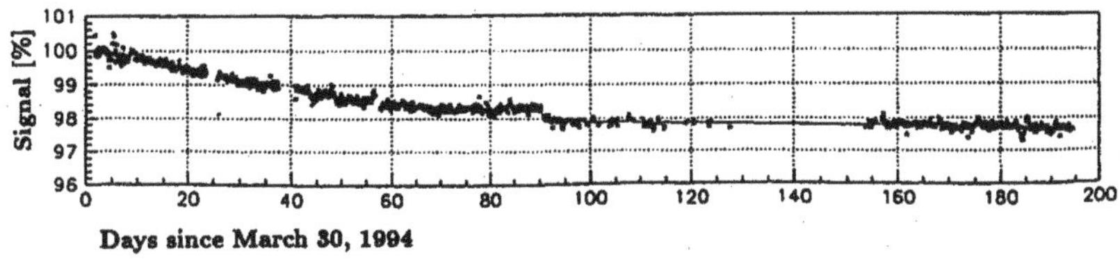

Figure 2: Decrease of the amplitude on the ionisation cell during the leakage in the cryostat valve. The amplitude before the leakage is 100%

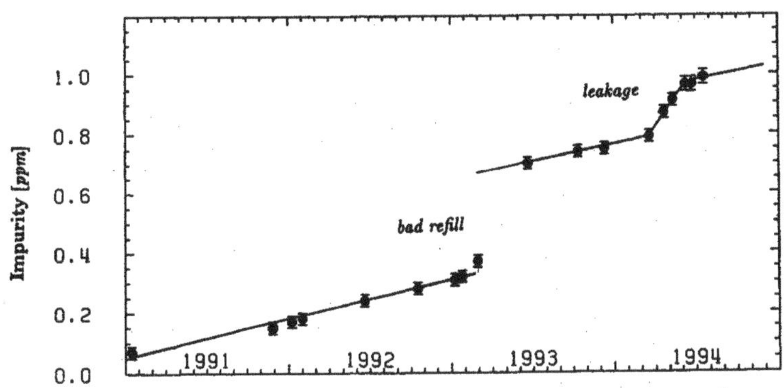

Figure 3: The polution in the LAr calorimeter as function of the time.

capacity is around $C_d = 5$ nF on average ($C_d^{max} = 16$ nF). The shaping time is chosen by 2.4 μs to optimize the signal to noise ratio. The noise charge of the entire chain is $Q_{noise} = 5e_0C_d + 5500e_0$. The measured electronic noise (1σ) is 34 MeV (10 MeV) for barrel (forward) part of the EM calorimeter and 28 MeV (18 MeV) for the HAD calorimeter.

The signals from the amplifier are digitized by ADC's and processed by Digital Signal Processor (DSP) which performs pedestal subtraction, noise suppression and electronic calibration correction.

The electronic calibration is done by pulsing 992 precise calibration pulse generators. The absolute precision of the calibration is of the order (0.1-0.2)%. A third order polynomial fit is performed according to $Q^i = \sum p^i_n(ADC^i_{counts} - pedestal)^n$, where Q^i is the pulsed charge and ADC^i_{counts} is the digital measurement. The parameters p^i_n and the pedestals are loaded into the DSP's to translate ADC counts into collected charges. The stability of the electronic calibration is checked at least every second day. Usualy a new calibration is necessary after a hardware change or done once per month to keep the calorimeter performance in good shape.

5. The energy calibration.

5.1 Constants for electrons and the electromagnetic energy scale.

The calibration constants for each wheel type were measured using an electron beam in the range of 3.7 GeV - 80 GeV [2]. By correcting effects of dead material, leakage and noise suppression cuts the absolute calibration constants C^{em}_{abs} were extracted. The numerical value of the constants C^{em}_{abs} is in the range (3.41–3.62) GeV/pC. The set of these constants, one for each wheel type, defines the *Electromagnetic Energy Scale.* The systematic uncertainty in the absolute scale is around 3% presently. The uniformity of the calorimeter response within a wheel is within ±1%. The deviation from linearity are limited to ±1%. The energy resolution of the EM calorimeter is $<\sigma/E> = 11.5\%/\sqrt{E} \oplus 1\%$ on average.

5.2 Constants for pions and hadronic energy scale.

The absolute calibration constants for the HAD calorimeter (*Electromagnetic Energy Scale*) are calculated using the absolute calibration constants for electrons and the inverse of the sampling ratio of the EM and HAD calorimeters as determined with MC simulations[3]. $C^{had}_{abs}(E) = (C^{had}_{MC}(E)/C^{em}_{MC}(E))C^{em}_{abs}$. The numerical value of the constants $C^{had}_{abs}(E)$ are in a range of (6.7–7.6)GeV/pC

The H1 calorimeter is a non-compensating calorimeter. A weighting procedure is applied to come to the final hadronic energy scale. The HAD calorimeter response was measured with pion beams in the range of 3.7 GeV - 205 GeV [3] at CERN. The pion signal is described by MC within an uncertainty of 3%. The energy resolution of the HAD calorimeter is $<\sigma/E> = 51\%/\sqrt{(E)} \oplus 1.6\%$.

6. Check of the energy calibration

Unfortunately at HERA no special calibration events like Bhabha scattering are available. But several methods provide a good check of the energy scale. Some of them are given below.

6.1 The radiative cosmic muons.

Passing the detector a part of the cosmic muons produce electrons. Their energy E and momentum p are measured in the calorimeter and the tracker. The measured and simulated E/p distribution are shown in fig. 4a for selected electrons with $p > 1$ GeV/c. The distribution demonstrates that the electromagnetic energy scale is verified better than ±1.5%. Fig. 4b gives a comparison of the E/p distribution for 1994 and 1993 data. The stability of the reconstructed calorimeter response is within ±1% during that period.

6.2 The double angle method (DA).

In NC events the energy of the scattered electron E_{DA} can be reconstructed by using the polar angles of the final electron and of the outgoing quark jet. Also the energy of the electron E_{el} is measured directly in the electromagnetic calorimeter. Fig. 5a demonstrates the distribution of the E_{el}/E_{DA} for $p_t > 10$ GeV/c. The value $< E_{el} >$ agrees with D.A. prediction better than 1%.

6.3 The p_t balance.

The direct verification of the hadronic energy scale at HERA can be done by

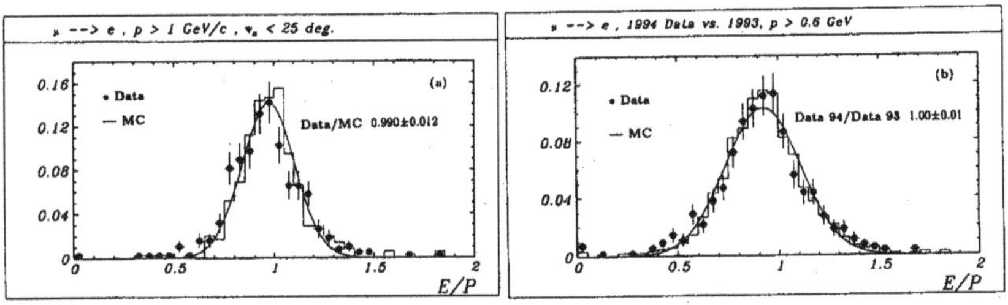

Figure 4: (a) Energy match of electrons generated by cosmic muons in the tracker and the LAr calorimeter; (b) Comparison of the data 1994 and 1993.

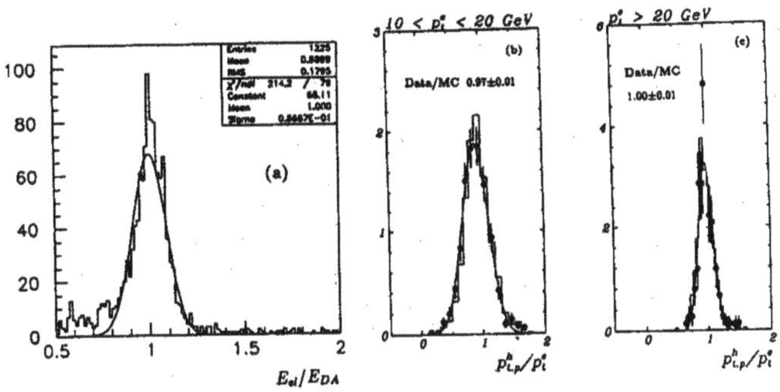

Figure 5: (a) Distribution of the E_{el}/E_{DA}. (b) The p_t^h/p_t^e ratio.

using the transverse momentum p_t^e of the scattered electron and the vectorial sum of the hadronic transverse momentum p_t^h which is measured in LAr calorimeter. Fig. 5b and 5c show the ratio p_t^h to p_t^e for $p_t^e > 10$ Gev/c and and $p_t^e > 20$ GeV/c and a comparison of the data with MC. The measured energy balance is compatible with simulation. Taking into account the uncertainty in the electromagnetic energy scale the hadronic energy scale is verified within ±5%.

7. The H1 calorimeter trigger

The signals of the 44352 calorimeter cells are grouped and summed in the 5009 trigger cells. The signals from the trigger cells pass through analog switches and are summed into 656 trigger towers (TT). The EM and HAD trigger towers separately pass through the analog threshold and form the 256 EM and HAD big towers (BT). The analog threshold for each TT is calibrated and set to 1.2 GeV during the data runs. The signals of BT's are digitized by FADC's and are sent to the digital electronics. After applying look-up tables calorimeter trigger elements are created, such as for missing energy, transverse energy, an electron trigger etc. with three level of thresholds. The entire time for trigger decision is about 21.5 bunch crossings (BC) or 2.1 μs. Fig. 7a,b show the precision of the time t_0 determination. It is 20 ns for the entire trigger system over the full dynamic range.

Fig. 7a demonstrates the correlation between the response of the trigger

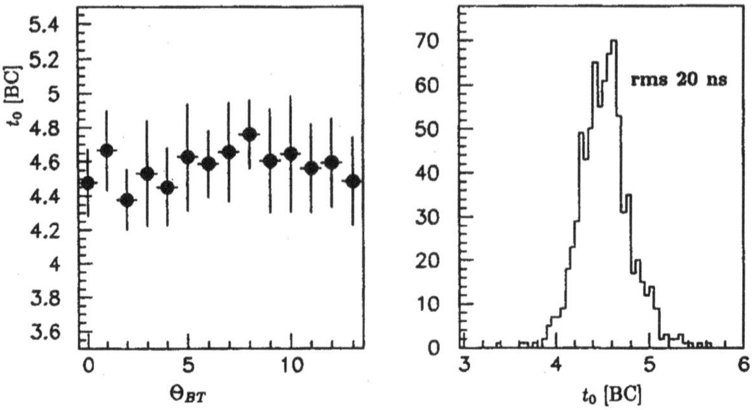

Figure 6: (a) Timing of the calorimeter trigger for different big tower index Θ_{BT}. (b) t_0 distribution for the central barrel. Time is given in bunch crossing (BC).

FADC's and the calorimeter ADC's. The linearity is good and no increase of the errors with energy is visible which indicates the absence of additional systematic errors. The efficiency of the trigger is presented in fig. 7b for all trigger towers. The 50% efficiency level is reached around 1 GeV.

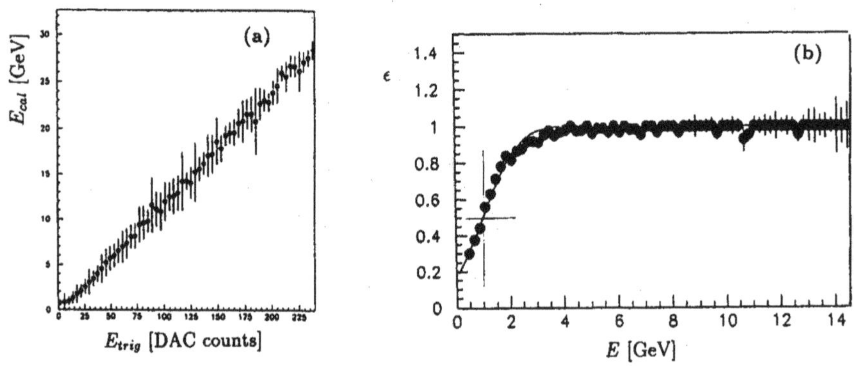

Figure 7: (a) The energy correlation between the trigger E_{trig} and the calorimeter E_{cal} signals. (b) Efficiency ϵ of the all trigger towers.

8. Conclusions

- The 'natural' decrease of the calorimeter signal due to pollution is 0.77% per year for period 1991–1993.
- The stability of the electronic calibration is around 0.1% during a month.
- The homogeneity and linearity are within ±1%. The resolution of calorimeters are $<\sigma/E>=11.5\%/\sqrt{E}\oplus 1\%$ for e's and $<\sigma/E>=51\%/\sqrt{E}\oplus 1.6\%$ for π's.

- The two presented checks of the electromagnetic energy scale verify it within ±1.5%.
- The hadronic energy scale is compatible with MC within ±5%.
- The analog threshold of the LAr calorimeter trigger is 1.2 GeV. It is well understood and calibrated.
- The precision of the t_0 determination is 20 ns for the entire trigger system.
- Both the LAr calorimeter and its trigger system were in full operational mode during the 93 – 94 run period and showed a good performance.

9. Acknowledgements

I am grateful to my colleagues of the H1 calorimeter group, especially to K.Borras and J.Gayler for a useful discussion and a critical reading of the draft.

10. References

1. H1 Calorimeter Group, *NIM* **A336** (1993) 460-498.
2. H1 Calorimeter Group, *NIM* **A350** (1994) 57-72.
3. H1 Calorimeter Group, *NIM* **A336** (1993) 499-509.

THE PERFORMANCE OF THE ZEUS CALORIMETER

JAMES A. CRITTENDEN
Physikalisches Institut, Universität Bonn
Nußallee 12, D-53115 Bonn, Germany

for

The ZEUS Calorimeter Group

ABSTRACT

The ZEUS experiment has now completed its third year of operation at the electron-proton collider HERA. The uranium/scintillator sampling calorimeter surrounding the inner tracking detectors has proven an essential component for the online triggering algorithms, for offline event-type identification, for kinematic variable reconstruction, and for a variety of physics analyses. This paper summarizes the experimental context, the operating characteristics, the calibration techniques, and the performance of the calorimeter during its first three years of operation.

1. Introduction

The uranium/scintillator sampling calorimeter for the ZEUS experiment has been operating at the electron-proton storage ring facility HERA since 1992. Its measurements of electromagnetic and hadronic shower energies, positions, and times play a central role at each of the three trigger levels, as well as in offline background rejection, event classification, reconstruction of kinematic variables, jet-finding, and a variety of other analysis tasks. It is the intention of this contribution to the conference to detail the present status of the calorimeter performance. After a review of the mechanical design, readout and calibration techniques, and selected test beam results, we concentrate on lessons learned from the first three years of operation.

2. Experimental Context

Fig. 1 depicts the general purpose, 4π-detector ZEUS designed to study the interactions of HERA's 820 GeV protons and 27 GeV electrons. The innermost component is a high resolution drift chamber (VXD) which provides transverse position measurements of track segments in a radial region from 11 to 15 cm. The central tracking detector (CTD) surrounds the vertex detector, its nine superlayers of drift cells of eight sense wires each extending from 16 to 85 cm in the radial coordinate. A superconducting solenoidal coil one radiation length (X_0) thick, providing a 1.4 Tesla field in the tracking region, is situated outside the CTD and in front of the barrel calorimeter (BCAL). In the forward (proton flight) direction a multilayer drift chamber/transition radiation detector (FDET) provides further tracking outside the CTD and in front of the forward calorimeter (FCAL). The endplate of the CTD and the

FDET are the dominant contributions to material between FCAL and the interaction point, consisting of $\simeq 1\,X_0$. In the rear (electron flight) direction there are a small angle scintillation hodoscope (SRTD) and a planar drift chamber (RTD) in front of the rear calorimeter (RCAL). The material between RCAL and the interaction point amounts to $\simeq 0.5\,X_0$, except for a narrow region near the beam pipe, where it ranges up to several X_0. Thus over most of the detector volume the material traversed by particles before entering the calorimeter is less than $\simeq 1\,X_0$, except for narrow regions near the edges of the calorimeters, where the total amount of material traversed by a straight trajectory can reach several X_0. Surrounding the uranium calorimeter is a backing calorimeter (BAC) consisting of wire chambers installed in slots in the iron magnetic field return yoke, and the muon tracking system (FMUON, BMUON, RMUON). The forward muon system includes toroidal magnets for an additional momentum measurement. Two planes of large-area scintillation counters (Veto Wall) behind RCAL provide rejection against upstream interactions of the proton beam with residual gas in the beam pipe. A luminosity measurement is provided by small-angle electron and photon calorimeters in the HERA tunnel in the electron flight direction. A silicon-strip leading-proton spectrometer installed in Roman pots and a forward neutron calorimeter furnish information on the hadronic final state at small angles in the proton flight direction.

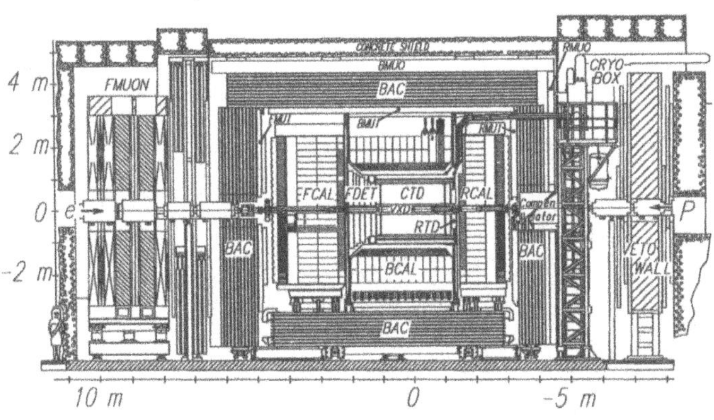

Figure 1: *The ZEUS detector*

The measurement of structure functions at the high Q^2-values attainable at HERA requires calorimetric measurements of the scattered electron, of the hadronic final-state formed from the struck proton constituent, and of particles emanating from the transition region of color compensation extending between the jet and the proton remnant. The latter component of the final state is often observed as large energy depositions in FCAL near the beam pipe. In contrast, such energy depositions are conspicuously absent in such candidate events for diffractive scattering as exemplified in Fig. 2 ($Q^2 \simeq 70$ GeV^2 !). While not all diffractive scattering events show such structure, the probability that a scatter from a colored object in the proton exhibit such an energy configuration is much less than the relative frequency with which the HERA experiments have observed such events. Particularly interesting from the point of view of calorimetry is the necessity of a low noise level in order to be able to identify events with "nothing" at forward angles. This event picture does not display signals with less than 60 MeV in electromagnetic cells and 110 MeV in hadronic cells. One observes an isolated 0.3 GeV deposit in the forward calorimeter which does not exceed the 0.4 GeV threshold used by the simple algorithm classifying this event as a candidate for diffractive scattering. This intriguing process involving a high momentum transfer scatter of the electron from an apparently colorless proton constituent is one of a wide variety of interactions which result in depositions of only a few GeV in the calorimeters.

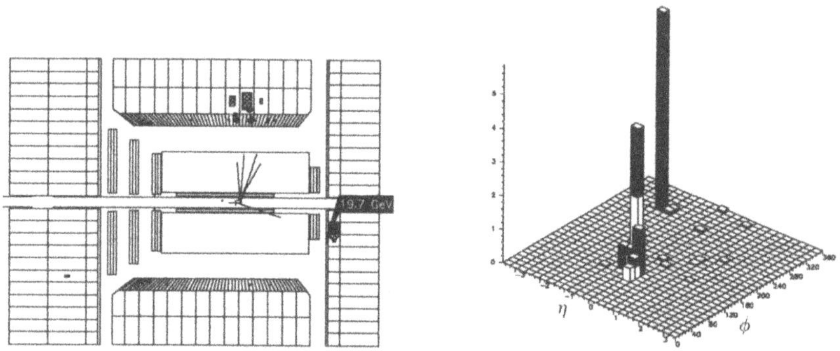

Figure 2: *Candidate Event for Diffractive Scattering at High Q^2. The picture on the left shows tracks and energy depositions in the upper (lower) half of the detector rotated around the beam axis to the upper (lower) half of the vertical plane. The picture on the right shows a lego plot of transverse energy (GeV) in bins of pseudo-rapidity and azimuthal angle. Dark (light) towers represent transverse energy in the electromagnetic (hadronic) sections.*

3. Mechanical Design

The 700-ton ZEUS calorimeter [1,2,3] consists of forward, central barrel, and rear sections, covering the polar angle ranges $2° - 40°$, $37° - 129°$, and $128° - 177°$. Each is made up of layers of 2.6 mm SCSN-38 scintillator and 3.3 mm stainless-steel-clad depleted-uranium plates. Each layer corresponds to $1.0\,X_0$ and 0.04 interaction lengths (λ_I). This uniformity in structure throughout the entire calorimeter and the distributed nature of the natural radioactivity of the uranium are decisive factors in our ability to maintain a precise calibration of the calorimeter. The choice of active and passive thicknesses results in a sampling fraction of 4% for electromagnetic and hadronic shower components (hence compensation [4]) and 7% for minimum-ionizing particles.

The wavelength-shifting optical readout components subdivide the calorimeters longitudinally into electromagnetic (EMC) and hadronic (HAC) sections. The EMC sections are $25\,X_0$ $(1.1\,\lambda_I)$ deep. The FCAL (BCAL) HAC sections are further segmented longitudinally into two sections, each of $3.0\,\lambda_I$ $(2.0\,\lambda_I)$ depth. RCAL has a single HAC section $3.0\,\lambda_I$ deep. The optical components subdivide FCAL and RCAL into towers of transverse dimensions 20 cm × 20 cm. The EMC sections are further divided vertically into four (two) cells in FCAL (RCAL.) Thus the transverse segmentation of the EMC cells is 20 cm × 5 cm in FCAL and 20 cm × 10 cm in RCAL. The front face of FCAL (RCAL) is 2.2 m (1.5 m) distant from the nominal electron-proton interaction point. BCAL consists of 32 wedge-shaped modules situated 1.2 m distant from the beam axis. Its EMC sections are divided by the optical readout components into 53 cells per module, each of transverse dimensions 24 cm × 5 cm, approximately projective in both the polar and azimuthal angles. The fourteen HAC sections in each module measure 24 cm × 28 cm in the transverse dimension and are projective only in the azimuthal angle. A 2.5° rotation of all the BCAL modules relative to the radial direction ensures that the intermodular regions containing the wavelength shifters do not point at the beam axis.

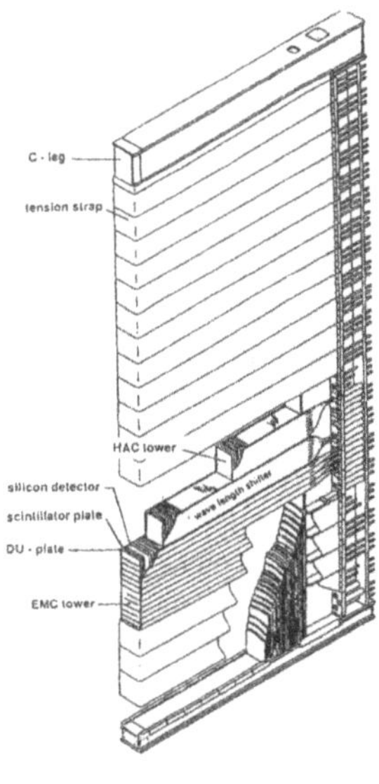

Figure 3: *Geometrical configuration of an FCAL module*

The light signal from each calorimeter cell is received by two phototubes which view the ends of wavelength-shifting light guides situated on opposite sides of the cell. This redundancy in the readout allows not only for a transverse position measurement within the cell, but permits the use of the left/right sum as an energy measurement independent of the impact point (which is important for the trigger), and further allows the separation of intrinsic shower fluctuations from those of photostatistical origin in beam tests. Light yield requirements imposed tight tolerances on the quality of the wavelength shifting light guides.[5] Since they occupy "crack" regions between modules, their thickness was limited to 2 mm, the minimum consistent with constraints on the attenuation length along the length of the active area of the light guides. This source of inhomogeneity due to the attenuation was corrected with a patterned reflective wrapping. The minimum light yield is 100 $\frac{photoelectrons}{GeV}$ and 1.2 $\frac{photoelectrons}{layer}$ for each phototube. Thus photostatistics contribute less than 7% to the energy resolution for electromagnetic showers, which is $18\%/\sqrt{E(GeV)}$.

Fig. 3 shows the an FCAL module as an example of the above-described construction geometry. Spaces are available at depths of $3\,X_0$ and $6\,X_0$ for insertion of the 3 cm × 3 cm silicon diodes (HES) which allow hadron-electron separation by sampling the electromagnetic shower near its maximum. These detectors were installed in RCAL (which has such a slot at a depth of $3\,X_0$ only) for the 1994 data-taking period. The FCAL HES is not yet installed.

4. Readout Technique

The phototube high voltages are set using the radioactivity of the depleted uranium. Tight tolerances during construction of the calorimeter allow the cell-to-cell variation of the response to electrons to be controlled at the 1% level when normalized to the uranium background level. The uranium calibration further determines the relative EMC/HAC phototube gains, ensuring compensation at the 1% level for shower energies exceeding 5 GeV. The compensation results in a hadronic energy resolution of $35\%/\sqrt{E(GeV)}$.

The 11836 phototube signals are each split five ways in the front end analog cards: to a current integrator averaging the input current over $20\,ms$ and providing the uranium calibration signal measurement, to a current sum node serving the first level trigger, to the two shaping-and-sampling paths of differing gains, and to a termination resistor. The shapers use precision components to define an effective shaping time constant of about 100 ns uniformly for all shapers to 0.2 ns accuracy.

The processing of the shaped signals is based on CMOS switched-capacitor chips developed for the ZEUS application.[6] The signals are sampled at the 10.4 MHz HERA bunch crossing rate such that a sample on each of the rising and falling edges be guaranteed. A sample on the baseline preceding the pulse serves as protection against pile-up effects. The slope-weighted sum of the two baseline-corrected samples on the pulse edges provides a jitter-independent measurement of the energy deposit and the difference provides a shower time measurement. Two gain scales, differing by a factor of 22.22 (5.00) in FCAL and BCAL (RCAL), and 12-bit digitization allow an effective dynamic range of 17 (15) bits. The RMS noise level is dominated by the signal from the uranium activity and is at a level of $\simeq$ 10 MeV per cell (see section 5). The gain scales are defined such that the high gain path saturates at 24 GeV in FCAL and 18 GeV in BCAL and RCAL. The low gain paths saturate at 530 GeV in FCAL, 380 GeV in BCAL, and 90 GeV in RCAL.

Gain and pedestal correction factors are necessary for each switched-capacitor cell to ensure that noise and linearity requirements are fulfilled. These are obtained via calibration runs exploiting the on-board digital-to-analog converter (DAC) on each 12-channel front-end analog card. A complete calibration data set consists of about 20 Mbytes. The sample calibration corrections and the charge and time reconstruction algorithms are performed in digital signal processors (DSP's) treating the 12-bit digitization results. The digitization occurs at a rate of 0.6 MHz in modules which receive the analog, multiplexed outputs of the analog cards over $\simeq$ 50 m of shielded twisted-pair cable. The 58-cell pipelines are clocked continuously at 10.4 MHz until a first-level trigger occurs, at which time they are stopped for 10 μs to transfer six samples for each gain path to the switched-capacitors in a buffer-multiplexer chip. Front end data acquisition then continues while the buffer-multiplexer data is clocked at 0.6 MHz to the digitizers. This double-buffering scheme achieves a deadtime of 1% for a 1 kHz first-level trigger rate and allows 5 μs for a trigger decision.

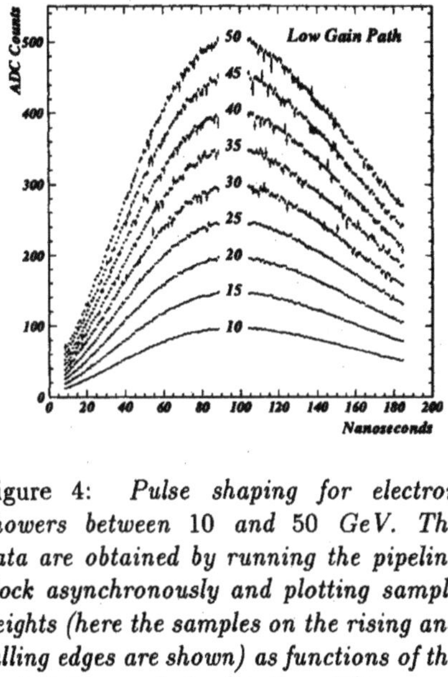

Figure 4: *Pulse shaping for electron showers between* 10 *and* 50 *GeV. The data are obtained by running the pipeline clock asynchronously and plotting sample heights (here the samples on the rising and falling edges are shown) as functions of the arrival time - clock edge time difference.*

The DSP's provide calibrated energy and time measurements to the second-level trigger transputer network, as well as to an event builder serving the third-level MIPS processor farm.

One means of measuring the pulse shape is provided by running the pipeline clock asynchronously and plotting sample heights versus the time difference between particle arrival time and pipeline clock edge. Fig. 4 shows the pulse shapes for test beam electrons between 10 and 50 GeV.

A second generation of switched-capacitor chips was installed during the shutdown preceding the 1994 data run. The new chips have been demonstrated to survive an order of magnitude greater radiation dose compared to the original generation, continuing to operate at doses up to 1000 Gray.[7]

5. Operating Characteristics

The operating characteristics of the calorimeter were determined in extensive test beam programs at FNAL and CERN [3,8] on both prototype and final modules. Fig. 5 shows the energy dependence of the resolution and the ratio of electron to pion signals, measured in FCAL prototype modules at CERN, documenting the values cited above. The uniformity across module and tower boundaries was thoroughly investigated as well. Nonuniformities at the level of 10% occur in narrow regions near cracks, resulting in an averaged nonuniformity under 1%. Of the total of 80 modules, six FCAL, four RCAL, and six BCAL modules were examined in test beams. One goal was the verification of construction tolerances as reflected in the cell-to-cell reliability of the uranium signal calibration. Fig. 6 shows an example of a result for 15 GeV electrons in FCAL and RCAL modules. These ten modules comprised 37% (28%) of all the electromagnetic cells in FCAL (RCAL). Cell-to-cell RMS variations of 1.3% (1.5%) were measured for these FCAL (RCAL) cells, to which the measurement uncertainty contributed 0.3%. All FCAL and RCAL modules were subsequently measured at DESY using cosmic ray muons, with such accuracy that the individual cell variations within the 1% RMS spread could be corroborated for the cells which had been measured at CERN. Of the 32 BCAL modules, 22 underwent extensive investigations with cosmic rays as well.[9]

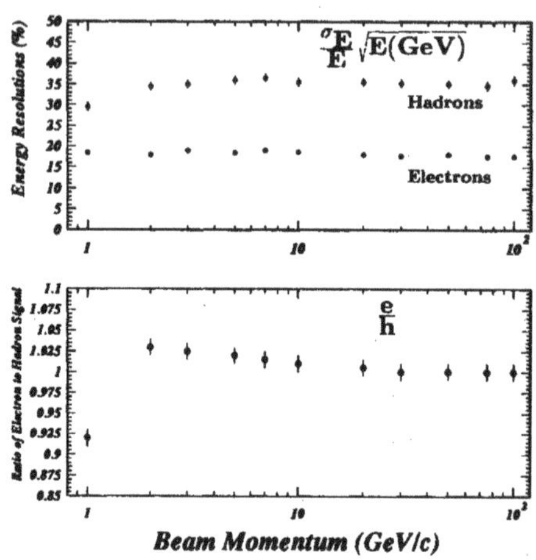

Figure 5: *Energy resolution and the ratio of electron to pion signals as a function of beam momentum, measured for the FCAL prototype modules at CERN.*

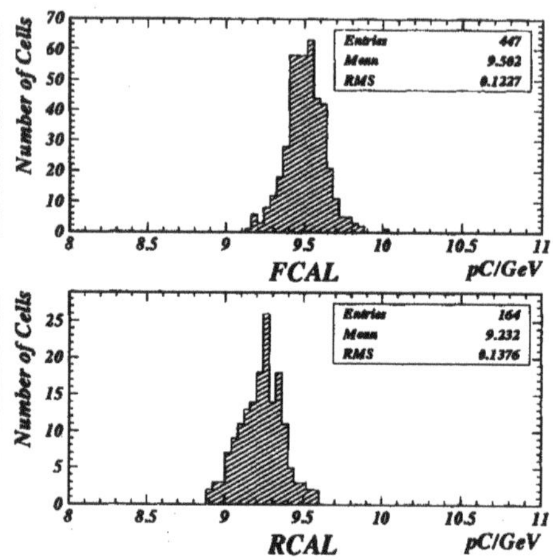

Figure 6: *Cell-to-cell calibration uniformity for the six FCAL and four RCAL modules measured at CERN. These modules comprise 37% (28%) of all the electromagnetic cells in FCAL (RCAL). RMS nonuniformities of 1.3% and 1.5% were measured.*

6. Calibration Methods

The natural radioactivity of the uranium provides the absolute calibration of the ZEUS calorimeter. Test beam studies furnished the calibration factors necessary to maintain the calibration at the level of 1%. During data-taking periods the uranium currents are recorded daily for use as offline normalisation correction factors. Fig. 7 shows measurements of the uranium current signal, averaged over 20 ms to remove statistical fluctuations, for each phototube of the central FCAL module. The nominal currents are chosen such as to ensure that the response of each cell to minimum-ionizing particles scale with the number of layers in the cell. The factor of five greater HAC uranium signals as compared to the EMC uranium signals is the result of the larger transverse cell dimension (20 × 20 cm versus 5 × 20 cm), the larger number of uranium/scintillator layers (80 versus 25), and a thicker stainless steel cladding (400 μ versus 200 μ) on the uranium plates. Deviations of the uranium current settings from nominal are shown in Fig. 8 for a recent run. Of the 11836 tubes, 126 deviated by more than 20% from the nominal setting and were excluded from the offline analysis. Variations over the period of a week are typically under 1%. Adjustments of the phototube high voltage values to reëstablish the nominal currents are undertaken once a month.

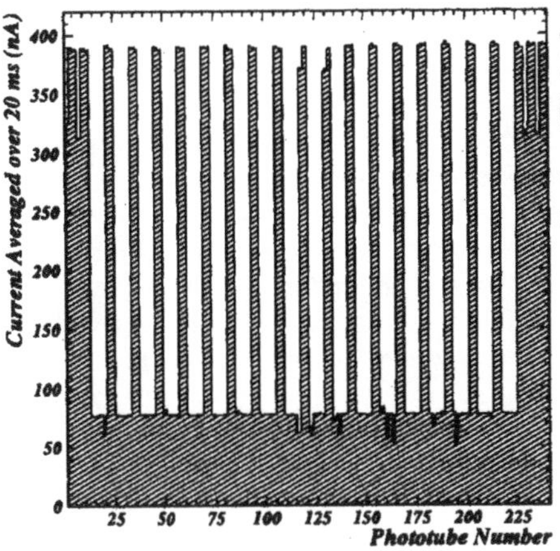

Figure 7: *Uranium activity calibration current values for the central FCAL module on 1 September 1994.*

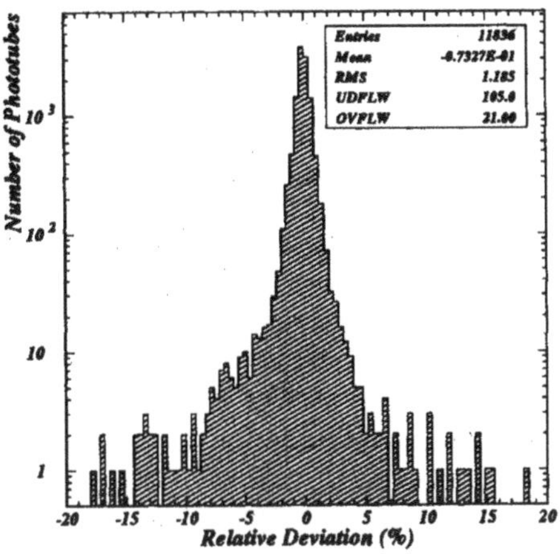

Figure 8: *Relative deviations from the nominal uranium activity calibration current values for the central FCAL module on 1 September 1994.*

As mentioned in section 4, the switched-capacitor readout is calibrated by means of on-board DAC's which provide levels at the inputs to the pipelines, allowing the measurement of gain and pedestal values for each capacitor. Calibrated charges are also injected at the inputs to the shapers. A complete calibration of the front-end electronics is performed once a week.

Strict rejection criteria based on the weekly front-end calibration and the daily measurements of charge injection and uranium currents are used to generate a list of channels to be excluded from the offline analysis. Typically $\simeq$ 2% of the 11836 readout channels are declared useless. The number of calorimeter cells for which *both* readout channels malfunction at any time is typically less than two.

Each phototube is equipped with an optical fiber permitting light injection into the wavelength-shifting light guide just in front of the photocathode. The LED and laser light sources provide periodic measurements of the gain, linearity, photostatistics, and phototube transit times.

Precise measurements of the structural integrity of the FCAL and RCAL modules are obtained by sliding a 1-mm-long 70 MBq ^{60}Co source in steps of 0.2 mm through tubes positioned along the length of a tower.[11] This method allows the verification of the positioning of each scintillator tile and of the wavelength-shifter assemblies. All modules were scanned before installation in 1991 and during shutdowns accessible modules in the beam pipe region have been scanned each year. The shapes of the ^{60}Co distributions have shown stability at the 2% level over a period of three years, consistent with the negligible radiation damage expected from dosimeter measurements.

7. Selected Applications in Physics Analyses

The 1994 HERA data-taking period, which ended on November 1, has provided the ZEUS experiment with 6.1 pb^{-1} of integrated luminosity, to be compared to 1.1 pb^{-1} in 1993 and 0.05 pb^{-1} in 1992. The rapid progress of the HERA accelerator performance means that our ability to study systematic effects in the calorimeter response continues to improve dramatically. The studies presented in this paper are based on the data recorded in 1992 and 1993, thus representing 15% of our entire data sample to date. We present here a selection of the applications of the calorimeter data, sorting them according to the use of the time, position, and energy measurements.

The shaping-and-sampling algorithm described in section 4 provides measurements of the phototube pulse time relative to the pipeline clock, which is synchronized to the accelerator RF. After corrections for time shifts of the bunch arrival time relative to the HERA RF, differences in the distance of each calorimeter cell from the interaction point, shower development times, and phototube transit times, subnanosecond resolutions are obtained.[10] Fig. 9 shows the time resolution for an FCAL EMC cell in an energy range where systematic effects begin to dominate the statistical and shower fluctuations. In general, one achieves a resolution of better than 1 ns for depositions greater than a few GeV per phototube.

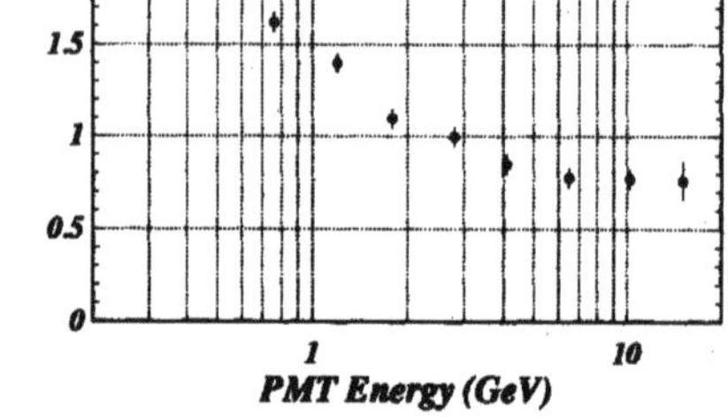

Figure 9: *Single phototube calorimeter time resolution for electromagnetic energy depositions in FCAL.*

This time information plays an important role in the rejection of triggers caused by particles produced in interactions of the proton beam with the residual gas in the beam pipe. These upstream interactions produce depositions in the rear of RCAL, which differ in time from depositions by particles produced at the electron-proton interaction vertex by 12 ns. Since the RCAL time measurement as calculated from all RCAL energy depositions is accurate to better than 1 ns even without all the offline corrections listed above, it provides a powerful tool for background rejection at the second trigger level.

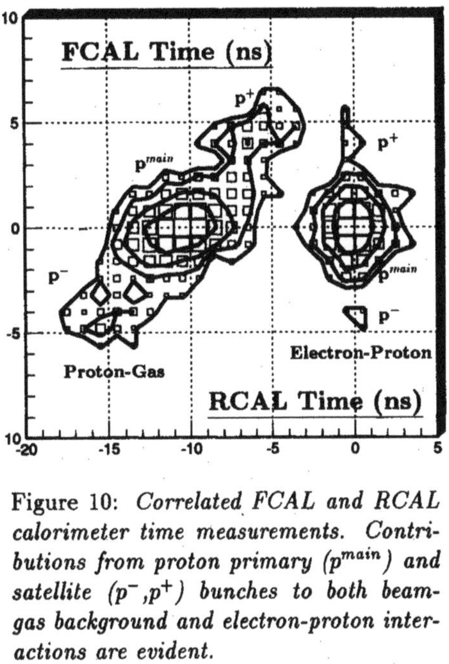

Figure 10: *Correlated FCAL and RCAL calorimeter time measurements. Contributions from proton primary (p^{main}) and satellite (p^-,p^+) bunches to both beam-gas background and electron-proton interactions are evident.*

Another contribution of the calorimeter event time measurements is the monitoring of irregularities in the HERA beams' RF structure. In 1993 a new RF system was activated for the proton beam, reducing the primary proton bunch RMS length from 40 cm to 20 cm. Some protons were trapped in satellite bunches which lead and trail the primary proton bunch by 4.5 ns. Since the population and lifetime of these satellite bunches varied dramatically from fill to fill, monitoring their behavior and correcting the data samples for triggers caused by their interactions was essential. Fig. 10 shows a contour plot of FCAL time versus RCAL time, illustrating the time structure of the contributions from proton primary and satellite bunches to both beam-gas background and electron-proton interactions.

The calorimeter event times are also used to determine the position of the interaction vertex along the beam line. While the resolution of 8 cm is an order of magnitude worse than that of the CTD, the systematics and acceptance are very different. In particular, the calorimeter measures the vertex position better than the CTD in events with tracks at small angles only.

The primary use for the calorimeter measurements of incident position is for that of the scattered electron produced in the neutral-current deeply inelastic scattering (DIS) process. The rate of the low-Q^2 electrons at small scattering angles falls steeply with scattering angle and the position cut defines the lower bound on the acceptance in Q^2. The best estimate of the calorimeter position resolution in this region is given by comparison to position measurements calculated from the HES silicon-diode detector data in RCAL, which have a resolution better than 3 mm. The comparison indicates a resolution of 1.3 cm vertically (via charge sharing between cells) and 0.8

cm horizontally (via the transverse light attenuation in the scintillator tiles), averaged over the position and energy distributions of the scattered electrons, which peak at small scattering angles and near the electron beam energy. The position resolution varies strongly with the position within a cell and reaches an accuracy of 3 mm in the vertical coordinate near cell boundaries.

We begin our discussion of the calorimeter energy measurement with the noise levels, which are dominated by the uranium signal contribution to fluctuations in the three samples used to reconstruct the charge. Fig. 11 shows for all 11836 channels the RMS width values from pedestal triggers recorded during empty bunches of a physics run. Two bands are observed for each calorimeter, each corresponding either to the EMC or to the HAC cells. In general, the noise levels are $\simeq$ 10 MeV in the EMC cells and $\simeq$ 15 MeV in the HAC cells, for which the uranium currents are greater. Also shown is the noise level measured with the high voltages off, corresponding to $\simeq$ 5 MeV.

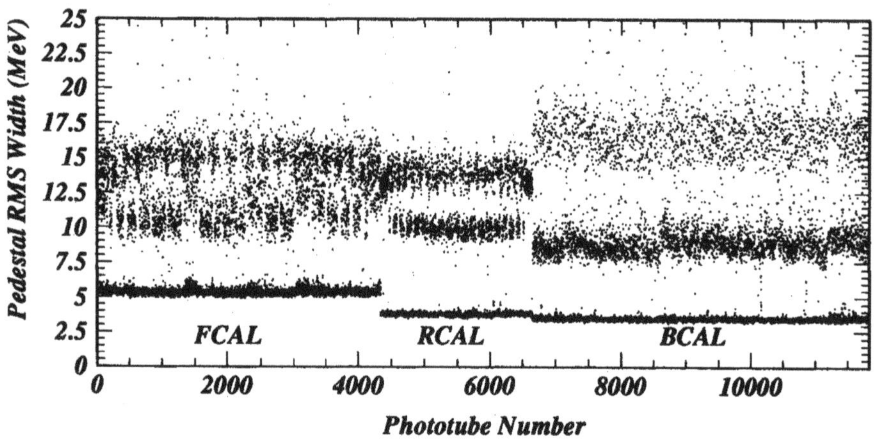

Figure 11: *Pedestal RMS width values for the 11836 phototubes as calculated from triggers recorded during empty bunches of a physics run. Bands at values of* 10 *MeV and* 15 *MeV show the levels for the EMC and HAC cells of each calorimeter. The values near* 5 *MeV were measured with the high voltages turned off and indicate the contribution from the electronics noise. Disfunctional readout channels (*$\simeq$ 2% *of all channels) are not excluded from this plot.*

For comparison, fig. 12 shows the RCAL signal reconstructed from proton beam halo muons, which deposit 330 MeV electron-equivalent-energy in the EMC section and 1100 MeV in the HAC section, as well as the pedestal distribution for the six-channel sum for a single RCAL tower, which has an RMS width of 36 MeV. There is a slight correlation in the noise from the two phototubes of the same calorimeter cell.

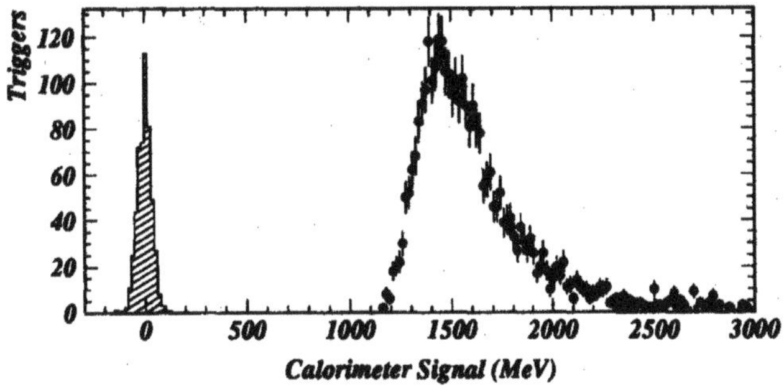

Figure 12: *A comparison of the pedestal signal summed over the six channels of an RCAL tower and the pedestal-suppressed signal for the entire RCAL from beam halo muons, which deposit* 330 *MeV electron-equivalent-energy in the EMC section and* 1100 *MeV in the HAC section.*

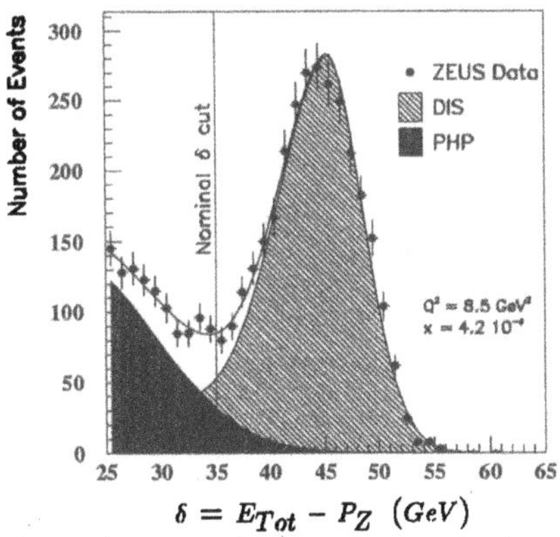

Figure 13: *The distribution in the variable* $\delta = E_{Tot} - P_Z$ *summed over the entire calorimeter for DIS candidate events in a bin of* x_{Bj} *and* Q^2 *where the background from photoproduction processes is significant.*

Fig. 13 shows how the calorimeter data contribute to the event-type selection by exploiting the longitudinal kinematic constraint for DIS events.[12] Shown is the distribution of the total energy measured in the calorimeter with the total longitudinal momentum subtracted. This quantity is insensitive to fluctuations in the energy measured in the forward direction and is kinematically constrained for DIS events to a value of twice the electron beam energy. For the present purpose of illustration, a region of x_{Bj} and Q^2 has been chosen such that the background from photoproduction processes, where the electron exits the detector through the beam pipe, is significant.

The DIS contribution is observed to peak near 45 GeV rather than at the value of 54 GeV expected if sources of signal loss are ignored. The kinematic constraint is compromised by energy loss of both the electromagnetic and hadronic final state in dead material between the interaction point and the calorimeter.

The energy-loss correction algorithms under development will not be discussed in this paper, however two observations are appropriate. The first is that studies using the proton beam halo muons mentioned above and comparing their signals with both the cosmic ray data and the test beam data verified the energy scale calibration of the calorimeter with an accuracy of 3%. Secondly, the SRTD, which covers an area of $4500\,cm^2$ around the beam pipe in front of RCAL, has proven to provide an accurate means of correcting for early showering. The depositions in these strips exhibit an anticorrelation to the calorimeter signal, as shown in fig. 14 for a sample of DIS electrons in which the scattered electron energy is constrained to the beam energy. The selection of these electrons exploits the overconstrained kinematics, deriving the electron energy from its angle and the hadronic final state. The distribution of selected energies has an asymmetric tail due to radiation by the initial state electron and to the resolution in the cut variables. The simple linear correction function depicted in the upper left-hand plot in fig. 14 results in the improvement indicated in the lower right-hand plot. The SRTD was installed during the 1993/1994 shutdown and such studies are just now becoming available.

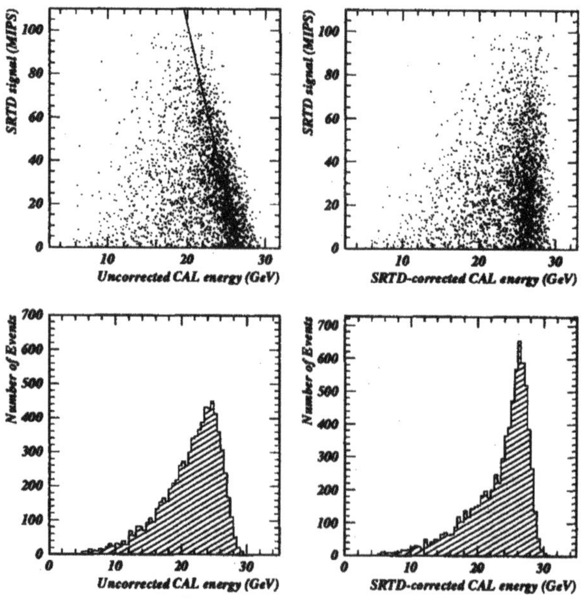

Figure 14: *A correction procedure for energy loss in material in front of RCAL using presampling by the SRTD for scattered electrons kinematically constrained by selection to the electron beam energy. The simple linear correction function derived from the anticorrelation of the SRTD and RCAL signals shown in the upper left-hand figure results in the improved spectrum indicated in the lower right-hand figure.*

8. Upgrades in Progress

A presampling detector component consisting of $20 \times 20\,cm^2$ scintillator tiles read out via wavelength-shifting fibers and covering the portions of the faces of FCAL and RCAL which are not shadowed by BCAL was approved early this year for installation during the 1994/1995 shutdown. The proposal for this detector was based on test beam studies performed in 1992 and 1993 which indicated that an energy correction algorithm was practical. In particular, it was shown that the correction could be made independent of specific knowledge of the presence of dead material in front of the presampler up to $4\,X_0$ and that an event-by-event correction brought substantially more improvement than an average correction based on the geometry of the dead material.[13] Further analyses of test beam data recorded in 1994 are in progress.

Preparations for the installation of the HES silicon-diode detectors in FCAL, complementing those installed in RCAL in 1994, are underway. Pilot modules will be instrumented for the 1995 data-taking period, with full instrumentation of one slot approved for the following year.

Initial simulation and test beam studies have been undertaken for barrel presampler and shower-max detectors. A proposal for this upgrade is in preparation.

A redesign of the beam pipe which will reduce the dead material in front of RCAL for small electron scattering angles has been concluded. Construction is underway and installation is expected for the 1994/1995 shutdown.

9. Summary

The performance of the ZEUS uranium/scintillator sampling calorimeter during the first three years of HERA operation has been consistent with design parameters and expectations gained from test beam measurements. The use of the natural radioactivity of the uranium has proven to be a simple and reliable calibration technique. To date no aging effects inconsistent with the design lifetime of ten years have been observed.

The calorimeter measurements of energy, position, and time play critical roles at each of the three trigger levels, in the event classification algorithms, and in the physics analyses. A spatial resolution of 1 cm and a time resolution of 0.5 ns are typical for the inelastically scattered electrons produced in neutral-current interactions at HERA. The hadronic and electromagnetic energy measurements are degraded at a level of up to $\simeq 20\%$ by material of thickness ranging up to several radiation lengths between the calorimeter and the interaction vertex in narrow angular regions. The use of presampling techniques in test beam measurements and in situ show results encouraging for energy correction algorithms. A full-scale presampler for the forward and rear calorimeters will be installed prior to the 1995 data-taking period.

10. References

1. The ZEUS Detector, Status Report 1993, DESY 1993.

2. M. Derrick et al., NIM A309 (1991) 77.

3. A. Andresen et al., NIM A309 (1991) 101

4. H.Brückmann et al., NIM A263 (1988) 136.
   R.Wigmans, NIM A259 (1987) 389.
   J.E.Brau and T.E.Gabriel, NIM A238 (1985) 489.

5. J.Hartmann et al., NIM A305 (1991) 366.
   B.G.Bylsma et al., NIM A305 (1991) 354.

6. A.Caldwell et al., NIM A321 (1992) 356.
   A.Caldwell et al., DESY 92-130 (1992).
   L.Hervas, Ph.D. Thesis U.A.Madrid, DESY F35D-91-01 (1991).
   W.Buttler et al., NIM A277 (1989) 217.
   W.Sippach et al., IEEE-NS 36 (1989) 465.

7. A.Caldwell et al., DESY 92-130, 2nd Annual Conference on Electronics for Future Colliders (1992)

8. A.Bernstein et al., NIM A336 (1993) 23.
   A.Andresen et al., NIM A290 (1990) 95.
   U.Behrens et al., NIM A289 (1990) 115.
   R.Klanner, NIM A265 (1988) 200.

9. U.Behrens et al., NIM A339 (1994) 498.
   I.Ambats et al., NIM A320 (1992) 161.

10. R.Schattevoy, Ph.D. Thesis U.Bonn, BONN-IR-94-08 (1994).

11. U.Behrens, et al., NIM A323(1992) 611.

12. M.Derrick et al., *Measurement of the Proton Structure Function F_2 from the 1993 HERA Data*, DESY 94-143, submitted to Z. Phys. C.

13. H.Grabosch et al., NIKHEF-H-93-11 (1993).

THE KLOE ELECTROMAGNETIC CALORIMETER

JULIET LEE-FRANZINI

Laboratori Nazionali di Frascati
Via Enrico Fermi 40 - 00044 - Frascati, Italy
Physics Department, State University of New York at Stony Brook
Stony Brook, New York 11794, U.S.A.

and

A. ANTONELLI[b], M. ANTONELLI[b], G. BARBIELLINI[g], M. BARONE[b], S. BERTOLUCCI[b], C. BINI[d], C. BLOISE[b], V. BOLOGNESI[c], F. BOSSI[b], P. CAMPANA[b], F. CERVELLI[c], R. CALOI[d], M. CORDELLI[b], G. DE ZORZI[d], G. DI COSIMO[d], A. DI DOMENICO[d], O. ERRIQUEZ[a], A. FARILLA[a], A. FERRARI[c], P. FRANZINI[d], F. GARUFI[d], P. GAUZZI[d], E. GERO[b], S. GIOVANNELLA[b], R. HAYDAR[b], M. INCAGLI[c], L. KEEBLE[b], W. KIM[f], G. LANFRANCHI[b], A. MARTINI[b], A. MARTINIS[g], S. MISCETTI[b], F. MURTAS[b], A. PARRI[b], A. PASSERI[e], F. SCURI[g], E. SPIRITI[e], L. TORTORA[e], S.WÖLFLE[b].

[a] *Dipartimento di Fisica dell'Università e Sezione INFN Bari,*
[b] *Laboratori Nazionali di Frascati dell'INFN, Frascati,*
[c] *Dipartimento di Fisica dell'Università e Sezione INFN Pisa,*
[d] *Dipartimento di Fisica dell'Università e Sezione INFN Roma I,*
[e] *Istituto Superiore di Sanità and Sezione INFN, ISS, Roma,*
[f] *Physics Department, State University of New York at Stony Brook,*
[g] *Dipartimento di Fisica dell'Università e Sezione INFN Trieste/Udine.*

ABSTRACT

A general purpose detector, KLOE, is under construction for operations at the Frascati ϕ factory, DAΦNE. Its central mission is the study of direct CP violation in K^0 decays, which places very stringent requirements on electromagnetic shower measurements in the 20–280 MeV/c region. We have chosen to use a lead–scintillator sampling calorimeter, EmC, consisting of very thin (0.5 mm) lead layers in which are embedded 1 mm diameter scintillating fibers. Much prototyping and testing has been done during its design, yielding, for the final EmC, an expected energy resolution of $\sigma(E)/E \sim 5\%/\sqrt{E\ (\mathrm{GeV})}$ and a time resolution of $\sim 66\ \mathrm{ps}/\sqrt{E\ (\mathrm{GeV})}$, with excellent linearity in the region of interest and with little dependence on incidence angle and entry position.

1. Introduction

The main goal of the KLOE[1] experiment at the DAΦNE ϕ-factory[2] in Frascati is to study CP violation in K^0 decays by measuring $\Re(\epsilon'/\epsilon)$ with an accuracy of $\mathcal{O}(10^{-4})$. Since $K_{L,S}$ neutral decays ($K_{L,S}\to\pi^0\pi^0$) produce four photons between 20 and 280 MeV, identification of these decays and rejection of the $K_L\to3\pi^0$ background[1] means that the electromagnetic calorimeter (EmC) plays a crucial role in this measurement. Vertex reconstruction of $K^0\to\pi^0$'s$\to\gamma$'s in KLOE is performed by measuring the arrival time of each photon the the calorimeter ,[3] because of the low speed of the K^0 (β_K ~0.2), requiring excellent timing accuracy of the EmC. In addition, the calorimeter should also provide a fast and unbiased first level trigger for the detector. Finally, it should have some particle identification capability, to aid rejection of $K_{\mu3}$ decays. The requirements for the KLOE EmC are: a) time resolution $\mathcal{O}(70\ \text{ps}/\sqrt{E\ (\text{GeV})}\,)$, b) energy resolution $\mathcal{O}(5\%/\sqrt{E\ (\text{GeV})}$, c) shower vertex resolution $\mathcal{O}(1\ \text{cm})$, d) full efficiency for γ's in the energy range 20–280 MeV, and e) hermeticity.

2. The KLOE Electromagnetic Calorimeter Design

The KLOE EmC is a very fine sampling lead-scintillating fiber calorimeter, with photomultipliers, PMs, read-out. The central part, barrel, approximates a cylindrical shell of 4 m inner diameter, 4.3 m active length and 23 cm thicknes. The barrel covers the polar angle region $49°<\theta<131°$and consists of 24 sectors with trapezoidal cross section, ~60 cm wide. Two end-caps, 4 m in diameter and 23 cm thick, close hermetically the calorimeter. Each end-cap, consists of 26 "C" shaped modules which run vertically along the chords of the circle inscribed in the barrel. At the two ends they are bent at 90°, becoming parallel to the barrel ends, to decrease the effects of the magnetic field on the PMs and to increase hermeticity. A quater cross section of the KLOE calorimeter is shown in fig. 1. In our EmC fibers run mostly transversely to the particle trajectories. This reduces sampling fluctuations due to channeling, resulting in improved resolution particularly important at low energies.

Each module of the KLOE EmC is built by glueing 1 mm diameter blue scintillating fibers between thin grooved lead plates, obtained by passing 0.5 mm thick lead foils through rollers of proper shape. The grooves in the two sides of the lead are displaced by one half of the pitch so that fibers are located at the corners of adjacent, quasi-equilateral triangles resulting in optimal uniformity of the final stack, see fig. 2. The grooves are just big enough to insure that the lead does not apply direct pressure on the fibers. Light travelling in the cladding is effectively removed because of the glue surrounding the fibers. The selected fiber pitch of 1.35 mm results in a structure which has a fiber:lead:glue volume ratio of 48:42:10 and a sampling fraction of ~15% for a minimum ionizing particle. The final composite has a density of ~5 g/cm^3 and a radiation length X_0 of ~1.6 cm, is self supporting and can be easily machined. The very small lead foil thickness ($< 0.1\,X_0$) results in a quasi–homogeneous structure and in high efficiency for low energy photons. Measurements indicate that the blue-green Kuraray SCSF-81, Bicron BCF-12 and Pol.Hi.Tech-46 fibers, satisfy our requirements for light yield, scintillation decay time and attenuation length.[4] Since the time resolution depends on the light yield, great care has been put in maximizing the efficiency of the light collection system and insuring uniform photocathode illumination. Each light guide consists of a mixing part and a Winston cone concentrator.[5] We are thus able to match the calorimeter elements to the PM photocathodes, with an area reduction factor of up to ~4, without losses, because of the small divergence angle, 22°, of the light travelling in the fibers.

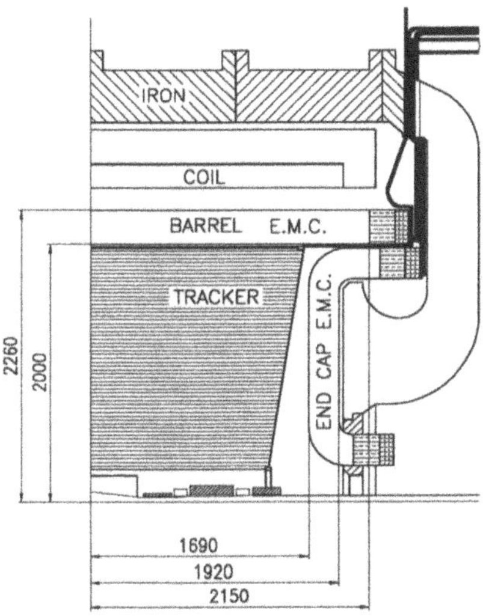

Fig. 1. Cross section of the KLOE Calorimeter

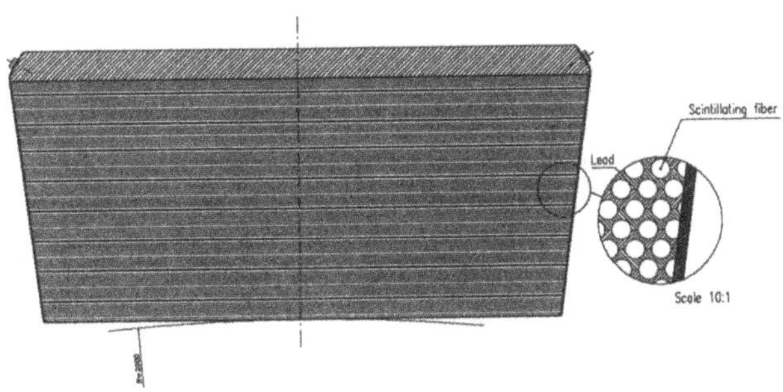

Fig. 2. Fiber and lead layout.

3. Construction of Prototypes

We built a *barrel* prototype module of dimensions $14 \times 24 \times 203$ cm^3 to establish construction procedures and perform realistic tests of the final EmC performance, see fig. 3a). The module depth of 24 cm is equivalent to a thickness of 15 X_0, consisting of 207 lead and scintillating fiber layers for a total of 21,450 fibers, the first 3.5 cm are Kuraray SCSF-81 fibers followed by 20.5 cm of Kuraray SCSF-38 fibers. It weighs 330 kg. The module was made in the Frascati machine shop. The grooved lead foils were produced with a home-made rolling machine starting from 0.5 mm thick lead foils. The grooved rollers, made of hardened steel and ground to shape by a sintered diamond tool, are mounted with ball bearings on a very rigid frame and are aligned and checked with a set of micrometers. Foils of almost 6 m length and 15 cm wide have been produced. The thickness uniformity obtained is around few tens of μm, the grooves deviate from a straight line by less than 0.1 mm per meter of foil length. The module was constructed by glueing the first lead foil to a 2 cm thick iron back plate carrying all the mounting fixtures. Bicron optical epoxy resin, BC 600 ml, was then applied in a precisely measured amount to the top surface of each lead foil. A layer of fibers was then laid down in the grooves and epoxy applied anew. Typically, 6–7 layers of fibers and lead were stacked in about one hour, before the glue started curing. The growing stack was then compressed under vacuum for at least 30 min to apply uniform pressure, and to squeeze out excess glue. Once the final thickness was reached the two ends were cut and milled. No fiber was damaged during the stacking phase, while four fibers were lost during the final milling. The calorimeter was wrapped in a 1 mm thick steel skin, in order to simulate the final design and provide protection for the module.

A prototype for the *end-cap* region, fig. 3b), consisting of two "C" shaped modules, was also built. The width of each module is 7 cm and they both have the same depth of 24 cm. Each section consists of a central straight piece (89–110 cm), two curved pieces with an inner radius of 7.5 cm, and two straight end pieces to which light guides are glued, the first 7 cm are Kuraray SCSF-81 fibers, the rest are Pol.Hi.Tech-46 fibers. It has been constructed by cutting the grooved lead foils and fibers to the desired lengths, incising at the two ends the central groove of each lead foil, leaving the central part uncut for a length of 89 cm, then bending the two adjacent modules.

The readout, identically for both prototypes, was organized into 22 elements for each side: five planes of four small elements, 3.5×3.5 cm^2, and two large rear elements, 6.5×7 cm^2, see fig. 3b). For the small elements, on each side the light was transported from the calorimeter through light guides to 1-1/8 inch diameter PMs, with an area concentration factor of $\sim$2.7. 1-1/8 inch Hamamatsu phototubes, type R1398 with a transit time spread of (< 1 ns) have been used. The rear elements were similarly coupled to 5 cm diameter PMs, with an area concentration factor of $\sim$2.5.

In order to investigate the effect of coarsening the readout granularity and using larger light guides, we took data with the *barrel* readout organized in four planes of three 4.2×4.2 cm^2 cells, leaving the last plane unchanged. This configuration is referred to as "coarse" granularity, see fig. 3a). Since the same PMs were used, the light guides had an area concentration factor of ~3.8.

4. Prototype Tests

4.1 Test Beams Layout

The calorimeter prototypes have been tested[6,7] in a beam at the Paul Scherrer Institut, PSI, in Villigen, Switzerland. The PSI machine is an isochronous cyclotron which accelerates protons to 600 MeV kinetic energy. A secondary beam of π's, μ's and e's in the momentum range 50–400 MeV/c was delivered to the test areas.

Two crossed scintillators in front of the calorimeter and a third counter, 4.5 m away from the calorimeter, provided a beam trigger, defined the beam position at the calorimeter and enabled $\pi/\mu/e$ identification by time of flight, see fig. 4.

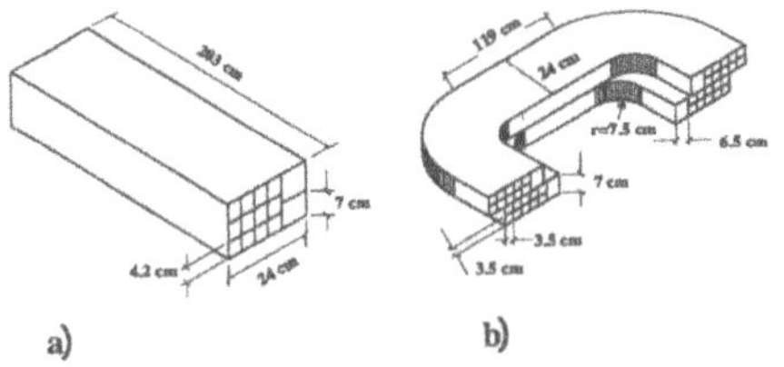

Fig. 3. a) *Barrel* and b) *end-cap*prototype.

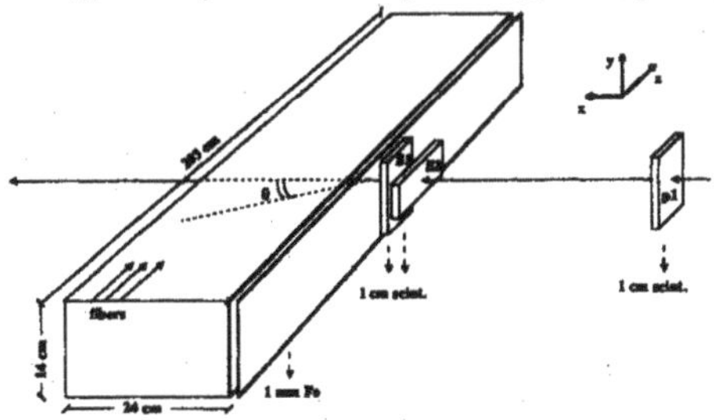

Fig. 4. Calorimeter test setup at PSI.

The calorimeter PM outputs were connected to active splitters through 5 m of RG-58 cable. The two outputs of each splitter were sent to CAMAC ADCs, to low threshold (5 mV) discriminators, the signal for 50 MeV being ~350 mV. The discriminator outputs acted as TDC stops. The trigger signal was used to generate the ADC gates and the reference time signal T_0 which supplied the "start" for the TDC system.

The *barrel* prototype has also been tested with tagged low energy photons (20–80 MeV) at the Frascati LADON facility.[8] At LADON photons are obtained by back scattering a laser beam from electrons stored in ADONE. A tagging system for the recoil electrons yields a determination of the photon energy with a resolution of ~1.7 MeV. The trigger was obtained by a coincidence from the two calorimeter ends, after adding all signals from each side. The trigger generated ADC gates and TDC "start"s which were cleared if there was no trigger from the tagging system within 200 ns.

4.2 Energy Response and Resolution

The coordinate system used in the data analysis is shown in fig. 4, where the definition of the incidence angle θ is also indicated. The origin was put in the center of the calorimeter surface facing the beam.

In order to study the energy response of the prototype we define the total visible energy E_{vis} as the sum over all elements read out on both ends, labelled A and B. To correct for the response of each channel we divided the signal $E_{adc,i}$ of channel i, by the calibration constants $K_{mip,i}$: $E_{vis} = (1/2)\sum(E_{A,i} + E_{B,i})$ with $E_i = E_{adc,i}/K_{mip,i}$. The calibration constants K_{mip}, in ADC counts, were obtained with cosmic rays and 400 MeV/c pions, both of which are minimum ionizing particles to a first approximation, incident at the center of each element. Clean samples of events selected for the calibration were ensured by means of triggering and software cuts. A gaussian fit to the pulse height distribution was performed for each channel to determine the peak position. The fit interval was chosen to be asymmetric with respect to the mean value, to take into account the asymmetric shape of the distribution.

We define 1 MIP as the visible energy deposited by a minimum ionizing particle in a 3.5×3.5 cm^2 readout element; our simulation[9] determines 1 MIP to be equivalent to an average energy deposit of 3.24 MeV in the active material. For the coarse granularity and for the larger cells of the last plane geometrical correction factors have been used to express the results in our calibration units. Pions of 400 MeV/c momentum are minimum ionizing particles only in the first plane of the calorimeter. We have therefore corrected for the increase of the specific ionization with calorimeter depth and for straggling effects. The correction factors used for both granularities are given in table 1.

Table 1. Specific ionization correction factors.

Plane Number	1	2	3	4	5	6
3.5 × 3.5	1.00	1.03	1.04	1.06	1.10	1.14
4.2 × 4.2	1.01	1.04	1.05	1.09	1.14	–

Before reaching the calorimeter the beam had to pass through the trigger counters and the iron skin of the calorimeter itself. The total amount of material is equivalent to 3.6 cm of scintillator. Positrons lose ~7 MeV while the energy loss of pions and muons depends on their kinetic energy and was appropriately taken into account.

These corrections have been applied to the beam energy of the charged particles.

Energy Response and Resolution for Photons and Positrons

The following results were obtained with positrons and photons entering the center of the calorimeter front face ($z = 0$ cm, $\theta = 0°$). The distributions of the energy signal for photons ranging from 28 to 63 MeV are are displayed in fig. 5. From the very good symmetry of the line shape at the lowest energy of 28 MeV, we can conclude that at these energies, the calorimeter is still fully efficient.

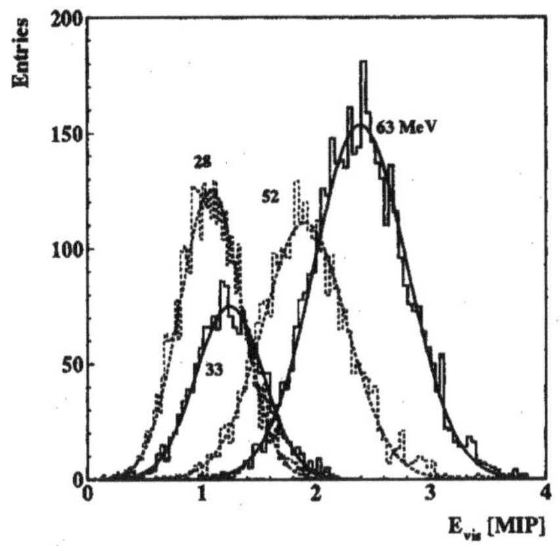

Fig. 5. Energy spectra for photons.

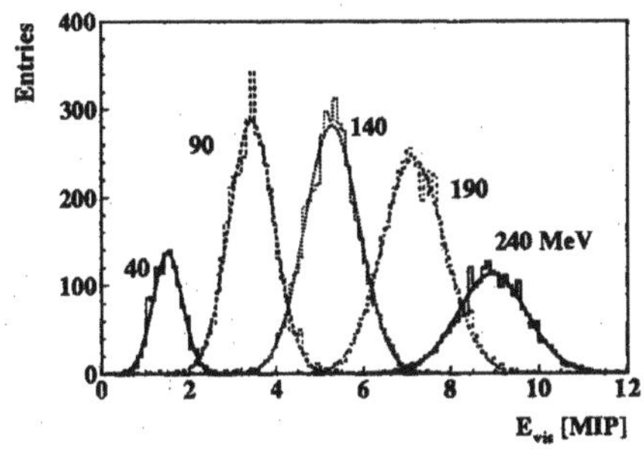

Fig. 6. Energy spectra for positrons.

For each spectrum a gaussian fit, leaving all three parameters free, has been performed; the mean of the fit was used as the average calorimeter response E_{vis} and the square root of the variance $\sigma_{E,vis}$ was used to compute the energy resolution. The plot of E_{vis}/E *vs.* the energy E have a maximum deviation from linearity of less than 3%, mostlty due to inaccuracies in tagging system. The line shapes for positrons in the energy range 40–240 MeV are shown in fig. 6 for *barrel* data with the coarse granularity. The linearity of the response is excellent with a deviation of at most 2%, again mostly from beam setting inaccuracies.

Fitting photon and positron data together we obtain a slope $s_{barr} = (36.8 \pm 0.2)$ MIP/GeV, see fig. 7. In the same plot we have also added data for the *end-cap* prototype exposed to positrons (triangle symbols) for which the slope is $s_{ecap} = (37.4 \pm 0.6)$ MIP/GeV. The average response to electromagnetic (e.m.) showers is proportional to the incident particle kinetic energy only if the shower is fully contained. Simulations[9] show that the depth of the prototype keeps leakage below the 2% level for the energy range of interest. The sampling fraction for e.m. showers can therefore be estimated as $f_{em} = E_{vis}/E = (36.8 \pm 0.2)$ MIP/GeV $\times$ 3.24 MeV/MIP $= (11.9 \pm 0.1)\%$.

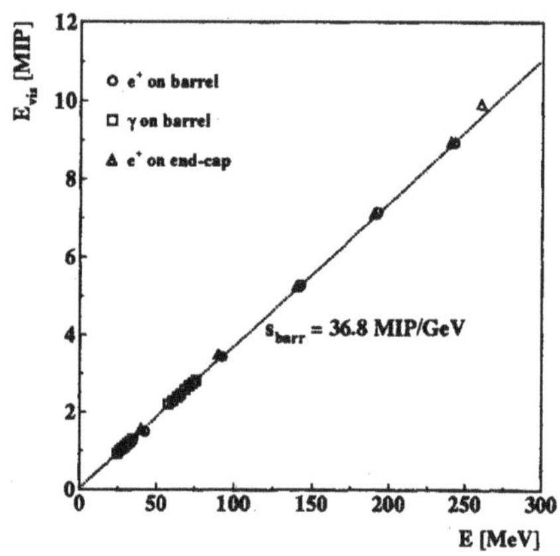

Fig. 7. Energy response for photons and positrons vs. kinetic energy.

The fluctuation of the ratio R between the energy signals, from each readout element, E_A and E_B at the two calorimeter ends, $R = E_A/E_B$, allows us to estimate the number of photoelectrons N_{pe} in each tube, from $N_{pe} = (\sqrt{2} \times \mu_R/\sigma_R)^2$, where μ_R and σ_R are the mean of R and the r.m.s. spread of its distribution respectively. For the *barrel* prototype the typical signal for a minimum ionizing particle at 1 m distance from the photomultiplier is ~65 pe. Results within 10% were obtained by measuring the PM gain directly. A lower light yield of ~50 pe/MIP is obtained in the *end-cap* prototype due to the worse quality fiber used in its construction. y resolution for positrons is $\sigma_E/E = (4.47 \pm 0.02)\%/\sqrt{E\ (\text{GeV})}$ and for photons $\sigma_E/ = (4.35 \pm 0.01)\%/\sqrt{E\ (\text{GeV})}$. The measured resolution for positrons and γ's are shown in fig. 8. The $1/\sqrt{E}$ dependence indicates that sampling fluctuations and photoelectron statistics dominate the resolution. The contribution of the latter has been evaluated from the light yield to be $1.6\%/\sqrt{E}$ and is therefore negligible. A similar analysis for the *end-cap* gives an energy resolution of $\sigma_E/E = (4.59 \pm 0.04)\%/\sqrt{E(\text{GeV})}$.

Energy Response and Resolution for Pions and Muons

The response for muons and pions in the momentum range 150–280 MeV/c, of interest for the KLOE experiment, has also been studied. Due to their low kinetic energy (50–180 MeV) these particles lose energy mostly by ionization and could stop in the calorimeter. The response *vs.* kinetic energy for both types of particles is identical. A linear fit of the energy response curve gives a slope s_μ=46.8 MIP/GeV, corresponding to a sampling fraction for muons of $f_\mu = E_{vis}/E =$ 46.8±0.4 MIP/GeV×3.24 MeV/MIP = 15.2±0.1% and an "e/μ" ratio of f_e/f_μ=0.79. The energy resolution does not scale perfectly with $1/\sqrt{E}$ because sampling fluctuations become larger near the end of the particle's range. However, an approximate parametrization of the resolution is given by $3.7\,\%/\sqrt{E}$.

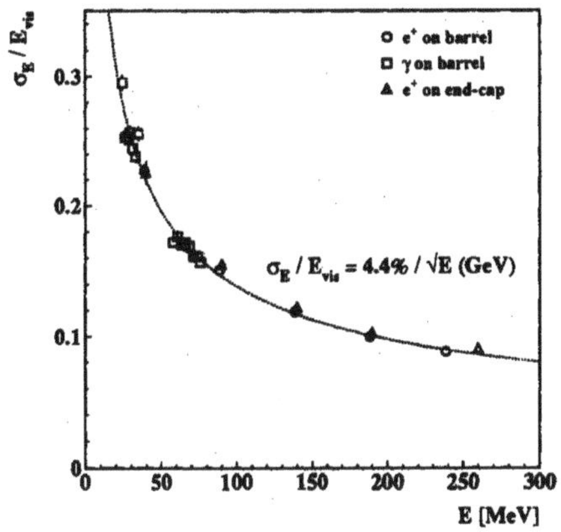

Fig. 8. Energy resolution for photons and positrons vs. kinetic energy.

Dependence on the z–coordinate

A z–scan along the module axis was performed with 200 MeV/c positrons at seven different z–positions from −80 to +80 cm. The z dependence of the signal is well represented as $E_{vis}(z) = E_{vis}(0)\cosh(z/\lambda)$, due to adding two exponentials, with an attenuation length λ=240 cm, while the resolution is constant.

Dependence on the Incidence Angle

The response of the barrel prototype as a function of the incidence angle, θ, has been studied using photons, positrons and muons at 30°, 45° and 60° incidence. Response and resolution do not depend on θ as expected for a quasi homogeneous calorimeter with very thin converter layers. For low energy photons. E <60 MeV, we observe a small worsening of the resolution, ~25%, at $\theta = 60°$. The curved parts of the end-cap prototype have been investigated by scanning the prototype with 200 MeV positrons incident at 45° at different positions, moving along the outer surface of the bent part. No appreciable variation was found in the scan.

4.3 Timing Performances

The particle arrival time in the calorimeter is given, for each PM, by a discriminator output, used as a TDC "stop". For charged particles, the common "start" of the TDC system was obtained from the beam defining counters. For photons we self triggered on the analog sum of all calorimeter signals, as mentioned previously. Time walk correction depending on pulse height were applied off-line. The times measured at each calorimeter end T_A, T_B are obtained by the energy weighted average over all elements: $T_A = \sum E_{A,i} T_{A,i} / \sum E_{A_i}$; $T_B = \sum E_{B,i} T_{B,i} / \sum E_{B_i}$. The average of T_A and T_B then defines the mean arrival time T. For photons we perform our analysis in terms of the time difference $\Delta T = (1/2) \sum E_i (T_{A,i} - T_{B,i}) / \sum E_i$, ΔT and T are the difference and sum of T_A and T_B and therefore have the same r.m.s. spreads, *i.e.* $\sigma_{\Delta T} = \sigma_T$ except for the fact that the "start" time jitter is cancelled event by event in ΔT while it has to be unfolded for the external trigger case.

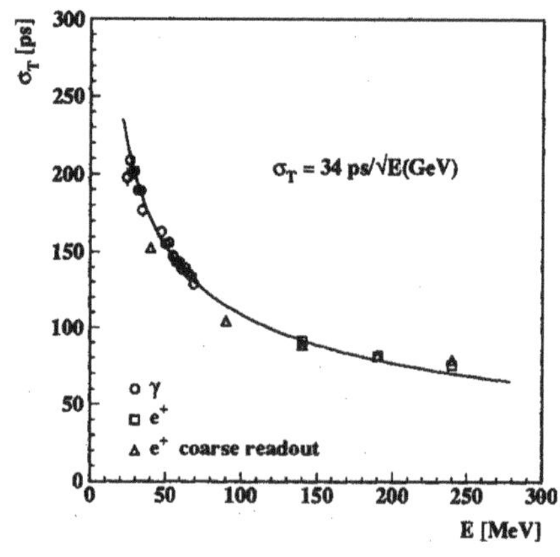

Fig. 9. Time resolution vs. energy.

Timing Resolution at Normal Incidence

The dependence of $\sigma_{\Delta T}$ on beam energy for 20–80 MeV γ's and of σ_T for 50–250 MeV positrons, at normal incidence on the calorimeter surface, are shown together in fig. 9, corresponding to a time resolution of $\sigma(T \text{ or } \Delta T)$=34 ps/$\sqrt{E\ (\text{GeV})}$. Fitting all measurements of $\sigma_{\Delta T}$ from 20 to 250 MeV gives $\sigma_{\Delta T}$=36 ps/$\sqrt{E\ (\text{GeV})}$, in excellent agreement with the result requiring unfolding of the "start" uncertainty. The measurements of the time resolution with a coarser readout granularity, (4.2×4.2 instead of 3.5×3.5 cm^2) triangle symbols in fig. 9, are indistinguishable although the light guides area reduction factor increases from 2.7 to 3.8, showing that light collection efficiency is unchanged. For the end-cap prototype we find a time resolution of σ_T =40 ps/$\sqrt{\text{E(GeV)}}$, with 50–200 MeV positrons.

Timing Resolution Dependence on Incidence Angle

The dependence of σ_T on θ was studied for all available beam energies. At 60° the resolution for 250 MeV positrons is ~25% worse than the result for normal incidence. The distributions are still gaussian. At lower energies the angular dependence of the resolution becomes negligible, being dominated by sampling fluctuations. The investigation of the curved portion of the end-cap prototype as described, found no dependence. Only for the most extreme conditions, when the shower effectively crosses the fibers twice, the time resolution deteriorates by ~20%.

Timing Resolution Dependence on z-Coordinate

All our measurements confirm that the time resolution scales as the number of photons to the $-1/2$ power. We therefore expect that the resolution is best at the center of the calorimeter. The z dependence of the time resolution is given by: $\sigma_T(z) = \sigma_T(0)\sqrt{\cosh(z/\lambda)}$. This has been verified with the data from the z-scan. The resolution at the two ends of the barrel module is ~1.05 times worse than at the center.

Measurement of the z-Coordinate

The z-coordinate of the entry point z_e obtained from the time difference ΔT and the effective light propagation speed in the fibers v_f is: $z_e = v_f \Delta T$. ΔT is plotted *vs.* the nominal z-position in fig. 10. From a fit to the data we obtain v_f=17.2 cm/ns, in agreement with the refractive index of the fiber core, n=1.6, and the bounce angle in the fiber.

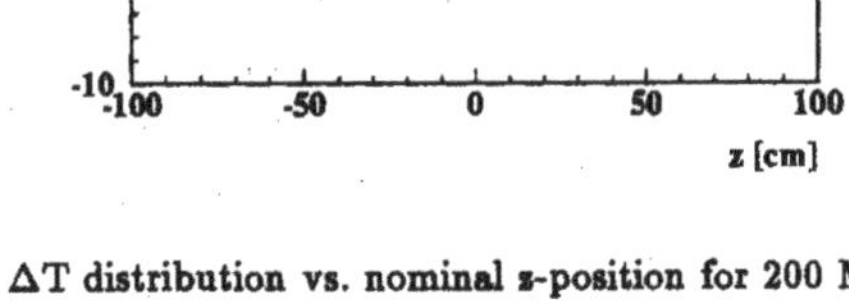

Fig. 10. ΔT distribution vs. nominal z-position for 200 MeV positrons.

The position resolution σ_z, given by $\sigma_z = v_f \sigma_{\Delta T}$ is σ_z=6 mm/$\sqrt{E\ (\text{GeV})}$ for the e.m. showers.

5. Particle Identification

$K_L \to \pi\mu\nu$ decay, $K_{\mu 3}$, constitutes the most significant background to the identification of $K_L \to \pi^+\pi^-$ decays. The calorimeter can provide some π/μ identification to improve the background rejection from kinematics.[10] π/μ identification can be obtained by time of flight, Monte Carlo studies[3] show that the time resolution of the KLOE calorimeter, combined with tracking information result in a $K_{\mu 3}$ rejection factors of 5–6, with 93% signal efficiency. π/μ identification can also be achieved by studying the energy signals and the first and second moments of the energy deposition distributions in the calorimeter. This method has been tested with data collected for pions and muons of 200–250 MeV/c, a $K_{\mu 3}$ rejection factor of 5 for is obtained while maintaining an efficiency of 95% for pions.

6. Module Ø

All the machinery for building the *barrel* calorimeter have designed and built by the KLOE calorimeter group, mostly at the Laboratori Nazionali di Frascati, LNF. The first full size *barrel* module, with dimensions of 23 cm×60 cm×430 cm and trapezoidal cross section, called module Ø, was then succesfully built, also at Frascati. The machinery has been turned over to a company which will assemble all modules of the *barrel* calorimeter. The front half of module Ø is built with Kuraray SCSF-81 fibers and the back half with Pol.Hi.Tech-46 fibers.

The module read out consists of five planes of 12 elements each, of ~4.4×4.4 cm^2, varying slowly across the module face. Each element is viewed through light pipes which provide mixing and terminate with a Winston concentrator coupled to mesh photomultipliers. The PMs are manufactured to KLOE specifications by Hamamatsu (R2021, mod.). KLOE needs to use these special tubes because they can operate in moderate magnetic field.[11] As shown in fig. 1, they are inside the return yoke where they see up 0.15 T inclined less than 25° with respect to the PM's axis.[12] These PMs have a measured quantum efficiency ~60% lower than conventional tubes like the ones used in the previous prototypes.[13]

Module Ø employs the final KLOE electronics. The PM base contains the voltage divider for the PMs working with grounded cathode to allow grounding of the tube shieldings and an A.C. coupled preamp which drives the cables carrying the signals outside of the magnet yoke. The signals enter three way splitters, providing input to a constant fraction discriminator, a driver for amplitude measurements and first level analog sums for trigger generation. All these functions are mounted on a 9U VME card, containing 30 complete channels. Signals from this card go to the KLOE designed 12 bit ADC's, 2 V×10 ns full scale, and 12 bit TDC's, 25 ps per count. A description of KLOE electronics can be found in ref. 14.

7. Conclusions

Module Ø was exposed to a test beam at PSI this past July. Analysis of the collected data is underway at present, preliminary results indicate that $\sigma_E/E \sim 5\%/\sqrt{E\ (\mathrm{GeV})}$ and $\sigma_T \sim 66\mathrm{ps}/\sqrt{E\ (\mathrm{GeV})}$. These results are precisely the expected resolutions taking into account the length of the full module (4.3 m) and the reduced quantum efficiency of the mesh PM's, halving the number of photoelectrons.

In sum, we have studied extensively the performance of both the *barrel* and *end-cap* prototypes of the KLOE calorimeter and confirmed the findings with module Ø. Aside from the measured energy and time resolutions quoted, we found the energy resolution to be independent of the incidence angle and we expect very little dependence of the time resolution on incidence. We have verified that the calorimeter can provide some added means to identify pions and muons. Finally, the KLOE electronics performed extremely well at the PSI test, fully to specifications and with very low noise. Therefore, we have succeeded in the design of a calorimeter which satisfy, with some safety margins the requirements of the KLOE experiment.

ACKNOWLEDGEMENTS

We thank W. Kluge and his group, the PSI management and technical staff for their hospitality, help and support during our work at PSI, thank the LADON group for their help and collaboration. We acknowledge the efforts of the SPECAS mechanical shop for the construction of the prototypes. We also thank A. Balla and G. Corradi and M. Santoni for help with the electronics, and especially M. Anelli and A. Rutili for their help in setting up and testing the calorimeter.

REFERENCES

1. The KLOE Coll. *KLOE, A General Purpose Detector*, LNF-92/019 (1992).
2. G. Vignola, *Proc. of the XXVI Int. Conf. on HEP*, ed. J. Sanford AIP (1992) 1941.
3. The KLOE Coll. *The KLOE Detector Technical Proposal*, LNF-93/002 (1993).
4. G. de Zorzi, *Proc. of 4th Int. Conf. on Calorimetry*, La Biodola, Italy (1993).
5. H. Hinterberger, R. Winston, Rev. Sci. Instr. 37 (1966) 1094; A. Di Domenico *et al.*, KLOE Notes 83 and 84.
6. S. Miscetti, *Proc. of 4th Int. Conf. on Calorimetry*, La Biodola, Italy (1993).
7. J. Lee-Franzini, *Proc. of 6th Int. Conf. on Advanced Detectors*, Elba, PISA, Italy (1994).
8. D. Babusci *et al.*, Nucl. Instr. and Meth. 305 (1991) 19.
9. S. Miscetti and A. Parri, KLOE note 111 (1994).
10. The KLOE Coll. *The KLOE Central Drift Chamber*, LNF-94/028 (1994).
11. C. Bini *et al.*, KLOE Notes 42, 50, 60, and 78; G. Barbiellini *et al.*, KLOE Notes 56 and 81.
12. F. Bossi, KLOE notes 54 and 91.
13. L. Keeble, G. Lanfranchi and M. Anelli, KLOE Notes 90, 96 and 109.
14. The KLOE Coll. *The KLOE Data Acquisition System*, LNF-94, (1994).

EM Crystals

Session Chair: Craig Woody (BNL)

TEST RESULTS OF A CeF_3 CRYSTAL MATRIX

T.BECKERS
VUB. Brussels/UIA, Antwerp, BELGIUM

M.SCHNEEGANS
LAPP, Annecy, FRANCE

P.DEPASSE, H.EL MAMOUNI, J.FAY, M.GOYOT, B.ILLE, P.LEBRUN, J.P.MARTIN, G.MAURELLI, M.REYNAUD, P.SAHNE, J.P.WALDER
Université Claude Bernard, Lyon, FRANCE

M.CHIPAUX, I.SORIC
CE Saclay, FRANCE

H.HILLEMANNS, T.KIRN, D.SCHMITZ, J.SCHWENKE
I. Physikalisches Institut, RWTH Aachen, GERMANY

V.COMMICHAU
III. Physikalisches Institut, RWTH Aachen, GERMANY

R.RAGHAVAN
Tata Institute, Bombay, INDIA

M.MATTIOLI, L.PACCIANI, S.PIRRO
INFN Roma, ITALY

E.AUFFRAY, I.DAFINEI, U.DYDAK, P.LECOQ
CERN, Geneva, SWITZERLAND

L.DJAMBAZOV, M.FELCINI, H.HOFER, P.LECOMTE, I.MELNIKOV, F.NESSI-TEDALDI, D.REN, U.RÖSER, G.VIERTEL, H.P.VON GUNTEN, S.WALDMEIER-WICKI
Swiss Federal Institute of Technology (ETHZ), Zurich, SWITZERLAND

Presented by Pierre LECOMTE
Swiss Federal Institute of Technology (ETHZ), Zurich, SWITZERLAND

ABSTRACT

A cerium fluoride matrix of nine longitudinally segmented towers, each approximately 42cm long, has been tested with electrons and pions from 10 to 150 GeV. Despite less than optimal geometry and crystal quality, an energy resolution of 0.5% has been obtained with PIN photodiodes readout. The performance of cerium fluoride makes it thus currently the most promising crystal for electromagnetic calorimetry. Among the crystals we received, several, up to 14cm long, had excellent light yield and radiation hardness. The longest crystal we received was 28 cm long and 3×3 cm^2 in cross section.

1. Introduction

The Crystal Clear Collaboration has conducted generic crystal research since 1990, in view of applications in high energy physics. Many of its members participated to the design and construction of the 11360 crystals BGO electromagnetic calorimeter for the L3 experiment at LEP. The collaboration established worldwide contacts with crystal growers and its work led to the selection of practical candidate crystals for calorimetry. Some early CeF_3 crystals were tested in a beam last year by the collaboration[1].

The Swiss Federal Institute of Technology, which also played a major role early on in the design and testing of the L3 BGO calorimeter, signed in 1991 a joint research agreement with the Shanghai Institute of Ceramics and with the Beijing Glass Research Institute to produce and test large CeF_3 crystals.

Both groups began this year a collaboration to evaluate the feasibility of a Cerium Fluoride calorimeter for the CMS experiment at the CERN-LHC. The results presented here are preliminary and were obtained from May till September 1994 in electrons and pions test beams at CERN, at energies ranging from 10 to 150 GeV.

2. Why Cerium Fluoride?

A comparison of the properties of the most likely crystal candidates (Table 1) for the CMS detector at LHC led us to test CeF_3. BGO is included in the table as a reference, but it is too slow, not sufficiently radiation resistant for the LHC environment, expensive and has a rather high temperature coefficient. Barium Fluoride suffers from unfavorable radiation length and two components emission, the one of interest being weak and in the far UV. Lead Tungstate offers compactness and possibly "low" cost, but it suffers from a high temperature coefficient, low light yield and its radiation hardness remains to be confirmed. Cerium Fluoride is at present the best candidate, assuming that it can be mass-produced at an acceptable price; thus we set to demonstrate that the energy resolution necessary at LHC, in particular to detect efficiently the Higgs decay into two photons, is achievable with a CeF_3 calorimeter.

Table 1: Properties of Crystals for Calorimetry at $p\bar{p}$ colliders, compared to BGO

Crystal Type and Properties				
$Bi_4Ge_3O_{12}$	BaF_2	CeF_3	$PbWO_4$	Parameter
7.13	4.88	6.16	8.28	$\rho[g/cm^3]$
2.15	1.49	1.62	2.30	n
1.12	2.06	1.68	0.85	$\lambda_r[cm]$
21.8	29.9	26.2		$\lambda_I[cm]$
2.33	3.39	2.63	2.19	Molière Radius [cm]
480	195-220/310	340/300	440/530	$\lambda_{max}[nm]$
300	0.8/630	30/5	2.2/9.9/39	$\tau[ns]$
100	16/100	55	4.4	Light Yield [% of BGO]
-1.55		0.14	-1.9	Temperature Dependence [%/°C]
10	$>10^4$	$>10^4$	?	Radiation Hardness [Gy]

3. Test Beams

Most of our tests were performed at the CERN-X3 beam, a low intensity unseparated tertiary beam providing muons, pions and electrons of energies ranging from 4 to 50 GeV. The beam optics gives 0.5% momentum resolution and there is a beam spectrometer consisting of four wire chambers and two bending magnets to confirm this resolution (Fig.1). The beam momentum spread was quadratically substracted from our results. Data were acquired from CAMAC by a PC, stored on disk and transferred at the end of every run via Ethernet to a Unix workstation, processing being monitored from an X terminal in the counting room.

We also obtained two days in the H4 beam to work with electrons in the 50-150 GeV range. This beam has an intrinsic resolution of 0.1-0.2% but its tuning is extremely sensitive to magnet settings; we have some indication of a variable pion contamination in our electron sample, a contamination to which we are sensitive because we used slow electronics for this test.

4. Test Matrix

Our test matrix consists of nine towers (Fig.2). A tower is segmented longitudinally into a 14cm long front crystal and into a rear part 28cm long. A rear part is made of two crystals glued together with Epotek 340. The cross section of the four corner towers is approximately $2\times2cm^2$; it is $3\times3cm^2$ for the other five towers.

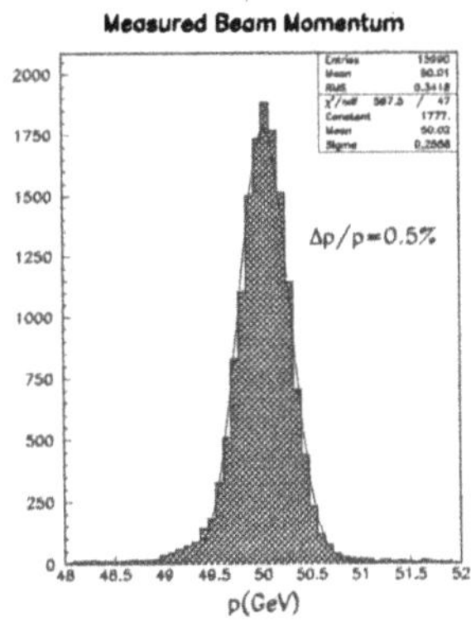

Fig.1 X3 beam momentum resolution at 50 GeV as measured by the beam spectrometer.

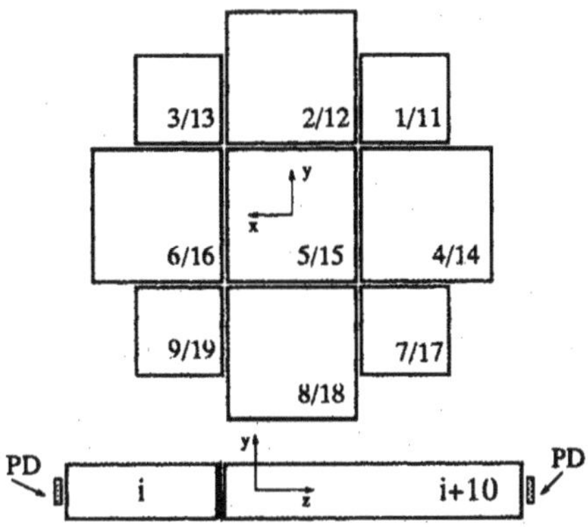

Fig.2 Test Matrix layout. Front crystals are numbered from 1 to 9, rear parts from 11 to 19.

The actual crystal cross sections vary from 20 to 21 mm in the 2 by 2 towers and from 27.5 to 32.7 mm in the 3 by 3 towers. The actual crystal lengths vary from 115 to 148mm (front crystals) and from 253 to 280mm (rear pairs of crystals). The actual tower depths vary from 23 to 25.5 X_0. All towers are aligned at the rear, the differences in length of up to 42 mm are all in the front.

The light yield varies significantly from crystal to crystal, the worst tower giving 280 photons/MeV and the best one 1080 photons/MeV, on the standard test bench of the Crystal Clear Collaboration, using an XP2020Q photomultiplier and with the wrapping adjusted to homogenize the light collection efficiency.

There is no temperature regulation nor monitoring but the matrix and its preamplifiers are enclosed in a double walled aluminum box to provide electromagnetic shielding.

This matrix is therefore far from optimum in terms of length, cracks between towers, light yield and temperature regulation.

5. Test Matrix Readout

Each tower segment is equiped with a $1cm^2$ windowless Silicon PIN photodiode Hamamatsu 3590-04, 300μm thick, with a quantum efficiency of 35-40% in the UV. The diodes were coupled to the crystal with a UV transparent, viscuous glue OKEN 6262A.

Most of the results presented were obtained with slow electronics (the same preamplifiers as used in the L3 BGO calorimeter[2], shaping amplifiers with a shaping time of 2 μs, coupled to CAMAC ORTEC 811 peak sensing ADC). The shaping amplifiers installed 3m away from the matrix had remotely controlled gains so that they could be quickly set for pions and electrons runs. We also tested three types of fast amplifiers developed by Lyon[3,4], with shaping times of 20-30 ns, coupled to LeCroy 2249 charge integrating ADC and verified that their performance is adequate for our needs.

6. Test Results

6.1 Light Yield, Noise, Calibration

50 GeV pions are aimed at the edge of the photodiode. The resulting spectrum (Fig.3 for slow electronics, Fig.4 for fast electronics) shows 3 peaks, respectively corresponding to the pedestal, the light in crystal and to the light in crystal plus energy directly deposited in the photodiode. Since a minimum ionizing particle produces a most probable signal of 22500 electrons in the photodiode[1], this gives the ADC calibration in electrons/channel. The calculated peak energy deposited in the crystal shown (#1) is 113.4 MeV, allowing us to convert our ADC calibration into MeV.

6.2 Matrix Electronic Noise

Using the above procedure, we measure for the best crystals 2.5MeV noise with slow electronics, 12.4MeV with fast electronics. Ideally, the matrix noise would be $\sqrt{18}=4.24$ times larger. In reality, if we compare the measured noise to the noise calculated from the best crystal, we have for the sum of nine towers 70 MeV instead of 11 MeV with slow electronics and 300 MeV instead of 53 MeV with fast electronics. This is not correlated pick up noise but it due to the poor light yield of several crystals, i.e some peripherial towers contribute in a major way to the total noise. In other words, the noise we have today in a matrix of nine towers with slow electronics and some relatively low light yield crystals is similar to what we expect with fast electronics if all the crystals were of the same quality as our best ones.

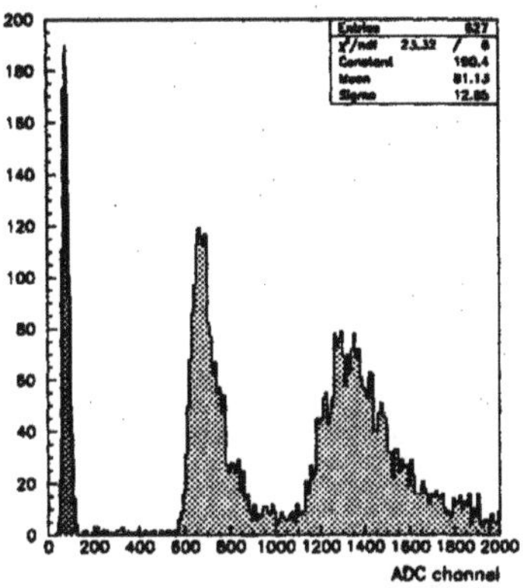

Fig.3 SLOW PREAMPLIFIER (2μs)
Noise=13 counts ⇒ 2.5MeV
Light =599 counts ⇒ 179e/MeV
Diode=663 counts

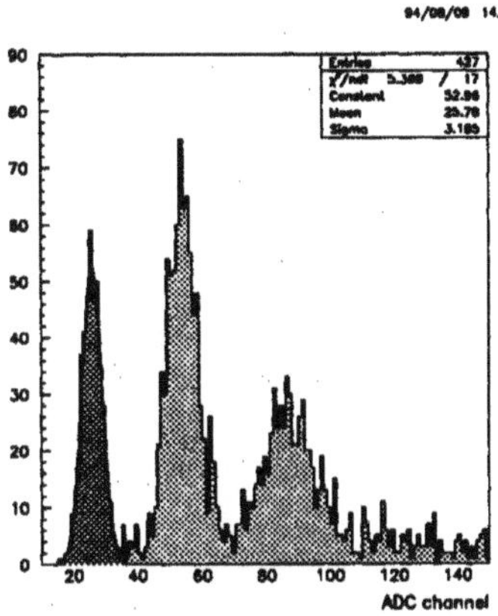

Fig.4 FAST PREAMPLIFIER (30ns)
Noise=3.18 counts ⇒ 12.4MeV
Light=29 counts ⇒ 174 e/MeV
Diode=33 counts

6.3 Single tower spectrum and energy resolution with photodiode readout

The 50 GeV electrons beam is aimed in turn at the center of each tower to obtain the front to back intercalibration of each tower. The resulting central tower spectrum is shown as an example in Fig.5; ignoring the high energy tail, we get from these distributions the energy resolutions plotted in Fig.6. The differences in resolution is due to dimensional problems: Tower 1 for example is misaligned by 2-3mm because of varying crystal cross sections. The energy dependence of the single tower energy resolution is shown in Fig.7.

6.5 Matrix energy resolution

The measured and Monte-Carlo simulated energy resolutions of the sum of nine crystals versus energy are plotted on Fig.8, the energy spectrum at 50 GeV in Fig.9. The 50 GeV point has been measured both in the X3 and H4 beam. The H4 data are slightly worse than simulation and X3 data, which we ascribe to pion contamination. The X3 data at 25 and 50 GeV closely reproduce our Monte Carlo simulation and the 10 GeV point is affected as expected by the electronics noise contribution.

6.6 Energy Resolution with slow and fast shaping

We have compared the resolutions of a single tower for 50 GeV electrons obtained with slow and fast electronics (Fig.10). The 2μs shaping time electronics gives 1.08±0.02% and the 30 ns shaping time electronics gives 1.09±0.02%.

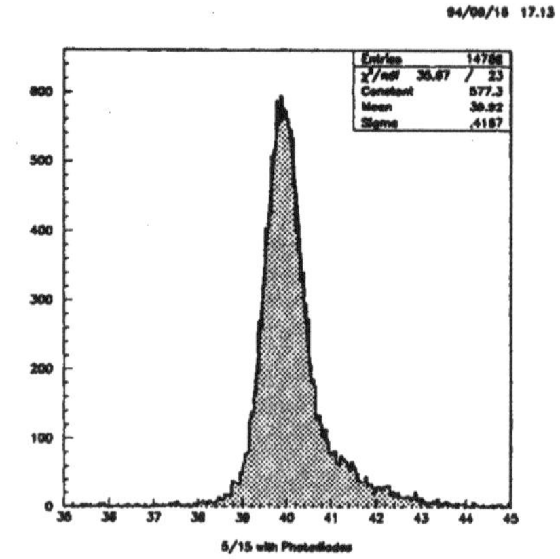

Fig.5 Central tower energy distribution for 50 GeV electrons; PIN photodiodes readout

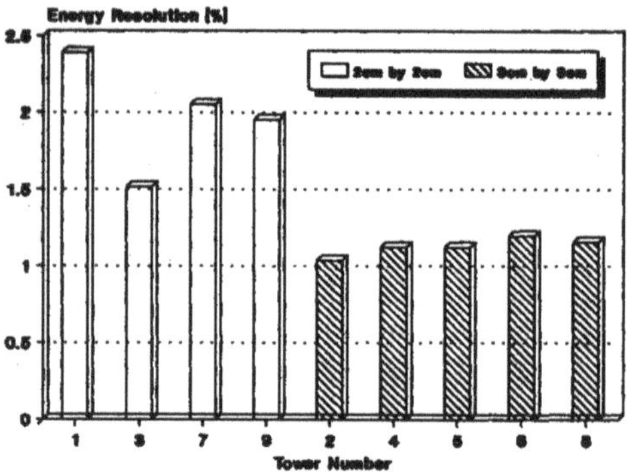

Fig.6 Measured energy resolution of individual towers for 50 GeV electrons

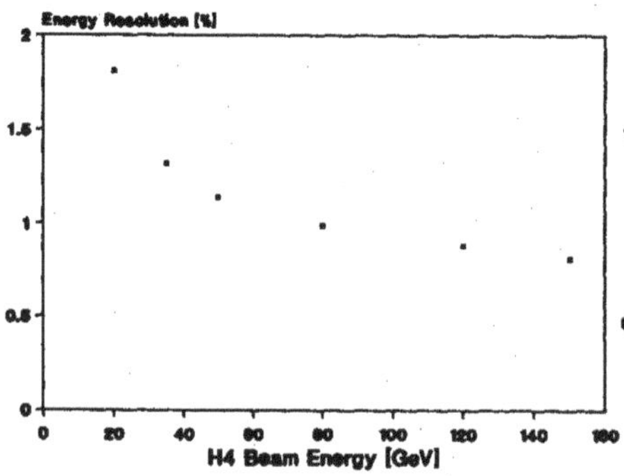

Fig.7 Energy dependence of the central tower resolution, measured in the H4 beam

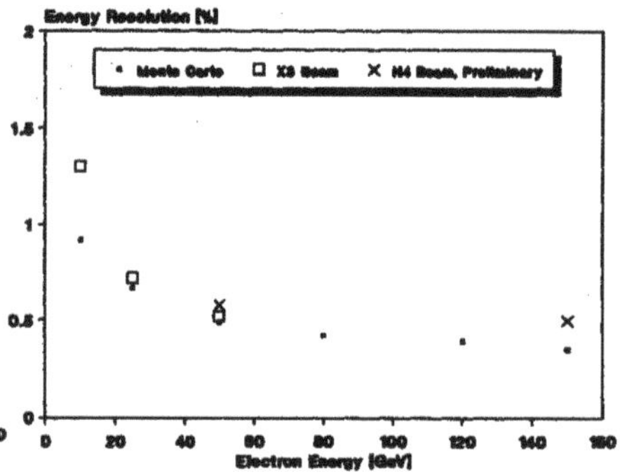

Fig. 8 Measured and simulated energy dependence of the resolution of the CeF_3 matrix (sum of nine towers, X3 beam momentum spread substracted)

6.7 Photomultiplier and Si Photodiode Readouts

To check the origin of the tail observed in the energy distribution, we have replaced the Si photodiodes on the central tower with vacuum photomultipliers. It can be seen from Fig.11 that the tail in the high energy part of the distribution originates in the PIN photodiodes and that both rear leakage and albedo contribute to the tail. Curing this will require optimisation of diode thickness, area, crystal light yield or the use of low gain avalanche photodiodes

6.8 Photodetector tests at 3 Tesla

We tested two Hamamatsu photosensors in a 3T magnetic field (Fig.12): a Si PIN photodiode with preamplifier and a R5189 UV sensitive vacuum phototetrode. The photodiode was oriented at 0, 45 and 90 degrees from the field and excited either directly with a ^{57}Co source or with the light from a CsI crystal/^{137}Cs source. The phototetrode was viewing a NaI crystal/^{137}Cs source at field orientations of 30 and 60 degrees. The photodiode/amplifier combination showed no detectable sensitivity to the field. The vacuum phototetrode, although having its gain affected by the field, is still usable at 3T.

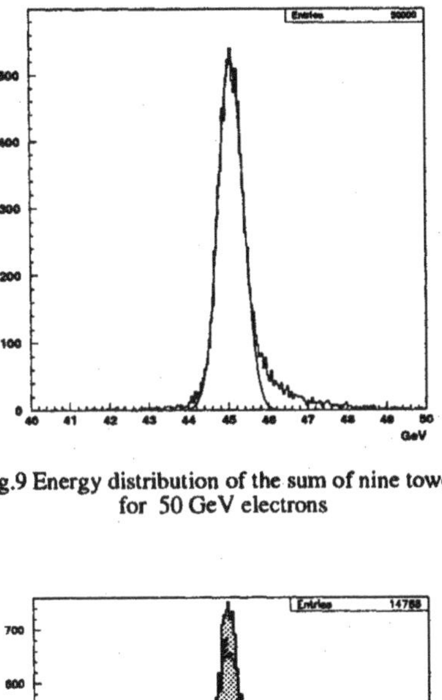

Fig.9 Energy distribution of the sum of nine towers for 50 GeV electrons

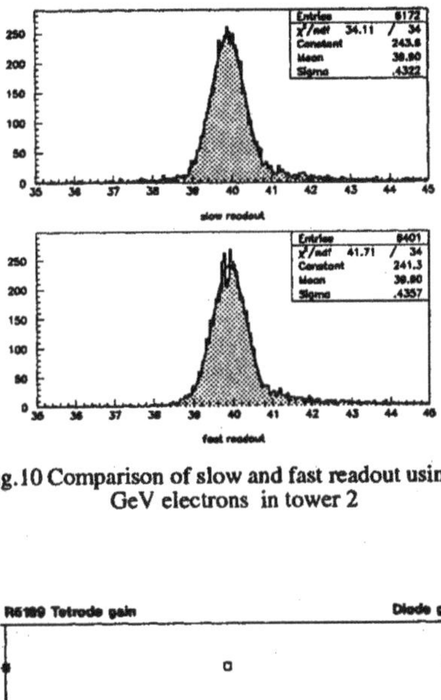

Fig.10 Comparison of slow and fast readout using 50 GeV electrons in tower 2

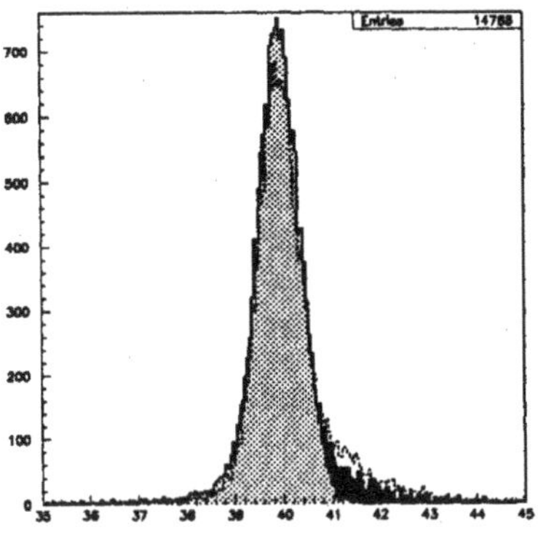

Fig.11 Energy distribution of 50 GeV electrons into one tower, with photomultiplers readout (light grey), PIN diodes readout (Dotted line) and PIN diode on front segment, Photomultiplier on back (dark grey).

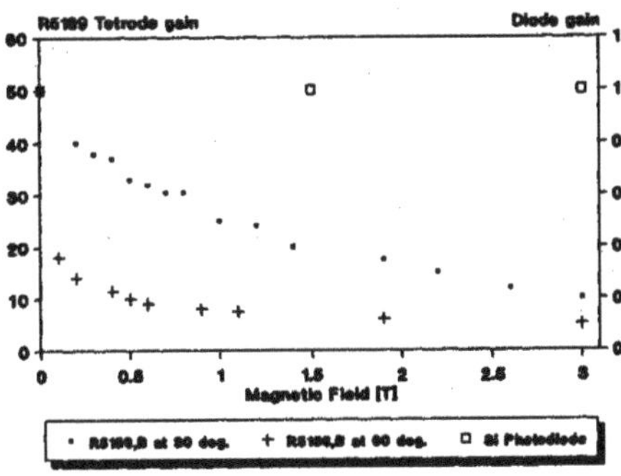

Fig.12 Photodetector gain dependence on magnetic field up to 3 Tesla

7. Summary and conclusion

We have demonstrated an energy resolution of the order of 0.5% at energies ≥ 50GeV in a small longitudinally segmented Cerium Fluoride matrix, using PIN photodiodes readout, under far from optimal conditions. **Consequently, from the performances point of view, Cerium Fluoride is presently the best material for EM calorimetry at LHC.**

We have studied the effect of calorimeter leakage and albedo on the solid state photodiodes and are considering ways of reducing the high energy tail.

We have also demonstrated adequate performance of our fast preamplifiers, in view of operation at LHC.

Large crystals 3×3cm^2 in cross section, up to 28cm long, have been produced and smaller, high quality crystals are available. Nevertheless, cost effective mass production of large, consistently high quality crystals remains the key issue for the use of Cerium Fluoride in a large scale detector at LHC.

8. Acknowledgments

We wish to thank P.Grafström, N.Doble, M.Haguenauer and J.Bourotte for their help in operating the X3 and H4 beams, the setting up of the experiment and data taking facilities, J.Spangaard for his assistance in operating the X3 beam spectrometer and our thanks also go to the technical staff of the participating institutions.

8. References

1. E.Auffray et al., *First results on large cerium fluoride crystals in a test beam,* Preprint **CERN/PPE 94-112.**

2. M.Goyot, *Performance of preamplifier-Silicon photodiode readout systems associated to large BGO crystal scintillators,* Nucl.Instr.Meth. **A263** (1988) 180.

3. M.Goyot, *IPN Lyon Development.*

4. J.C.Caldero and P.Sahuc, *Shaping amplifiers for photodiode readout,* IPN Lyon internal note.

Beam test results of $PbWO_4$ crystal calorimeter prototype

G.A. Alexeev, A.V. Dolgopolov, S.V. Donskov, V.A. Kachanov, V.Yu. Khodyrev, V.A. Medvedev, Yu.D. Prokoshkin, P.M. Shagin, A.V. Singovsky, V.L. Solovianov, S.M. Stepoushkine, V.P. Sugonyaev
Institute for High Energy Physics, Protvino, Russia

A.A. Fiodorov, M.V. Korzhik, O.V. Missevich
Iistitute of Nuclear Problens, Minsk, Belarus

J.P. Peigneux, J.P. Vialle
Laboratoire d'Annesy de Physique des Particules, Annecy, France

Abstract

Beam tests of an EM calorimeter prototype made of $PbWO_4$ crystals produced with a significant improvment of growth technology have been performed at CERN SPS. The measured energy resolution is $\sigma_E/E = 2.78\%\sqrt{E} \oplus 0.45\%$.

1 Introduction

Lead Tangstate crystals (PWO) are considered as a promising material for electromagnetic calorimetry since more than 2 years [1, 2]. Today significant progress has been made both in basic crystal properties investigations [3, 5, 7, 8] and in specific applications to electromagnetic calorimetry [4, 6].

Among interesting results obtaned so far where first a good energy resolution measured in beam tests at CERN and KEK with 3×3 matrices of $20 \times 20 \times 180$ mm^3 crystals [4, 6] and second, a significant variation of crystal characteristics along the ingot axis [8]. This variation together with a rather strong temperature dependence of the light yield (about $-2\%/^\circ C$) would spoil PWO calorimeter performances. As a result, one may expect a much better performance with uniform and temperature stabilized crystals.

After optimisation of the crystal growth procedure, precise stoichiometry control and doping with Nb, a new set of crystals has been produced. Laboratory tests did not show any significant crystal properties variation from top to bottom of ingots.

A new EM calorimeter prototype made with the new set of crystals have been tested during the summer 1994 at the X1 beam of CERN SPS. The first results of these tests are presented below.

2 Measurement conditions

The test calorimeter was built as a mairix of 5×8 cells. Each cell was made of $20 \times 20 \times (180 \div 200)$ mm^3 PWO crystal wrapped with 100 μm TYVEK paper. Two specially shaped plastic pieces where attached to the front and back faces of each crystal to hold an optical fiber of the monitoring system light distributor and a photomultiplier respectivelly. These pieces where kept together by streched $50\mu m$ mylar strips. One layer of $50\mu m$ black PVC tape was inserted between cells for cell-to-cell optical isolation. The total gap between crystals in matrix was less than $300\mu m$.

Philips PM1911 photomultipliers (PM), optically connected to the crystals by Dow Corning Q2-3067 couplant where used for light detection.

The 5×8 cells matrix, surrounded by massive Cu plates with pipes for water circulation used for temperature stabilisation, was put in a dark box. The top raw of crystals was equiped with temperature sensors. All measurements where performed at a temperature of (14.5 ± 0.2) $°C$.

Signals from PM where digitized by 12 bit charge ADC with 100 ns gate. A monitoring system based on red LEDs triggered by a temperature stabilized pulser was used to control PM gain and ADC stability to a precision of 0.2%.

3 Test results

3.1 Energy resolution

The characteristics of the EM calorimeter prototype where measured with 10, 20, 50 and 70 GeV electrons. The momentum spread of the beam was $\sigma_p/p \simeq 0.3\%$. The calibration of calorimeter cells was done in two steps. First, all cells where irradiated by a narrow 50 GeV electron beam with a spot size of 2×2 mm^2. The peak positions obtained from a Gaussian fit of the amplitude distribution for each cell was used as a first approximation for the minimization procedure to determine the optimal calibration coefficients.

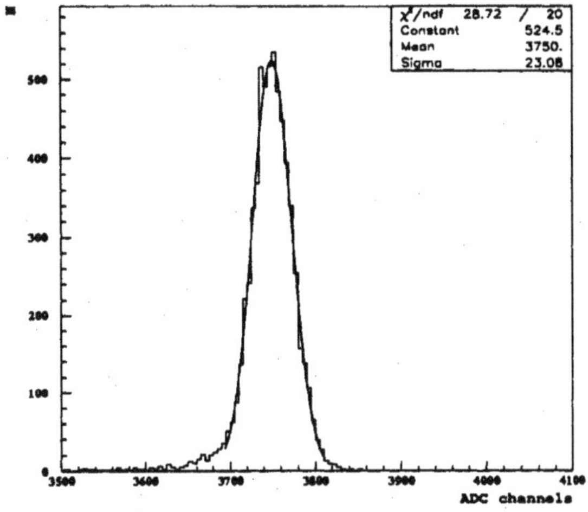

Figure 1: Calibrated sum of 5 × 5 cells measured in the 70 GeV electron beam. The beam spot size is 6 × 6 mm^2.

The calibrated sum of the signals from 5 × 5 cells irradiated by 70 GeV electrons is presented in fig. 1. The energy resolution for the different electron energies measurements was obtained by a Gaussian fit. It is shown in fig. 2 as a function of energy. The beam momentum spread is quadratically subtracted. The energy resolution can be parametrized as:

$$\sigma_E/E \;=\; (2.78\ \pm\ 0.05)\%/\sqrt{E}\ \oplus\ (0.45\ \pm\ 0.02)\%$$

3.2 Temperature dependence of the light yield

The light yield temperature dependence has been measured after the energy scan with the temperature stabilization system been switched off. The crystals temperature increased from 14.34 to 17.44 $°C$ during 4 hours. The slow temperature rise ensured the absence of significant temperature gradients along the crystals. This was also controlled by the temperature sensors attached to front and back sides of one cell. The matrix was continuously irradiated with 50 GeV electrons. The temperature dependence of one crystals light amplitude is presented in fig. 3. It is well fitted by a linear function with a slope $(-2.10 \pm 0.06)\%/°C$.

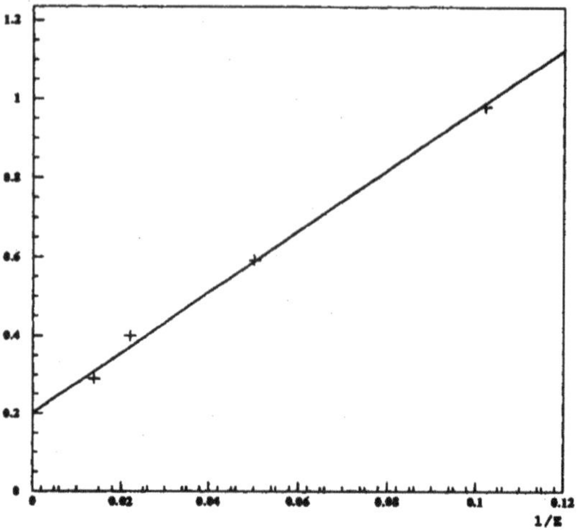

Figure 2: $(\sigma_E/E)^2$ versus electron beam energy. Straight line is for $a/\sqrt{E} \oplus b$ dependence.

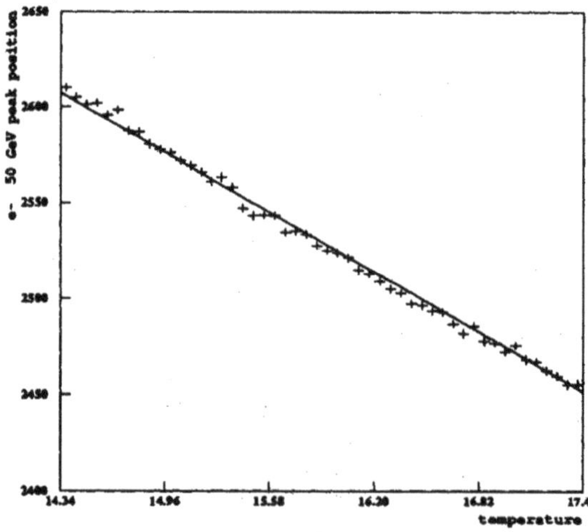

Figure 3: The temperature dependence of 50 GeV electron signal in one cell.

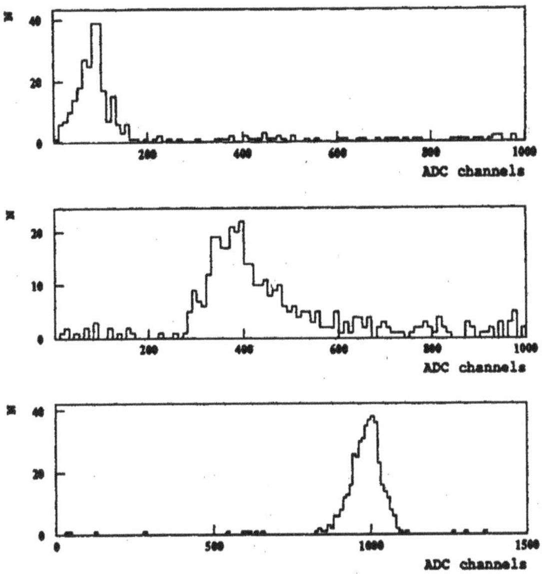

Figure 4: Amplitude distributions for one cell with PIN diode readout: signal from MIP in the crystal, signal from MIP in crystal and diode and signal from the 5 GeV electron respectively.

3.3 Light yield

The light yield of PWO crystals was measured by several groups mainly in laboratory conditions using low energy photons from radioactive sorces. To compare it with the light output obtained with EM showers, the PM at one crystal has been replaced by a Hamamatsu PIN diode S3590-03. The signals from minimum ionizing particle (MIP) passing through the crystal, crystal and diode and the amplitude spectrum from 5 GeV electron are presented in fig. 4.

Comparing to the number of e-hole pairs created by a MIP in the $300\mu m$ PIN diode (25000 [9]) the number of electrons created in a diode by an EM shower is 16 photoelectrons/MeV. Taking into account a geometrical matching factor 1/4 ($1cm^2$ diode and $4cm^2$ crystal) and a mean quantum efficiency of the diode $\sim$ 50% (the mean value for PWO emission spectrum), the PWO crystal light output is estimated to be $\simeq$120 photons/MeV.

4 Conclusions

Beam tests of an EM calorimeter prototype made of PWO crystals grown with improved technology have been performed. The better crystal uniformity and good temperature stabilisation result in a significant improvement of the calorimeter performance.

The energy resolution obtained is close to the design parameters of CMS [10] and ALICE [11] EM calorimeters. This result together with the good radiation hardness [8] indicates that PWO crystals may be used in LHC experiments.

References

[1] V.G. Baryshevsky *et al.*, Nucl. Instr. Meth. **A322** (1992) 231.

[2] M. Kobayashi *et al.*, Proc. CRYSTAL 2000 International Workshop, Chamonix, France, Editione Frontieres,(1993) 375.

[3] V.A. Katchanov *et al.*, IEEE Conference Record, Nuclear Science Symposium and Medical Imaging Conference, San Francisco, USA, (1993) 146.

[4] O.V. Buyanov *et al.*, Nucl. Instr. Meth. **A349** (1994) 62.

[5] M.V. Korzhik *et al.*, preprint LAPP-EXP 94-01, LAPP, Annecy.

[6] S. Inaba *et al.*, preprint KEK 94-105, submeted to Nucl. Instr. Meth.

[7] M.V. Korzhik *et al.*, presented at MRS'94 Meeting, April 1994, San Francisco, USA.

[8] A. Fyodorov *et al.*, to be presented at IEEE conference, Nuclear Science Symposium and Medical Imaging Conference, USA, (1994).

[9] Particle Data Group, Phys. Rev. D50 (1994) .

[10] Letter of Intent by the CMS collaboration, CERN/LHCC 92-2 (1992).

[11] Letter of Intent by the ALICE collaboration, CERN/LHCC 93-5 (1993).

A New CsI(undoped) Endcap Photon Veto System for BNL E787: A Study of the Decay $K^+ \to \pi^+\nu\bar{\nu}$

I-H.Chiang[(1)], E.Garber[(1)], T.Inagaki[(2)], K.Ino[(3)], S.Kabe[(2)], S.Kettell[(1)], M.Kobayashi[(2)], T.K.Komatsubara[(3)], Y.Kuno[(2)], K.K.Li[(1)], L.S.Littenberg[(1)], T.Morimoto[(3)], T.Nakano[(4)], K.Omata[(3)], T.Sato[(2)], T.Shinkawa[(2)], S.Sugimoto[(3)], K.Tauchi[(3)], A.Yamashita[(1)] and Y.Yoshimura[(2)]

[(1)]*Brookhaven National Laboratory (BNL), Upton, NY 11973*

[(2)]*National Laboratory for High Energy Physics (KEK), Tsukuba 305, Japan*

[(3)]*Institute for Nuclear Study (INS), University of Tokyo, Tokyo 188, Japan*

[(4)]*Osaka University, Osaka 560, Japan*

ABSTRACT

A new CsI(undoped) calorimeter has been installed for the endcap photon veto system of BNL E787, a search for the decay $K^+ \to \pi^+\nu\bar{\nu}$. This new endcap provides substantially more light output than the previous Pb-scintillator endcap and has better timing resolution.

1. Introduction

The experiment E787* at the Alternating Gradient Synchrotron (AGS) at BNL is searching for the decay $K^+ \to \pi^+\nu\bar{\nu}$[1]. The standard model predicts this branching ratio to occur at $\sim 10^{-10}$. A measurement of this branching ratio is particularly interesting, as it is sensitive to the CKM matrix element V_{td} and the theoretical uncertainties are not very large ($\sim$7%) [2].

The signature for the decay $K^+ \to \pi^+\nu\bar{\nu}$ is a single charged pion track, so a search for this decay at a sensitivity of $\sim 10^{-10}$ requires a very efficient veto of the photons from the copious $K^+ \to \pi^+\pi^\circ$ ($K_{\pi 2}$) decay, a large rejection of muon tracks by requiring the observation of the decay chain $\pi^+ \to \mu^+ \to e^+$, and good reconstruction of the energy, range and momentum of the charged track .

The two π° photons from $K_{\pi 2}$ decay at rest have energies between 20–225 MeV, but due to unavoidable gaps and dead material in the detector it is necessary to veto energy deposits of even lower energy. Essentially, the whole E787 detector is used to veto photons. The photon veto has to be efficient in the very high rate environment necessary to measure a 10^{-10} branching ratio, yet not introduce a large inefficiency caused by accidental vetoing of good events. This requires good timing resolution in almost all detector systems.

A diagram of the E787 detector is shown in figure 1. The previous version of the detector has been described elsewhere[3]. The new endcap system is described below.

*E787 collaboration: BNL, Princeton University, TRIUMF, KEK, INS and Osaka University

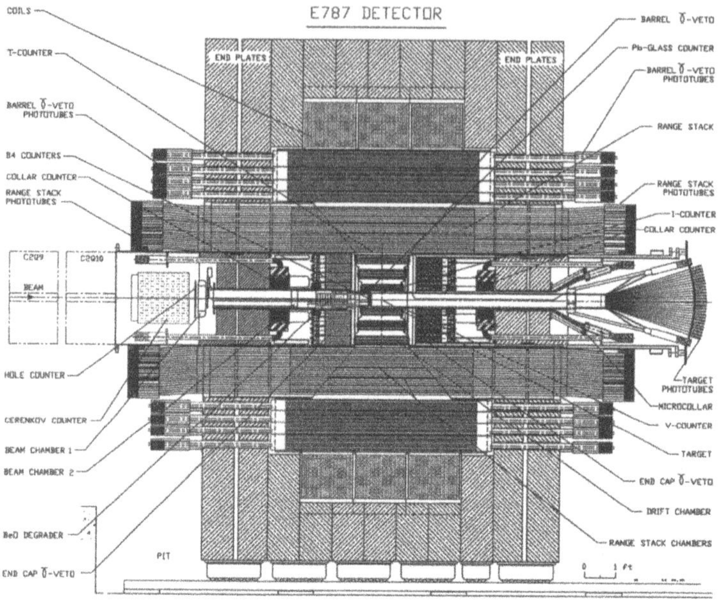

Figure 1: The E787 detector.

2. CsI Endcap Photon Veto System

The endcap is in a particularly high rate environment near the beam. The old endcap was constructed as a Pb-scintillator sandwich with a visible energy/ deposited energy of 30%. The light output was only a couple of photoelectrons per MeV of deposited energy (pe/MeV) and the timing resolution for 3 MeV photons was greater than 2.5 ns. In addition, due to the large amount of dead material (Pb) the old endcap was more likely miss photons that underwent photonuclear interactions.

The width of the veto window should be as narrow as possible to reduce losses from accidental hits. It is also necessary to be efficient for detecting signals riding on the tails of earlier hits. The CsI provides substantially more light than the old endcaps and should provide much better timing resolution, but since it is slower, pulse recognition of signals on the tails of earlier signals is more difficult. The slow component of scintillation light is suppressed through the use of optical filters and the pulses are recorded with transient digitizers to obtain good rejection of overlapping signals. The design of the endcap is to achieve sub-nanosecond timing resolution for 1 MeV signals.

2.1. CsI Crystals

The endcaps consist of 143 CsI(undoped) crystals arranged in four concentric rings in each endcap (75 crystals in the upstream endcap and 68 crystals in the downstream endcap). The crystals are right-pentagonal-polyhedrons (see figure 2a). The dimensions of the Ring 1 crystals are A=7.8–8.0 cm, B=6.1 cm and C=25.0 cm; the dimensions of the Ring 2–4 crystals are A=9.9–11.2 cm, B=8.6 cm and C=25.0 cm. They were grown by Crismatec using the Kyropoulos method. The ingots were 16"–19" in diameter and needed to be greater than 4" high in order to cut crystals for Ring 2–4. The number of Ring 2–4 crystals cut from one ingot varied from 1–3. These large crystals generally have good light output, a mean of 255 pe/MeV for full photocathode coverage (QE=16%).

CsI is a dense scintillator that is relatively fast with reasonably good light output. [4] The density is $\rho = 4.5$ g/cm^3; the radiation length is $\chi_0 = 1.86$ cm; the Moliere radius is $R_M = 3.8$ cm. Undoped CsI has at least two components of scintillation light, the fast component has a time constant of ~45 ns and the slow component has a time constant of ~600 ns. The time constant of the fast component measured with these large crystals is slightly larger than for smaller crystals. The fast component has a peak emission at ~310 nm, while the slow component peaks at ~450 nm. The ratio of fast to slow components is a measure of the quality of the crystal and is frequently quoted as F/T, where F (Fast) is the integrated charge in a 100 ns ADC gate and T (Total) is the integrated charge in a 1 μs gate. The crystal light output and F/T ratio are shown in figure 2(b). The light output is measured with a 5" PMT with a UV glass window. With this PMT the entire crystal surface is covered by the photocathode and the measured light yield varies from 180–310 pe/MeV. There is a strong correlation between F/T and npe/MeV.

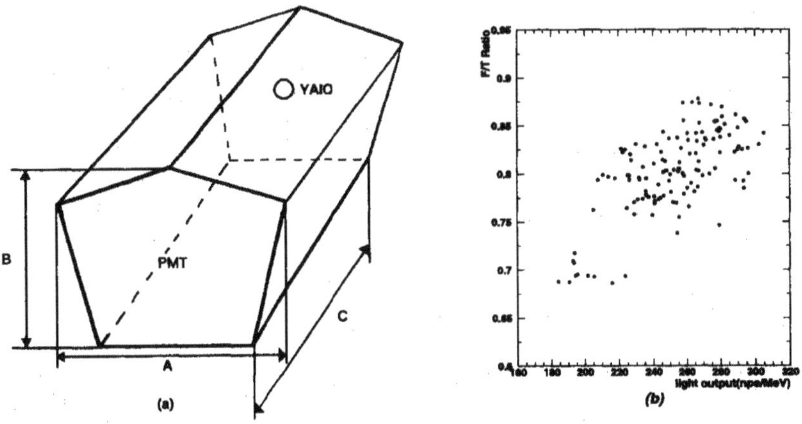

Figure 2: (a) Sketch of CsI crystal. (b) Plot of the F/T ratio vs the crystal output (for full PMT coverage).

The crystal size and shape were designed minimize the amount of dead material in the endcap. Each crystal is wrapped with 0.008" Teflon (in 3 layers) and 0.0015"

Al-Mylar. This assembly is then wrapped with 0.0015" Monokote to protect the inner wrapping and to provide optical isolation. Increasing the Teflon beyond 0.008" increases the light yield only slightly, while adding more dead material. The open end of each crystal has a white Noryl flange (with holes for the PMT and for a quartz optical fiber) glued to it with a whitened epoxy (Stycast 1266 with MgO, 1:1 by weight). The white flange and white epoxy were found to increase the light output by about 15% relative to a black flange with clear epoxy. This flange is used to mount the PMT to the crystal and the crystals onto the aluminum support plate of the endcap. The material between the crystals and the target has been minimized — only the wrapping and a 0.188" lucite endplate (the inner wall of the endcap, inside of Ring 1, is 0.031" Al). The lucite cover helps to provide a seal of the crystal volume; dry nitrogen is continuously flowed into the endcap to keep the humidity low (5-15%).

2.2. High Field Fine-Mesh Phototubes

The PMTs are directly coupled to the crystals in order to maximize the light collection, so they must operate with high gain in a 10 kG magnetic field. They must have high quantum efficiency (QE) at a wavelength of 310 nm (need UV glass or quartz window) and large photocathode area. Fine-Mesh PMTs were designed for this experiment by Hamamatsu. They are high gain, 19-stage PMTs with 14.5 cm^2 (2" PMT) and 32.2 cm^2 (3" PMT) photocathodes and UV glass windows. The gain is greater than 2×10^7 at **B**=0 kG, with typical gains of 1–2$\times10^6$ at 9.6 kG. The gain drop in the E787 detector is $\sim$ 40, at **B**=9.6 kG. They have reasonably fast rise times, high QE (average QE=17%), but not very good single photoelectron resolution. The ring 1 crystals use 2" PMTs (R4721mod) and the larger crystals use 3" PMTs (R5065mod).

Each PMT has an Kenko U330 filter glued onto it. The PMT/filter combination then has a 3 mm thick cookie of Sylgard 527 silicone gel formed onto it. The surface is slightly convex, to allow good optical coupling to the crystal and to allow for removal of the PMT from the crystal. The F/T ratio is typically 80% without the filters and 86% with the filters. There is a loss of ~15% of the fast light and a suppression of the slow component by almost an order of magnitude with the filter.

The PMTs are mounted to the crystals with a spring loaded locking mechanism that keeps a constant pressure of 7 pounds on the cookie.

2.3. Electronics

Signals from the CsI PMTs are amplified (PS776 $\times$10 amplifiers) and then split three ways (0.5:0.4:0.1). The first output goes to an 8-bit, 500 MHz (256 ns range) CCD transient digitizer[5] for pulse recognition and precise timing, the second output goes to a 11 bit ADC (LRS4300B) for energy calibration, and the third output is multiplexed by four and is then sent to a high range ADC, a high-range CCD and to a discriminator. The discriminator output is used for the photon veto requirement of the L0 trigger. The discriminated signal is also sent to a 1 ns least count, (2μs range) TDC (LRS1789) as a check on the CCDs and for measuring rates. The high-range ADC and CCD are used due to the large dynamic range of the photons of interest.

An average pulse shape from the CCDs is shown for one of the crystals in figure 3(a). (The signal has been delayed with 250 ns of RG8x cable and 150 ns of RG58 cable.) A fit to the pulse shape is overlayed on the histogram, the rise time is 9.9 ns and the fall time is 44.1 ns.

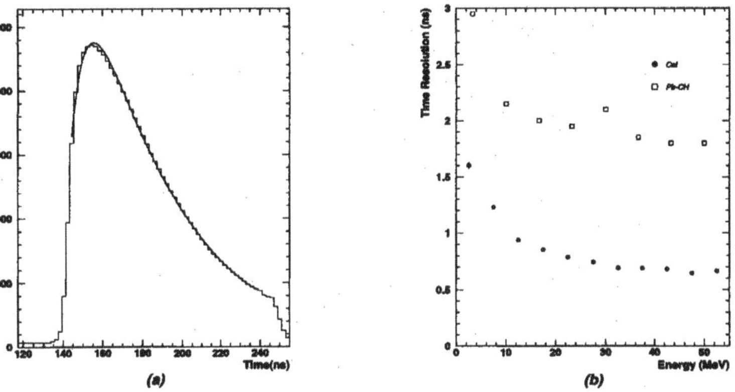

Figure 3: (a) An average pulse shape in the low-range CCD for one crystal. (b) Timing resolution for old and new endcaps as a function of energy.

2.4. Monitoring System

The PMT gain is monitored with three separate techniques 1) a Xe lamp , 2) $YAlO_3$:Ce(^{241}Am) light pulsers[6], and 3) calibration triggers from $K^+ \rightarrow \mu^+\nu$ ($K_{\mu2}$) decays. The Xe lamp (Hamamatsu L2436) output is monitored by 2 PIN photodiodes and the light is distributed by quartz fibers to all 143 crystals and several reference PMTs. The Xe light is broad band the spectrum is cut by a UV filter to match the light spectrum of CsI scintillation; the pulse width (FWHM~120ns) is considerably longer than that of the CsI (FWHM~35ns), but has been shown to be stable over time. The resolution is 2–3%. Over the one month of running the Xe light output was stable to ±0.4%. Both the Xe lamp and reference PMTs are in temperature controlled environments. These temperatures, as well as the temperatures of the crystal and PMT environments, are monitored once per AGS spill (every 4 seconds). The Xe lamp is triggered by a clock synchronized to the AGS RF to flash once between each AGS spill. This provides a continual monitoring of the PMT gains.

The YAlO light pulsers are glued to the CsI crystal on the end facing the PMT. They scintillate at a wavelength that is well matched to the CsI and have a similar pulse shape (FWHM~50 ns). They are calibrated to the ^{137}Cs peak between AGS runs; their typical equivalent energy is 4–9 MeV. Special runs, triggering on the YAlO signals, are made to set the PMT HV and then about once per week as an additional check of the gain calibration.

The energy calibration is accomplished by comparing the energy deposited by muons from $K_{\mu 2}$ decays in data and Monte Carlo simulations. The $K_{\mu 2}$ muons are mono-energetic (152 MeV) and, neglecting corner clipping, stop in the CsI; however, due to the variation in kaon stopping position in the target, the peak is quite broad. Special triggers are made with $K_{\mu 2}$ decays with the muon stopping in the endcap. These calibration triggers are accumulated continuously with the $\pi\nu\bar{\nu}$ data, and special calibration runs are made every few weeks. This energy calibration is verified by reconstructing the π^0 from $K_{\pi 2}$ decays, where one photon converts in the barrel veto and one converts in the endcap. The current level of agreement is better than 10%.

The timing calibration is done by comparing the time of photons in the endcap with the π^+ in the range stack from $K_{\pi 2}$ decays.

3. Initial Performance

The new CsI endcap calorimeter was installed for the 1994 AGS run. Data on the performance has been collected and analyzed and some preliminary results have been obtained.

Rates are measured by counting random hits in the TDCs (for energies above ~10 MeV) and from special runs where a small number of endcap crystal signals were sent to the long baseline Transient Digitizers [7] (500MHz, 8 bit, 6μs range). The rates in the inner rings (Ring 1–2) on the upstream endcap are quite high. The number of kaons entering the target (K_B) has been raised from K_B=300 k/spill (the effective AGS spill length is $\sim$ 0.8–1 sec) to K_B=1.2 M/spill, with the goal to reach K_B=3 M/spill. With K_B=1.2 M/spill, the incident kaon rate measured at the beam Cherenkov counter is C_K=6 M/spill and the incident pion rate is C_π=1.5 M/spill. At this intensity the rate per crystal (2 MeV threshold) in the upstream endcap Ring 1 is 850 kHz; Ring 2 is 650 kHz, and the other Rings (including downstream) vary from 150–250 kHz. The rate (R) dependence on energy threshold (E) is fit with a sum of two exponentials

$$R(E) = A_1 e^{-E/E_1} + A_2 e^{-E/E_2} \tag{1}$$

with E_2=40 MeV and 1.3 MeV< E_1 <3.5 MeV. The total rate in the upstream endcap is estimated to be 15.4†(3.6) MHz with a 2 MeV(30 MeV) threshold. The fraction of endcap hits above 30 MeV which are in coincidence with C_K is 76% and the fraction in coincidence with C_π is 17%. The total rate in the downstream endcap is 50%(60%) of the upstream endcap. The average number of crystals hit per event is 2 (for thresholds above 5 MeV).

The calibration of the detector is continuing, but some preliminary results on the endcap timing show that the resolution is less than 1 ns for all $K_{\pi 2}$ decays. The timing resolution as a function of photon energy is shown in figure 3(b). We see that for energies above 10 MeV the resolution is better than 1 ns and at all energies we

†assuming a multiplicity of 2

have 2–3 times better resolution than with the old endcap. We expect even further improvements as the calibrations are refined.

4. Conclusions

Two new photon veto endcaps using undoped CsI crystals have been constructed and installed into the E787 detector. We have measured an improvement of a factor of 2–3 in the endcap time resolution. Further improvements are expected. Calibration of the detector is continuing and a measurement of the veto efficiency for $K_{\pi2}$ photons should be available soon.

5. Acknowledgements

We gratefully acknowledge J.R. Cuccia Jr., W. Lenz, G. Munoz, and H. Ratzke for their valuable technical assistance. This work was supported in part by the U.S. Department of Energy, under contract number DE-AC02-76CH00016, and in part by the Ministry of Education, Science and Culture of Japan through the Japan-U.S. Cooperative Research and Development Project in Energy and Related Fields.

6. References

1. M.S. Atiya *et al.*, *Phys. Rev. Lett.* **70**(1993)2521; *Phys. Rev.* **D48**(1993)R1.
2. G. Buchalla and A. Buras, *Nucl. Phys.* **B412** (1994) 106.
3. M.S. Atiya *et al. Nucl. Instr. and Meth.* **A321** (1992) 129.
4. Z. Wei and R. Zhu, *Nucl. Instr. and Meth.* **A326** (1993) 508; C.L. Woody *et al.*, *IEEE Trans. Nucl. Sci.* **37** (1990) 492; S. Kubota *et al.*, *Nucl. Instr. and Meth.* **A268** (1988) 275.
5. D. Bryman *et al.*, *IEEE Trans. Nucl. Sci.* **38** (1991) 295.
6. M. Kobayashi *et al.*, *Nucl. Instr. and Meth.* **A337** (1994) 355.
7. M. Atiya *et al.*, *IEEE Trans. Nucl. Sci.* **36** (1989) 813; *Nucl. Instr. and Meth.* **A279** (1989) 180.

THE KTeV PURE CsI CALORIMETER

RONALD E. RAY
Fermi National Laboratory
Batavia, IL 60510, USA

ABSTRACT

KTeV is currently building a state-of-the-art pure CsI electromagnetic calorimeter with a sophisticated digital readout. The CsI array is expected to have better than 1% resolution over a dynamic range of 2 - 80 GeV. The design of the CsI array is driven by the difficult physics goal of attempting to measure the CP violation parameter Re(ϵ'/ϵ) to 1 part in 10000 in a high-rate neutral beam environment. The physics requirements and their impact on the final design will be discussed.

1. Introduction

In spite of the fact that kaon physics is a very mature field, it remains an area of intense study at laboratories around the world. The continuing interest in kaon physics is due to the fact that the system lends itself to incisive experiments yielding results with definitive physics interpretations and because of the ever-present possibility of new discoveries. The origin of CP violation is still a subject of intense interest. Insight into this issue can be obtained by measuring the quantity Re(ϵ'/ϵ) which parameterizes CP nonconservation in $K^\circ \rightarrow \pi\pi$ decays. A nonzero value of ϵ' would provide definitive evidence for the presence of *direct* CP violation and would support the Standard Model interpretation of CP violation arising from a non-zero phase in the CKM matrix. A value of zero for ϵ' would be ambiguous, however, and could support either *Superweak*[1] models or the CKM model where ϵ' is equal to zero because of an accidental phase cancellation. The latter could occur for a top quark mass in the range of 200-300 GeV [2].

The most recent results on ϵ'/ϵ come from Fermilab experiment E731[3], which has obtained a result consistent with zero, and CERN experiment NA31[4], which has obtained a positive result 3σ away from zero. This is a clear discrepancy which has resulted in new, more accurate experiments being mounted at both locations.

The KTeV (<u>K</u>aons at the <u>TeV</u>atron) collaboration at Fermilab* is presently constructing a new detector to begin operation at the beginning of 1996. The state-of-the-art detector, along with a new beamline producing intense fluxes of high-energy kaon decays, will extract the CP violating parameter Re(ϵ'/ϵ) with an accuracy of 1×10^{-4}, almost an order of magnitude improvement over current sensitivity. Additionally, the detector will be sensitive to rare four-body K_L decays with branching ratios as small as 10^{-11}, two orders of magnitude more sensitive than current limits.

*Chicago, Colorado, Elmhurst, Fermilab, Osaka, Rice, Rutgers, San Diego, UCLA, UCSD, Virginia, Wisconsin

2. Design Considerations for the KTeV Electromagnetic Calorimeter

2.1. Resolution and light output

The value of Re(ϵ'/ϵ) can be determined by measuring the double ratio of decay rates

$$\frac{\Gamma(K_L \to \pi^\circ\pi^\circ)/\Gamma(K_S \to \pi^\circ\pi^\circ)}{\Gamma(K_L \to \pi^+\pi^-)/\Gamma(K_S \to \pi^+\pi^-)} = 1 - 6Re(\epsilon'/\epsilon). \quad (1)$$

The electromagnetic calorimeter must accurately identify and count the number of neutral kaon decays to $\pi^\circ\pi^\circ$ while controlling systematic uncertainties. The most important contribution to systematic errors which are relevant to the calorimeter are the semileptonic background due to residual $\pi e\nu$ decays where the electron is mis-identified as a pion; $3\pi^\circ$ background where two photons are lost in the detector or fuse in the calorimeter; accidental activity which can cause biases which might differ for K_L and K_S events; and finally, energy scale uncertainties which can bias the event count because of the close relationship between reconstructed vertex position and cluster energies.

The calorimeter performance specifications required in order to measure ϵ'/ϵ to an accuracy of 10^{-4} have been determined largely from experience with the E731 lead glass calorimeter where electromagnetic calorimetry limitations which contribute to the systematic errors listed above have been thoroughly understood. Improved electron resolution and π-e rejection will reduce the contribution of residual semileptonic decays. Improved photon energy and position resolution will reduce backgrounds from $3\pi^\circ$ events by allowing for tighter mass cuts and by providing more efficient identification of overlapping clusters. Good shower containment and light transmission will reduce the difference in response between photons and electrons which is one of the ingredients necessary to fully understand one's absolute energy scale.

To accomplish these goals an energy resolution of 1% and a position resolution of 1 mm at the mean photon energy of 15 GeV is required. Non-Gaussian tails can spoil good energy resolution, so a line shape as nearly Gaussian as possible is required. The energy response of the calorimeter must be linear, i.e. $E_{true} = E_{meas}^{(1-\alpha)}$ where $\alpha < 0.5\%$. Finally, the components near the beam holes must be radiation hard to 10 kRads.

Pure Cesium Iodide (CsI) crystals were chosen to replace the old lead glass blocks for the KTeV project. CsI was chosen because of its light output, speed, radiation hardness, transparency and cost. Additionally, large quantities of CsI (doped with thallium) have been grown in the recent past, providing an existence proof that large scale manufacturing was possible. The calorimeter is shown in Figure 1 and the basic properties of Pure CsI can be found in the literature [5].

The calorimeter consists of 3100 pure CsI crystals, each 50 cm (27 $X_\circ$) long and read out with a Hamamatsu PMT at one end. The crystals come in two transverse sizes; 25 mm × 25 mm (small) and 50 mm × 50 mm (large). The large crystals are a cost cutting measure to reduce the number of channels. The position of the boundary between large and small crystals (0.6 m from the center of the array) was determined from the K° mass resolution as a function of the extent of the small block region.

The 50 cm length of the crystals has provided a significant challenge to the three vendors producing the crystals; Bicron, Crismatec and Horiba. Horiba has devised a process which results in single 50 cm long crystals of high quality and uniform properties along the crystal length. Bicron and Crismatec, unable to grow such long crystals, provide high quality glued pairs.

The design parameters of the CsI array were arrived at only after extensive prototyping and Monte Carlo studies using GEANT. The energy resolution of the array is found to depend primarily on shower leakage, uniformity of light output along the length of the crystals and the thickness of the material in which the crystals are wrapped. A crystal length of 50 cm (27 X_o) was chosen to minimize tails in the energy distribution due to leakage out the back of the crystal. For 20 X_o crystals, even 16 GeV showers were found to develop significant low-side tails. At 25 X_o the tails are still present, though greatly reduced. By 27 X_o the resolution as a function of crystal length begins to level out.

Because of the intrinsic longitudinal fluctuations in the development of electromagnetic showers, the calorimeter resolution degrades quickly with increasing deviations from uniform response along the crystal length. To preserve good resolution, each crystal is required to have a uniform response along its length to within 5%. Uniform response is not an intrinsic property of a 50 cm long CsI crystal. Uniformity is obtained with a compensational wrapping technique where, to first order, the half of

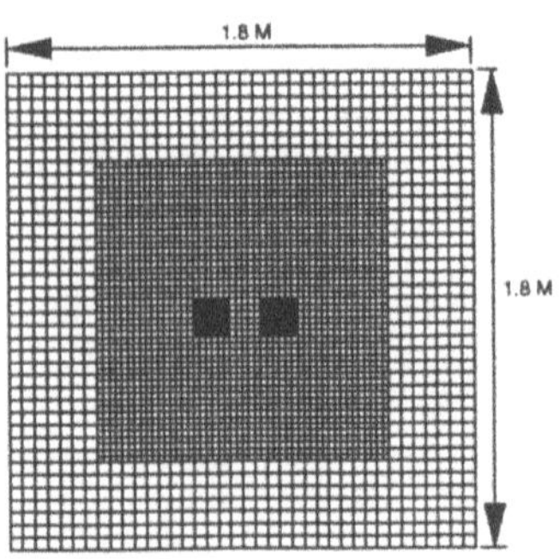

Fig. 1. The KTeV CsI array

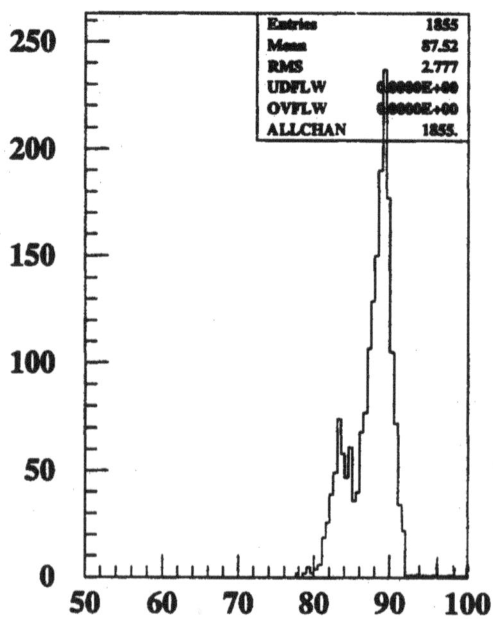

Fig. 2. Fast-to-total ratio for crystals currently in hand

the crystal farthest from the PMT is wrapped with aluminized mylar and the half closest to the PMT is wrapped in black mylar. By varying the amount of reflective material each crystal can be individually tuned. To minimize the amount of dead material between crystals, 0.005 inch thick mylar is used for wrapping.

The light output from pure CsI has both fast and slow time components with different emission spectrums[5]. Slow light is a source of accidental energy in the calorimeter and must be minimized. The crystals received from the vendors all have a fast to total (fast + slow) ratio of >70%. By placing a broad band filter (UG-11 glass), with a peak transmission of 90% at 300 nm for normally incident light and a near-zero transmission band between 400 and 650 nm, the fast-to-total ratio of the crystals can be boosted to a minimum of 80% with the peak of the distribution at 87%. The fast-to-total ratio for the crystals currently in hand, read out through filters, is shown in Figure 2.

Pure CsI has sufficient light output that photo-electron statistics will barely contribute to the overall energy resolution, even at the bottom of our dynamic range (2 GeV). The large/small crystals are read out through a 5 mm thick RTV cookie and a 1 mm thick filter into a 1.5/0.75 inch PMT resulting in an average of 30/20 p.e. per MeV. The number of p.e./MeV for both the small and large crystals currently in hand, when read out through a cookie, filter and appropriate size PMT, is shown in Figure 3.

2.2. Radiation hardness

The dominant limitation on the ability of E731 to calibrate the lead glass calorimeter was imposed by light absorption and radiation damage effects. CsI has significantly better transparency and response to radiation exposure than lead glass. Nevertheless, sufficiently large doses of radiation will degrade the transparency of CsI, reduce the overall light level and, most importantly, degrade the uniform response of the crystals over their 50 cm length. During the KTeV experiment, the crystals near the beam holes can expect an exposure of 7-10 kRads. The exposure falls off rapidly with distance from the beam. It is therefore important to identify a few hundred crystals which are radiation hard to 10 kRads and place them near the beam holes. In Figure 4 the effect of radiation damage on energy resolution is shown. The crystal was exposed to 8 kRads in several steps. At each dose level the uniformity response was measured and convoluted with longitudinally segmented GEANT showers to obtain a resolution. The GEANT showers spanned an energy range from 2 to 64 GeV. This particular crystal is relatively radiation hard, though not all of the KTeV crystals perform this well after irradiation.

CsI exhibits a wide range of susceptibility to radiation damage. Some crystals degrade significantly after 1/2 kRad while others remain unaffected after many kRads. The source of radiation damage in CsI is largely unknown. A correlation between radiation hardness and fast-to-total ratio was once suspected, but no such relationship is apparent in our data. KTeV's CsI vendors all grow large ingots from which they cut many crystals. The crystals from a given ingot tend to have similar radiation hardness properties. Ingots will be identified which have produced full length crystals which have been have been exposed to ^{60}Co and found to be radiation hard. The batches of crystals from these ingots will then be placed around the beam holes. During data taking, the uniformity of all 3100 crystals will be monitored continuously by cosmic ray muons passing transversely through the crystals and triggered by a longitudinally segmented hodoscope array above and below the calorimeter. In the event that the uniformity of a crystal degrades significantly, the reflectivity on the front face of the crystal can be increased in an attempt to boost the light collected from the half of the crystal farthest from the PMT.

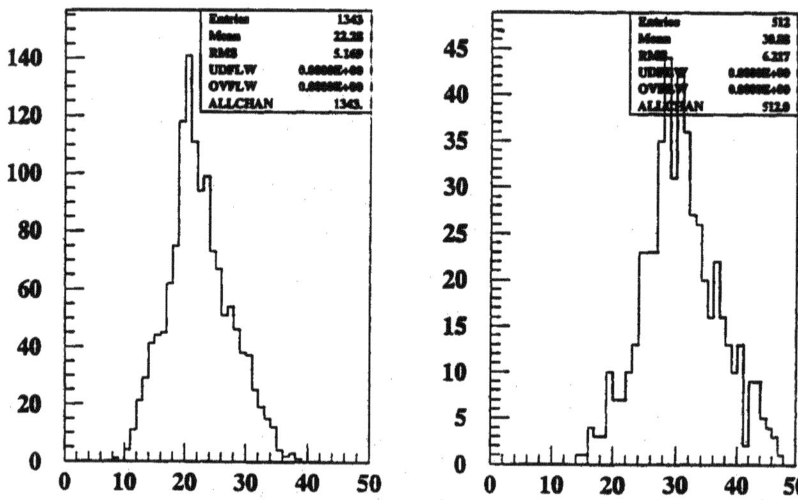

Fig. 3. Photo-electrons/MeV detected in small (left) and large (right) crystals currently in hand

3. Digital readout

A new digital readout technology has been developed at Fermilab which will be used to digitize the anode signals from the CsI PMTs at a rate of 53 MHz (the RF structure of the primary proton beam). The device resides in the base of the PMT, has 8 bits of resolution and a dynamic range of 17 bits. The device consists of an 8-bit commercial flash ADC and a custom current splitter implemented in an ASIC. The current splitter integrates binary weighted divisions of the input charge (I, I/2, I/4,...I/512), identifies the highest range which overflows and Grey codes that range as a 4-bit exponent. The integrated signal in the first non-overflowing range, or the remainder of the charge not accounted for in the exponent, is sent to the flash ADC. The 53 MHz sampling rate allows 18.8 ns slices to be recorded prior to the arrival of a CsI pulse, during the fast part of the pulse, and during the slow tail of the pulse.

The pre-samples provide a means for detecting accidental energy deposition in the calorimeter. Details of this device can be found in J. Whitmore's contribution to these proceedings.

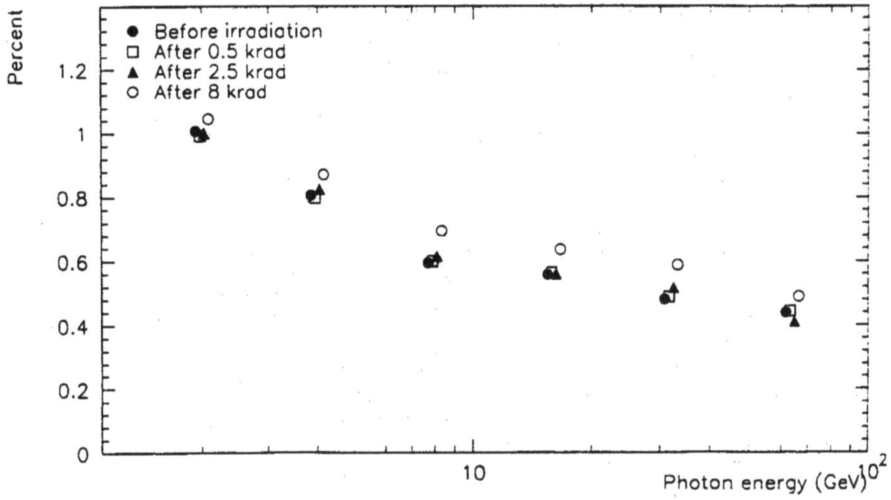

Fig. 4. Simulated energy resolution as a function of photon energy using the measured uniformity response before and after radiation exposure for a relatively radiation hard CsI crystal

4. Acknowledgements

Thanks to Aaron Roodman and Peter Shawhan, both from Chicago, for several of the plots appearing in this report.

5. References

1. L. Wolfenstein, *Phys. Rev. Lett.* **13** (1964) 562.
2. B. Winstein and L. Wolfenstein, *Rev. Mod. Phys.* **65**, No. 4, (1993) 1113.
3. L. K. Gibbons *et al.*, *Phys. Rev. Lett.* **70** (1993) 1203.
4. G. D. Barr *et al.*, CERN-PPE/193-168 (1993)
5. Particle Data Group, *Phys. Rev.* **D50** (1994) 1261.

KTEV CSI TEST BEAM RESULTS AT CERN

R.S. Kessler
University of Chicago
The Enrico Fermi Institute, 5640 South Ellis Avenue,
Chicago, Illinois 60637, USA

ABSTRACT

A 25 channel prototype of the KTeV CsI calorimater was brought to a test beam at CERN. Results are presented on the CsI energy resolution, π/e separation, and a uniformity measurement with muons.

1. Introduction

The KTeV CsI calorimeter[1,2,3] is designed to detect photons and electrons from Kaon decays at FNAL. One of the KTeV goals is to measure ϵ'/ϵ with a precision of 10^{-4}, which requires an energy resolution of $\sim 1\%$ for 15 GeV photons. It is also planned to search for the rare decay $K_L \to \pi^0 e^+ e^-$ with a single event sensitivity of 10^{-11}. To reject $K \to \pi e \nu$ related backgrounds at this level, a pion rejection of several hundred is needed from the CsI calorimeter.

To test the performance of this device, we exposed a 25 channel array at the CERN X1 test beam facility. We studied electron momenta of 5,10,20,30,40 and 60 GeV/c (Sec. 2), a 40 GeV/c pion beam (Sec. 3), and transverse muons to measure the response uniformities. (Sec. 4).

2. Setup

The setup (Fig. 1) consists of 25 CsI crystals, $5 \times 5 \times 50$ cm^3. The crystals were custom wrapped to have $\sim 5\%$ non-uniformity along their 27 X_0 length, and light yields ranging from 20-40 photo-electrons/MeV. Each crystal is viewed by a single 1.5" Hamamatsu phototube.

The readout electronics is perhaps the most unique feature of this system. The Digital PhotoMultiplier Tube Base, or DPMT[4,5], was specially designed at FNAL to give high precision digitization over a large dynamic range. The DPMT consists of two major components: (i) the charge integrating encoder [QIE] determines in which of the 9 ranges the signal lies, and (ii) the FADC determines the *mantissa*. The digitized signals were stored in a CAMAC FIFO, and then read out by a PC computer.

The FADC is an 8 bit device, which determines the readout precision. Combining the 8 bit FADC with nine ranges, the total dynamic range is 17 bits. In the lowest range, 1 FADC count corresponds to 13 fC, or ~ 1 MeV. In the highest range, 1 FADC count is ~ 3.3 pC, or ~ 250 MeV. The pedestal [RMS] noise is < 1 count. Thus for a 25 channel sum, the observed noise is about 5 MeV.

In addition to the wide dynamic range and low noise, the DPMT operates with four pipelined capacitors to provide a dead-timeless readout. The digitization frequency for this test setup was 10 MHz (for KTeV it will run at 53 MHz) so that each ADC gave the integrated charge in a 100 ns time *slice*. Since the test beam is asynchronous with respect to the DPMT, we only select events that arrived in a 10 ns window at the beginning of an ADC time slice.

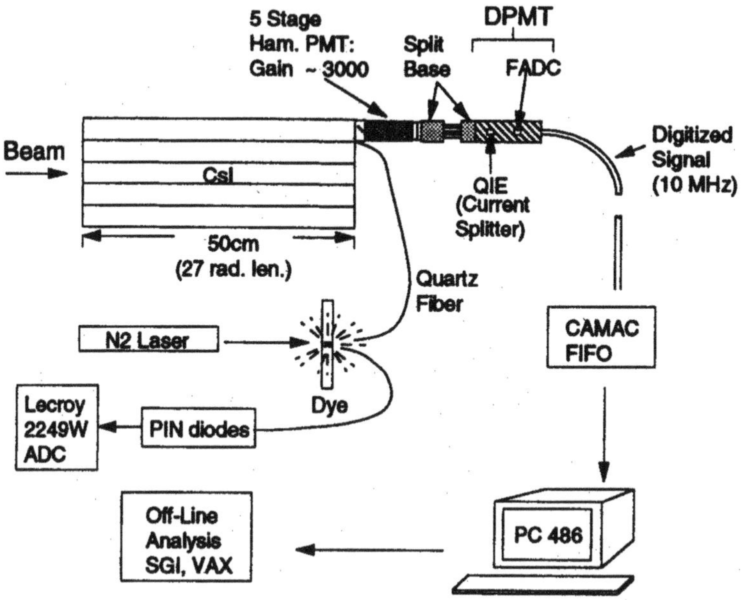

Figure 1: CsI setup for CERN beam test.

3. Electrons in CsI

The CsI raw data are shown in Fig. 2 for 20 GeV electrons incident on the central block. Note that as the range switches between 7 and 8, the FADC switches between the high and low end.

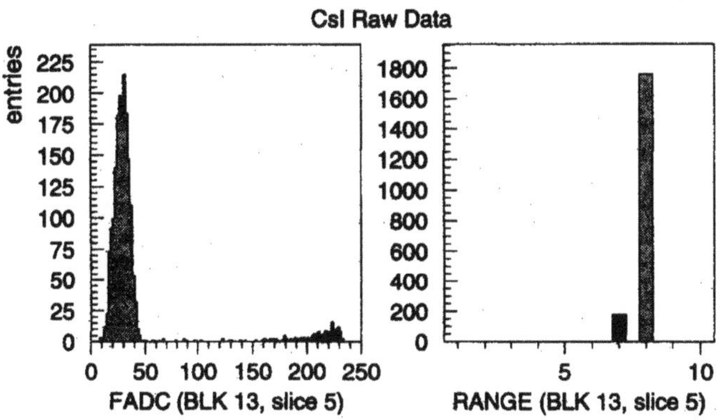

Figure 2: CsI raw data for 20 GeV electrons. The FADC and range are shown for the in-time slice of the central block (i.e., BLK 13).

To determine the energy resolution of the CsI calorimeter, there are two important calibration steps. First the DPMT calibration converts the FADC and range into a single number, which is the integrated charge. The second step is to convert charge to energy for each CsI block.

For the DPMT calibration, each crystal was exposed to a quartz fiber that delivered light from a laser flash (Fig. 1). Each laser flash was also viewed by two PIN diodes, whose signals were digitized by a Lecroy 2249W ADC. The laser light level was varied over the entire dynamic range of the DPMT (using filters), so that the DPMT values could be related to the PIN diode ADCs. Further details about the laser calibration can be found in these conference proceedings.[6]

To convert charge to energy, the matrix inversion technique was used. For this calibration we used independent data from a position scan at 20 GeV/c. Each of the inner nine blocks were illuminated to provide a good calibration for each block.

Using the calibration procedure described above, the E/p spectra are shown in Fig. 3. The beam was incident on the central block for the e^- energy scan. The flat tails ($E/p < 0.95$) resulted mainly from the few percent pion contamination of the electron beam. The resolution is estimated by a Gaussian fit to each E/p spectrum, and is shown in Fig. 4a (solid circles). The Monte Carlo resolution is also shown in Fig. 4a (open circles) using perfect calibration constants for both the DPMT and the charge $\rightarrow$ energy conversion. The contribution from the beam momentum spread (open triangles) is based on simulations of the X1 beam line optics and includes multiple scattering effects; however, radiative effects were not included in this simulation. The radiative effects in the beam may contribute to the low side tails that appear in the region $0.95 < E/p < 0.98$.

In Fig. 4b, the beam contribution has been subtracted from the E/p resolution to obtain an estimate of the *intrinsic* CsI energy resolution. The CsI energy resolution is below 1% for the range of electron momenta studied in this test.

4. Pions in CsI

After the electron energy scan we studied π/e separation using 40 GeV/c pion and electron beams. For this test, only the secondary targets were changed to select particle type; the beam optics remained unchanged so that the pion and electron energy scales could be compared. To get sufficicient rate, the collimators were opened; this resulted in a 2.5% (RMS) momentum spread compared with the $\sim 1/2\%$ σ_p/p used for the e^- energy scan.

The pion and electron E/p spectra are shown in Fig. 5a, where the total energy was determined from the inner 9 blocks. Since there is some pion contamination in the e^- sample, the e^- efficiency is taken to be 100% at $E/p = 0.9$. As the E/p cut is moved above 0.9, the e^- efficiency drops and the π^- rejection increases. Fig. 6 shows the π^- rejection as a function of the e^- efficiency. With 95% e^- acceptance the π^- rejection is $420 \pm 35(stat)$ at 40 GeV/c. Since the E/p resolution was 2.5%, we extrapolated to the case with better E/p resolution by replacing the e^- spectrum in Fig. 5 with a Gaussian of variable width. Fig. 6 shows that extrapolating to 1% E/p resolution, the pion rejection estimate increases to $\sim 700 \pm 75(stat)$ with 95% e^- efficiency.

Fig. 5b shows the same π^- E/p spectrum, but the E/p axis is re-scaled to show the MIP strucure (~ 400 MeV) in detail.

The muon contamination was estimated by looking at the E/p spectra in the CsI blocks outside the central block where the pion beam was centered. Assuming that

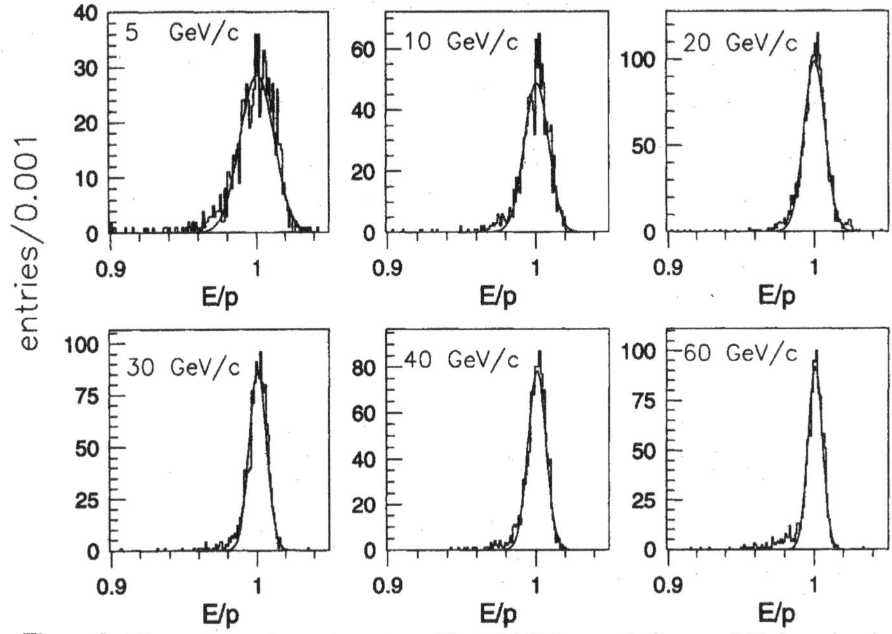

Figure 3: E/p spectra vs. beam momentum. The total CsI energy is the sum of 25 channels using one 100 ns ADC slice.

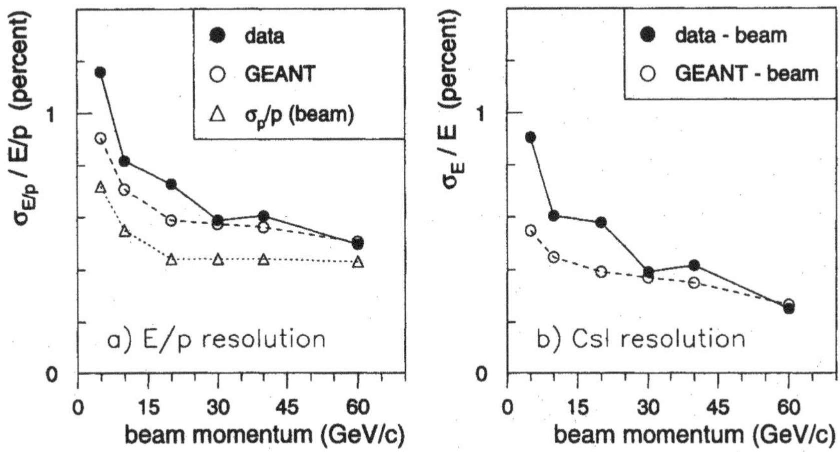

Figure 4: a) E/p resolution vs. beam momentum, and b) the intrinsic CsI resolution with the beam contribution subtracted.

the muon halo is distributed uniformly over the array, we estimate that the muon content of the pion beam was no more than a few percent. Unfortunately we have no way to measure the e^- contamination of the pion beam; such a contamination would make our pion rejection estimates too low. Thus we make a conservative estimate by not correcting for muon or electron contamination. A GEANT/FLUKA[7] based E/p spectrum for 40 GeV/c π^- is compared with the data in Fig. 7. This shows the MC/data ratio as a function of E/p. At $E/p \sim 1$ the MC spectrum has a deficit, suggesting that there may be a small e^- contribution in the data.

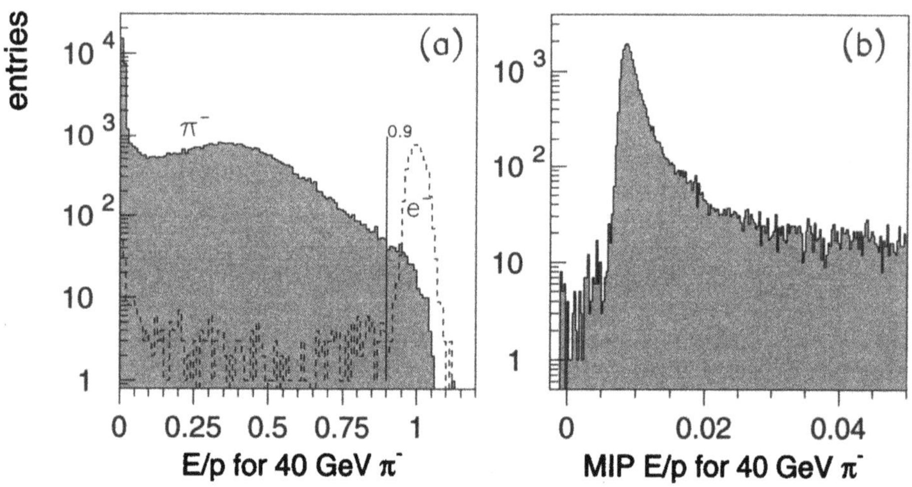

Figure 5: (a) E/p spectra for π^- (shaded) and e^- (dashed) at 40 GeV/c, and (b) same spectrum for π^- MIP peak. The energy sum is over the inner 9 blocks, using one ADC slice. The line at $E/p = 0.9$ defines 100% e^- acceptance.

5. Muons in CsI

Near the end of the experiment the CsI uniformities were measured using transverse beam muons. The entire CsI setup was rotated 90^0, and the beam collimators were closed to select the muon halo from a 70 GeV/c pion beam. An 8 channel scintillation hodoscope behind the CsI was used to measure the longitudinal position of the muon along the CsI axis; i.e., Z_{PMT} = distance from PMT. Fig. 8a shows the muon ADC of the central block, for one Z_{PMT}-bin. Using a Landau fit to find the peak for each Z_{PMT}-bin, the uniformity is shown in Fig. 8b. The response varies by $\sim 5\%$ over the length of this crystal. The measured uniformity for each crystal was used in our GEANT simulation of electrons.

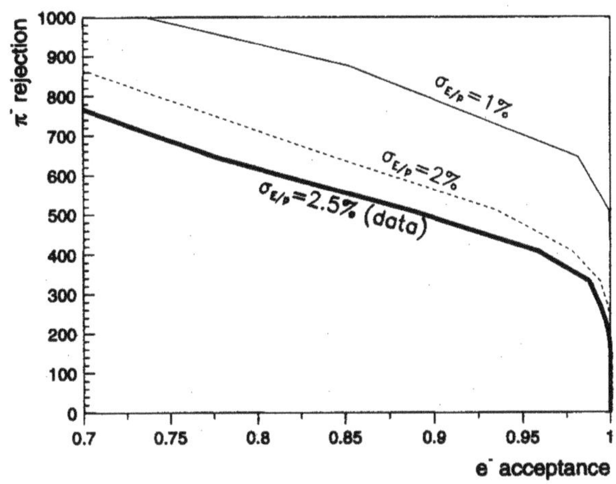

Figure 6: π^- rejection vs. e^- efficiency at 40 GeV/c. The thick solid line is the data. The thin lines are extrapolations based on replacing the e^- E/p spectrum with a Gaussians whose widths are shown in the figure.

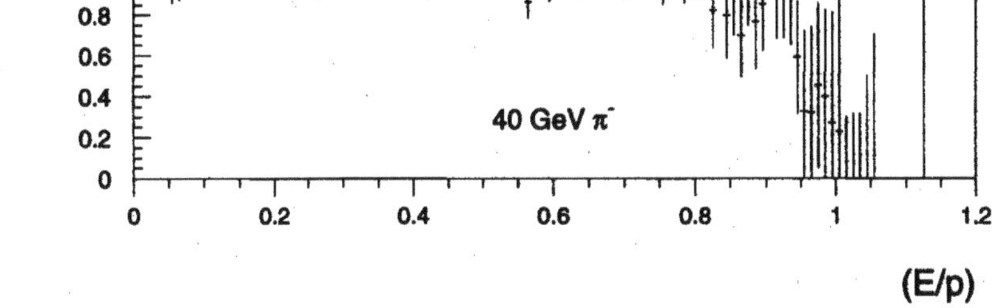

Figure 7: MC/data ratio for 40 GeV/c π^-. The MC program usedGEANT/FLUKA for pion shower generation.

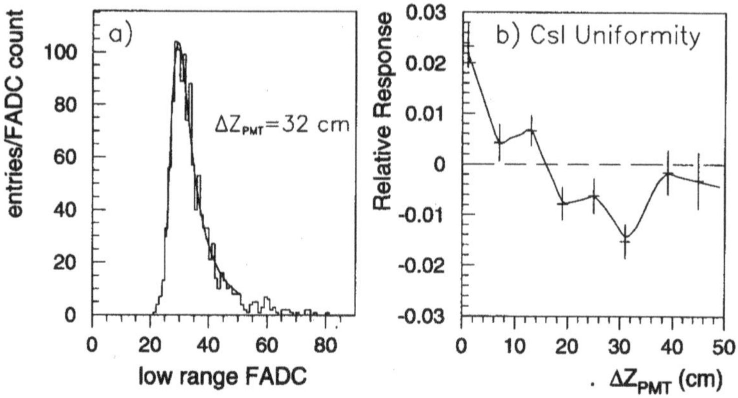

Figure 8: a) ADC for transverse muons. The peak at 30 FADC corresponds to 30 MeV. b) the relative response (deviation from average) vs. distance from the PMT for the central block in the array.

6. Conclusion

We have successfully completed the first KTeV CsI system test using a test beam at CERN. After subtracting the beam RMS from the E/p resolutions, the CsI resolutions are better than 1% from 5–60 GeV/c. Comparing 40 GeV/c pions and electrons, we observe a π^- rejection of $420 \pm 35(stat)$ with 95% e^- efficiency and a 2.5% RMS beam spread; extrapolating to 1% E/p resolution gives a pion rejection of $700 \pm 75(stat)$. Using beam muons transverse to the CsI array, the relative responses were determined to better than 1% every 6 cm along the crystal axis. Finally, the large dynamic range of the DPMT readout system is illustrated by our measurements that vary from 30 MeV muon energy deposits, to 60 GeV electron showers.

We are grateful to the CERN staff for their hospitality, especially to Per Grafstrom for his guidance in using the X1 beam line.

1. B.Winstein, *A Pure CsI Calorimeter for the Study of Kaon Decays*, Heavy Scintillators for Scientific and Industrial Applications, (Edition Frontieres, 1993), p. 279.
2. A.Roodman, *Radiation Damage and Scintillation Uniformity in Pure CsI*, Heavy Scintillators for Scientific and Industrial Applications, (Edition Frontieres, 1993), p. 479.
3. R.Ray, *The KTeV Pure CsI Calorimeter*, these proceedings.
4. R.J.Yarema et. al., *A High Speed, Wide Dynamic Range Digitizer Circuit ...*, Presented at the 6'th Pisa Meeting on Advanced Detectors, 22-28 May, 1994, La Biodola, Isola d'Elba, Italy. Also see references within.
5. J.Whitmore, *A High Speed Digitizing Photomultiplier Tube Base for the KTeV CsI Calorimeter*, these proceedings.
6. H.Nguyen, *The KTeV PMT Laser Monitoring System*, these proceedings.
7. *GEANT Detector Description and Simulation Tool*, CERN Program Library Long Writeup W5013. Also see FLUKA references [2,3,4,5,6] within.

THE ELECTROMAGNETIC CsI(Tl) CRYSTAL CALORIMETER FOR THE *BaBar* DETECTOR AT PEP-II

Gerald Eigen
Lauritsen Laboratory of High Energy Physics
Caltech
Pasadena, Ca. 91125, USA
E-mail: Gerald@CITHEX, Gerald@SLACVM

ABSTRACT

The design of the *BaBar* CsI(Tl) crystal calorimeter is shown, light yield and noise measurements are presented for different readout schemes with photodiodes and the expected performance of the calorimeter is discussed.

1. Motivation

The study of *CP* violation at an asymmetric *B*-factory benefits considerably from excellent calorimetry, because both *CP* eigenstate decays with π^0's can be well reconstructed and branching ratios for modes like $B^0 \to \pi^0\pi^0$ can be measured with high precision. The study of various *CP* eigenstate decays provides important information for understanding the mechanism for *CP* violation, as Standard Model predictions can be tested in various modes. The measurement of $B(B \to \pi^0\pi^0)$ is important for estimating contributions from penguin processes which may effect the *CP* asymmetry in $B \to \pi^+\pi^-$.

At the $\Upsilon(4S)$ the average γ-multiplicity from $B\bar{B}$ events is 11. Most photons originate from π^0's. Despite the boost of $\beta\gamma = 0.56$, the majority of them lies at rather low energies, *i.e.* more than 50% are below 0.2 GeV (see Figure 1a). Photons from $B \to \pi^0\pi^0$ on the other hand populate the entire energy spectrum up to 4.5 GeV (see Figure 1b). Monte Carlo studies indicate that even in modes with high-energy π^0's the photon energy resolution (σ_E) rather than the angular resolution is the dominant component determining the *B* mass resolution. Thus besides high acceptance, the electromagnetic calorimeter must have the best-possible energy resolution from 10 MeV up to the beam energy in order to minimize backgrounds and achieve high *B* meson reconstruction efficiencies.

The target energy resolution in *BaBar* is $\sigma_E/E = 1\%/\sqrt[4]{E(GeV)} \oplus 1\%$. Figure 2 shows both the energy resolution and momentum resolution for *BaBar* and CLEO II.[1] Also shown is a measurement of the ^{22}Na 1.275 MeV γ line in an $18X_0$ long CsI(Tl) crystal read out with a large-area Si-photodiode (PD). The increase expected in σ_E/E below 0.01 GeV is caused by electronic noise. The constant of 1% reflects rear leakage, non-uniform light collection and calibration effects. It should be noted that above 0.1 GeV the photon energy resolution is considerably worse than the momentum resolution. Therefore in modes containing π^0, σ_E is the determining component in

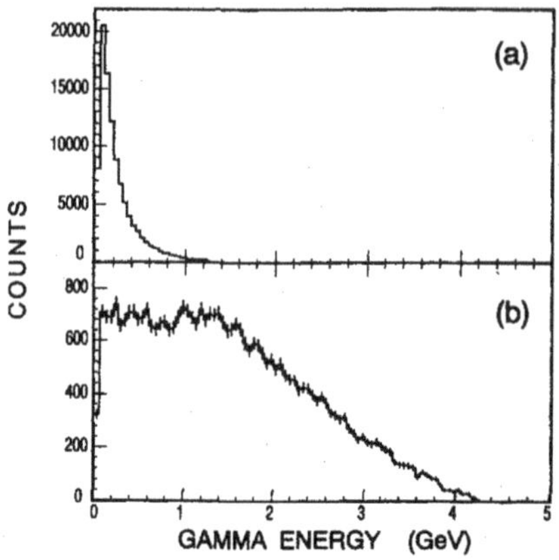

Figure 1. Photon energy spectrum in (a) generic B decays and in (b) $B^0 \to \pi^0\pi^0$ events

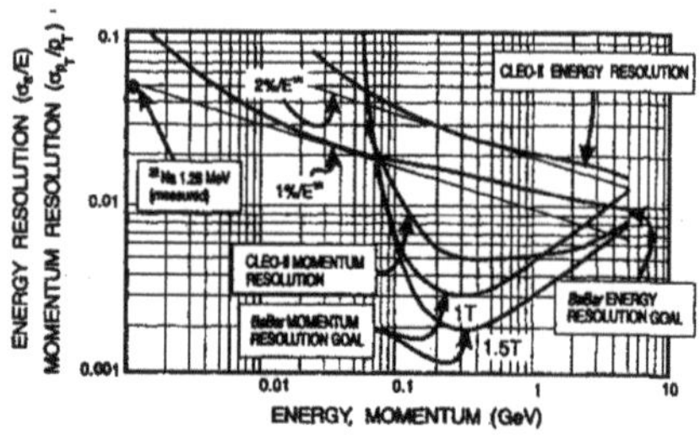

Figure 2. Energy and momentum resolution for *BABAR* and CLEO II (at $\theta = 90^0$)

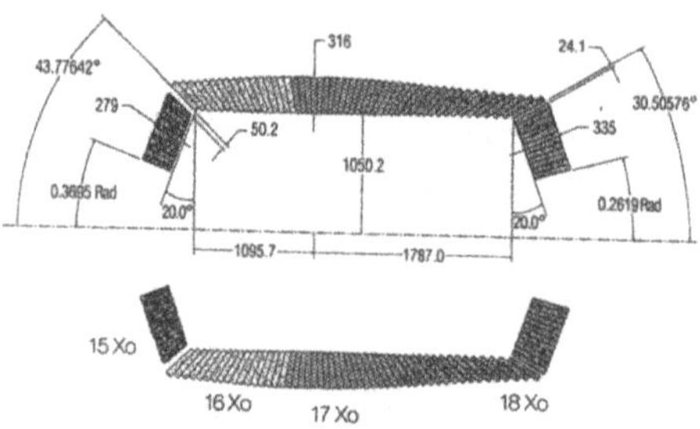

Figure 3. Configuration of the *BABAR* CsI(Tl) calorimeter

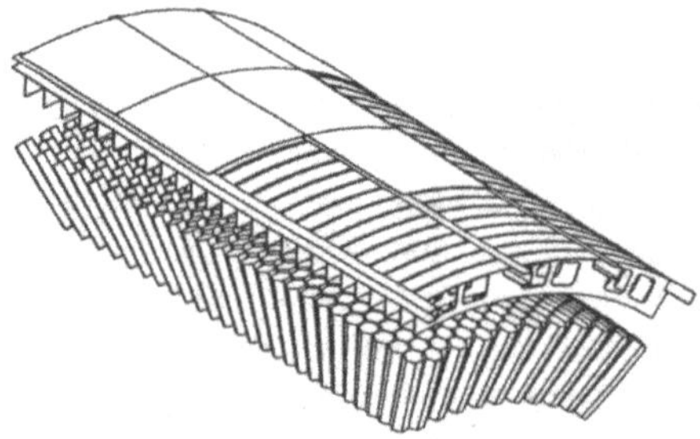

Figure 4. Schematics of the mechanical support structure of the barrel

the B-mass resolution. Angular resolutions are expected to be $\sigma_\phi < 10$ mr for $E_\gamma =$ 0.1 GeV at $\theta = 90^0$ and $\sigma_\theta \leq \sigma_\phi \times \sin\theta$. By minimizing the inactive material in front of the calorimeter we aim to retain a high detection efficiency down to 0.01 GeV.

After a detailed review of a liquid krypton calorimeter, a scintillating fiber calorimeter and a CsI(Tl) crystal calorimeter a committee decided in favor of the CsI(Tl) calorimeter on grounds of best performance and smallest ramifications on other detector components and on PEP II. CsI(Tl) has a high intrinsic light yield of 65,000 γ's/MeV (at T = 20^0 C);[3] the emission spectrum peaking at 550 nm is well-matched to the quantum efficiency of Si PIN diodes; the radiation length of $X_0 = 1.86$ cm keeps crystal lengths reasonably short and the moliere radius of 3.8 cm provides sufficiently high segmentation. In addition, one can build on the experience from CLEO II and Crystal Barrel.

2. Calorimeter Description

Figure 3 shows the configuration of the *BABAR* CsI(Tl) calorimeter.[2] A total of 10,000 CsI(Tl) crystals are arranged in a barrel (7,300 crystals) and two angled endcaps (2,700 crystals), covering a total volume of 8.1 m^3. The calorimeter is located directly behind the particle identification (PID) system. The inner radius lies either at r = 0.9 m or at r = 1.05 m depending on whether a DIRC or a silica aerogel PID system is selected. The angular coverage ranges from 260 mr in the forward direction to $\pi - 370$ mr in the backward direction, which is the region not shadowed by beam elements. Both barrel and endcaps have projective geometry. Favorite crystal shapes are truncated hexagonal pyramids with a diameters of ~ 4.5 cm in front and ~ 5.5 cm at the rear. Crystal lengths vary from 18 X_0 in the forward endcap to 15 X_0 in the backward endcap to balance shower leakage versus costs (see Figure 3).

Hexagonal crystals have several advantages over conventional trapezoidal crystals: a 25% higher light yield,[5] less background as showers are contained in fewer crystals, higher crystal growing yields using the Bridgeman technique and a need for fewer different crystal shapes thus yielding lower crystal prices. In the endcaps for example, only one crystal shape is necessary. In the barrel, however, a single shape would introduce too large gaps between crystals. Another option of two longitudinally segmented crystals is under study with respect to cost savings, improvements in e/π separation and split-off rejection from charged particles.

A support structure with good mechanical integrity is necessary for holding and positioning the crystals which weigh about 36.5 tons. The basic idea is to support the weight from the back by a strong cylindrical Al structure where a modular design is favored. As shown in Figure 4 the strong back is made of cylindrical ribs which are attached to longitudinal support bars.[4] Each rib will hold a ring of crystals. The support bars are then covered with removable plates providing access to the rear of the crystals where the PD's and preamplifier are located. A thin inner Al cylinder closes off the calorimeter volume, providing electric shielding as well as temperature and humidity stabilization. Individual crystals are held by a light structure made of honeycomb or some other composite material to minimize the inactive material between crystals. The support structure must also provide mounting and attachment points for the PID and central drift chamber systems.

3. Light Collection and CsI Crystal Readout

For protection against humidity the crystals are coated with teflon AF. To achieve a highly efficient light transport the crystals are wrapped in a white diffuse reflector and the front face is covered with a good mirror. Figure 5 shows the relative light yield of an 18 X_0 long crystal versus the thickness of a PTFE film reflector. Several other diffuse reflectors of different thicknesses including Kodak paint, Bicron paint and Tyvec have been tested. The best performance so far has been obtained with a 1.6 mm thick PTFE film coating. This wrapping, however, is impractical because of the thickness and its performance under pressure: PTFE film flows and eventually clings to the crystal surface. This impacts the total reflection and thus produces a decreased light output. We have observed a 10% light reduction over a 6 months period. An alternative reflector is Tyvec, a stiff material, which produces 2/3 of the light yield obtained with 1.6 mm thick PTFE film with just two 150μm thick layers. Other coating options consist of $BaSO_4$ or TiO_2 paints.

Due to tapered shapes and light attenuation the response is not a priori uniform along the crystal. To obtain uniform light response the crystal surface is usually roughened. This is a tedious procedure which increases costs. A more elegant solution consists of producing of a pattern of white and black dots adapted for each crystal shape, which is reliably reproducible in an automated process and which can be integrated into the wrapping or final coating.

The primary option for light collection consists of a fluorescent flux concentrator (FFC)[6] which is read out by two silicon PIN photodiodes (PD's) as shown in Figure 6. The FFC combines two function: wave length shifting and light concentration. It consists of a 3 mm thick acrylic plate containing a laser dye which efficiently absorbs the scintillation light and shifts it to higher wave lengths. The FFC is also a wave guide transporting the shifted light until it is collected by one of two rectangular 0.3×3 cm^2 PD's located on adjacent sides of the FFC. The remaining sides are covered with a highly-efficient white diffuse reflector to keep the light trapped until it reaches the PD's. To maintain total reflection, the FFC is separated from the crystal by an air gap. A white diffuse reflector is also located on the back to increase the collection efficiency by giving the traversing light a second pass through the FFC. Figure 7 shows the photon energy spectrum of the ^{22}Na γ lines measured in an 18 X_0 long crystal with this read out.

Compared with our secondary option, a direct readout with two large-area PD's,

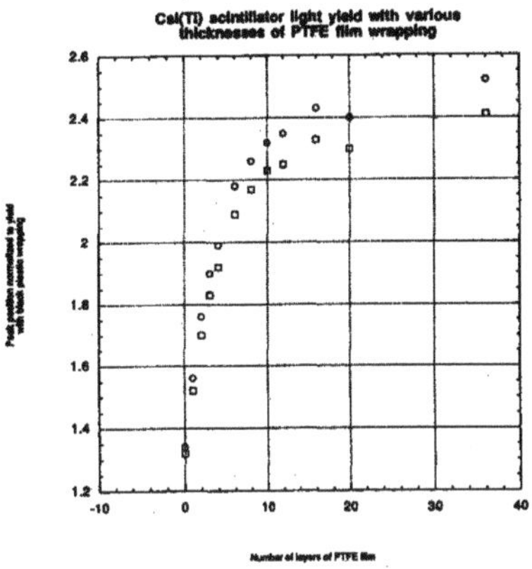

Figure 5. Relative light yield vs PTFE film thickness; circle: 511 keV line, squares: 1275 keV line

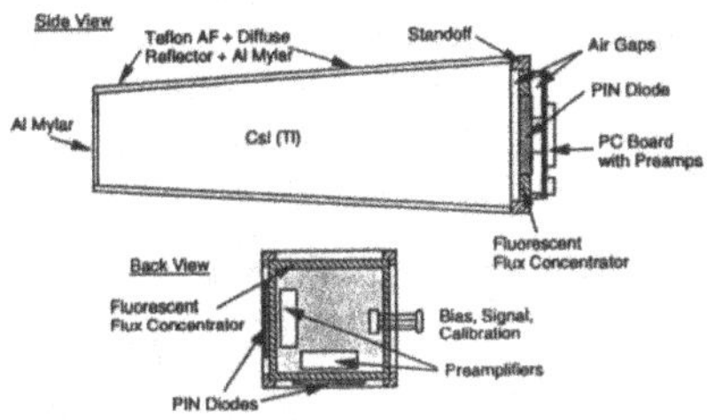

Figure 6. CsI(Tl) readout assembly with an FFC and a PD

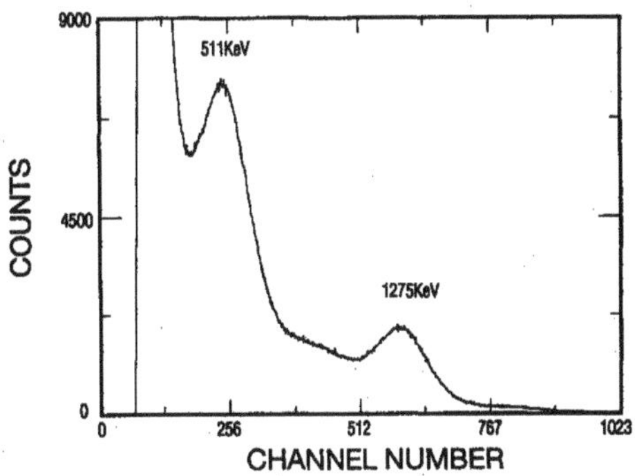

Figure 7. Energy spectrum of ^{22}Na measured in an $18X_0$ CsI(Tl) crystal with FFC/PD readout

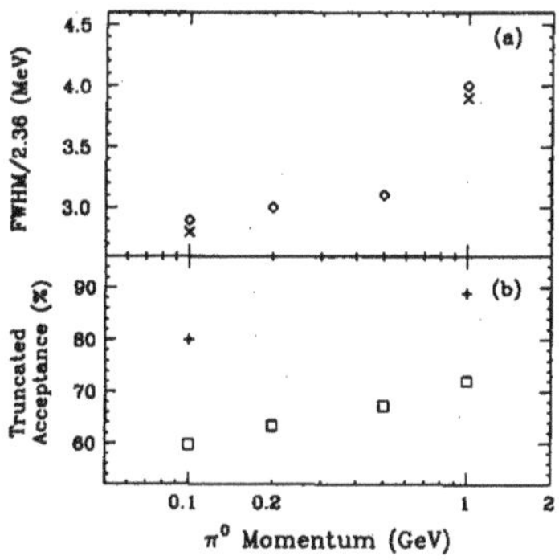

Figure 8. (a) Expected mass resolution and (b) truncated acceptance for single monoenergetic π^0's *vs* π^0 momentum

the FFC/PD readout is less sensitive to light attenuation in the crystal as most light is collected after the first pass through the crystal. Also non-uniformities in the crystal are less important. The absolute light yield, however, is reduced by a factor of 2.5 because of a lower light collection efficiency. But the capacitance of small rectangular PD's is also smaller than that of large-area (2.8×2.8 cm^2 or 1.8×1.8 cm^2) PD's used in the direct readout. Therefore, the signal-to-noise ratio for both readout schemes is rather similar. Tables 1-2 summarize light yield measurements for direct readout using one 2.8×2.8 cm^2 PD and for FFC/PD readout using red laser dye from BASF and one 0.3×3.0 cm^2 PD. The shaping times were set to 3 μs and 6 μs, respectively. The target equivalent noise energy (ENE) is 100 keV. With just one PD, the achieved ENE's are 55 keV for direct PD readout and 70 keV for FFC/PD readout.

Table 1: Light Yield and Noise of full Size Crystals with direct PD readout

Sample	Rear Face [cm^2]	Light Yield* [pe]	Light Yield** [pe]	Light Yield*** corrected [pe]	noise [keV]
Kharkov	6×6	13890	9860	45260	54.6
Budker	6×6	13010	9790	44950	59.5
Crismatec	5.8×5.8	10810	9420	40430	60.0
Horiba	4.5×4.5	13525	11580	29920	51.6

* Coverage of rear face with PD and PTFE film.
** Coverage of rear face with PD, remaining area is uncovered.
*** Yields from (**) corrected for the area of crystal

Table 2: Light Yield and Resolution of full Size Crystals with FFC plus PD readout

Sample	Rear Face [cm^2]	Light Yield [pe]	Resolution	Depletion layer [μm]
Kharkov	6×6	4300	0.079	200
Budker	6×6	4000	0.079	200
Crismatec	5.8×5.8	3810	0.093	200
Crismatec	5.8×5.8	4480	0.067	300

A low-noise charge-sensitive preamplifier is placed directly behind the PD. The intrinsic noise has to be 150-200 e and the slope has to be less than 2 e/pC. The PD's and preamplifiers are separately shielded. The signals are digitized outside the calorimeter but on the detector at an accessible place. A dynamic range of 17-18 bits is required to cover the entire energy range from 100 keV (noise) to 12 GeV (highest possible γ energy). An effective resolution of 10 bits is needed. The signal is sampled and digitized several times to determine its amplitude. The preferred shaping time lies in the $1-4$ μs range which fixes the digitization rate to 1-10 MHz.

For redundancy, two independent read outs are kept up to the ADC. The readout will be equipped with an electronic calibration system. The needs for an additional calibration and monitoring system with radioactive sources are investigated.

4. Expected Performance

The performance of the calorimeter has been tested with a full GEANT simulation.[2] The actual geometry was approximated with 18 X_0 in the forward endcap, 17 X_0 in the barrel and 16 X_0 in the backward endcap. It was further assumed that a 150μm thick Al layer is placed between crystals. To define showers the energy in An array of 5×5 crystals is summed. An incoherent noise of 100 keV per crystal is included and a constant of 0.5% is added in quadrature to account for non-uniform light collection and intercalibration uncertainties. The resolution is defined by FWHW/2.36. Simulations of photons have been done for different energies, different polar angles and different materials in front of the calorimeter. Furthermore, several bench mark modes have been studied, including $B^0 \to \rho^0\pi^0$, $B^0 \to \pi^0\pi^0$ and $B^0 \to J/\psi K^{*0}$.

The Monte Carlo prediction of the photon energy resolution at $\theta = 90^0$ lies slightly below our target resolution, except for the 5.3 GeV energy point which is largely affected by rear leakage. Inactive material in front of the calorimeter has little effect on the energy resolution but degrades the efficiency, especially for low energy photons. For 0.05 GeV energy photons, the additional material of a fast RICH degrades the efficiency by about 10% while for a 1 GeV photon the degradation is only 2%. The photon energy resolution also becomes worse at smaller polar angles because of front leakage from the staggered crystals. For example, the energy resolution for a 0.1 GeV photon at $\theta = 90^0$ and at $\theta = 30^0$ degrades by about 40% (from 2.0% to 2.8%) while the degradation is $\sim$ 20% (from 1.4% to 1.7%) for a 1 GeV photon. Placing a 150μm of Al between the crystals degrades the energy resolution by about 10%.

Figure 8 shows the π^0 mass resolution and acceptance versus momentum for π^0's from generic B decays. The crosses represent data with no inactive material in front of the calorimeter, while open symbols show data for a total of 0.2 X_0 at normal incidence. The inactive material has little effect on the resolution but significantly degrades the efficiency. Similar results have been obtained from the study of exclusive final states. In $B^0 \to J/\psi K^{*0}$, for example, the π^0 mass resolution is 3.3 MeV and the efficiency is just 60%. In $B^0 \to \pi^0\pi^0$ both resolution and efficiency are decreased by a factor of $\sim$two. Including overlapping π^0's helps to increase the efficiency to 68% while degrading the resolution to 7.6 MeV. With this technique a B mass resolution of 43 MeV and a reconstruction efficiency of 50% have been obtained for $B^0 \to \pi^0\pi^0$.

5. Radiation Hardness

The radiation environment at PEP II is considerably higher than that at existing

e^+e^- colliders. Scaling CESR luminosities to those expected at PEP II yields about 1.5 krad/year at $r = 45$ cm and 0.5 krad/year at $r = 100$ cm. The PEP II masking system prevents deposition of large amounts of radiation in the interaction region from accidental beam loss. Including a safety factor > 10, these radiation levels provide no problems for crystals in the barrel. For crystals in the endcaps, however, the situation is different. In order to clarify this issue we have started a dedicated radiation hardness study program.

6. Conclusion

The CsI(Tl) crystal calorimeter provides a state-of-the-art detector which is well suited for the physics goals at an asymmetric B-Factory. The study of multiphoton final states in Crystal Barrel and of exclusive B and τ decays with π^0's in CLEO II has profited from excellent electromagnetic calorimetery. The *BaBar* calorimeter is expected to have an even better performance than CLEO II and Crystal Barrel due to higher light yield, lower noise and less shower leakage. In order to complete the *BaBar* detector by the end of 1998, the first batch of CsI(Tl) crystals needs to be ordered in spring of 1995. To finalize the design on crystal sizes and geometry of the support structure, a final choice on the particle ID system needs to be made. In addition, more optimization studies are necessary to settle issues like crystal shapes, longitudinal segmentation, direct PD or FFC/PD readout, radiation hardness for crystals in the endcaps and source calibration. The final design will be tested by a prototype in an electron beam. The next step, however, consists of completing the technical design report before the end of January 1995.

7. Acknowledgements

I would like to thank the *BaBar* calorimeter group for useful discussions.

8. References

1. Y. Kubota et al., *Nucl. Instr. Meth.* **A320**, (1992), p. 66.
2. D. Boutigny et al., *The BABAR Coll.*, Letter of Intent, **SLAC-443** (1994), p. 147.
3. J.D. Valentine et al., *Nucl. Instr. Meth.* **A325**, (1994).
4. C. Wuest, *private communication* (1994).
5. W.J. Wisniewski, *private communication* (1994).
6. I. Holl et al., *IEEE Trans. Nucl. Sci.* **35**, (1988), p. 105.

EM Liquid and Others

Session Chair: Bernard Aubert (LAPP Annecy)

FIRST RESULTS FROM AN ELECTROMAGNETIC ENDCAP ACCORDION PROTOTYPE FOR ATLAS

RD3 COLLABORATION
Alberta, Alma-Ata, Annecy, Barcelona, BNL, CERN, Grenoble, Madrid, Marseille, Milano, Montreal, Orsay, Paris, Saclay, Stockholm, Victoria

Presented by

LAURENT SERIN
Laboratoire de l'accélérateur Linéaire, 91405 Orsay, France

ABSTRACT

Studies of various physical processes expected at LHC have shown the need for similar performances in the barrel and endcap electromagnetic calorimeters. For this reason it was decided to extend the idea of "accordion" shaped electrodes to the endcap. A prototype with a "fan" geometry was designed and built by the RD3 collaboration. This detector was exposed to electron beams in the energy range 20-287 GeV at the CERN SPS. First results are close to those obtained with the 2m-barrel : an energy resolution with a sampling term of $10\%/\sqrt{E}$ and a local constant term of 0.4% was achieved.

1. Introduction

The endcap region at LHC will have to perform in a harsh environment. The radiation dose received will be very high (about 2 MRads and 10^{14} neutrons/cm^2 per year at high luminosity at the hotest point) so that a liquid argon calorimeter, a radiation hard technology, is a good candidate. Potential problems of ion build-up could appear at large rapidity and are under study. Several beam tests have already demonstrated that the relatively slow response of such detectors can be solved using fast readout scheme associated with accordion geometry [1], as required in the high rate environment of LHC. In order to take advantage of these developments, the accordion concept was revisited and adapted to the endcap region. Physics processes expected at LHC (for instance $H \to \gamma\gamma$, $H \to ZZ^* \to 4$ leptons) put strong requirements on performances :

- A coverage up to $\eta = 3$ to increase the acceptance to signal.
- A position resolution of about $\sigma = (6/\sqrt{E(\text{GeV})} \oplus 0.15)$ mm
- A good energy resolution of $\sigma/E = 10\%/\sqrt{E(\text{GeV})} \oplus 1\%$. The electronic noise will be dominated by pile-up. As the particles in the endcap have higher energy than in the barrel, the sampling term contribution is less important. On the other hand the constant term has to be as good or better.

The principles of this detector with a fan geometry chosen by the ATLAS experiment will be exposed in a first part. Then the prototype built by the RD3 collaboration will be described and the testbeam results presented.

2. The electromagnetic endcap in ATLAS

A few different ways have been proposed to adapt the accordion geometry to the endcap [2]. The solution retained by ATLAS consists of two concentric wheels, each built with identical accordion shaped absorbers piled up in ϕ which are interleaved by readout kaptons. The wave crests run radially from the beam axis and the wave amplitude grows with the radius (the angle of the fold decreases from 120° to 50° within a wheel).

The entrance face of the calorimeter will be located at 3m60 from the interaction point and will extend from an internal radius of $\simeq$ 30 cm to an external radius of 196 cm. This corresponds to a pseudorapidity coverage from 1.4 to 2.4 (2.4 to 3.2) for the outer wheel (inner wheel). The granularity of the calorimeter is obtained in the η direction by suitably cutting the copper electrodes in strips, and in ϕ by grouping consecutive kapton circuits to the same readout channel.

In the outer wheel the first compartment (about 5 X_0 in depth) will be equipped with an integrated preshower with fine grained η strips (typically $\Delta\eta \times \Delta\phi = 0.003 \times 0.1$). In the two other compartments (resp 14 and 8 X_0 in depth) the granularity is 0.025×0.025 (twice coarser in ϕ in the last one by grouping six readout kaptons). The granularity of the inner wheel will be 0.05×0.05 in the first two samplings (9 X_0 in depth each) and 0.05×0.1 in the last one (9 X_0).

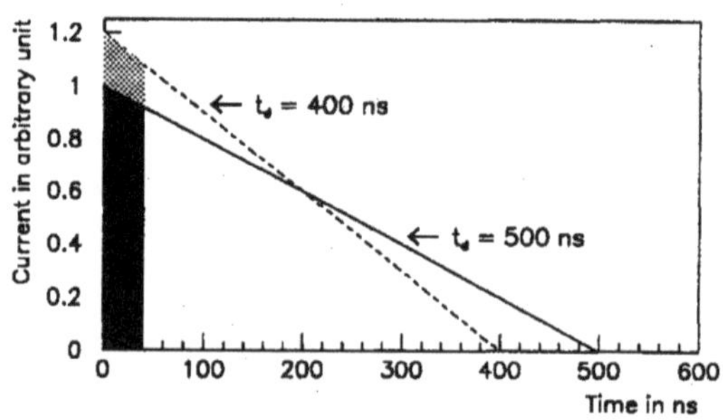

Fig. 1. Current signal in the calorimeter for two values of drift time. The gray area shows the difference of collected signal at short shaping

By construction the endcap exhibits a good hermeticity and a complete ϕ symmetry. Only a circular crack will exist between the two wheels. As The amplitude of the wave grows with the radius the response modulation induced by the accordion in ϕ will be also radius dependent.

In order to obtain a constant stochastic term of the energy resolution it was chosen to maintain the density, sampling fraction and sampling frequency of the calorimeter constant as a function of the radius. This was achieved by using suitably increasing lead thicknesses. As a consequence the gap is opening with the radius. This means that at constant high voltage the drift time (t_d) will depend on the radius. On figure 1 are shown the triangle current signal for two values of t_d. If we integrate over a long period the signal will be identical (the area of the triangle is constant). But using short shaping, constant response can be obtained only provided t_d is also constant. This is achieved by increasing the high voltage radially by groups of two strips.

3. The endcap electromagnetic prototype

3.1. *Description of the mechanics*

The endcap prototype is conceived as 1/6th of the full ATLAS inner wheel with strips pointing to a vertex at 2.8 m from the entrance face. It is divided into three compartments of 9, 9 and 6 X_0 in depth and covers the range 2.16 to 2.89 in pseudorapidity. The granularity is $\Delta\eta \times \Delta\phi = 0.03 \times 0.05$ and twice coarser in η in the third sampling. The lead absorber thickness increases radially from 1.3 to 2.3 mm as well as the gap from 1.85 to 2.65 mm.

The high voltage was distributed by cell pairs in radius, independently in the front and the middle samplings (The middle and back samplings have the same high voltage).

3.2. *Readout Chain*

The signals from kapton electrodes are grouped together in electronic cells on motherboards at the calorimeter front and back faces. The technology used to read out the signal is the so called "0T" [3] which consists in a 50 Ω transmission cable sending the calorimeter signal to a warm current preamplifier outside the cryostat. Obvious advantage is the absence of active electronics in the cold, especially in the high radiation environment at LHC. The preamplifier signal is shaped with a risetime of 30 ns (between 5% and the peak) to the current triangle. This value has been chosen in order to minimize the quadratic sum of electronic and pile-up noises. The shapers were followed by Track and Hold and 12 bits charge integrating ADC's.

The electronic noise was measured using pedestal events. The incoherent noise varies between 45 to 56 MeV as a function of radius, resulting from the capacitance increase. The coherent noise was estimated to about 6 MeV. Using better shielding it could be slightly improved at the level of 4 MeV as with the "2m" barrel prototype. The cross talk between adjacent electronic cells was found at the level of 1 %. The dispersion of the cell peaking times is less than 500 ps after having unfolded the known dependence of peaking time with capacitance.

The calibration was done by simulating a voltage pulse of the calorimeter outside and sending it to an injection resistor located on the motherboards. This calibration system is well suited to the "0T" technology because only passive elements are placed in the liquid.

4. Test beam results

During two periods, the electromagnetic endcap prototype was exposed to electrons of energy 20 to 287 GeV in the H8 beam line of the CERN-SPS.

The prototype was housed at the back of the cryostat, built to hold electromagnetic and hadronic modules, mounted on a computer controlled platform whose rotation in η and ϕ allowed the beam to hit the prototype with an incidence angle corresponding to particles originating from the LHC nominal interaction point. In order to minimize the amount of material in front of the calorimeter, the prototype was separated from the entrance of the cryostat by 1.4 m of low density foam leading to a total amount of inactive material of about 1 X_0.

The beam line was equipped by four scintillating counters used for triggering and three multiwire proportional chambers used to extrapolate the particle trajectory to the calorimeter front face. The resolution on the impact point given by the chambers is about 300 μm in the two views.

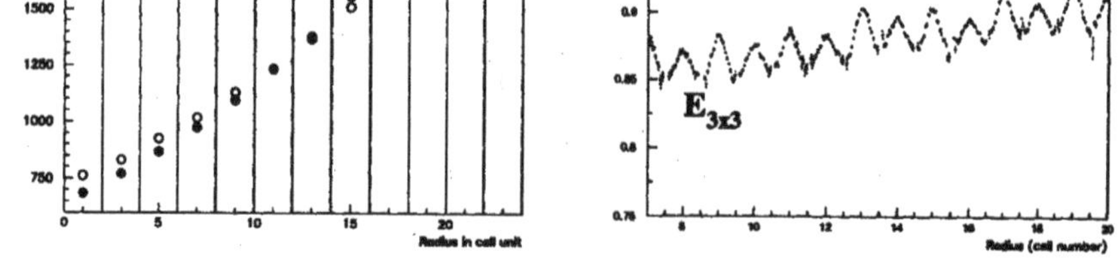

Fig. 2. High voltage (left) and normalized reconstructed energy (right) as a function of radius in cell unit

4.1. Setting of the high voltage

Initial values for each radius were given by Monte Carlo simulation with detailed geometry description including charge collection and non-uniformity of the electric field. An electron scan was performed as a function of the radius in order to check the accuracy of the values. On figure 2 is shown the high voltage used as a function of the radius before and after adjustment. The slope modification of about 5% was adjusted in order to have a flat response in each sampling as a function of the radius. The disagreement comes probably from approximations of the drift velocity used in the simulation. On the same figure is also presented the radial dependence of the energy reconstruted in a cluster after adjustment. Three different effects can be seen (for a 3x3 cluster) :

- The energy containment within the cluster varies with the radius (the cell size is constant in $\eta \times \phi$ not in cm^2).
- The lateral shower leakage depends on the impact point within the cluster. The amplitude of this parabola is about 2 % relative to the edges of the cell.
- The residual high voltage odd-even effect ($\sim 1\%$) due to the high voltage pairing distribution (It is adjusted with respect to the center of cell pairs). In this configuration the odd cell has higher voltage than needed so the signal is overestimated. Similarly the even cell signal is underestimated.

4.2. Energy response

The signals of the calorimeter are first corrected for the electronic pedestals and gains measured every few hours. The shower energy was reconstructed in a symmetric cluster $\Delta\eta \times \Delta\phi = 5 \times 3$ cells. The energy is corrected for the residual high voltage effect by applying odd-even factors and for the lateral energy leakage. As the amount of radiation lengths of Ar is not constant as a function of ϕ within a cell, due to the accordion structure, the response presents a modulation whose amplitude is radius dependent. Nevertheless the amplitude ($\leq \pm 1\%$) and the variation with the radius remain small, and are corrected for to achieve a uniformity of 0.3 %. Moreover a front/back factor is used to take into account the longitudinal leakage (the calorimeter is 25 X_0 depth) and for possible small misadjustements between the high voltage of the front and middle samplings.

4.3. Energy resolution and linearity

After the corrections explained before, the calorimeter resolution is extracted for each beam energy by fitting a gaussian to the energy distribution. On figure 3 are shown the experimental points as a function of the beam energy which can be parametrized at $\eta = 2.4$ by :

$$\frac{\sigma}{E} = \frac{(10.5 \pm 0.5)\%}{\sqrt{E}} \oplus (0.27 \pm 0.09)\% \oplus \frac{0.532 \pm 0.030}{E}$$

where E is in GeV and the beam energy dispersion is removed. The stochastic and constant terms do not depend on radius and agree with the expectations from previous barrel test beams. The 1/E term is higher than the expected from electronic noise and could be attributed to the experimental setup. (Not using events which have started their showers in the low density foam gives a 1/E term compatible with the electronic noise but increases significantly the statistical errors).

The linearity response was found to be better than ±1% over the 20-287 GeV energy range whatever is the position.

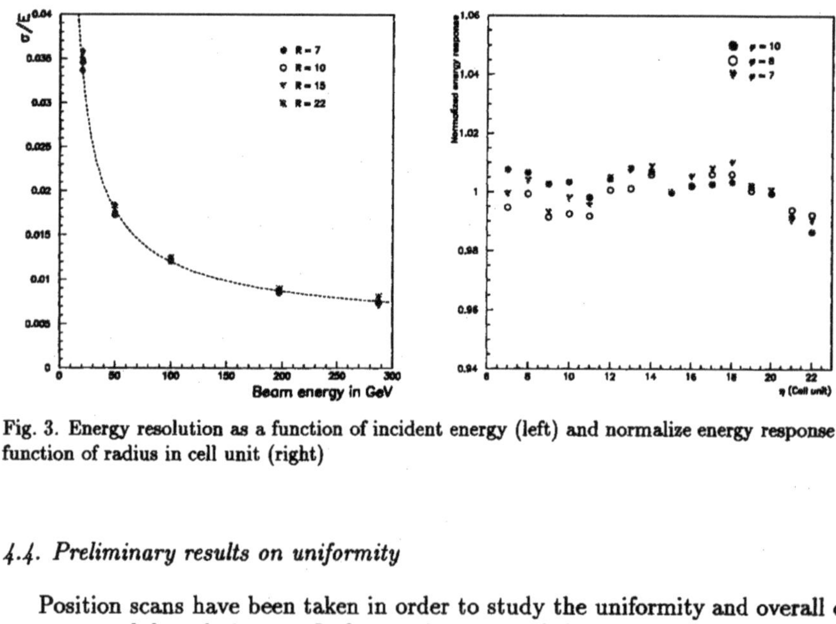

Fig. 3. Energy resolution as a function of incident energy (left) and normalize energy response as a function of radius in cell unit (right)

4.4. *Preliminary results on uniformity*

Position scans have been taken in order to study the uniformity and overall constant term of the calorimeter. In figure 3 is presented the normalized energy response as a function of the radius and for three ϕ rows after having corrected by a slope for cluster energy containment variation. All the cells are within ±1% and agree well between the three rows. Understanding of the systematic effects which can be clearly observed allows for improvement. Nevertheless the dispersion already corresponds to a 0.7 % overall constant term.

4.5. Position Resolution

The position resolution is obtained by comparing the energy weighted barycenter in the first sampling to the extrapolated beam chamber impact point. In the η direction the well-known S-shape has been corrected. After having unfolded the beam chamber resolution, the results are :

$$\sigma_\phi = \frac{(4.9 \pm 0.5)}{\sqrt{E}} \oplus (0.28 \pm 0.03) \oplus \frac{(23.7 \pm 2.8)}{E} \text{ mm}$$

$$\sigma_\eta = \frac{(4.3 \pm 0.5)}{\sqrt{E}} \oplus (0.19 \pm 0.04) \oplus \frac{(28.7 \pm 2.7)}{E} \text{ mm}, \quad \eta = 2.7$$

At smaller η the sampling term is worse because the cell size is larger. The 1/E term comes from multiple scattering in the front face of the cryostat which is 1.4 m from the calorimeter. Such a term will not exist at LHC nor with a well-suited croystat.

5. Conclusion

The test beam results have well validated the accordion fan concept for the endcap. The performances obtained already fulfill the ATLAS requirements and improvements are foreseen. The high voltage distribution has been shown not to be a problem. Moreover a radial calibration of the endcap (as was done in testbeam with the electron scan) can be done in one day of running at LHC at 10^{33} cm^{-2} s^{-1} with an accuracy of 0.3 % using $Z \to e^+e^-$ decays[4].

Acknowledgements

I would like to thanks D. Fouchez and A. Chekhtman for their help in preparing this talk and A.C Schaffer for having carefully read this text.

References

1. B Aubert et al. Performance of a liquid argon electromagnetic calorimeter with an Accordion geometry. *Nucl. Inst. and Meth.* **A309**(1991) 438
   B. Aubert et al. Performance of liquid argon calorimeter with fast readout. *Nucl. Inst. and Meth.* **A321**(1992) 467
2. B. Aubert at al. RD3 Status Report and further R&D for Calorimetry, CERN/DRDC/93-4
3. R. L. Chase et al. Transmission line connections between detectors and frontend electronics in liquid argon calorimetry. *Nucl. Inst. and Meth.* **A330**(1993) 228
4. Radial Uniformity of the spanish fan electromagnetic endcap, L. Serin, ATLAS Internal Note CAL-N0-25

Performance of an Accordion Electromagnetic Calorimeter with Liquid Krypton

David Lissauer
Brookhaven National Laboratory Address
Upton, New-York 11973, USA
E-mail: Lissauer@bnl.gov

For the Noble Liquid Calorimeter Collaboration
University of Arizona, Brookhaven National Laboratory, Columbia University, University of Milano, University of Pavia, University of Pittsburgh, University of Rochester, University of São Paulo, University of Tel-Aviv, and University of Washington

ABSTRACT

Beam test results of the liquid krypton electromagnetic calorimeter with a projective accordion type electrode structure are presented. The electrode had a fine segmentation in the front to enhance π^0 rejection and pointing. The test was carried out at the H4 line at the CERN SPS with e^- beams between 20 and 200 GeV. Preliminary results of energy resolution, linearity, μ response and the dependence of the energy resolution on the amount of inactive material in front of the calorimeter are presented.

1. Introduction

The physics requirements of the LHC require the development of a fast, stable, finely segmented calorimeter with excellent energy resolution and a small constant term at high energy. For example experiments designed for LHC[1] and SSC[2] to search for the Higgs via its decay into two photons, $H \to \gamma\gamma$, require very good energy resolution and background rejection. The aim of this test has been: a) to bring the energy resolution of an EM accordion calorimeter[3,4] into the range of about $5.5\%/\sqrt{E}$ with a small constant term; b) to achieve good position and pointing resolution by using a finely segmented first section that will act as an integrated pre-shower[2] detector; and c) to study the energy resolution dependence on the amount of inactive material in front of the calorimeter.

To improve the energy resolution we have taken an obvious step of increasing the sampling fraction by replacing the liquid argon (LAr) with the next heavier noble liquid, liquid krypton (LKr). Keeping the same volume (i.e. depth) of the calorimeter allows one to achieve a higher sampling frequency and a higher sampling fraction. A non projective accordion has been tested by us at BNL [5,6,7,8]. In this test we have built a projective accordion module with an integrated pre-shower section.

2. Experimetal setup

The test of the calorimeter was conducted at the H4 beam line at the SPS at

CERN, in which negative beams of electrons, in the momentum range from 20 to 200 GeV/c, as well as pions and muons are available. The calorimeter test setup is schematically shown in Fig. 1a. Two scintillation counters S1 and S2 of dimensions 2.5×2.5 cm^2 and 3×3 cm^2 respectively were used to define the beam direction and time t_0 for the test run. In front of the calorimeter were placed a set of four veto scintillators covered with 6 mm thick Pb plates in order to reject those events in which the shower process starts upstream of the calorimeter. The veto counters define the beam spot size of 2×2 cm^2 for the calorimeter. Two Cathode Strip Chambers[9] (CSC) were used to locate the particles in the x-y direction on the face of the calorimeter The CSC position resolution are nominally better $70\mu m$.

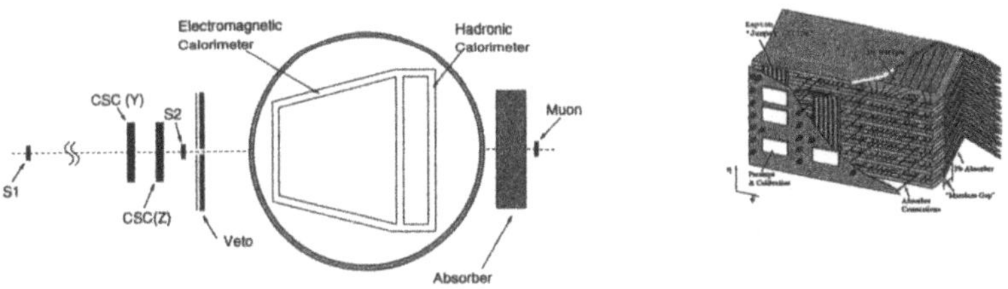

Fig. 1. a) Experimental setup for the test run. b) Isometric view of the calorimeter module.

2.1. *The Calorimeter Module*

The shape of the test calorimeter is a projective cylindrical wedge with a front face of dimensions 60×37.5 cm^2 and depth of 55 cm which corresponds to $25X_0$ with liquid krypton. An isometric view of the accordion showing the electrical connections, and internal electronics is shown in Fig. 1b. The absorber plates are made of 1.0 mm Pb plates sandwiched between two sheets of 0.2 mm stainless steel. Stainless steel provides both the mechanical strength as well as a better surface than lead. There is a nominal 2 mm liquid gap between the absorber and the electrode. A small hadronic module of an EST parallel plate design followed the electromagnetic module.

2.2. *Readout Electronics*

The readout chain used in the experiment is show in Fig. 2. Both the pre-amplifier and the calibration circuits are located as close as possible to the signal electrodes and are operated at cryogenic temperatures [10]. The high voltage connections to the electrodes are distributed over busses located on the electrode boards. The motherboard contains routing for all output signals, calibration signals, and low voltage power. Transmission lines of 50 Ω characteristic impedance terminated at both ends connect the preamplifier output to an intermediate amplifier and cable driver located on the exterior wall of the cryostat. The intermediate amplifier has a switchable voltage gain of 2 or 9 (the gain is fixed at 9 for the lower dynamic range front (strip) section and for the hadronic section), and drives differentially a 100 Ω shielded twisted pair cable 35 meters long. The signals are received in the counting room by Variable Gain Amplifiers (VGA).

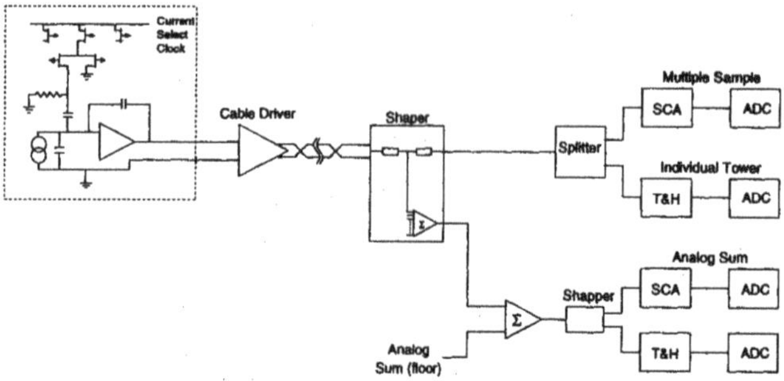

Fig. 2. Electronic readout chain.

The signals arriving at the VGA are split into two. One path takes the signal to a 40 ns, bipolar shaper/amplifier and the other path is input to a current sum used in the trigger. The most important control feature of the VGA is the adjustment of channel gain within a range of ±20% through a voltage-controlled capacitance in the amplifier. The voltage is set through a 12-bit DAC driven by a on-board microprocessor. This feature of the VGA, in combination with the calibration system discussed below, permits equalization of gains to a precision of a 0.2%. All shaped signals are fed into track and hold units and digitized by a 12-bit fastbus ADC.

A non-loading, high input impedance circuit ("High Impedance Pick-off") branches off some of the individual channels which together with the analog sums are sampled every 25 ns and stored in a switched capacitor array, thus simulating the mode of data acquisition likely to be employed at the LHC. A total of 32 samples per channel were kept for offline analysis. The status of the SCA readout is discussed by J. Parsons presentation in this conference.

One of the important advantages of liquid ionization calorimetry is stability and uniformity of response. In this test we used a calibration system developed at BNL that is based on the distribution of DC current to individually selected channels. The system thus permits the calibration of individual channels. With this technique an interchannel gain uniformity of better than 0.2% has been achieved.

2.3. Cryogenics and Purification System

A specially made bathtub surrounded the calorimeter modules with the goal of minimizing the amount of krypton required. The cryostat is the same as was used in our previous test in 1992[6]. The krypton was cooled and liquified using a liquid nitrogen heat exchanger at the top of the cryostat.

We used the same krypton as in 1992 but ran it through an Oxysorb filter before liquefying. We measured the stability of the signal and the lifetime of electrons in the krypton using a small ionization chamber and a ^{241}Am source (alpha cell) immersed in the same krypton as the calorimeter. Fig. 3a shows the response of this alpha cell as well as a curve resulting from the convolution of the associated electronics impulse response with the krypton current signal. From the calculation a drift velocity of 3.8 mm/μs at the nominal voltage of 10KV was derived. The total charge that the source deposit in the alpha cell was measured several times each day during our two week run, no measurable change in the charge during this time was found to the 1% level. We emphasize that no special treatment of the materials that entered the cryostat other than alcohol cleaning was needed and there was no need to purify the krypton once it was liquified.

3. Experimental Results

3.1. Calorimeter Response

Fig. 3b displays the shaped signals from the calorimeter taken at different high voltage settings. As the applied high voltage is raised, the electron signal reaches a limiting value. This can be seen both in the pulse shape and in amplitude. In liquid krypton, however, the drift velocity is already large at very low voltages and even at 50 V a pulse with 30% of the maximum amplitude can be seen. The data we are reporting in this paper were taken at 2000 V.

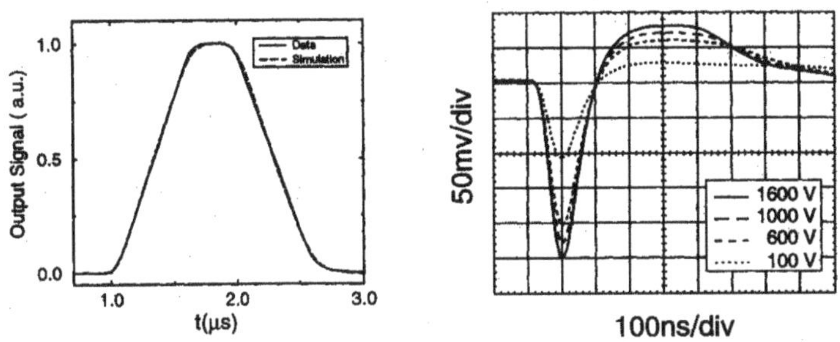

Fig. 3. a)Alpha Cell response. b)Online signal waveforms from the calorimeter with different voltage settings

3.2. Energy Resolution

To determine the energy resolution of the calorimeter, we define an active region to be a 6×6 tower configuration around the position of the incident particle. The energy was corrected for ϕ modulation, energy loss in the front of the calorimeter as well as energy leakage at the back. Minimal corrections are needed to compensate for the amount of energy deposited before the active part of the calorimeter, due to the presence of a passive massless gap. Using the corrected data, we show in Fig. 4a the energy spectrum obtained for 200 GeV electrons. The distribution of pedestal events are shown in the figures, and their width obtained from Gaussian fits are σ=240 MeV for 30 towers. Thus the equivalent noise for a single readout tower is 43 MeV/tower for the low-low gain setting. The rms value divided by the mean of the spectrum in Fig. 4a energy resolution of $5.5\%/\sqrt{E}$, after the noise contribution is subtracted in quadrature. There is no correction for the momentum spread of the beam

The fitting procedure was repeated for each of the eight beam momenta at which data were taken from 20 to 200GeV/c. In Fig. 4b, we show these data, along with a line fitted to the function

$$\frac{\sigma_E}{E} = a \oplus \frac{b}{\sqrt{E}}.$$

The fitted values of the parameters are a= (0.27±0.2)% and (b=0.055±0.004) $\sqrt{\text{GeV}}$.

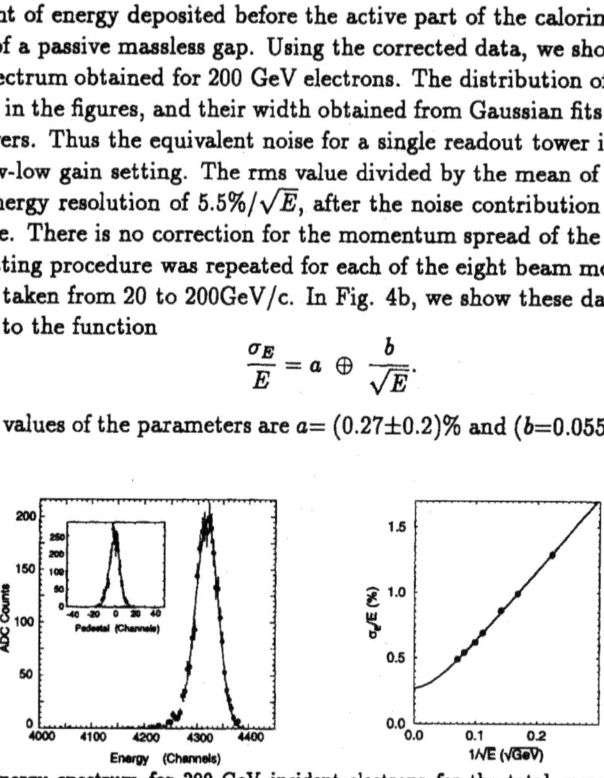

Fig. 4. a)Energy spectrum for 200 GeV incident electrons for the total energy sum. b)Energy Resolution as a function of $1./\sqrt{E}$

The linearity of the calorimeter response, over a range from 20 GeV to 200 GeV, is defined as the ratio of the fitted mean energy to the nominal beam energy. The relative deviations from a straight line as a function of beam momentum have been measured and the system is found to be linear over this energy range to within 0.1%.

3.3. *Energy resolution as a function of inactive material*

In a realistic electromagnetic calorimeter for a collider experiment, inactive material in front of the active calorimeter is inevitable. In this test, the amount of inactive material in front of the active calorimeter is $1.2X_0$ in the regular setup. To study the calorimeter response to additional inactive material, lead plates of various thicknesses were placed immediately upstream of the cryostat. The total amount of inactive material varied from $1.2X_0$ to $5.2X_0$.

A 50 GeV electron beam was used in this study. The total observed energy is plotted as a function of total inactive material in Fig. 5a. The resolution is determined from energy distributions where weights for each longitudinal sections were optimized for each case. It is clear that the addition of inactive material of $2.0X_0$ thickness results in a non-recoverable energy resolution.

3.4. *Position and Pointing resolution*

The shower position is measured with the transverse segmentation of the calorimeter for each longitudinal section. In the front section, three strips centered around the one with highest energy are used in the position calculation, while in the towers, 3×3 towers centered around the tower with highest energy are used. Fits to the curves are then used to correct for the non-linear relation between the impact point and simple energy weighted position. The position resolution is the width of the difference between the chamber position and position measured by the calorimeter. The noise contribution to the position measurement is negligible at the energies under study.

Fig. 5b shows the position resolution as a function of the beam momentum for the strips and middle longitudinal segment. A fit of the form

$$\sigma = a/\sqrt{E} \oplus b$$

gives the parameters a=2.7 mm $GeV^{1/2}$,b=0.24 mm for the strips, and a=4.4 mm $GeV^{1/2}$ and b= 0.3 mm for the middle longitudinal segment.

Using the positions measured at the three different depths, the angle of the calorimeter shower is also measured. In this calculation the average shower centroid in the longitudinal direction for each longitudinal segment is assumed according to the Monte Carlo simulation. The θ angular resolution obtained can be parameterized as

$$\sigma_\theta = (41 \pm 0.4/\sqrt{E} \oplus 2.0) mrad$$

3.5. *Response to Muons*

The calorimeter muon response is displayed in Figures 6a. The muon energy was

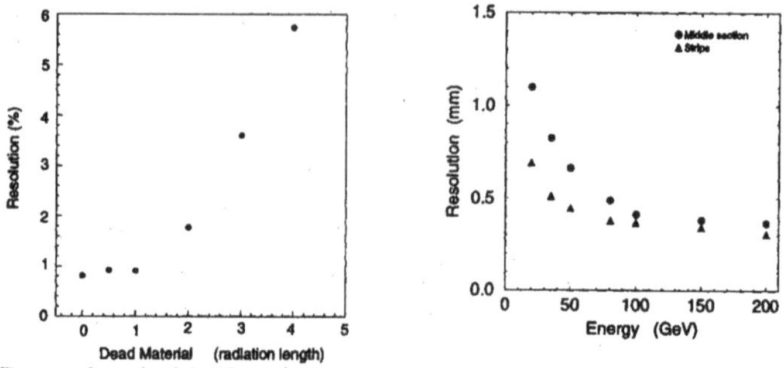

Fig. 5. a)Energy deposited in the calorimeter as a function of the additional amount of material infront of the calorimeter. b) Position resolution for the Strip section and Middle section.

defined as the energy deposited in a 2×1 tower geometry. The pedestal (in units of energy) is displayed along with the data and illustrates the signal to noise one can achieve with this system.

The suppression of the response of an electromagnetic shower relative to that of a minimum ionizing particle is usually called the e/μ ratio, which we define as the ratio of the electron response to the muon response (peak) for a given amount of energy lost in the calorimeter. The muon total energy loss was computed using the GEANT simulation program. Using the linear response curve obtained in the electron runs, the electron response for the same amount of energy loss was calculated. The e/μ ratios for liquid krypton is 0.83 ± 0.01.

3.6. *Timing Resolution*

The timing performance of the calorimeter was measured by discriminating the shaped signal from the middle section analog sum. The discriminator used has been described earlier [5]. The timing was measured in a TDC which was started by the beam scintillators and stopped by the output of the constant fraction discriminator (CFD). The online information is corrected offline for difference in cable lengths using the algorithm of Ref [5]. The timing resolution achieved during the test for electrons of 120 GeV is 260 ps. This resolution was limited by the scintillator timing resolution of 190ps and largely by the shaper (second stage) noise.

3.7. *Photon Data*

In addition to the the electron data we have also taken data with a tagged photon beam. The energy of the photon were between 35-80 GeV depending on the setup. The analysis of the data is ongoing, in Fig. 6b we show a typical event dump where one can immediately see the advantages of the integrated preshower for γ/π^0 separation.

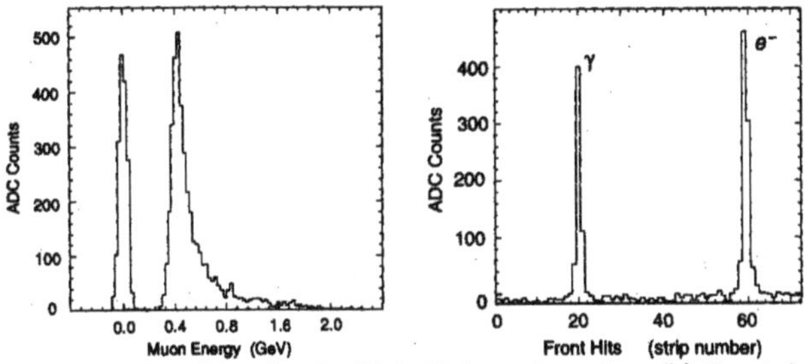

Fig. 6. a) Muon Energy deposition for the High High configuration. b) Event dump showing the energy profile of a photon and the electron from the tagged beam

The photon signal in the preshower is narrow and well separated from the noise. This will enable us to do a detailed shower shape analysis. General features of the data are in good agreement with the Monte Carlo (MC) but detailed analysis and comparison is now ongoing.

4. Discussion and Conclusion

We have tested successfully a projective liquid krypton accordion calorimeter with an integrated preshower. We found: i) The krypton signal was stable over the two weeks of the run, with no purification needed. ii) The energy resolution obtained was in good agreement with the MC expectation, showing that the accordion design can achieve resolution below 6% with a very small constant term. iii)Pointing resolution of the order of 40 $mrad/\sqrt{E}$ can be obtained with an integrated preshower. iv) Preliminary result on the shower shape analysis of photons look very encouraging. v) Inactive material does not deteriorate the calorimeter response significantly below $2X_0$. vi) The passive massless gap in the absorber minimizes the need for correction as a function of longitudinal shower development.

5. Acknowledgments

Our design of the accordion electrode structure has relied on the pioneering work of the RD3 collaboration. We would like to thank the the many people at CERN who were more than helpful during the test. In particular we would like to thank the teams from RD3/ATLAS, NA48 who have given us special help during the run. We also acknowledge the loan of the krypton from the Budker Institute for Nuclear Physics.

This research was supported in part by the U.S. Department of Energy under contract DE-AC02-76CH 00016. Part of this work is supported by the SSC closeout funds.

6. References

1. ATLAS, Conceptual design report
2. GEM experiment Technical Design Report, GEM TN-93-262 (1993) unpublished.
3. B. Aubert et al., Nucl. Inst. and Methods A309(1991)438.
4. B. Aubert et al., Nucl. Inst. and Methods A321(1992)467.
5. O. Benary et al., Nucl. Inst. and Methods, A344,(1994)78.
6. O. Benary et al., Nucl. Instr. and Methods, A344(1994)363
7. O. Benary et al., Nucl. Instr. and Methods, A350(1994)131
8. O. Benary et al., Nucl. Instr. and Methods, A349(1994)367
9. G. Bencze et al., BNL Report 60837, (1994) Nucl. Instr. and Methods (in Press)
10. V. Radeka and S. Rescia, Nucl. Inst. and Methods A265 (1988) 228

THE LIQUID KRYPTON CALORIMETER OF THE NA48 EXPERIMENT AT CERN FOR A PRECISION MEASUREMENT OF ϵ'/ϵ IN CP VIOLATING $K^0 \to 2\pi$ DECAYS

Presented by:
MAURO SAVRIE'
I.N.F.N. and Ferrara University
via del Paradiso, 12
Ferrara, 44100 Italy

NA48 Collaboration:
CAGLIARI, CAMBRIDGE, CERN, DUBNA, EDINBURGH,
FERRARA, MAINZ, ORSAY, PERUGIA, PISA, SACLAY,
SIEGEN, TORINO, VIENNA

ABSTRACT

The NA48 collaboration at CERN has constructed a prototype of an electromagnetic calorimeter using 180 litres of liquid krypton as active medium. It has a completely new readout structure consisting of CuBe2 metal ribbons as electrodes. The resulting structure is characterized by a very good homogeneity and fine granularity. Initial current readout technique is used with a shaping time of 50 ns. The detector has been tested with an electron beam in an energy range between 10 and 120 Gev. Results concerning energy, space and time resolutions are presented.

1. Introduction

A big effort has been made to understand the origin of CP violation, since its discovery in 1964 by J. Cronin and collaborators[1]. The upgrading in both beam and detector techniques permits nowadays a considerable improvement in the accuracy of measurements of the parameters describing CP violation. Recent results from NA31[2] and E731[3] experiments give $\mathrm{Re}(\epsilon'/\epsilon) = (2.0 \pm 0.7) \times 10^{-3}$ and $\mathrm{Re}(\epsilon'/\epsilon) = (7.4 \pm 5.2 \pm 2.9) \times 10^{-4}$ respectively. The NA48 collaboration is preparing a new experiment[4], whose aim is the determination of $\mathrm{Re}(\epsilon'/\epsilon)$ with a precision of 2×10^{-4}, measuring the double ratio of the decay rates of K_S and K_L into two charged and neutral pions. The four decays modes ($K_L \to \pi^0\pi^0$, $K_L \to \pi^+\pi^-$, $K_S \to \pi^0\pi^0$, $K_S \to \pi^+\pi^-$) will be concurrently detected using nearly collinear K_S and K_L beams, coincident in space at the detector position and with intensities ten times larger than the previous experiment[5]. K_S and K_L are distinguished by tagging the proton which produces the K_S[6].

2. The prototype calorimeter

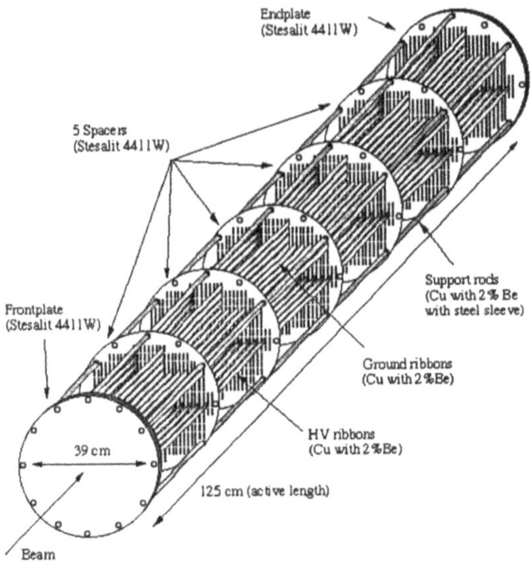

Figure 1: Prototype structure

2.1 The liquid krypton calorimeter

The heart of the experiment is a fast quasi-homogeneous electromagnetic calorimeter, with liquid krypton as the active medium, for the detection of photons from π^0 decays. The performance requirements are: a very good energy resolution ($(3 \div 4)\%/\sqrt{E}$) with one radiation length of equivalent material in front; spatial resolution better than 1 mm for photons above 25 GeV in order to have a good two photon separation; time resolution of 0.5 ns or better for photons above 10 GeV and high rate capability (a singles rate of about 1 MHz is expected). The readout structure consists of towers, with fine granularity, given by 13500 cells of 2×2 cm^2 each. The cross section of the sensitive area is about 2.6 m in diameter and the length is equivalent to 27 X_0 of liquid krypton. The properties of krypton, as well as the cryogenics and the purification system, are described elsewhere[7].

2.2 Structure of the prototype

The prototype (fig. 1) is a barrel with 39 cm in diameter and 27 X_0 in depth, segmented in 184 cells 2×2 cm^2 in size, filled with about 200 l of liquid krypton. One cell is defined by 3 copper-beryllium electrodes with ribbon shape, 40 μm thick, 18 mm wide and 1268 mm long. The electrodes are 1 cm apart on the horizontal plane and they are vertically separated by a 2 mm gap (see fig. 2). Due to the high elastic modulus of the material, the correct mechanical tension of the electrodes is achieved by means of two accordion-like springs at the extremities of each ribbon, formed by

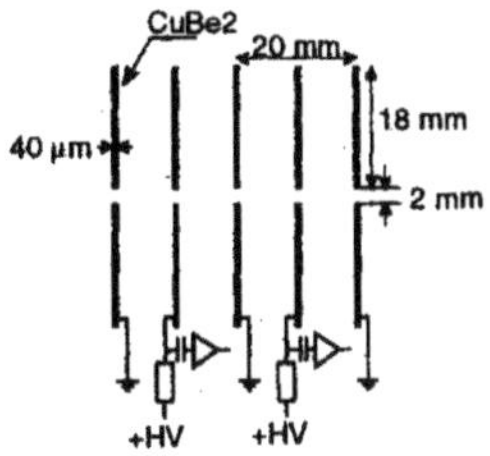

Figure 2: Readout structure

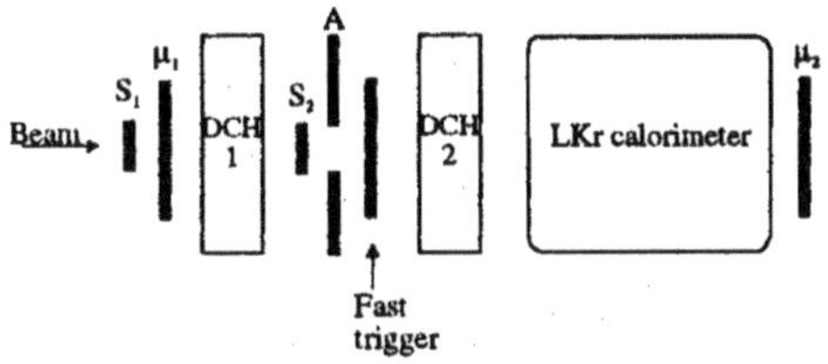

Figure 3: Test setup

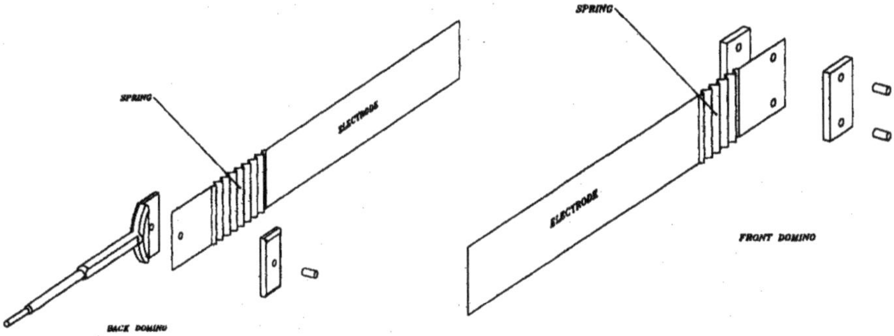

Figure 4: CuBe2 ribbon with connections

passing them through two cogwheels. The upstream spring is 15 mm long with a peak to peak amplitude of about 2 mm and with an elastic constant of about 160 gf/mm. The back spring is 27 mm long with the same amplitude and an elastic constant of the order of 100 gf/mm (see fig. 4). The electrodes are stretched, with a tension of 250 gf, between two STESALIT 4411W* end plates 50 mm thick. Both springs act as a sort of gimbal, reducing the sensitivity to the positioning of the electrodes and compensating for the deformation of the overall structure, when cooled down to liquid krypton temperature and of the two end caps, when mechanically loaded (the total load for the big calorimeter will be about 7 tons). One connecting block (domino) on each end of the electrodes, provides their fixation to the end plates. The front one has a dove tail shape and it is made of STESALIT 4411W, like the two end plates, in order to present a homogeneous material distribution to the incoming particles. The back domino is sketched in fig. 4. Its pin is inserted into a hole, machined in the downstream flange, fixed with a nut and provides the regulation of mechanical tension, as well as signal, H.V. and ground connections. The ribbons are passed

*The STESALIT 4411W is an epoxy resin reinforced with glass fibers. It has been chosen because of its good mechanical, electrical and cryogenic properties. Moreover it does not pollute the krypton.

through five slotted spacers of 5 mm thick STESALIT 4411W. Spacers are located along the detector about every 20 cm and are displaced by 5 mm on the left and on the right alternatively, allowing the electrodes to make a zig-zag of about 50 mrad. The zig-zag helps to reduce the effective length over which the ribbon is stretched to 208 mm. Moreover it is essential for a good uniformity response of the cell, as will be discussed later. The stiffness of the structure is provided by twelve copper beryllium (2%) rods running, from the front to the back plates, parallel to the detector axis. The materials and their thicknesses used in the prototype, will be the same in the real calorimeter.

2.3 Readout chain

In one cell the collecting electrode (anode) is the central one and it is connected to a positive H.V. supply bus via a 10 $M\Omega$ resistor, while the cathodes are at ground. The current induced at the anode by a drifting charge in the static electric field of the cell:

$$I = q \times v_d \times \frac{E_s}{V_0} \tag{1}$$

is read by a charge integrating preamplifier based on a design by V. Radeka[8]. It works in the liquid krypton directly coupled to the electrode through a H.V. blocking capacitor[†]. The technique of the initial current readout[9] is used to minimize the effects of the shower fluctuations. The signals from the preamplifiers are then sent to a $(CR)^2 - (RC)^4$ shaping amplifier. The resulting pulse has a peaking time of 50 ns and an undershoot lasting for $\simeq 2.8\ \mu s$, that corrensponds to the drift time at $V_0 = 5$ kV. Pulses are digitized by 12 bit ADC's (LRS2282B). Eight channels are equipped with new 10 bits, 40 MHz FADC's of the same kind as will be used in the real experiment[4]. Preamplifiers are mounted on a multi-layer mother board, holding also the calibration system that simulates an initial current pulse with a DMOS-FET switch and an RC stage. The linearity and the stability of the readout chain were measured to be better than 0.2%.

3. Test beam and results

3.1 Test beam and setup

The prototype has been tested in the H4 beam line at the CERN SpS with electrons of energies between 10 and 120 GeV. The momentum bite of the electron beam was $\sigma(E)/E = 0.5\%/\sqrt{E}$. The energy of the beam was known to $\Delta E/E \simeq 25\%/E \oplus .5\%$ with a reproducibility of $\Delta E/E = 3.2\%$. The test setup is shown in fig. 3. The trigger was defined by three scintillators with a diameter slightly smaller then the dimension of the entrance window of the cryostat. The impact point of the incoming particle was given by two drift chambers with 4 sense wire planes each and an intrinsic resolution of 100 μm.

3.2 Energy resolution

[†]produced by ASC Corp., Ogallala, Nebraska,USA , 13000V, 3nF±10%

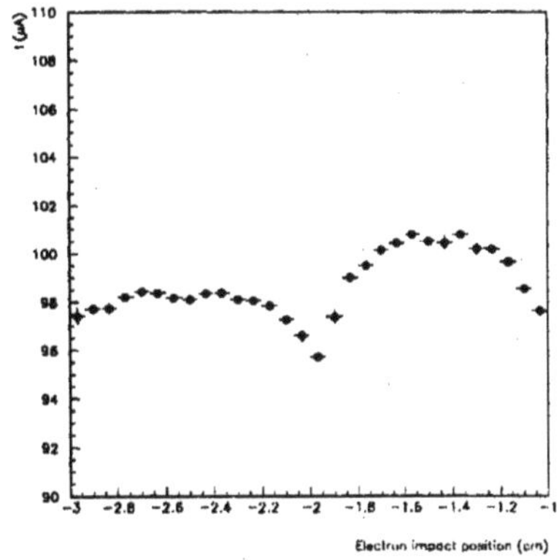

Figure 5: L/R asymmetry for kapton foil prototype

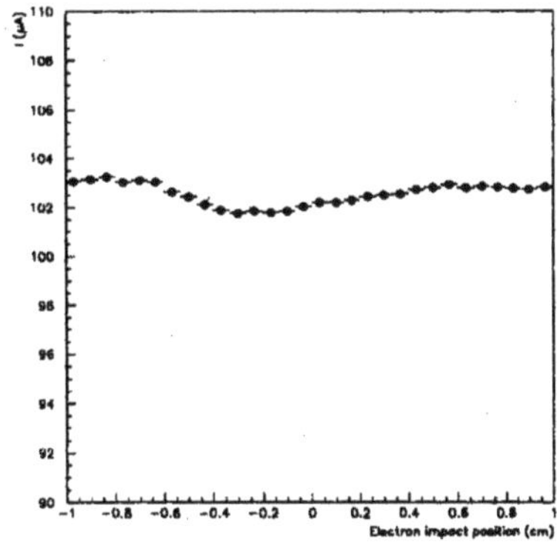

Figure 6: L/R asymmetry for ribbon prototype

With the previous prototype, with straight electrodes etched in copper cladded Kapton foils[7], the energy resolution was strongly affected by a non-uniformity in the response of the cells arising from mechanical tolerances of the cell gap and from the presence of the anode. Concerning the first effect, from equation (1), the induced current:

$$I = q \times v_d \times \frac{E_s}{E_s \times d} \propto \frac{1}{d} \tag{2}$$

depends on the gap width d. A Monte Carlo simulation[10] predicts that the gap dimension has to be defined at the 1% level or better, in order to keep the constant term in the energy resolution below .5%. This effect caused a large ($> 2\%$) left-right asymmetry that could not be corrected by the electronic calibration, because the two half-cells are read by the same preamplifier. The presence of the collecting electrode resulted in a sharp dip at the anode position, due to particles impinging on the cell at a distance from the anode, which corresponds to a drift time shorter then the clipping time. Both effects disappear in the case of a uniform charge distribution in the cell, which is not the case here. The typical shower profile (fig. 7), shows a very steep transverse developement, with about 50% of the energy deposited in one cell. Both effects have been seen with the previous prototype, (fig. 5). With the new ribbon readout structure, the problem was solved because a much better and stable geometry was obtained. Moreover, the dip was reduced by the 50 mrad tilt of the electrodes that produces a sort of smearing of the shower, therefore simulating a more uniform charge distribution within the cell (See fig. 6). The energy resolution is obtained as a gaussian fit to the energy distribution, summing the currents of a 11×11 shower box around the maximum. This gives a stochastic term of $3.5\%/\sqrt{E}$ with a noise contribution of ~ 40 MeV and a constant term of $\sim .4\%$. The resolution as a function

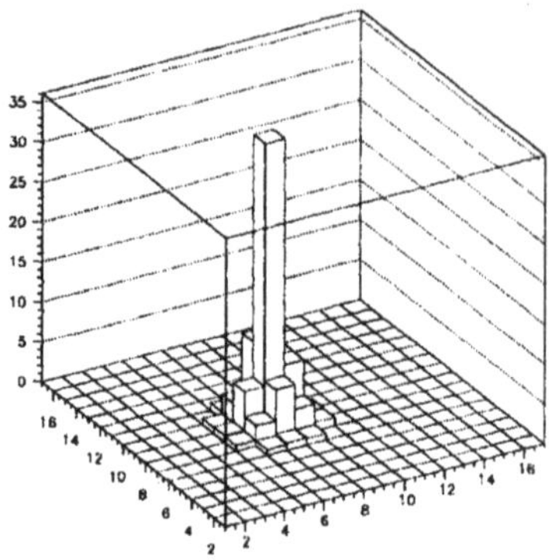

Figure 7: Shower profile at 120 GeV

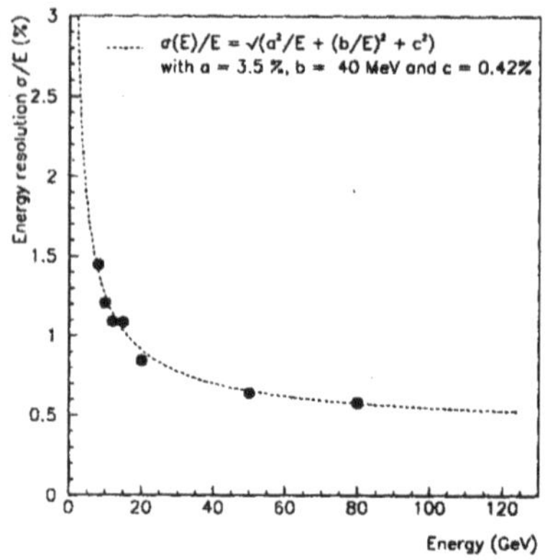

Figure 8: Energy resolution

of the energy, after calibration and impact point correction, is shown in fig. 8.

3.3 Space and time resolutions

The spatial resolution has been obtained comparing the center of gravity of the shower, with the impact point position given by the drift chambers. It is $4.1\%/\sqrt{E}$ mm for x and $4.4\%/\sqrt{E}$ mm for y with a constant term of 0.5 mm, well in agreement with Monte Carlo predictions[11]. A good time resolution is needed in order to separate K_S and K_L decays in coincidence with the tagging system. The time resolution was determined with data of the 40 MHz FADC system. The time of the signal is defined as the time at which the leading edge of the FADC pulse reaches half the maximum of its amplitude. It is calculated as a weighted mean of the times measured in a 3×3 box around the maximum. The time resolution is better than 300 ps at 50 GeV.

4. References

1. J. H. Christenson et al. *Phys. Rev. Lett.* **13** (1964) 138.
2. G. D. Barr et al. *Phys. Lett.* **B317** (1993) 233.
3. Winstein and Wolfenstein *Rev. of Mod.Phys.* **65** (1993) 1125.
4. G. Barr et al. *CERN SPSC* **90-22** (1990) 253.
5. H. Burkhardt et al. *NIM* **A268** (1988) 116.
6. P. Grafström et al. *NIM* **A344** (1994) 487.
7. V. Fanti et al. *NIM* **A344** (1994) 507.
8. B. Aubert et al. *CERN/DRDC* **90-31** (1990) p5.
9. C. Cerri et al. *NIM* **227** (1984) 227.
10. B. Gorini in *Diploma Thesis Pisa* (1993).
11. P. Debu, E. Mazzuccato *NA48 internal note* (1993).

ICARUS: A SECOND GENERATION PROTON DECAY AND NEUTRINO DETECTOR AT THE LNGS

P. Cennini, S. Cittolin, G.Maurin, A. Placci, J.P. Revol, C. Rubbia, W.H. Tian
CERN, CH-1211 Geneva 23, Switzerland

D. Dzialo Giudice, X. Li, S. Motto, P. Picchi
Lab. Naz. di Frascati dell'INFN, via E. Fermi 40, Frascati (Roma), Italy

P. Boccaccio
Lab. Naz. di Legnaro dell'INFN, via Romea 4, Legnaro (PD), Italy

F. Cavanna, E. Olejarczyk, G. Piano Mortari, M. Verdecchia
Dipartimento di Fisica e INFN, Università dell'Aquila, via Vetoio, Coppito (AQ), Italy

D. Cline, W. Hong, S. Otwinowski, J. Park, H.G. Wang, M. Zhou
Department of Physics, UCLA, Los Angeles, CA 90024, USA

A. Bettini, F. Casagrande, S. Centro, C. De Vecchi, A. Pepato, F. Pietropaolo, S. Ventura
Dipartimento di Fisica e INFN, Università di Padova, via Marzolo 8, Padova, Italy

P. Benetti, E. Calligarich, S. Cesana, R. Dolfini, A. Gigli Berzolari, F. Mauri, L. Mazzone, C. Montanari, M. Negrini, A. Piazzoli, A. Rappoldi, G.L. Raselli, D. Scannicchio, M. Terrani, C. Vignoli
Dipartimento di Fisica e INFN, Università di Pavia, via Bassi 6, Pavia, Italy

L. Periale, S. Suzuki
ICGF del CNR di Torino, corso Fiume 4, Torino, Italy

ABSTRACT

The ICARUS physics programme is unique in that it addresses several of the most fundamental issues of modern high-energy physics: proton decay, atmospheric neutrino, long baseline neutrino (with a neutrino beam from CERN) and solar neutrino studies. These will involve the construction and operation of a very large argon TPC (several thousand tons) in the underground Gran Sasso Laboratory. We have developed the detector technique to the point that we now feel ready to start the construction.

We report on the mechanical and cryogenic aspects of the dewar, the argon purification techniques and results, the characteristics of the multiplane internal detector, the analog front-end and the digital readout electronics.

A recent important development concerns the possibility of doping the liquid argon with TMG to reach a better energy resolution.

Presented by F. Mauri

1. Introduction

The aim of the ICARUS project[1] is to build a multi-kton liquid argon TPC to be put in the Gran Sasso underground laboratory (LNGS) to perform a rich physics programme. The detector is homogeneous, fully sensitive with no dead time and provides high quality images of ionizing particles (like an electronic bubble chamber) as well as good energy and dE/dx measurements.

The priorities of the ICARUS programme are the following two fundamental issues:

(a) The stability of the nucleon, which is the only way to access phenomena at the energy scale of Grand Unification. From LEP measurements we know that the electromagnetic, weak and strong running coupling constants do not cross in a single point, unless we invoke some intermediate symmetry breaking scale[2]. If minimal SUSY provides such a breaking, then the favourite proton decay channel will be $p \to \nu + K^+$. Thanks to its large sensitive mass and to its spatial and energy resolution capabilities, ICARUS is an ideal device for nucleon decay detection, in particular for those channels which are not accessible to Cherenkov detectors, like $p \to \nu + K^+$. In a five year running time we can reach a limit of 10^{33} years for most of the decay channels.

(b) The nature of neutrinos, in particular the question of the neutrino mass. This is being investigated in ICARUS both through the study of atmospheric neutrinos and through long baseline studies with a CERN neutrino beam (three orders of magnitude improvement in the length of baseline with respect to the present accelerator experiments).

Even though ICARUS is not optimized to study solar neutrinos, the issue remains extremely interesting and therefore solar neutrinos are also an important part of our programme.

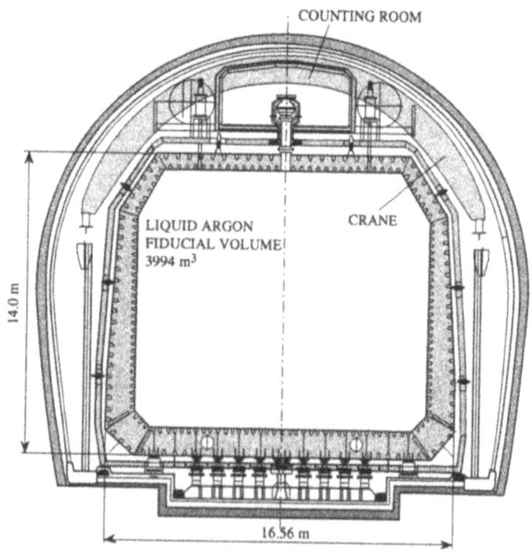

Fig. 1: Frontal view of the ICARUS detector dewar.

Finally astrophysical and cosmological neutrinos from supernova or any unexpected event which might occur during the lifetime (≈10 years) of the experiment will merit all our attention.

2. Detector dewar

In planning the feasibility study of the ICARUS cryostat, we have paid special attention to the choice of the dewar geometry in order to maximize the useful liquid argon volume. A parallelepiped efficiently matches the geometry of the readout chambers and fits the LNGS Hall C geometry and the facility structures. The frontal view of the detector dewar is shown in fig. 1 and the main design parameters are summarized in Tab. 1.

External dimensions	29.18 x 18.5 x 15.22 m^3
Internal dimensions	26.45 x 13.8 x 11.8 m^3
Outer vessel weight	≈ 530 ton
Inner vessel weight	≈ 1100 ton
Internal volume	4205 m^3
Volume occupied by LAr	3995 m^3
Total LAr mass	5591 ton
Active volume	3500 m^3
Active LAr mass	4800 ton
Project pressure	1.5 bar abs
Hydrostatic pressure	1.6 bar
Working temperature	90.5 K
Maximum drift length	2.1 m
Maximum drift time	1.4 ms
Drifting electric field	500 V/cm
Maximum high voltage	105 kV

Tab 1: ICARUS main design parameters.

The cryostat is composed of two vessels, the inner and the outer one, structurally independent and connected by support devices and spacers. The volume occupied by liquid argon (LAr) is about 4000 m^3, while the active LAr mass, i.e. the mass seen by the readout chambers, is about 4800 ton.

3. Argon purification

The free electrons produced by ionization in liquid argon must be able to drift over distances of the order of metres (2.1 m of maximum drift length), without substantial capture by electronegative impurities. This implies that the contamination of electronegative impurities must be kept below 0.1 ppb.

One of the first results obtained by the ICARUS R&D programme[3] was that of being able to successfully obtain liquid argon at this level of purity using simple and reliable purifiers.

Our goal is to fill the 5 kton detector in about 3 months, corresponding to about 50 ton of ultrapure liquid argon per day; this can be obtained using the liquid phase purification system[4] made of several small units, similar to those used in our R&D tests, running in parallel. The ultrapure liquid argon in the dewar, even in absence of any leak, can be contaminated by outgassing of the walls and of the various materials in contact with the gaseous argon at the top of the dewar. To keep the liquid argon at the necessary purity, we have designed and successfully tested at CERN for our 3 ton prototype a vapour recirculation/purification system, that can be scaled up to the 5 kton detector.

4. Readout chambers and electronics

The performance of the multiplane wire chambers readout have been well tested in our 3 ton prototype[5] at CERN. Nevertheless their use presents some weak points due mainly to the mechanical structure. Our present design for the 5 kton detector foresees the use of large-size 3D image chambers, based on the multi-layer printed-board circuit technology, which present an intrinsically simpler and safer structure and with performance comparable to that of the wire technique[6].

In total there are 3 double readout chambers with 5 mm electrode pitch and a number of readout channels of the order of 56,000 (30,000 in collection mode and 26,000 in induction mode).

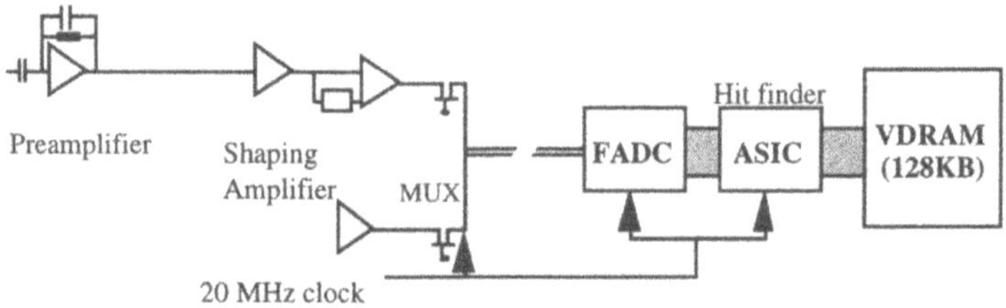

Fig. 2: Front-end block diagram.

The electronic design has to face the following criteria:

- a signal-to-noise ratio of 10 is necessary for dE/dx measurements (collection plane), while a $S/N \geq 6$ is enough for tracking (induction plane);

- to distinguish a minimum-ionizing particle, while ensuring no overflow when a shower is detected, 10-bit FADCs are needed;
- given a drift velocity of 1.5 mm/μs, corresponding to a maximum drift time of 1.4 ms, a sampling time of 400 ns is required to efficiently reproduce the signal shape.

The block diagram of one front-end electronic channel is shown in fig. 2.

5. The liquid argon doping test

In pure liquid argon, only a small part of the deposited energy results in free electrons, especially at high dE/dx and low electric fields. Part of the deposited energy ends up in the form of scintillation light, coming from direct excitation of the Ar discrete levels or due to the electron-ion recombination, and so is lost in our measurement. This effect implies a non-linear response of charge collection with respect to the dE/dx.

Recently we successfully tried to convert the scintillation light to free electrons by doping the liquid argon with photosensitive molecules. We have chosen TMG (Tetramethylgermanium) as dopant because of its high photo-absorption cross-section and its good solubility in liquid argon.

The main results of the test are[7]:

- we observe no degradation of the electron lifetime;
- the spatial resolution remains unchanged, provided that the dopant concentration exceeds 3.5 ppm. At lower concentrations, the electrons emitted by TMG molecules are sensibly far from the original track, which implies a worse spatial resolution;
- there is a considerable linearization of the charge yield for heavy ionizing particles, as measured by muons and protons near their stopping points (fig. 3);

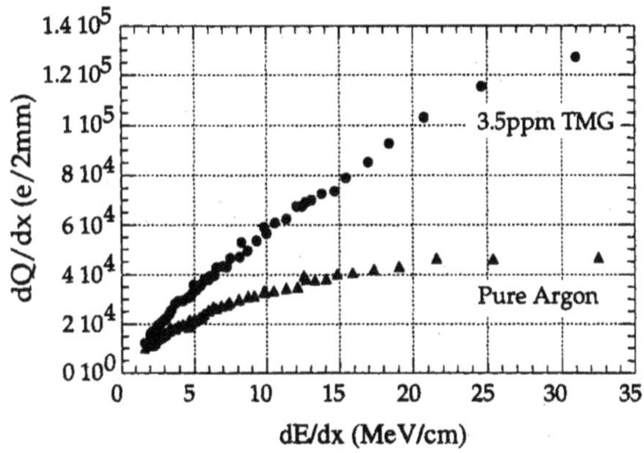

Fig. 3: Charge yield vs. energy density in pure LAr and with a TMG concentration of 3.5 ppm (electric field of 350 V/cm).

- there is a noticeable increase (≈30%) of the free charge also for minimum ionizing particles (fig. 4).

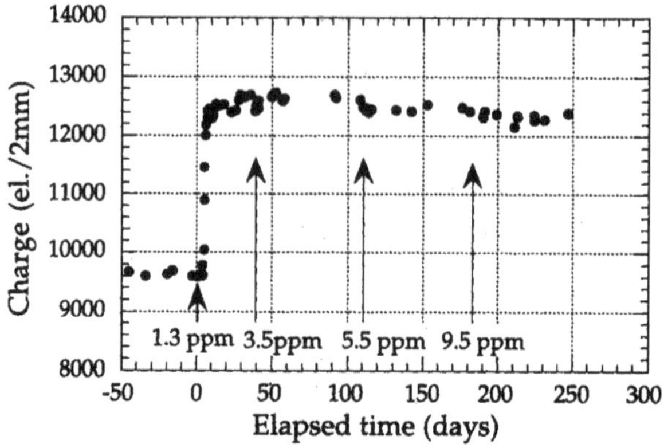

Fig. 4: Collected charge in different doping concentrations for m.i.p. (electric field of 350 V/cm).

6. Conclusion

A scientific and technical proposal has been completed in May 1994; the LNGS scientific committee has approved the lines of research and has recommended the continuation of the engineering designs.

The future programme foresees:

- the detailed engineering of the 5 kton detector;
- the realization of an intermediate mass prototype to optimize the main parameters of the final detector.

7. References

1. ICARUS Collaboration, "ICARUS II, a second generation proton decay experiment and neutrino observatory at the Gran Sasso Laboratory", proposal vol. I LNGS-94/99-I (22 September 1993) and vol. II LNGS-94/99-II (21 May 1994).
2. A. Zichichi, *Rivista del Nuovo Cimento* Vol. 2 N. 14 (1979) 1.
3. A. Bettini et al., *Nucl. Instr. and Meth.* **A305** (1991) 177.
4. P. Cennini et al., *Nucl. Instr. and Meth.* **A333** (1993) 567.
5. P. Benetti et al., *Nucl. Instr. and Meth.* **A332** (1993) 395.
6. P. Cennini et al., *Nucl. Instr. and Meth.* **A346** (1994) 550.
7. P. Cennini et al., "Improving the Performance of the Liquid Argon TPC by Doping with Tetra-Methyl-Germanium", submitted to *Nucl. Instr. and Meth.*

THE BROOKHAVEN NATIONAL LABORATORY E852 LEAD GLASS CALORIMETER SYSTEM

S. Teige
Department of Physics, Indiana University
Bloomington, IN 47405, USA
For the E852 Collaboration

ABSTRACT

A 3053 channel lead glass calorimeter has been built for BNL experiment 852, a search for mesons with exotic quantum numbers. This search places stringent requirements on the resolution of the calorimeter and requires that it be used as part of the experimental trigger. To satisfy these requirements quality control procedures were established and implemented and custom electronics were designed and built. This paper discusses the measurements made on the lead glass blocks and phototubes as part of the quality control procedures. The custom electronics designed and built for the calorimeter include the ADCs and a trigger processor system. The ADCs have 12 bit resolution with a 4 microsecond conversion time, 0.025% linearity and 250 nanosecond reset time. The trigger processor is capable of determining the total energy deposited in the calorimeter and the effective mass of the photons hitting it within 1 microsecond of the end of data transfer to the processor. Trigger processor performance is discussed as are preliminary experimental results.

1. Lead Glass Detector System Overview

The E852 Lead Glass Detector (LGD) consists of an array of 3045 $4 \times 4 \times 45$ cm type F8-00 lead glass blocks stacked in a 43×71 rectangle with four corner blocks and four central blocks removed. Cerenkov light produced by interacting photons and electrons is measured by type FEU-84-3 Russian made phototubes. The tubes are powered by a computer controlled, Cockroft-Walton type base. The tube/base assemblies are held in an aluminum and soft iron structure referred to below as the cellular wall. The entire assembly is carried by a transporter to allow each module to be placed in an electron beam for calibration. The pulses from the phototubes are digitized by custom made, 12 bit ADCs. A laser based illumination system is used to monitor the detector.

The glass, phototubes and bases are described in detail in reference 1 but a brief description follows.

The transverse dimentions of each block were determined by an automated measuring system. The average block size was 40.0068 millimeters with a standard deviation of 24.0 microns. The high precision machining of the blocks allowed them to be stacked tightly in the array. The space between two typical blocks (as determined by a feeler guage) is less than 100 microns.

The quantum efficiency and noise rate was measured for each tube. The average quantum efficiency was 21 % with a standard deviation of 3 %.

The bases are computer controlled Cocroft-Walton voltage multipliers. This design allows the system to operate with a small power consumption and with minimal cabling not associated with the signal.

1.1. The Transporter

The array of lead glass blocks is contained in a frame made of I-beams and steel tube. This frame also supports a 1 cm thick soft iron front cover that serves as magnetic shielding, the monitor system, the "cellular wall" (both discussed below) and a light tight rear enclosure. This inner frame is carried by an outer frame, also made of steel tube which in turn rides on wheels on a precision rail system. Vertical and horizontal motion, for calibration purposes, is acomplished by computer controlled, high torque stepping motors. The transporter position was determined by rotational position transducers connected to the speed reducers. Comparison of mechanical position measurements and the position measurements derived from these transducers showed that the transducers produced repeatable results in agreement with the independent mechanical measurement to within 0.5 mm.

1.2. The Cellular Wall

In order to hold the phototube/base assemblies at the rear of the lead glass blocks an aluminum and soft iron assembly was constructed. This "cellular wall" must serve two purposes: mechanical support and alignment of the phototubes and their bases and magnetic shielding of the tubes.

The design selected to achieve these goals consisted of a sandwich of aluminum plates supporting steel tubes between them. Two aluminum plates, 1/2 inch thick, had 3053 (43 by 71) 39 mm holes bored in them. The holes were on 40 mm centers. Each hole in one of the plates recieved a press fitted machined soft iron tube. When all the tubes were installed, the second plate was placed atop the array of tubes. The tubes were machined to have a slip fit into the upper plate. The outer edges of the plates were secured to aluminum blocks to provide a structure with sufficient rigidity for it to be moved. Further rigidity was obtained by filling the inter-tube gaps with epoxy. The epoxy was poured in several stages and allowed to cure between stages. 59 gallons were required. Additional magnetic shielding was provided by three layers of μ-metal installed in the front half of the soft iron tube.

1.3. The Monitoring System

The gain of the phototubes was monitored by a laser based monitoring system. A nitrogen laser was used to excite a small cylindrical scintillator with an emission peak at 410 nm. Light from this scintillator illuminated a bundle of optical fibers.

12 of these fibers launched light into the corners of three sheets of plexiglas mounted directly in front of the lead glass blocks. Sufficient light was scattered out of the plexiglas and onto the photocathodes to give a pulse with an amplitude similar to a typical data event. Another four fibers carried light to an RCA8575 photomultiplier tube in a temperature controlled box outside the fringe field of the analysis magnet. The signals from this tube were used to measure the pulse to pulse variation of the laser output. Table 1 shows the performance of this system over a period of 24 hours.

Average PMT Signal	Average Monitor Signal	Ratio	Change
235.21	2314.27	0.10163	—
234.93	2310.26	0.01068	+0.05 %
237.67	2325.38	0.10220	+0.51 %
237.79	2321.67	0.10242	+0.21 %
239.68	2330.97	0.10282	+0.39 %
238.30	2323.15	0.10257	-0.24 %
238.79	2329.91	0.10248	-0.08 %
239.52	2335.24	0.10256	+0.07 %

2. The Analog to Digital Converters

Because of the large dynamic range of the signals (approximatly 5 MeV to 15 GeV) in this experiment and the need to use digitized pulse heights in the trigger, custom designed and built analog to digital converters were used. The ADCs have 12 bit resolution with a 4 microsecond conversion time, 0.025% total non-linearity over the full scale and a 250 nanosecond reset time.*They are packaged as 32 channels per single wide FastBus module.

Baseline pre-sampling is used to supress low frequency noise. The output of these ADCs is unchanged when a 1 Volt peak to peak, 1 kHz noise signal is superimposed on a typical phototube signal.

There are two channels of programmable thresold discriminator per input channel. One of these is programmed to some low threshold and is used for zero supression during readout by the data acquisition system. The other channel is set to some higher threshold and is used to select channels to be read by the trigger processor. The state of these discriminators is presented on the auxillary backplane of the FastBus crate after a front panel strobe. These states are available as early as 100 ns after the end of the signal gate so that they can be used for fast pre-trigger applications.

These ADCs can be reset by a front panel pulse to within 1 channel of the pedestal any time before 1 microsecond after the leading edge of the signal gate. The reset time is 250 ns. After this "last chance for reset" conversion begins and is completed 3 microseconds later.

*Compare to the LeCroy 4300 FERA with an 11 microsecond conversion time, an 11-bit range and a 1 microsecond reset time

The output is a 32-bit word containing the digitized integrated charge, a 16-bit user programmable address and two bits giving the discriminator states. Readout can be accomplished via a custom readout controller (reading only the channels with the "high" discriminator channel set, or through standard FastBus operations through, for example, a segment interconnect.

3. The Trigger Processor

E852 required the the trigger be able to select events with an η in the final state. Since the η has a large branching ratio into 2γ or $3\pi^0$, it was nesseccary to design a trigger processor that could quickly evaluate the effective mass of a system of photons.

The effective mass squared of n photons is given by

$$m^2 = (\sum_{i=1}^{n} E_i)^2 - (\sum_{i=1}^{n} \vec{p_i}) \cdot (\sum_{i=1}^{n} \vec{p_i}). \tag{1}$$

Since photons deposit their energy over several blocks in the lead glass array, the energy and momentum of the $i'th$ photon are determined by sums over "clusters" of blocks in the detector. The sum over photons in Eq. (1) can then be written as a sum over all blocks in the detector as

$$m^2 \approx 2(\sum_{k=1}^{N} C_k PH_k)(\sum_{k=1}^{N} PH_k \delta_{k_x}) - (\sum_{k=1}^{N} PH_k q_{k_y})^2 - (\sum_{k=1}^{N} PH_k q_{k_z})^2 . \tag{2}$$

where C_k, δ_k and q_k are constants that depend only on the address of a block.

The sums in Eq. (2) can be quickly evaluated using commercially available multiply/accumulate circuits. The q_k terms are approximated as integers and downloaded to the processor. An EPROM is used to store these constants. It is also interesting to note that the first sum in equation 13 is the total photon energy in the LGD. It has already been shown that this quantity is a powerful selector of well contained all-neutral events.

Figure 1 shows a block diagram of the device. The processor is connected to the readout controller (ROC) boards by 4 ribbon cables. During a readout sums corresponding to the 4-vector components od Eq. (2) are evaluated as are the largest pulse height and the number of blocks over threshold.

4. Performance

Figures 2 and 3 show the effective mass of two photon events with and without a trigger processor requirement. Figure 2 shows the decays $\pi^0 \to \gamma\gamma$ and $\eta \to \gamma\gamma$ in the ratio of $\approx 20 : 1$. Figure 3 shows the observed 2-photon effective mass distribution when the trigger processor was configured to select events with a photon system with an effectiver amss greater than the mass of the π^0 .

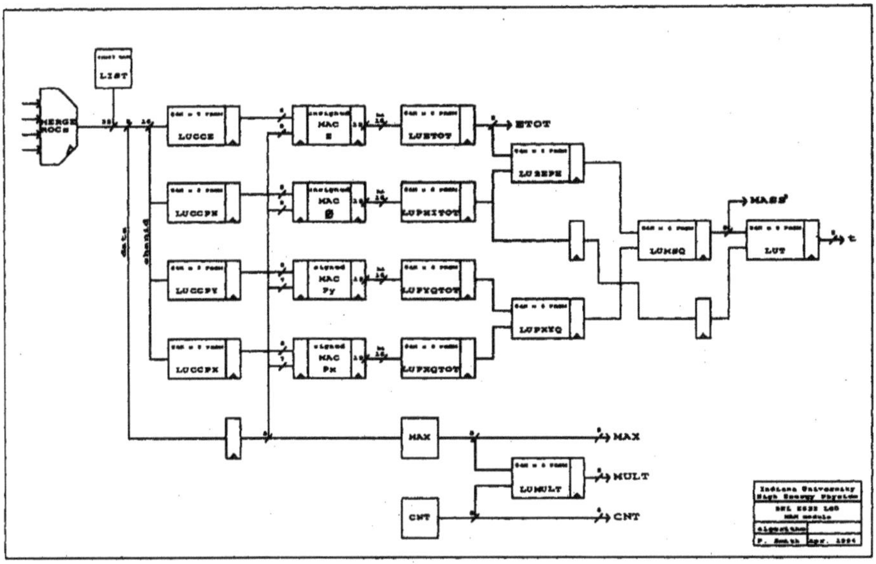

Figure 1: A block diagram of the trigger processor. There are 4 crates of ADCs, each with its own readout controller. These controllers are connected to the trigger processor by ribbon cables allowing the crates to be read out in parallel. The 32-bit ADC word is split into address and data parts. The 12-bit address part is used as the address for 4 EPROMS that are loaded with the constants of Eq. (2). The constants from these EPROMS are multiplied by the data part of the ADC word and added to a stack by the multiply and accumulate (MAC) circuits. After readout is complete, the sums in the MAC stacks are used as addresses in another stage of EPROMs. These EPROMS return functions of the values of the sums from the MACs. For example, the value of P_x^2 is computed from the value of P_x. These function values, in turn, serve as addresses for subsequent EPROMs until the effective mass squared is evaluated. A calculation similar to this one can evaluate the momentum transfer to the photon system. The maximum pulse height found is also calculated as is the number of blocks read. The number of photon clusters is correlated with the number of blocks read and the maximium puse height found. This correlation is evaluated externally to the trigger processor by a commercial MLU.

5. References

B. Brabson *et al.* Nucl. Instr. and Meth. **A332** (1993) 419-443

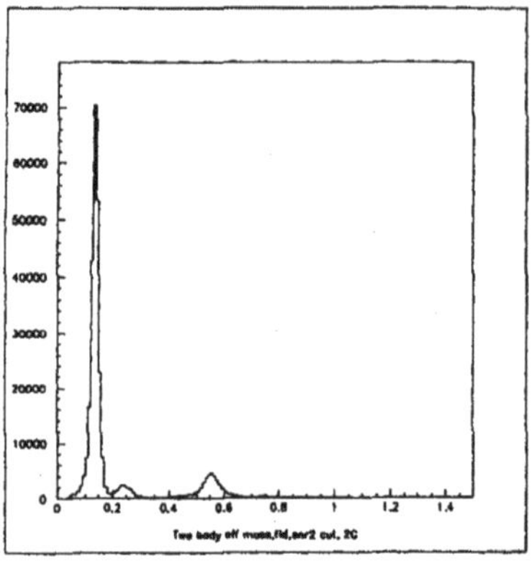

Figure 2: The two photon effective mass distribution without a trigger processor requirement.

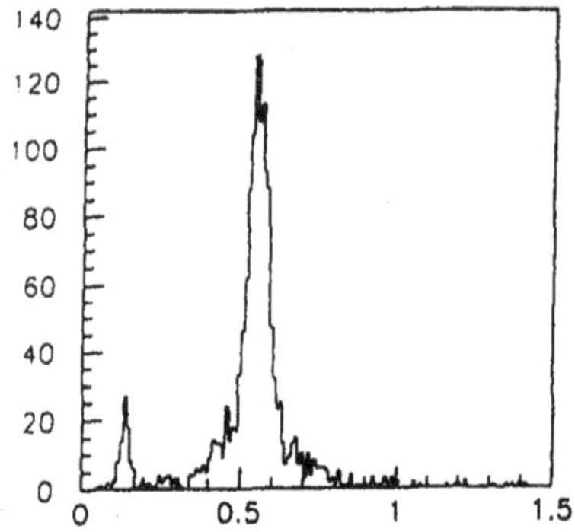

Figure 3: The two photon effective mass with the trigger processor configured to select events with an effective mass greater the the mass of the π^0

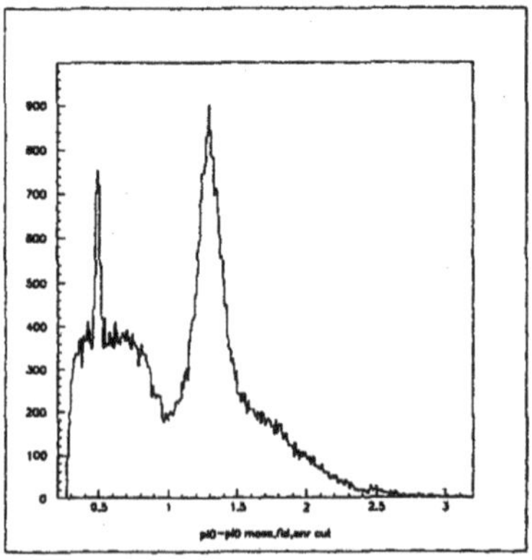

Figure 4: The $2\pi^0$ effective mass distribution from 4 photon events. The structure near 0.5 GeV/c^2 is due to $K_s \rightarrow 2\pi^0$ and has a width of 0.015 GeV/c^2.

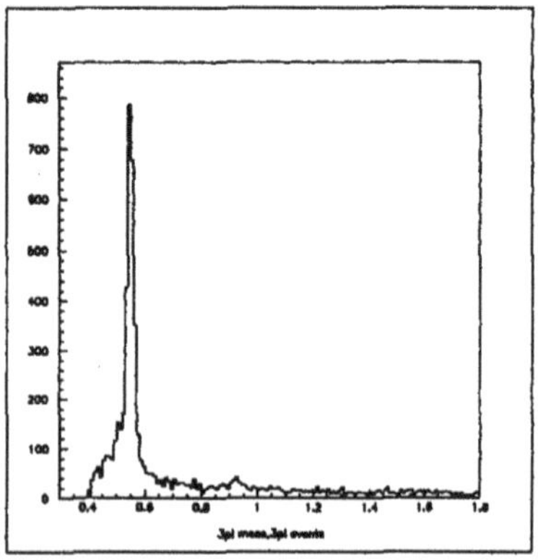

Figure 5: The $3\pi^0$ effective mass distribution from 6 photon events. The structure near 0.55 GeV/c^2 is due to $\eta \rightarrow 3\pi^0$ and has a width of 0.030 GeV/c^2. Evidence for $\eta' \rightarrow 3\pi^0$ is also seen.

THE NOMAD ELECTROMAGNETIC CALORIMETER

Cosenza [a]- Firenze [b]- Moscow [c]- Padova [d]- Pavia [e]- Pisa [f] collaboration

Presented by Mauro Mezzetto
INFN, Sezione di Padova, Via Marzolo 8, 35131 Padova, Italy

The NOMAD experiment [1] searches for $\nu_\mu \leftrightarrow \nu_\tau$ oscillations in the wide band neutrino beam at CERN SPS.

The search is based on the observation of τ production via ν_τ charged current interactions: $\nu_\tau + \text{"}q\text{"} \to \tau + X$. In order to select such events kinematical cuts such as missing P_T angular correlations etc. are used. To be effective, these cuts require very good resolution on energy and momenta.

The experiment is sensitive to most ($\sim$ 84%) of the decay channels of the tau, and is particularly suited to detect its decay into one electron and two neutrinos.

The role of the electromagnetic calorimeter (ECAL) is a crucial one in the electron identification and energy reconstruction. Moreover, ECAL detects and measures with high precision γ, π^0 and ρ^0.

1. The Electromagnetic Calorimeter Requirements

The ECAL has to fulfil the following requirements:

(i) To work properly inside the NOMAD magnetic field: $\vec{B} = 0.4\,T$ perpendicular to the detector longitudinal axis.

(ii) An energy resolution $\sigma/E \leq 10\%/\sqrt{E\,(GeV)}$.

(iii) To give a linear response over the full dynamic range from about 50 MeV to 100 GeV. A low noise electronic amplification is used to guarantee the low energy threshold.

(iv) A uniform response over its full 3×3 m^2 surface. To this purpose no mechanical substructure has been used in assembling the detector, except for an external frame, requiring a very careful machining of the lead glasses.

(v) An e/π rejection factor better than 10^{-3}. To this purpose a preshower (PS) made of 1.5 X_0 of lead followed by 2 planes of proportional tubes is used in conjunction with the ECAL; it also measures the coordinates of the γ's hitting the ECAL.

[a] L. La Rotonda, M. Valdata-Nappi.
[b] G.Conforto, E.Iacopini, A.Marchionni, F.Martelli, M.Veltri.
[c] S.N.Gninenko, A.V.Kovzelev, S.V.Volkov.
[d] M. Baldo-Ceolin, F. Bobisut, F. Cassol, A. Cavestro, D. Gibin, A. Guglielmi, M. Laveder, M. Mezzetto, A. Sconza and L. Visentin.
[e] G. Ambrosini, P. Cattaneo, C.Conta, R.Ferrari, M.Fraternali, A.Lanza, M.Livan, D.Orestano, F.Pastore, E.Pennacchio, R.Petti, G.Polesello, A.Rimoldi, V.Vercesi.
[f] D.Autiero, A.Cardini, V.Cavasinni, T.Del Prete, A.De Santo, V.Flaminio, G.Renzoni, C.Roda.

2. The Electromagnetic Calorimeter

The electromagnetic calorimeter consists of 875 TF1-000 lead-glass Čerenkov counters, forming a matrix of 35 rows and 25 columns. Each counter is $\sim$ 19 radiation lengths deep and has a rectangular cross-section of $79 \times 112\,mm^2$. The glass characteristics are summarized in table 1.

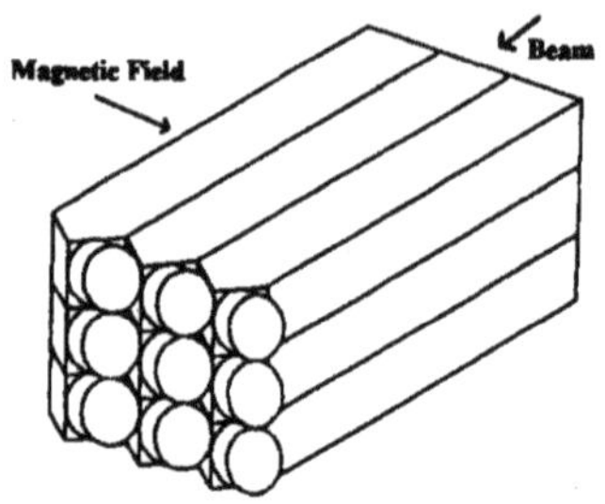

Fig. 1. Layout of the glasses.

Table 1: Characteristics of the TF1-000 lead glass

X_0	2.67 cm
ρ	3.85 g/cm^3
n	1.65 (at $\lambda = 589$ nm)
R_m	3.6 cm

Photomultipliers in general do not work efficiently in magnetic field while detectors having a small number of amplification stages (triodes and tetrodes) can operate at $B = 0.4\,T$ if their axis is at an angle $\theta \leq 45°$ with respect to magnetic field. To match this condition the glasses are cut wedge shaped, see Fig. 1. The amount of light reaching the cathode with this layout has been evaluated through a MonteCarlo simulation. The result shows a 20% loss with respect to a geometry where the photodetector is placed on a squared back side perpendicular to the longitudinal axis of the glass.

The number of photoelectrons per GeV in these conditions was measured to be $N_{phe} \simeq 1400$.

Hamamatsu 3" R2186 tetrodes are used as photodetectors. They provide a gain $G \simeq 40$ (at 800 V) for $B = 0.4\,T$ parallel to their axis, which becomes $G' \simeq 0.8G$ with $B = 0.4\,T$ at 45° with respect to their axis.

The signals are amplified and shaped by a low noise electronic chain developed from the former project of Delphi experiment [2].

A 12 bit FASTBUS peak sensing ADC, developed at CERN, records the amplified signals with a deviation from linearity lower than 0.3%.

Two blue LED's on each block are used both to monitor the stability of the detection chain and to convert the calibration constants measured with $B = 0$ into the ones needed when working in magnetic field [3].

A fast timing signal extracted before the shaping circuit is used to avoid overlapping of uncorrelated events occurring in the same beam spill. It can also be used as calorimeter trigger.

3. Linearity

Extensive tests with electrons and pions have been performed at CERN PS and SPS beams. The energy response to electrons has been measured in the range 1 - 80 GeV. The linearity and the deviations from the linear fit are reported in Fig. 2. The calorimeter results to be linear within 1 % in the full energy range.

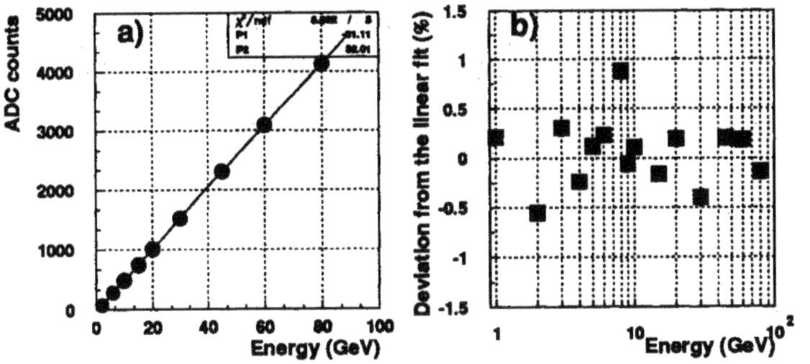

Fig. 2. a) Linearity of the ECAL response to electrons between 1 and 80 GeV; b) percentual deviations from the linear fit.

4. Uniformity

The uniformity of the ECAL response was tested by measuring the ADC counts in a multiplet of towers in several horizontal, vertical and angular scanning.

As an example the result of a horizontal scanning is shown in Fig. 3, the horizontal direction being the one asymmetric with respect to the tetrodes, see Fig. 1. The ECAL response turned out to be uniform at the level of 1% or better.

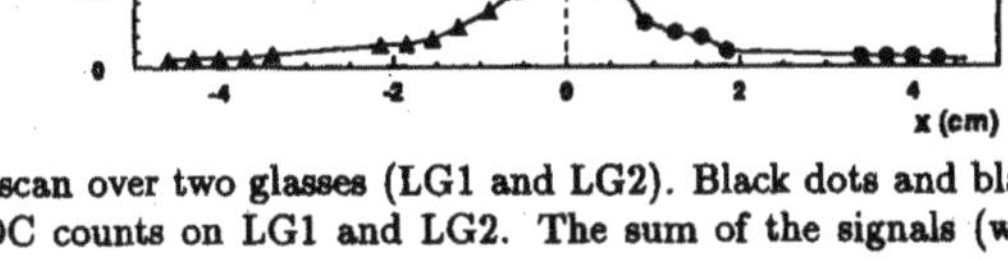

Fig. 3. Horizontal scan over two glasses (LG1 and LG2). Black dots and black triangles represent respectively the ADC counts on LG1 and LG2. The sum of the signals (white dots) is constant within 1%.

5. Energy Resolution

The intrinsic energy resolution of the ECAL has been measured in a matrix of 3×5 towers ($33 \times 40\, cm^2$) using electron beams with energies between 1 and $80 GeV$ hitting the center of the central tower. Fig. 4 shows the distribution of the measured

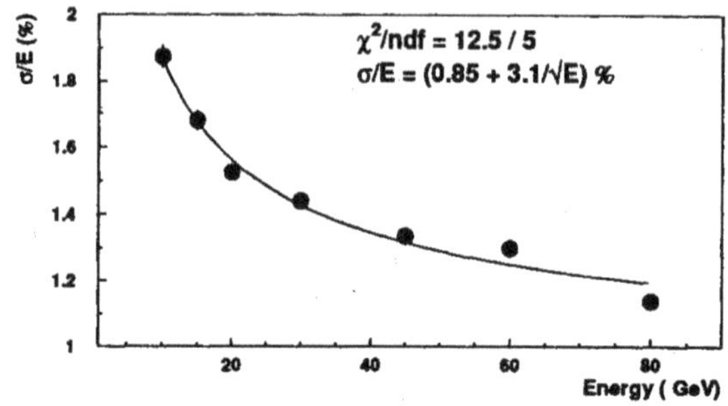

Fig. 4. Energy resolution $\sigma/E(\%)$ as function of E.

$\frac{\sigma}{E}$ values from 10 to 80 GeV ($B = 0$), where σ is the r.m.s. of the signal distribution at each energy and E is its mean value. Two contributions are subtracted from the measured σ: the momentum byte of the beam: $\Delta p/p = 1\%$, and the equivalent noise of the electronics, directly measured by the pedestal width of each tower. A fit to the measured points done with the function

$$\frac{\sigma}{E} = a + \frac{b}{\sqrt{E}} \tag{1}$$

gives $a = (0.85 \pm 0.03)\%$ for the constant term and $b = (3.1 \pm 0.16)\%(\sqrt{GeV})$ for the statistical one.

The equivalent noise for all the 875 towers measured with the ECAL inside the NOMAD magnet is shown in Fig. 5, its average value being $\sim 13\ MeV$.

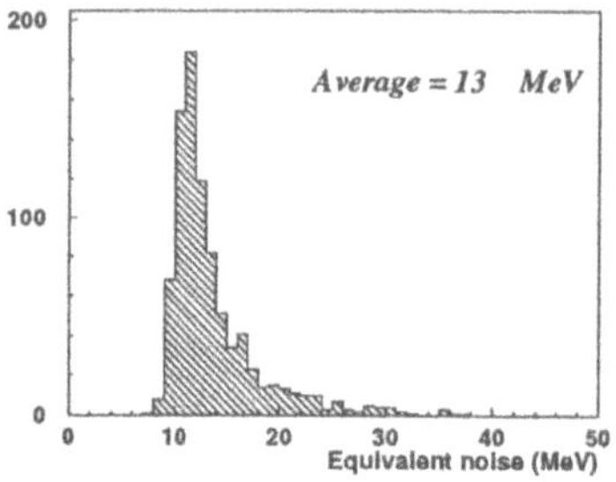

Fig. 5. Equivalent noise of the pedestals for all the ECAL towers.

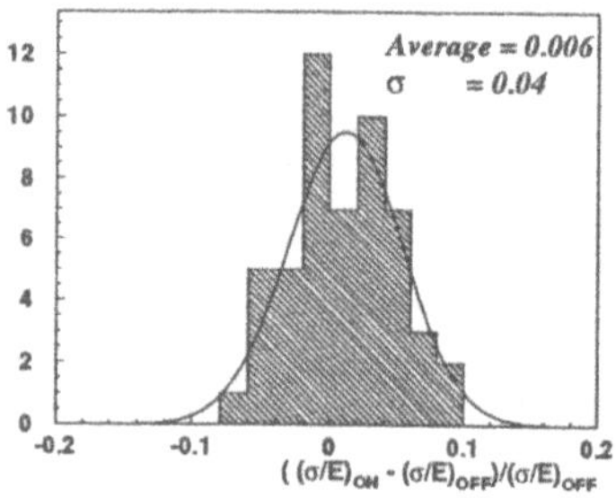

Fig. 6. Fractional variation of $\frac{\sigma}{E}$ with the magnetic field .

In order to detect possible effects of the magnetic field on the energy resolution, $\frac{\sigma}{E}$ was measured in a sample of 52 towers using 10 GeV electrons, once with $B = 0.4\,T$ and then with $B = 0$. The measured fractional variation of $\frac{\sigma}{E}$, shown in Fig. 6, assures that the magnetic field does not affect the energy resolution.

6. Operation Inside Magnetic Field

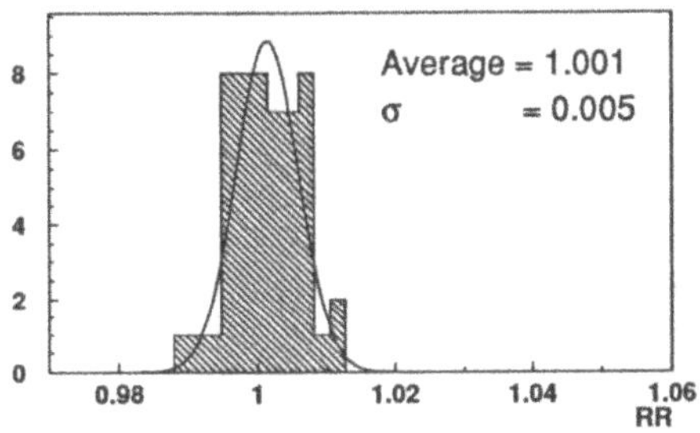

Fig. 7. RR values as measured in a 52 towers sample.

LED light pulses are used to trace the gain variations of the tetrodes in the magnetic field. If R_{ele} (R_{LED}) is the ratio of the electron (LED) signal measured with and without the magnetic field: $R_{ele} = (e^-)_{B=0.4T}/(e^-)_{B=0}$, $R_{LED} = (led)_{B=0.4T}/(led)_{B=0}$, one can define the double ratio RR as:

$$RR = \frac{R_{ele}}{R_{LED}}. \tag{2}$$

RR has been measured on a sample of 52 towers using 10 GeV electrons. The distribution of the experimental results, shown in Fig. 7, gives $\langle RR \rangle = 1.001$ with a r.m.s of 0.005, showing that LED light can measure the gain variation of each tetrode within 0.5%.

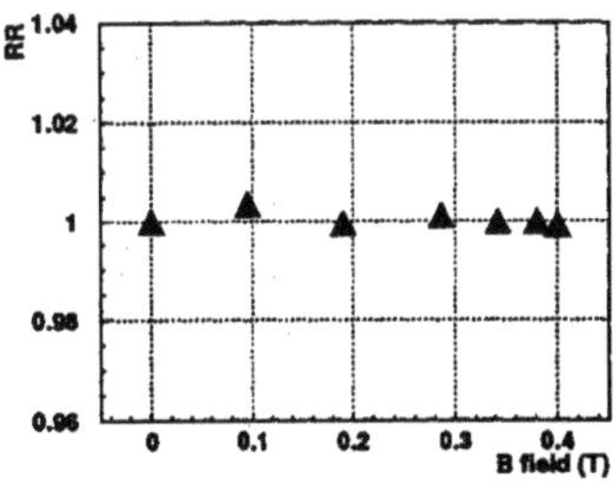

Fig. 8. RR values as function of magnetic field.

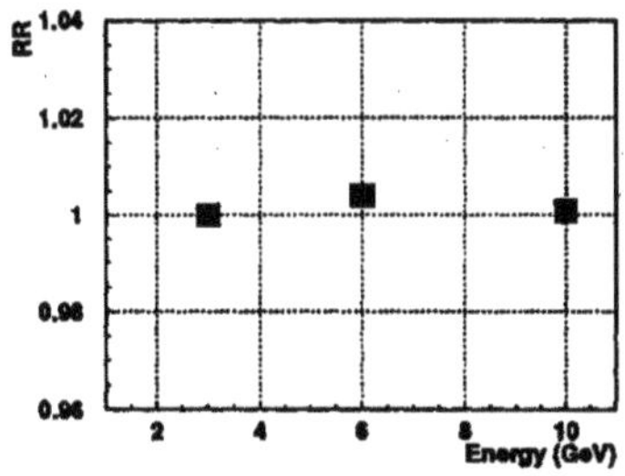

Fig. 9. RR values as function of electron energy.

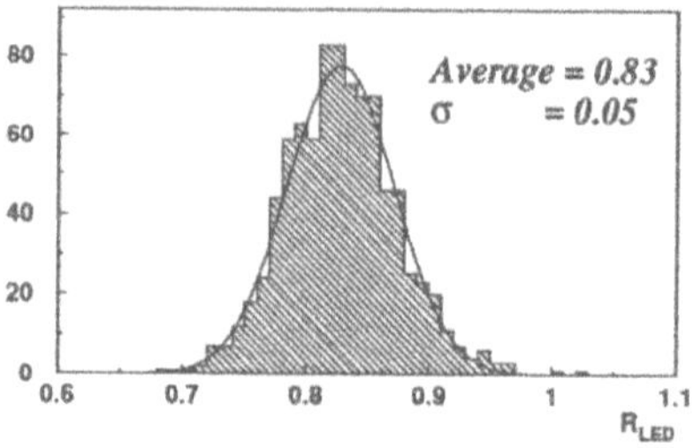

Fig. 10. R_{LED} measured for all the towers inside the NOMAD magnet.

Fig. 8 and 9 illustrate the RR independence both from magnetic field and energy variations, assuring the quality of R_{LED} as R_{ele} estimator.

The gain variation of each tetrode R_{LED} can be measured directly inside the NOMAD magnet, as shown in Fig. 10. In this way any disuniformity in direction and absolute value of the magnetic field is included in the correction.

7. Electron/Pion Separation

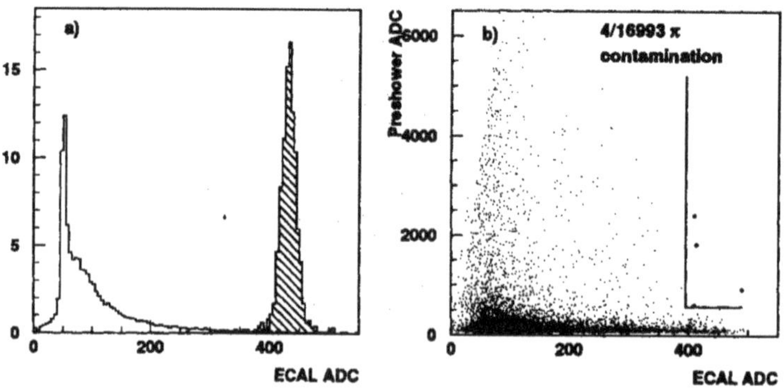

Fig. 11. a) 6 GeV electron (shaded) and pion (white) signals in ECAL; b) preshower vs. ECAL 6 GeV pion signals; the solid lines define the 90% efficiency region for detecting electrons.

The pion rejection factor has been measured with a multiplet of towers exposed to e^- and π^- beams at several energies. Fig. 11a) shows the energy deposition of e^- and π^- in the calorimeter; Fig. 11b) shows the scatter plot of 6 GeV π^- PS versus ECAL signals, only 4 events out of 17000 fall into the region corresponding to a 90% acceptance for e^-.

Fig. 12 summarizes the results obtained as a function of energy.

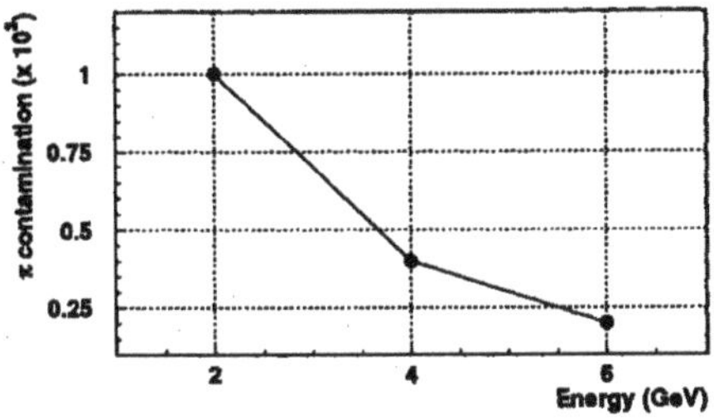

Fig. 12. π rejection factor (90% electron efficiency) as function of π energy.

8. Conclusions

The results of the tests on the performances of the NOMAD electromagnetic calorimeter show that the experiment requirements are completely fulfilled.

Magnetic field effects are well understood and do not affect its performances.

The ECAL response is linear and uniform within 1% in the 1-100 GeV range.

The intrinsic energy resolution is $\sigma/E = (0.85 + 3.1\sqrt{E(GeV)})$ % while the equivalent noise of electronic chain is lower than 15 Mev on average.

Finally, the ECAL signals combined with the preshower allow for an e/π rejection factor at the level of 10^{-3} or better.

Furthermore it has to be added that the ECAL installed in NOMAD in April 1994 took data satisfactory during the entire neutrino run (15 april - 21 october).

9. References

1. Neutrino Oscillation MAgnetic Detector (NOMAD). CERN-SPSLC/91-21 SP-SCP261 (1991) and CERN SPSLC/93-19 SPSLC M520 (1993).
2. G. Barichello et al., *Nucl.Instr.Meth.* **A254**(1987) 111.
3. See D. Autiero in this conference.

EM Scintillator

Session Chair: Sebastian White (BNL)

SHASHLIK CALORIMETRY
AN ELECTROMAGNETIC CALORIMETER + PRESHOWER FOR LHC.

LUDWIK DOBRZYNSKI
LPNHE, Ecole Polytechnique, Palaiseau, France

ABSTRACT

We report here the last test beam results obtained by RD36 collaboration with a Shashlik towers 6 x 6 matrix. The calorimeter uniformity is studied on a 20 x 20 cm^2 area (16 towers). After correction, the remaining non-uniformity of the response is ≈ ±0.3% on the largest part of the towers. The measured energy resolution is defined by: a stochastic term = 8.5%/√E (E in GeV), a noise = .310/√E (E in GeV), and a constant term = 0.5%. The achieved angular resolution when both the calorimeter and the preshower responses are used is equal to 50 (mrad) / √E (E in GeV). These performances are at the needed level for a LHC electromagnetic calorimeter . Results from data in a 3 Tesla magnetic field are also reported.

1. Introduction

The benchmark for an electromagnetic calorimeter at LHC is the detection, at high luminosity, of an intermediate mass Higgs boson by way of its two photon decay.

In the Standard Model there is an important mass range ($90 < m_H < 130$ GeV) in which only this channel is detectable. In the Minimal Super symmetric extension to the Standard Model this channel allows exploration of a large fraction of the parameter space inaccessible to LEP II (√s = 190 GeV). To perform well in detecting the Higgs two photon decay the electromagnetic calorimeter needs excellent energy resolution and good rapidity coverage. Efficient rejection of neutral pions and the measurement of the angle of incidence of the photon in the r-z plane ("photon pointing") are also required[1].

Using the design parameters of the CMS Shashlik electromagnetic calorimeter:

- Stochastic term = 9%/√E (GeV),
- Constant term = 1%,
- Noise = 300/E (MeV),
- Angular resolution = 70mrad/√E (GeV)

a signal with a statistical significance of more than 5 sigma expected from a Standard Model Higgs for $m_H > 85$ GeV can be observed as shown on figure 1 (result of a single Monte-Carlo experiment for 10^5 pb^{-1} and illustrates how SM Higgs of masses 90, 110, 130 and 150 GeV would look after background subtraction. For more details of this analysis see reference [2] and references therein.

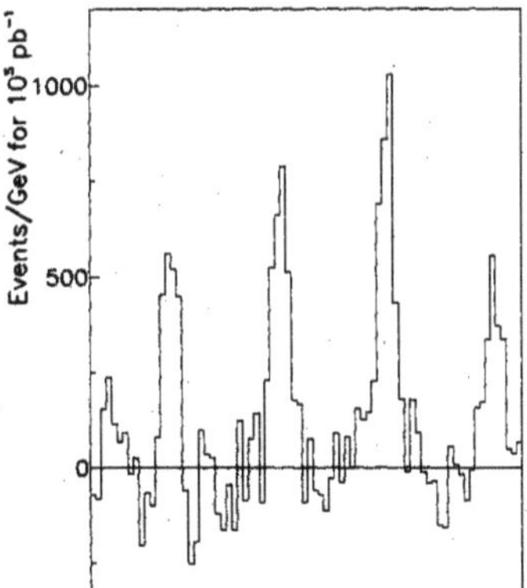

Figure 1: *Result of a single Monte-Carlo experiment for 10^5 pb^{-1} with Higgs peaks at 90, 110, 130 and 150 GeV.*

2. RD36 Shashlik Prototypes

Shashlik calorimeter is based on a technique to readout the scintillation light of a lead/scintillator plate sampling calorimeter with the use of WLS fibres running perpendicular to plates through holes in the plates. This technique is now proposed for many high energy experiments [3]. Different kind of towers were built and studied during the last 2 years. Their parameters are summarized in table 1. They differ by size, WLS fibre choice and mainly assembly technique (stainless steel rods running through additional holes, four stainless steel straps (50 and/or 100μm thick) welded on a stainless steel front and back plate or aluminum foils (50-125 μm thick) glued on the lead tiles and on a front and rear holding piece). The towers were wrapped either with aluminum or aluminized mylar. To separate lead from scintillator either ordinary white paper or tyvec was used.

	NONET (parallelipedical)	PROJECTIVE (projective η=0)	New IHEP (projective η=0)	Glued towers (at large η)
Number of towers	9	16	36	6 [4]+7 [5]
Tower lateral size	47 x 47 mm^2	52 x 52 mm^2	42 x 42 mm^2	trapezoidal size
Number of planes	75	75	70	80
Total depth	27.5 X_0	27.5 X_0	25 X_0	30 X_0
Radiation length	16.9 mm	16.9 mm	16.9 mm	16.9 mm
Moliere radius	34 mm	34 mm	34 mm	34 mm
Lead thickness	2 ± .005 mm	2 ± .05 mm	2 ± .002 mm	2 ± .002 mm
Scintillator type	Polystyrene + .5% POPOP +2%Para-Terphenyl	Polystyrene + .5% POPOP +2%Para-Terphenyl	Polystyrene + .5% POPOP +2%Para-Terphenyl	Polystyrene + .5% POPOP +2%Para-Terphenyl
Scint. thickness	4 ± .05 mm	4 ± .05 mm	3.74 ± .01 mm	3.74 ± .01 mm
WLS fibres	Y7	K27 and Y7	Y11	Y7
Number of fibres	25	36	36	36
Fibre pitch	9.5 mm	9.5 mm	7mm	7mm
Fibre diameter	1.2 ± .03 mm	1.2 ± .03 mm	1.0 ± .03 mm	.83 ± .03 mm
Hole Φ in scint.	1.3 mm	1.3 mm	1.2 mm	1.0 and 1.2 mm
Hole Φ in lead	1.5 mm	1.5 mm	1.2 mm	1.0 and 1.2 mm
Front fibre Ends	Aluminized	Loops	Aluminized	Aluminized
Light yield	12 γ / MeV	12 γ / MeV	20 γ / MeV	not measured

Table 1: *Main parameters of CMS prototype Shashlik tower.*

All these towers were placed in a high energy electron beams (10-150 GeV) at CERN SPS equipped with a 10x10mm^2 silicon PIN photodiode readout (S3590 Hamamatsu) followed by a low noise amplifier designed at INR [6]. Typically we have taken data with a shaping of 30-40 ns and a gate width of 160-200ns [7]. Light yield response of the towers was for the two first set of prototypes 12 γ/MeV. For the NEW IHEP one, we achieved a light yield of 20 γ/MeV. This increase comes from several factors: better quality of the construction , use of multiclading fibres, use of special tyvec for lead/scintillator separation.

Figure 2 shows the standard setup we had in the beam: a matrix of calorimeter towers with in front a preshower detector containing 2 planes of silicon strip detectors. Each plane was build out of 4 wafers (6 x 6 cm^2 each) with 2X_0 (1X_0) absorber in front. The preshower was readout by a 16-channel Amplex-SICAL signal processor[8]. Each detector was connected to a printed board circuit containing two AMPLEX's. The analog signals from the strip detectors are used to find the barycenter of the shower and to correct for the energy lost in the absorber material.

In what follows, we will report the results obtained with a matrix of 6 x 6 projective towers (NEW IHEP).

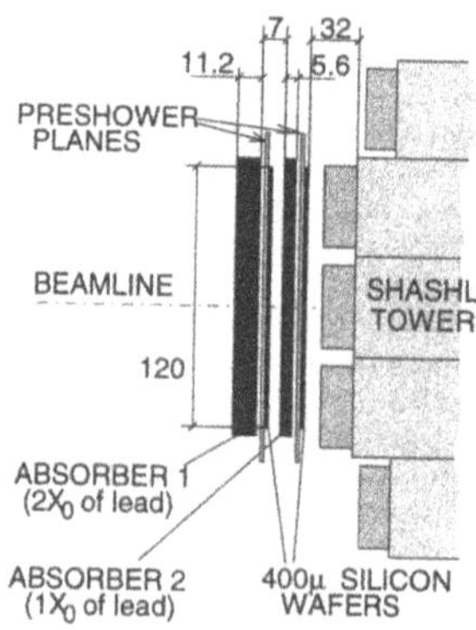

Figure 2: *Test beam setup. Shashlik towers with preshower in front.*

One of the main objective of the data (5 10^7 triggers) taken with this matrix was to study the uniformity response of a Shashlik type detector and to study the influence of the uniformity correction on the constant term of the energy resolution.

3. Main results

3.1. *Quality of the towers*

To appreciate the quality of the towers, we have measured their response to a Cobalt 60 radioactive source. Data have been taken every 2 cm from the front to the rear of the tower. The towers were readout with a PMT connected to a multimeter. Figure 3 gives the mean light response when the response was averaged between x=6 and 36 cm (see figure 4). The dispersion for all the towers (except for tower 33 and 36 for which we have used .83mm diameter Y7 fibres) is ±7%. When the longitudinal profile measurements of each tower were normalized to the mean light response shown on figure 3, one obtains the results presented in figure 4. It can be underlined the small dispersion of these measurements and so the good quality of the towers.

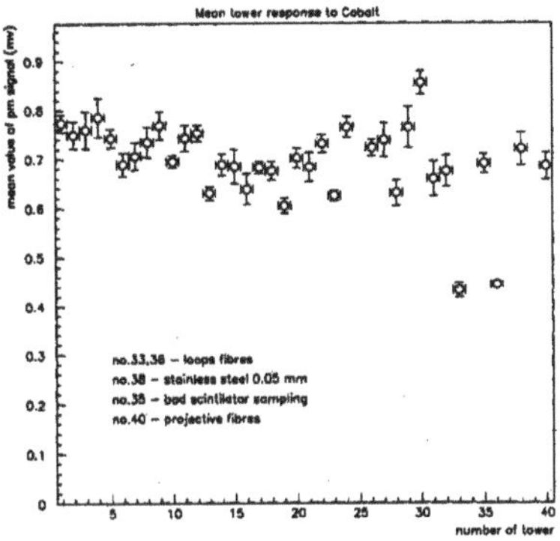

Figure 3: *Mean light response for the 36 NEW IHEP towers.*

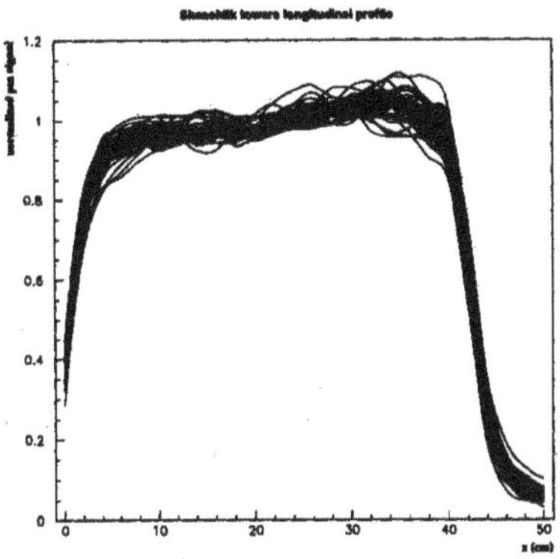

Figure 4: *Longitudinal light response for the 36 NEW IHEP towers.*

3.2. *Single tower response*

The quality of the towers was also investigated with 80 GeV electrons. Figure 5 shows the mean response for 24 towers out of the 6x6 matrix. One can remark the good symmetry of the response. Figure 6 gives a y slice of this response for a range of x = ±10 mm. Data for the 24 towers have been superimposed. The small ondulations apparent on the data are due to the WLS fibres. The distance between fibres is 7mm..

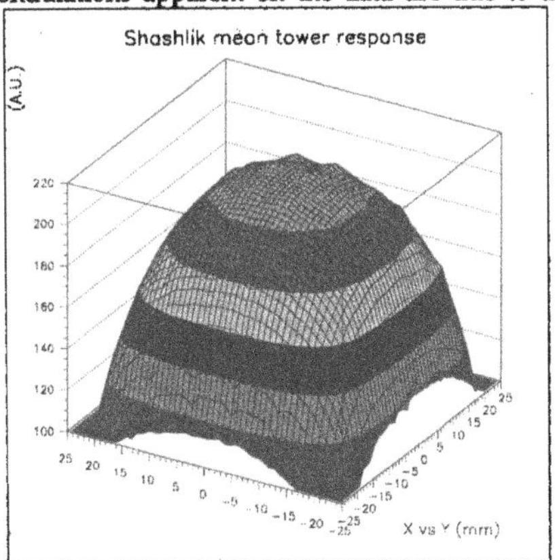

Figure 5: *Shashlik mean tower response obtained by averaging the response of 24 towers exposed to 80 GeV electron.*

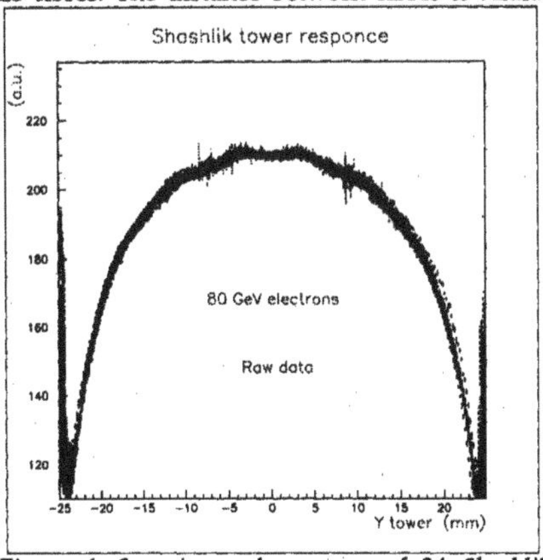

Figure 6: *Superimposed response of 24 Shashlik towers exposed to 80 GeV electrons. Data for x = ±10 mm. Similar results were obtained for a x slice when y =±10 mm.*

The tower transversal profile indicates a tower size of 47.7mm some what larger as the physical size of the towers (42mm). This is due to the projective shape of the towers and the size we observe is the one at the shower maximum. Again the dispersion looks small. It can be appreciated on figure 7 where we have plotted the deviation (in percent) of each point (250 per tower) of figure 6 to the mean value at each point. All the data have a mean dispersion of ±0.72%. For a x slice when y = ±10 mm the mean dispersion is ±0.90%. The difference can be explained by the fact that our setup is tilted by 3 degree in x (angle between the incident electron and the fibre axis) and is fully projective in y where the channeling is more important. From this study we also tried to estimate the dead space between towers. From the x profiles, we found some loss in energy measurement over 0.2mm. For y, the losses were larger (no tilt) and spread over 0.5-0.8mm.

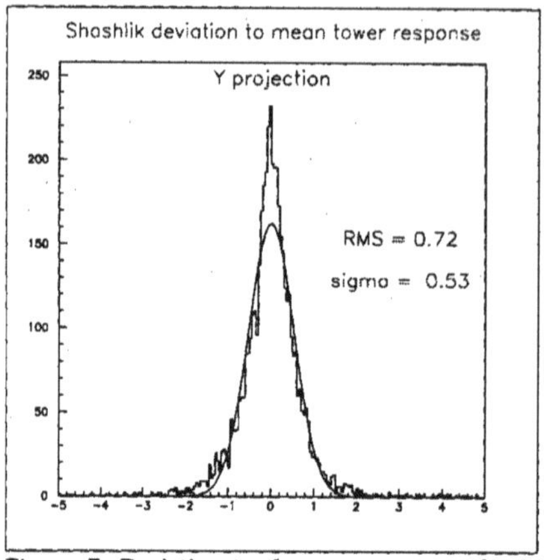

Figure 7: *Deviation to the mean response for 24 Shashlik towers exposed to 80 GeV electrons.*

3.3. *Uniformity of the calorimeter response*

Uniformity both in lateral and longitudinal direction is the crucial issue for the calibration of a calorimeter involving more than ten thousands towers. Radiation damage both on the scintillator and on the fibres will deteriorate the longitudinal response of the detector. This last effect can be corrected once total dose is known.

The lateral non uniformity is mainly caused by the tower geometry, non-uniform reflectivity at the side of the tiles and due to our choice to have the fibres parallel to each other implying that the distance between the fibres and scintillator edges changes form the front to the rear of the tower. This last effect can be compensated by an adequate choice of the fibre attenuation length. More generally, as it will be shown an adequate correction is required. There also exists local non uniformity around fibres due to Cerenkov photons. But this problem can be avoided by tilting the tower with few degrees with respect to the incident beam angle. The non uniformity will be studied on the shower reconstructed energy what implies to sum the deposited energy in 9 towers.

Figure 8 gives the averaged response of the 16 towers for which deposited energy can be reconstructed by the use of the tower which contains the highest deposited energy and the 8 surrounding towers collecting the leaking energy. Figure 9 shows the superposition of the response for 16 individual towers when x slices for a range of y = ±10 mm are studied.

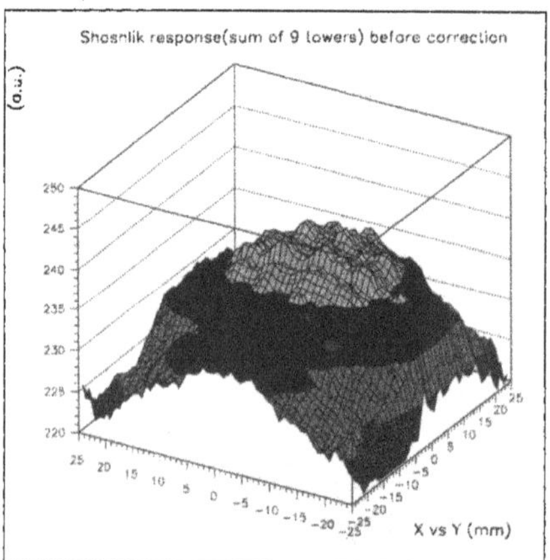

figure 8: *Mean response (Σ of 9 towers) for the 16 central towers of the 6x6 matrix exposed to 80 GeV electrons. The dispersion of all the data is ±3%.*

Part of the dispersion is due to differences in calibration when the sum of 9 towers is considered, but it is clear that a 2^{d} order polynomial shape can be adjusted to all the responses. A deviations of 0.71% (0.72%) to the mean value at each point, for x (y) slices is obtained, confirming the deviations from tower to tower described in section 3.2. We have parametrized the x (y) slices of the tower responses by a second order polynomial:

$$f(E) = a_1 \{1 - a_2 (x - a_5)^2\}$$

where a_1 is the tower response at its center, a_2 the global non uniformity correction and a_5 the tower center position. The quadratic term represent the main correction to be applied to the data. The corrected results for the same 16 towers are shown in figure 10. The overall dispersion is now small (≈0.3%).

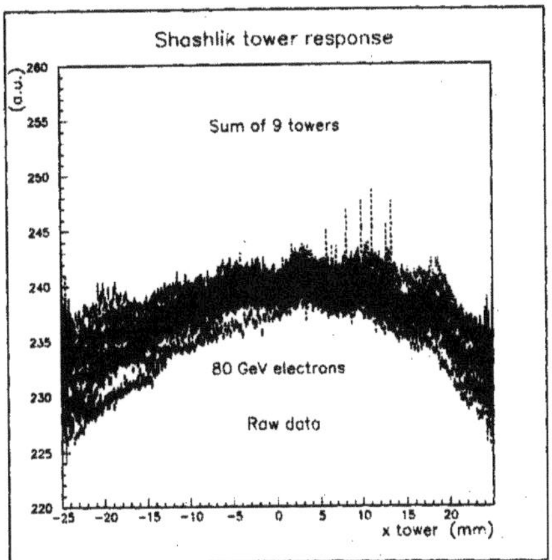

Figure 9: *Superimposed x slice profiles for 16 towers exposed to 80 GeV electrons when y = ±10 mm. y slice profiles present the same shapes(not shown). The dispersion of all the data is ±3%.*

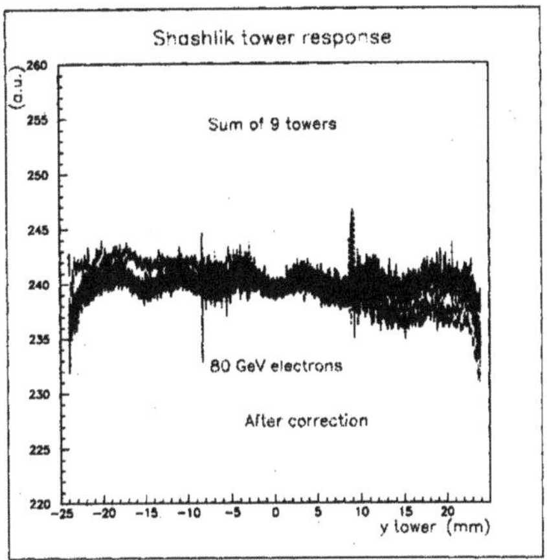

Figure 10: *After uniformity correction: superimposed x slice profiles for 16 towers exposed to 80 GeV electrons for y = ±10 mm.The dispersion of all the data is ±0.3%.*

3.4. *Energy resolution*

For precise energy measurement, we have implemented an additional correction to overcome the fibre effect which is clearly visible on figure 10. If particles hits calorimeter in normal incident angle, there are clear fibre effect of Cerenkov lights, scintillation effects in fibres and shower leakage due to air gap between fibre and scintillator. We have parametrized this local effect with the following function:

$$g(x) = f(E)\ \{1-a_3 \cos\{2\pi/a_4(x-a_5)\}\}$$

a_3 is the local non uniformity correction coefficient, a_4 gives the fibres distance. For our analysis and correction we have used the x coordinate given by beam chambers ($\sigma_{x,y} \approx 300\mu m$).

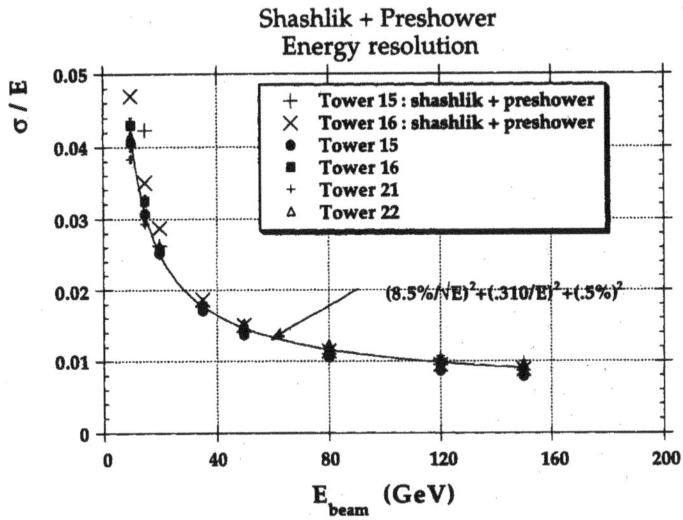

Figure 11: *Energy resolution at tower center (20 x 20 mm²) for 4 Shashlik towers and 2 Shashlik towers + preshower from the 6x6 matrix set up.*

Results for 4 different towers are shown in figure 11. Results with preshower for 2 towers are superimposed.

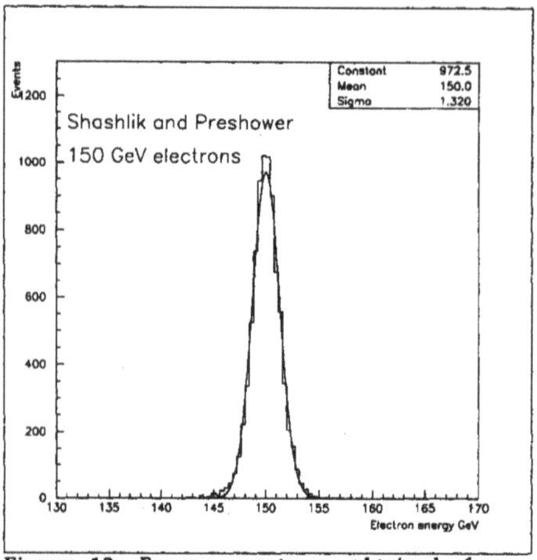

Figure 12: *Energy spectrum obtained from Shashlik + preshower (150 GeV electrons).*

In the Shashlik+Preshower detector the energy measured in the silicon wafers is properly weighted and added to the energy deposited in the towers. Figure 12 shows an energy spectrum obtained with 150 GeV electrons. It perfectly fits a gaussian distribution giving a resolution of 0.88%. At low energy (<35 GeV) the resolution of the bare Shashlik is not fully recovered when the preshower is used (fig. 11). This effect could partly come from the way the lead/scintillator sampling starts for our prototypes. In the present prototypes we have started the tower with an absorber tile which prevents the low energy component of the shower existing behind the preshower, to reach the first scintillator tile.

For all the presented data, we succeeded to reduce the noise term to 300 MeV for the sum of 9 towers and to get a constant term of ≈ 0.5%. Data are consistent with a statistical term of 8.5%.

3.5. *Response in magnetic field*

The Shashlik and preshower detector have been placed at the center of EHS magnet. The energy response is shown in figure 13for magnetic field varying from 0 to 3 Tesla. Signals are normalized to that at 0 Tesla. Signal increases like $\sqrt{B}$. At 4 Tesla (CMS design), signal increase is expected to be 13% compared to 0 Tesla. These measurements confirm similar observation for scintillator sampling calorimeters placed in magnetic field.

For CMS, it was important to check how magnetic field affects the energy resolution. Figure 14 shows the results for the bare calorimeter after standard uniformity correction and for an area of 20x20 mm^2 in the tower center. No special degradation is observed between no field and a 3 Tesla field. The constant term is ≈0.6-0.7% and the statistical term is equal for both configuration to 8.8%.

These results are conserved when the preshower is used (figure 15) like in the case presented in the previous section when the setup was studied without magnetic field.

For the energy resolution study in magnetic field all electrons are not hitting the central tower at the same point. At 150 GeV, the vertical displacement is only ≈2mm, but at 10 GeV the field displaces the impact point by about 45mm what brings it in an other tower. The uniformity corrections are then no more valid and this explains why the low energy point deviate from the fitted curves on figure 14.

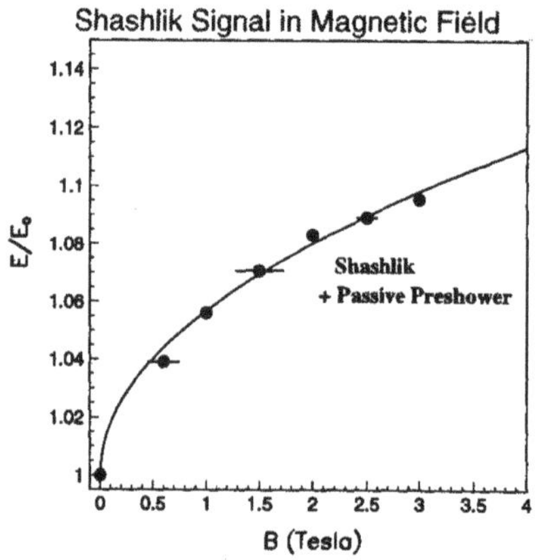

Figure 13: *Variation of the energy measured in the Shashlik with the preshower in front of it, as function of the magnetic field.*

3.6. *Position and angular resolution*

For the H-->γγ process at high luminosity the measurement of the angle of emission of the photons is required. In order to reconstruct this angle the photon shower position has to be measured at two depth in the electromagnetic calorimeter. In the foreseen design, this is accomplished by introducing a preshower detector. Two planes of silicon strip with a pitch of 2mm placed after $2X_0$ and $1X_0$ absorber will give the x and y of a first point, the second one is obtained from the shower barycenter in the calorimeter itself.

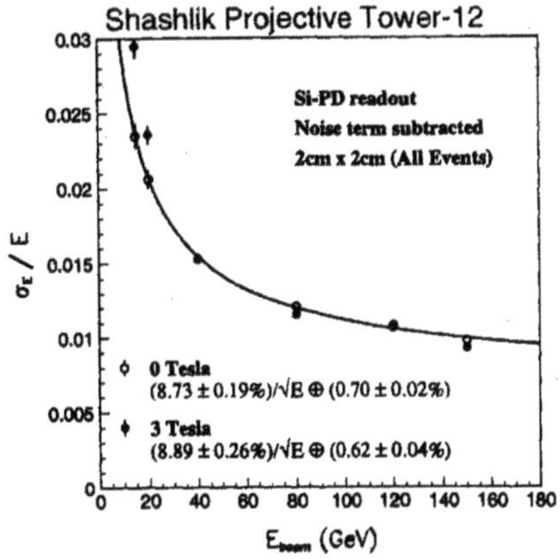

Figure : 14 *Energy resolution for a projective Shashlik tower setup for 0 and 3 Tesla magnetic field.*

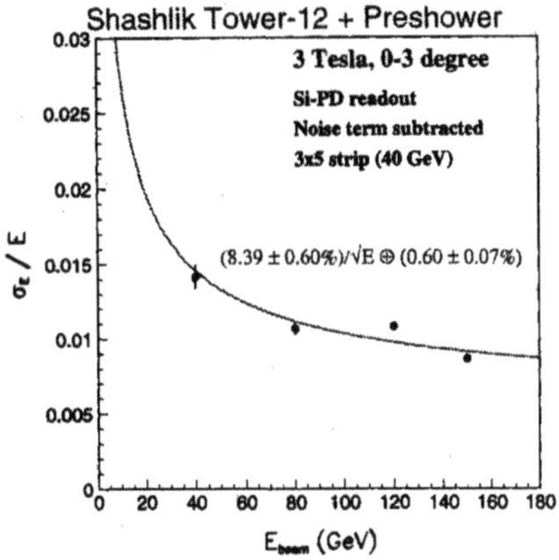

Figure : 15 *Shashlik + preshower energy resolution for 3 Tesla magnetic field.*

Figure 16 shows the position resolution obtained with the preshower when 80 GeV electrons are hitting the detector. We obtained, after unfolding of the beam chamber contribution, a resolution of ≈300µm. The resolution varies as function of the energy. It was found to follow: $\sigma_{x,y}$ (mm) = $2/\sqrt{E} + .2$ where E is in GeV. Figure 17 gives the measured position resolution for the NEW IHEP towers. The mean result corresponding to the tilted direction is as foreseen slightly worse (σ = ≈1.2mm) than the one for the non tilted direction (σ = ≈0.8mm). These results lead for the Shashlik+Preshower detector to an angular resolution of:

$$\sigma_\theta \text{ (mrad)} = \approx 50/\sqrt{E}. \qquad \text{(E in GeV)}$$

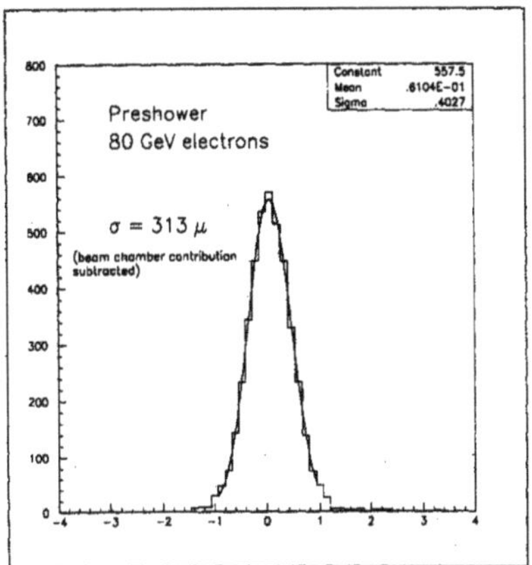

Figure 16: *Preshower position resolution for 80 GeV electrons after unfolding of the beam chamber resolution.*

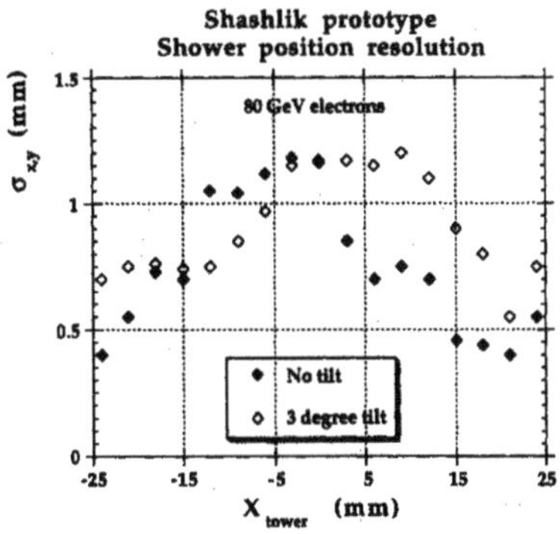

Figure 17: *Shashlik shower position resolution for the non tilted and the 3 degree tilted direction.*

4. Prospects and conclusions

New techniques have been developed to read out the light from lead/scintillator sampling calorimeters. These techniques involve the use of wavelength shifting optical fibres. The use of optical fibres enables fine lateral segmentation to be achieved with a minimum of dead space. In addition, it was expected and now confirmed that such calorimeter can be built at a relatively low cost. Several new projective prototypes have been build and tested in electron beams at CERN. The light yields are in excess of 20 photons per MeV. The uniformity of the response has also been investigated on an area of 20x20cm^2. A mean dispersion of less than ±1% has been achieved. Prototype towers were also studied in magnetic field (up to 3 Tesla). An increase of the calorimeter response of ≈10% has been observed. No degradations of the performances of the calorimeter have been found. The energy resolution measured on projective towers (42x42mm^2 at front face) from our last 6x6 matrix prototype is:

$$\frac{\sigma}{E} = \frac{.08}{\sqrt{E}} \oplus \frac{.3}{E} \oplus 0.005 \qquad (E \text{ in GeV})$$

Another goal of RD36 is to study the preshower detector which will be placed in front of the calorimeter. The preshower consists of several layers of lead radiators followed by Silicon strip detectors. The aim of the preshower is to measure accurately the shower position and to provide some separation between single photons and π^0. Together with the position measurement in the calorimeter, the preshower allows a measurement of the shower direction. In 1994, tests were made with a 3 X_0 thick preshower containing 2 planes of Silicon, covering a transverse area of 12x12cm^2 A good position resolution (≈300 μm at 80 GeV) has been obtained. Using the shower barycenter position deduced from energy deposited in the calorimeter and the position measured in the preshower we achieved an angular resolution of:

$$\sigma_\theta(\text{mrad}) = \frac{50}{\sqrt{E}} \qquad (E \text{ in GeV})$$

It was shown that the energy loss in the lead radiators can be measured in the Silicon and corrected for, so that the energy resolution of the Shashlik calorimeter is only slightly degraded when the preshower detector is added, even when the detector is put in a high magnetic field.

The measured performances of a Shashlik+Preshower detector are at the expected level needed for LHC. In 1995, we will test a new fast analog electronics, developed initially by CERN ECP for the RD2 experiment and modified for the specific requirements of the preshower detector. This electronics includes a 32 channels, 128 slots analog memory running at 40 MHz. Radiation hard versions of the principal components will be tested as well. As far as the Shashlik towers are concerned, further research and development are essential before building a full scale projective detector. Mainly detailed studies on radiation hardness, on multibundle large scale matrix will be continued.

References

[1] C. Seez *et al*, Proc. Large Hadron Collider Workshop, Aachen, 1990, eds. G. Jarlskog and D. Rein, CERN 90–10, vol. III, 474

[2]CMS Letter of Intent, CERN/LHCC 92-3, LHCC/I1, October 1992

[3] H. Fessler et al. Nucl. Inst. and Meth. A228(1985)303 and A240(1985)284;
B. Lohr et al. Nucl. Inst. and Meth. A254(1987)26
G.S. Atoyan et al. Nucl. Inst. and Meth. A320(1992) 144
T. Camporesi et al. Stic calorimeter for Delphi. Proceedings of the Florida IEEE Conference (1992)
CERN-DRDC Proposal 50: A combined Shashkik+Preshower detector for LHC. CERN/DRDC93- 28(1993)
S. White The Phenix em calorimeter. IV International Conference on Calorimetry in High Energy Physics. La Biola d'Elba, Italy(1993)
E. Kistenev report for this conference on the Phenix electromagnetic calorimeter

[4] Tower build at Ecole Polytechnique (France)

[5] Towers build by a russian industrial company under the supervision of INR(Moscow)

[6] Shashlik readout. CMS-TN/93-96

[7] Shashlik calorimeter. Response to high energy electrons. NIM A 344(1994) 57-63
Shashlik calorimeter. Beam test results. NIM A 348(1994) 74-86
Multi-Bundle Shashlik calorimeter prototypes. Beam-test results. CMS-TN/94-197. To be pub. in NIM
New test beam results of Shashlik and preshower prototypes. CMS-TN/94152
Shashlik calorimeter. Beam test results. CMS-TN/93-66
Position and angular resolution with the Shashlik calorimeter. CMS-TN/93-65

[8] E. Beuville at all. Nuclear Physics B. (Proc. Suppl.) 23A(1991) 198

MULTI-BUNDLE SHASHLIK CALORIMETER FOR CMS.

EVGENI GUSCHIN
Institute for Nuclear Researches of Russian Academy of Sciences
7a, the 60-th October annivs. prospect, 117312
Moscow, Russia

ABSTRACT

The beam-test results for two- and three-bundle shashlik tower prototypes are described. We found that the spatial resolution, the uniformity of energy response, the calorimeter reliability and hermeticity and also two showers separation are improved in multi-bundle design approach.

1. Introduction.

Shashlik calorimeter[3] is lead-scintillator sandwitch type calorimeter read-out by WLS fibers. As the option for the electromagnetic calorimeter for the CMS experiment at LHC[1,2] it should satisfy the main requirements for LHC experiments : high rate capability, compactness, good energy and spatial resolution, etc[2,4]. The main advantage of such calorimeter type is the much lower cost in comparison with a homogenious (total absorption) calorimeters. The cost of it is essentialy defined by high precision machining of lead and scintillator tiles, that is necessary for high hermeticity and uniformity, and depends mostly on the number of towers and much less on the its volume. From the other side, a calorimeter cell size should be small enough to provide the required spatial resolution and low pile-up noise. As a compromise, the multi-bundle tower design was proposed and tested.

2. Multi-bundle tower design.

Among the 'shashlik' prototype projective towers[6] tested at CERN two towers were manufactured as shown on the fig.1-3. The fibers collected in bunches share the light signal from calorimeter tiles, leading to two and three readout cells in the same tower. The separation of fibers is done in one direction, that fits the requirement of more precise coordinate measurement in the plane of pseudorapidity for CMS experiment. Since all the fibers in the tower capture a light from the same scintillator tiles, a point-like light source will give signals in all channels. This may be considered as a cross-talk between channels. However its value depends on the light source position so it can be used for coordinate measurements. As a drawback, the pile-up noise may increase. More simulation studies are still necessary to define the optimal tower dimension for LHC experimental conditions. The Monte-Carlo study of the spatial resolution and two showers separation in the nonprojective tower was done in the work[5]. The present towers were of projective pointing geometry with square cross-section. The dimensions of front and rare faces of the towers was 52.5 *mm* and 64 *mm*. The fibers were of a 1.2 *mm* diameter. Placed in a 6×6 matrix, with

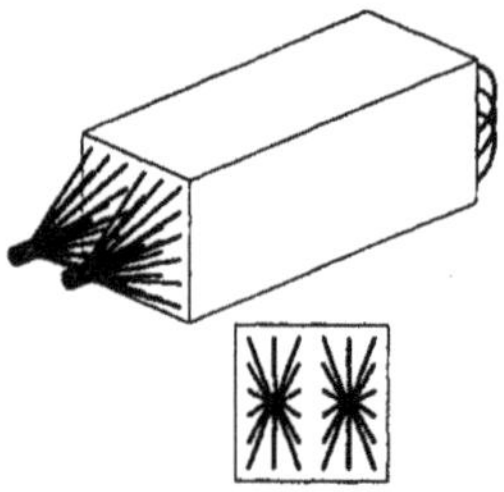

Figure 1: Two-bundle tower.

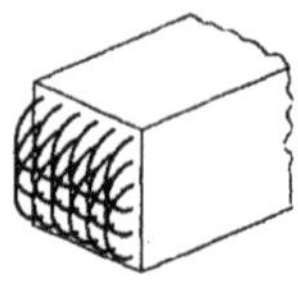

Figure 2: Front view.

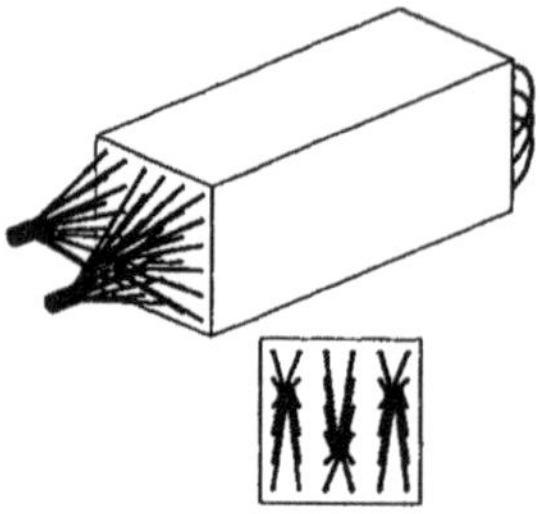

Figure 3: Three-bundle tower.

distance between fibers of 9.5 mm, they run parallel to each other and tower axis. Three-bundle tower was equipped with Y-7 type of fibers, and two-bundle one with K-27 fibers.

3. Experimental setup.

The program of test measurements was performed on SPS H2 test beam at CERN. It consisted of coordinate scan by 40 GeV electrons beam in horizontal plane at the middle of the tower. The tower to be tested ("t") was situated between two towers as shown on the fig.4. The tower axis was parallel to the beam. Beam particle coordinate was measured by the beam chamber with an accuracy of 300 μm. To detect the light from WLS fibers HAMAMATSU-3590-05 Si-photodiodes were used, equipped with low noise hybrid amplifiers[7,8]. The shaping of $\tau_{RC-CR} = 35 \div 40ns$ was used. The photodiode sensitive area was 9×9 mm^2, providing a capacitance of $25 \div 30pF$ at the input of amplifier.

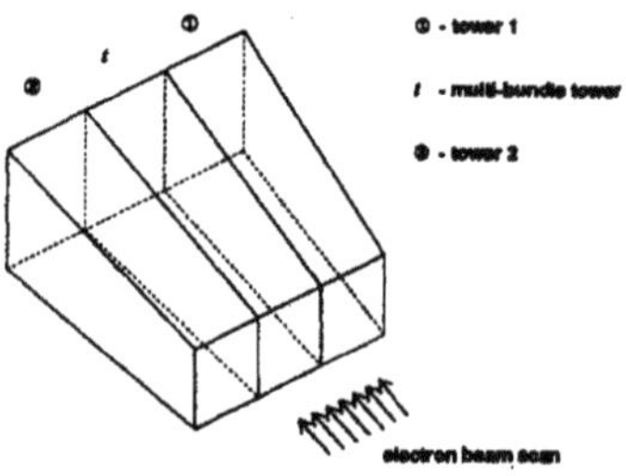

Figure 4: Experimental setup.

The electronic noise was measured to be $130MeV$ per readout channel for 3-bundle tower, and about $170MeV$ per channel for 2-bundle tower.

4. Experimental data and method.

4.1. Channels intercalibration.

The average responses of multi-bundle towers channels are shown on the fig.5-6 for a transversal position scan of the tower with a $40GeV$ electron beam. In the case of the 2-bundle tower, calibration required that the integrals of both channel responses over the tower transversal coordinate are equal. For the 3-bundle tower the left and the right channels were intercalibrated the same way. Central channel was tuned to minimize the lateral nonuniformity of the tower. The total response after calibration was uniform within ±1% through the tower.

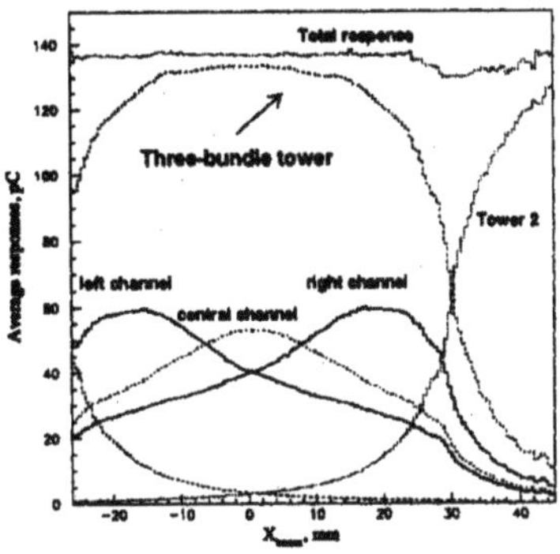

Figure 5: Channels responses in 3-bundle tower.

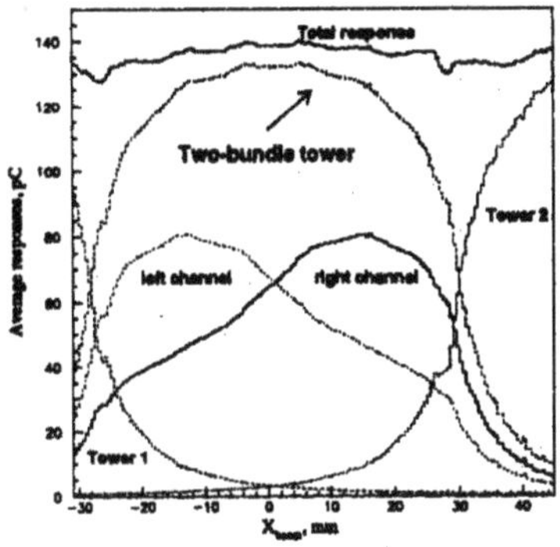

Figure 6: Channels responses in 2-bundle tower.

4.2. Self-correction of fiber effect in multi-bundle tower.

The scatter-plot on fig.7 shows the amplitude of each channel of 3-bundle tower vs. x_{beam}-coordinate. The events were selected by y_{beam} to be in ± 1.5mm around fibers. The response of channel containing the fiber was distorted.

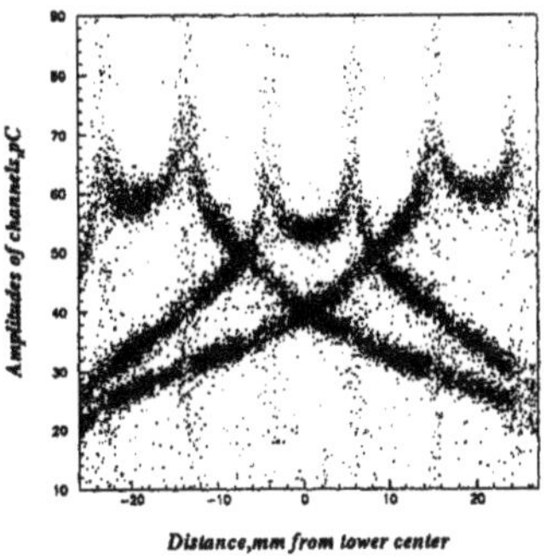

Figure 7: The coordinate scan across fibers in 3-bundle tower. Responses of channels are superimposed.

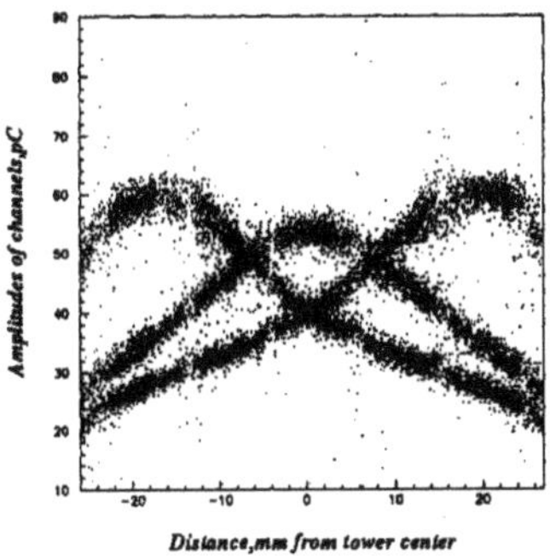

Figure 8: Channels responses in 3-bundle tower after correction.

The signals in the adjacent channels are not perturbed allowing the correction. The redundancy of multi-bundle readout leads to the strong correlation of channel responses[9]. The correlation functions were defined by shower shape and by the light collection conditions and were practically independent on the shower energy. The corrected channels responses are shown on the fig.8.

4.3. Shower spatial resolution.

The electromagnetic shower coordinate in a calorimeter can be determined from the center of gravity of the response of adjacent towers. In general the amplitude of the response of each tower should be fitted to the shower shape. The cross-talk between readout channels in the case of multi-bundle tower leads to a distortion of the shower shape. This makes the analysis of spatial resolution more complicated than for a towers with a single readout.

The details of the method we used for the reconstruction of the shower position were described in the work[9]. The improvement in the spatial resolution achieved with the multi-bundle readout is shown on the fig.9. For comparison the resolution obtained for a single-bundle tower with and without electronics noise is also shown. The mean value of the spatial resolution in the 2-bundle tower was $830 \pm 140\mu m$. The contribution of calorimeter itself into the spatial resolution after unfolding the electronic noise contribution ($330\ \mu m$) and the beam chambers resolution ($300\mu m$) was $690 \pm 125\mu m$. In the case of 3-bundle tower the mean value of the spatial resolution was $760 \pm 125\mu m$ including the contribution of the electronic noise ($290\ \mu m$) and of the beam chambers. The contribution of the calorimeter itself was $620 \pm 130\mu m$.

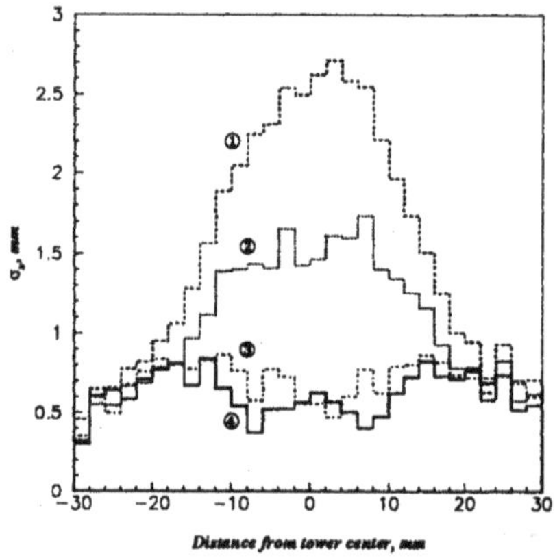

Figure 9: Spatial resolution vs. x_{beam} coordinate. (1) 2-bundle tower with only the sum of amplitudes used (identical to the tower with single channel read-out). (2) The same as (1), but electronics noise and beam chamber resolution are subtracted. (3) 2-bundle tower. (4) 3-bundle tower.

4.4. Recovery of dead channel.

If the correlation functions are known, a channel signal can be restored by using the adjacent signals. This improves the calorimeter security what is important for experiments at LHC that will have a duration of over 10 years. We found[9] that in the case of 2-bundle tower the channel signal can be restored with precision of better than $2GeV$ for $40GeV$ shower energy. The spatial resolution with one channel dead is the same as for single bundle tower. The precision of restoring the right or the left channel signal in the case of 3-bundle tower is better than $1GeV$, for central channel it is about $300MeV$. The spatial resolution in the case of central channel dead is not realy changed. In all cases the precision of restoring the channels responses was mostly defined by the electronics noise.

4.5. Two showers separation.

The redundancy of multi-bundle readout allows the rejection of double shower events. For this we used the same methodics as for 'fibers' effect correction. We simulated a double shower events by taking the sum of two single showers responses. Thus, the electronic noise was also taken twice. The separation efficiency as a function of the distance between the centers of gravity of two showers is shown on the fig.10 (with the efficiency of single shower event acceptance larger than 98%).

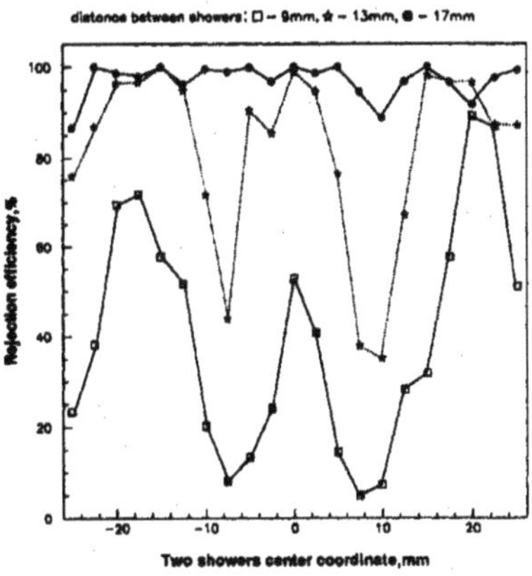

Figure 10: The rejection efficiency of double shower events in the 3-bundle tower.

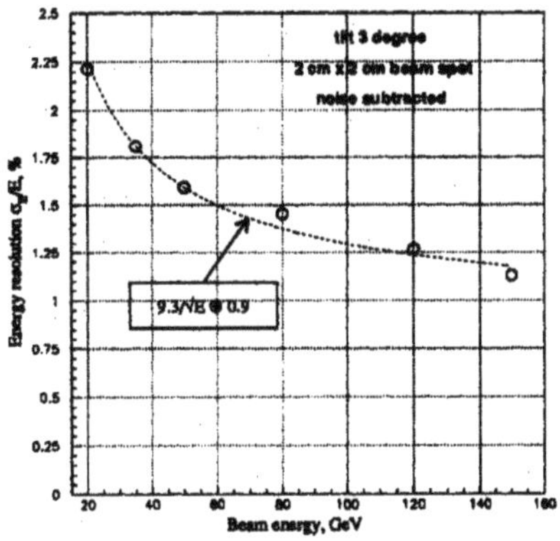

Figure 11: The energy resolution in the 3-bundle tower.

4.6. Energy resolution.

The energy resolution study for the 3-bundle tower was performed in the SPS-H4 test beam at CERN. The tower studied was installed inside the matrix of 4×4 regular towers. The calorimeter was tilted by 3 degree with respect to the beam direction in the horizontal plane. For the energy measurements we used the total response of 9 towers. The beam spot size was $2 \times 2\ cm^2$. We applied only a small correction of the total response nonuniformity around the fibers rows position in the vertical (nontilted) direction. The result is shown on the fig.11. It is not different from one for a regular tower[6].

5. Conclusion.

We examined the performance of 'shashlik' calorimeter that can be achieved in a multi-bundle design approach. Our experimental results show that:

- the spatial resolution at the tower center in multi-bundle towers is improved by 2-3 times in comparison with the same type of tower with a single channel readout.
- the lateral uniformity of the energy response in 3-bundle tower is improved in comparison with two and one bundle cases.
- the correlation of channels responses in multi-bundle tower provides a possibility to correct the calorimeter response in the vicinity of fibers as well as nuclear counting effect in Si-photodiode.
- the reliability of a calorimeter performance and the two showers separation are improved by mutli-bundle design.
- all the performances mentioned above are mainly limited by electronic noise.

6. Acknowledgements

The work described in this paper is a common contribution from the RD-36 collaboration laboratories. We are grateful to our colleagues from RD-5 collaboration, in particular to G. Bencze, F. Szoncso, C. Wulz and also to M. Haguenauer and J. Bourotte who made our running conditions at H2 and H4 beams possible and easy. We are grateful to the staff of SPS for providing assistance during our tests.

7. References

1. *CMS Letter of Intent.* CERN/LHCC 92,1 October 1992.
2. *CMS Status Report and Milestones.* CERN/LHCC 94-**, May 1994
3. H.Fessler et al., *Nucl.Instr. and Meth.* **A228** (1985) 303 and *Nucl.Instr. and Meth.* **A240** (1985) 284,
   B.Loher et al., *Nucl.Instr. and Meth.* **A254** (1987) 26,
   G.S.Atoyan et al., *Nucl.Instr. and Meth.* **A320**(1992) 144.
4. J.Badier et al. *Nucl.Instr. and Meth.* **A348** (1994) p.p. 74.
5. E.Guschin,Yu.Musienko,I.Semenjuk. INR-1005/94 preprint (Institute for Nuclear Researches,Moscow,1994).
6. J.Badier et al., CERN–CMS TN/93-108 (1993).
7. E.Guschin et al., CERN–CMS TN/93-108 (1993).
8. A.Karar, CERN–CMS TN/93-96(1993) and *Proceedings of the IV International Conference on Calorimetery in High Energy Physics* (La Biodala, Isola d'Elba,Italy,September,19-25,1993).
9. J. Badier et al. CERN-CMS-TN/94-197, (1994), *submitted to Nucl.Instr. and Meth.*.

TEST OF A HIGH RESOLUTION ELECTROMAGNETIC CALORIMETER WITH WAVELENGTH SHIFTING FIBER READOUT†

WILLIAM BROWER
Physics Department, University of California, San Diego, 9500 Gilman Dr. La Jolla, CA 92093, USA

Presented at the V International Conference on Calorimetry in High Energy Physics, Brookhaven National Laboratory, Upton, NY, September 25 - October 1, 1994

ABSTRACT

We have constructed two compact, high resolution electromagnetic calorimeters suitable for the next generation hadron colliders. The calorimeters consist of stacks of very thin lead and scintillator plates which have no physical boundaries between the readout cells. Holes in these plates accommodate a high density of wavelength shifting fibers used for the readout. By constructing the calorimeters with high sampling frequency we are able to use modest sampling fractions (~ 12%) while maintaining excellent energy resolution and compactness. Extensive data was collected during May 1994 in the H4 beam line at CERN with 6-150 GeV e^- and 225 GeV μ^+. The test beam program was designed to accurately measure many properties of each calorimeter including crosstalk, energy resolution, spatial uniformity, signal linearity and the light yield.

1 Introduction

SPAKEBAB is a collaboration between Texas Tech University and the University of California at San Diego and is now part of the RD36 program at CERN. We attempt to tune the available parameters of sampling calorimeters in order to develop one with small sampling fraction and excellent energy resolution. A small sampling fraction is desirable to maintain compactness and thus limit both the lateral and longitudinal electromagnetic shower size. The energy resolution of sampling calorimeters at low energies is dominated by sampling fluctuations. The sampling fluctuations can be controlled by changing the sampling fraction and/or the sampling frequency. It is therefore possible to compensate for a small sampling fraction by increasing the sampling frequency[1]. Two calorimeter prototypes were designed to meet the electromagnetic calorimeter requirements imposed by experiments at the next generation hadron colliders. They are best suited for high energy environments where excellent electromagnetic energy resolution is needed such as at the Large Hadron Collider (LHC) at CERN.

† Collaboration between UCSD (W. Brower, A. Domingues, C. Duba and H.P. Paar) and Texas Tech University (O. Dubois, O. Ganel and R. Wigmans).

2 Description of Prototypes

The prototypes are lead and scintillator plate sampling calorimeters with wavelength shifting fiber readout. They are based on designs previously researched[2][3] but have notable improvements including higher sampling frequency, smaller fiber pitch and finer lateral segmentation. While the similar and more general characteristics are outlined first, the readouts and other differences between the two prototypes are discussed later.

Our goal was to create a sampling calorimeter whose energy resolution would satisfy the guidelines provided by the proposed experiments at the large hadron-hadron colliders. Typical requirements for the energy resolution, parameterized as

$$\left(\frac{\sigma_E}{E}\right)^2 = a^2 + \left(\frac{b}{\sqrt{E(\mathrm{GeV})}}\right)^2,$$

ask that the energy resolution at $\sim$ 50 GeV should be $\leq$ 1%. EGS4[4] was used to estimate the sampling fraction and sampling frequency needed to achieve a stochastic term suitable for the requirements. Using various plate materials, sizes and thicknesses, we settled on the two prototypes described below.

The prototypes were constructed from alternating 19×19 cm^2 plates of lead and scintillator. The scintillator plates (SCSN-81[5]) are 1.0 mm thick. One prototype, called the 35-Mil, has lead plates of 0.89 mm (35 mils) thickness while the other, the 25-Mil, has lead plates of 0.64 mm (25 mils) thickness. The lead plates[6] were cut from rolls of lead sheet with an office-style paper cutter. The lead sheets are made from a lead alloy with a fractional composition of 0.53% Tin, 0.074% Calcium and 99.396% Lead. Each of the lead and scintillator plates have 1539 (19×81) holes drilled in them as shown in figure 1. The 1.27 mm (50 mil) diameter holes were drilled mechanically on a numerically controlled press. Diamond-tipped drill bits were used to drill the scintillator plates. The holes are arranged on a square lattice with center to center spacing of 4.0 mm. Wavelength shifting fibers (Y-7[5]) of 1.0 mm diameter are inserted through these holes.

It is important to emphasize that no material was placed between the lead and scintillator plates as they were stacked. Some designs, with larger fiber pitch and thicker scintillator plates, place reflective white paper on each side of each scintillator plate to help capture light for readout. Our design, with high sampling frequency (and therefore a large number of plates) would require many such sheets of paper, each with 1539, 1.0 mm diameter holes in them. Also, because our scintillator plates are only 1.0 mm thick, there was concern that when a wavelength shifting fiber was inserted through the stack, some paper may cover the crucial cylindrical area of contact between scintillator plate and fiber. It is through this area that all light generated inside the scintillator plates can get passed into the wavelength shifting fibers and ultimately to the readout. Because of the difficulty involved with manufacturing these sheets of paper and the possible catastrophic blocking of signal, it was decided to not install reflective paper.

Although the readout sections of the 25-Mil and 35-Mil prototypes differ in the number of readout cells, the basic cell design is the same. The fiber bundles are coupled, through a plastic hexagonal light mixer, to green-extended photomultipliers. The purpose of the light mixers is to reduce the sometimes large nonuniformities that exist on the photocathode surfaces[7]. The light mixers, photomultipliers, μ-metal shields, photomultiplier sockets and bases and the support structure are all housed in steel tubes. Table 1 highlights the main characteristics of each prototype.

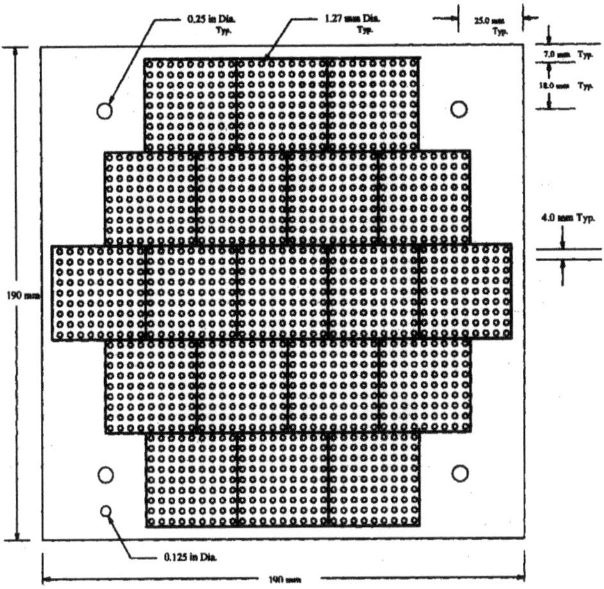

Figure 1: Hole pattern in scintillator and lead plates. The four large holes in the plate's corners are for guiding tie rods. The medium sized hole in the lower left corner defines the plates' orientation. The 19 squares shown are only to indicate which fibers were grouped to form readout cells in the 35-Mil calorimeter.

The 35-Mil prototype is composed of 183 lead plates, 182 scintillator plates and 1539 wavelength shifting fibers. A 9×9 square matrix of 81 fibers is bundled together and coupled, through a plastic hexagonal light mixer, to an eight-stage, 3/4-inch diameter, green-extended photomultiplier (Philips XP1981). There are 19 such fiber bundles and light mixers each with their own photomultiplier. There is a pressure contact between the light mixers and photomultipliers and a small air gap between the light mixers and the machined surfaces of the wavelength shifting fiber bundles.

The 25-Mil prototype is composed of 278 lead plates, 277 scintillator plates and 1141 wavelength shifting fibers. The wavelength shifting fibers are arranged in a nearly circular pattern about the center of the lead and scintillator plates. Since the density of holes in the plates is 1 hole per $0.16\ \text{cm}^2$, the approximate radius of the circle of wavelength shifting fibers is $\sqrt{1141 \times 0.16/\pi} = 7.62$ cm. All of these fibers are coupled, through a plastic hexagonal light mixer, to a ten-stage, 2-inch diameter, green-extended photomultiplier (Philips XP2201B). There is a pressure contact between the light mixer and photomultiplier and a small air gap between the light mixer and the machined surfaces of the wavelength shifting fiber bundles. The lead plates in this calorimeter are thinner than in the 35-Mil prototype to increase the sampling frequency. Since the thickness of the scintillator plates remains 1.0 mm, the sampling fraction is larger in this prototype. Both of these features help to improve the energy resolution.

	35-Mil	25-Mil
Number of lead plates	183	278
Number of Scintillator plates	182	277
Thickness of lead plates [mils]	35.0	25.0
Thickness of lead plates [mm]	0.889	0.635
Thickness of scintillator plates [mm]	1.00	1.00
Wavelength shifting fiber diameter [mm]	1.00	1.00
Center to Center fiber spacing [mm]	4.00	4.00
Effective radiation length (X_0) [cm]	1.27	1.53
Total instrumented length [cm]	34.5	45.4
Total instrumented length [X_0]	27.1	29.7
Number of readout fibers	1539	1141
Number of readout towers	19	1
Number of fibers/tower	91†	1141

Table 1: Characteristics of the 35-Mil and 25-Mil prototype calorimeters. The † is to indicate that 10 of the 91 fibers bundled in each readout tower do not pass through the body of the calorimeter but only serve to ensure a close-packed, hexagonal fiber bundle at the readout end.

3 Experimental Results

The results presented at this conference are preliminary. The test beam data was taken in the H4 beam at CERN by triggering on a multiple coincidence of small scintillation counters upstream from the detectors. By choosing which sets of the variously sized trigger counters define a trigger, we were able to easily control the particle flux and beamspot size. The "large beamspot" condition exposed an area of about 20×20 mm^2 while the "small beamspot" condition exposed an area of about 2×2 mm^2.

3.1 Energy Resolution

Using the "small beamspot" we performed an energy scan to measure the energy resolution of the 25-Mil prototype. Figure 2 shows the measured energy resolution plotted against $1/\sqrt{E}$. In the common parameterization of energy resolution, the stochastic term scales with energy as $a/\sqrt{E}$. The stochastic term appears as a straight line through the bottom right corner of this plot.

Depending on the parameterization of this data, we find a stochastic term of $5.5-6.0\%/\sqrt{E}$. The deviation from $1/\sqrt{E}$ scaling is $\sim 0.7\%$ at high energies. This deviation is called the constant term of the energy resolution and arises from sources like the finite momentum resolution of the beam, nonuniformities in response, shower leakage and the light attenuation in the fibers. We took data dedicated to measure the last of these contributions and find that the light attenuation in the wavelength shifting fibers can completely account for our measured constant term. We rotated the 25-Mil prototype so that the fiber axis was at 90° to the beam line. We then scanned with a 50 GeV electron beam along the side of the calorimeter. The response of the calorimeter was measured as a function of distance to the photomultiplier and shows that the attenuation length in the fibers is 1.6 m.

The longitudinal shower shape will fluctuate from event to event. Because of a finite attenuation length, identical particles that deposit energy close to the photomultiplier will give an

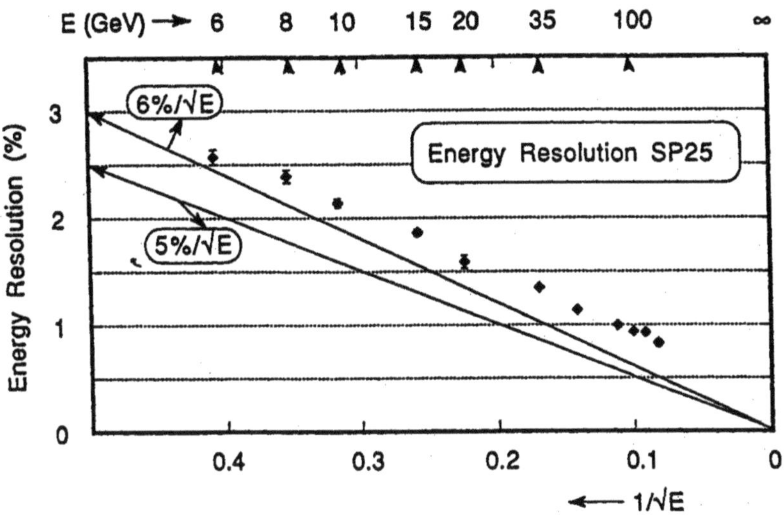

Figure 2: The energy resolution plotted as a function of the electron energy, which is plotted on a scale linear in $E^{-1/2}$. For comparison, the results for $\sigma/E = c/\sqrt{E}$, with c = 5% and 6% (the straight lines) are given as well. Data for the 25-Mil calorimeter exposed to the small beamspot.

enhanced signal compared with those that deposit energy far from the photomultiplier. We expect this effect to lead to signal variations on the order of $X_0/\lambda_{att} \sim 0.9\%$. This was confirmed with a detailed EGS4 study which included the measured attenuation length of the wavelength shifting fibers. At high energies, the dominant contribution to deviations from $1/\sqrt{E}$ scaling varies from $\sim 0.6\%$ to $\sim 0.9\%$. There is evidence from our EGS4 studies that the "constant" term in the energy resolution is not constant with energy. By mirroring the ends of the wavelength shifting fibers, we can expect to reduce the main contribution to the constant term of the energy resolution.

The calorimeter response is linear with incident electron energy. Results from the 25-Mil calorimeter show the average calorimeter signal per unit energy varies by less than 1% over the range from 6 to 150 GeV.

3.2 Spatial Uniformity

If the spatial uniformity of the calorimeter is not good, the excellent energy resolution that is measured is not very useful. To measure the spatial uniformity, we scanned the 35-Mil calorimeter with 50 GeV electrons over an area of 8×8 cm^2 using the large beamspot trigger. Wire chambers in the beam upstream from the calorimeter were used to measure the impact point on the front face of the calorimeter to within ± 0.3 mm. Figure 3 shows the average calorimeter signal (zero suppressed) as a function of the electron impact point in steps of 0.5 mm along the vertical axis. The underlying shape of this curve is due to incomplete shower containment. The "spikes" appear at precisely a pitch of 4.0 mm and correspond to the position of the wavelength shifting fibers. It is clear that light generated in the fibers contributes to the total signal at the few percent level.

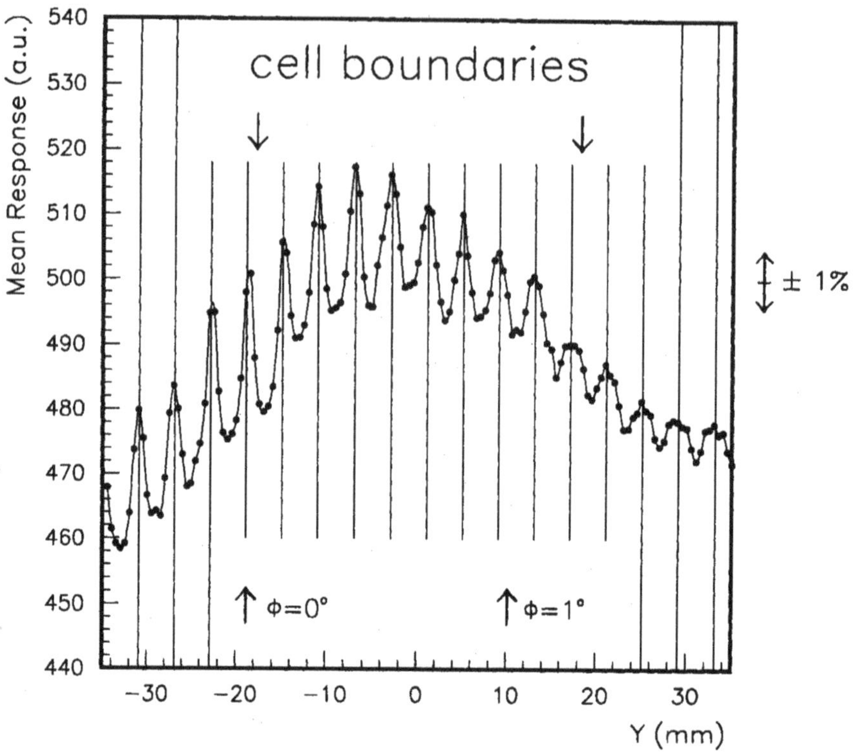

Figure 3: The average calorimeter response as a function of the impact point of the electrons. Results of a vertical scan in steps of 0.5 mm across the 35-Mil calorimeter.

The table on which the calorimeters were mounted could not be translated vertically. Because of this, a step in the vertical direction is really a change of angle, ϕ, with a rather large lever arm. A move of 1° in ϕ corresponds to a move of 29 mm in the vertical direction. It is seen that as the angle between the fiber axis and the beam direction increases, the effect due to light generation in the fiber decreases. This result is well understood and documented by the SPACAL collaboration[7]. They found, as do we, that the effect disappears at an angle of $\sim 3°$.

Also evident from figure 3 is the complete absence of edge effects in these calorimeters. Since there exists no physical boundary between the readout cells, there is no nonuniformity introduced when crossing or nearing the boundary. This feature allows for simple construction and a very hermetic device. It poses the question of crosstalk since only the attenuation length in the scintillator plates and the high fiber density prevents light from migrating laterally from where it was produced (see section 3.3). After unfolding the effect of incomplete shower containment, the overall spatial uniformity of the average calorimeter signal in 4×4 mm^2 cells is $\sim 0.4\%$.

Similar results from the 25-Mil calorimeter were found except the "spikes" at 4.0 mm pitch contribute only at the fraction of a percent level. We believe that the increased sampling frequency of the 25-Mil calorimeter compared to the 35-Mil is the cause of this reduced effect. We performed a grid scan of the 25-Mil prototype with 50 GeV electrons covering an area of 3×3 cm^2. The data was grouped into 81 units of 4×4 mm^2 and shows uniformity to $\pm 0.6\%$ over this area. It should be noted that this result has not been corrected for effects due to incomplete containment which we expect will further improve the spatial uniformity.

3.3 Crosstalk

The crosstalk is important in this type of calorimeter since there are no physical boundaries between the readout cells. If light generated inside the imaginary boundaries of one cell migrates to another cell, the effect is that the apparent lateral shower shape broadens and it becomes more difficult to distinguish individual, closely spaced showers.

We sent 225 GeV muons into the center of the 35-Mil calorimeter in order to measure the lateral migration of light into other readout cells. Muons provide an excellent measure of the crosstalk since they will deposit energy locally only along their path and, at this high energy, multiple scatter very little. With this data we find 78% of the calorimeter signal is deposited in the central readout cell (3.6×3.6 cm^2) and 20% is deposited in the ring of readout cells surrounding this central cell (see figure 1 for the cell structure). From this, we can deduce that the light travels approximately 2.3 cm before being absorbed in a wavelength shifting fiber.

In order to relate this measurement to lateral electromagnetic shower shapes, we ran detailed EGS4 simulations where we adjusted the average lateral distance light could travel in the calorimeter before being absorbed by a wavelength shifting fiber. We examine the fraction of signal deposited in the central cell and in the inner ring of cells surrounding this cell as a function of the average distance that the light is allowed to travel. When the light is not allowed to travel laterally (equivalent to ignoring any crosstalk effect) the simulations show that 80% and 12% of the signal from a 50 GeV electron shower will be deposited in the central cell and in the ring surrounding it, respectively. When the average distance the light is allowed to travel is set to 2.0 cm, the fractions become 68% and 25%, respectively. Our experimental data from 50 GeV electron showers indicate that 51% and 38% of the signal is deposited in the central cell and in the ring surrounding it, respectively. Using the results from the simulation, we conclude that the light is allowed to travel on average about 2.5 cm before being absorbed by a wavelength shifting fiber. This is in excellent agreement with the result obtained from the muon study of the crosstalk. The effect of the crosstalk can be incorporated into an effective Molière radius by increasing it from 2.6 cm to 3.6 cm.

3.4 Lightyield

The lightyield was determined by measuring the degradation of the energy resolution after installing neutral density filters in front of the photocathode of the photomultiplier. The filters, whose attenuation factor is precisely measured from the test beam data itself, reduce the number of photons that strike the photocathode. Fluctuations in the now reduced number of photoelectrons degraded the stochastic term of the energy resolution in a known way and have a large effect at low energy. We measured the energy resolution of the 25-Mil calorimeter with 6, 8, 10 and 15 GeV electrons after installing first one, and then a second filter, each with an attenuation factor of about 3. The results show that the 25-Mil calorimeter produces 1300 ± 90 photoelectrons per GeV deposited energy.

Thus, photoelectron statistics contribute $2.8\%/\sqrt{E}$ to the energy resolution implying that sampling fluctuations contribute about $5.0\%/\sqrt{E}$ to the stochastic term in good agreement with predictions made with EGS4.

4 Conclusions

We have demonstrated that with modest sampling fractions ($\sim$ 12%), and hence very compact calorimeters, one can obtain excellent energy resolutions by increasing the sampling frequency. The energy resolution has a stochastic term below $6\%/\sqrt{E}$, of which less than $3\%/\sqrt{E}$ comes from photostatistics. Deviations from $1/\sqrt{E}$ scaling at high energy are at maximum 0.7%. The attenuation length of the wavelength shifting fibers is the dominant contribution to this constant term. The spatial nonuniformity of response is shown to be better than $\pm 0.6\%$ over a 3×3 cm^2 area and the linearity of response is better than 1%. There is no evidence that any advantage is gained by optically isolating the readout towers. This leads to a very inexpensive and simple design which includes the ability to easily vary the calorimeter granularity and minimize its dead space.

5 Acknowledgments

We would like to thank our colleagues from RD36 and CMS, and in particular J. Bourotte, L. Dobrzinsky, M. Haguenauer, R. Tanaka and C. Seez, who made the described test beam possible. We are grateful to N. Doble, who provided us with particle beams of excellent quality. This project was carried out with financial support from the U.S. Department of Energy, the National Science Foundation and the Texas National Research Laboratory Commission.

6 References

1. M. Livan, V. Vercesi and R. Wigmans, *Scintillating-fibre Calorimetry*, and references therein, to be published in *Nucl. Instr. Meth.*
2. B. Loehr, et. al., *Nucl. Instr. Meth.* **A254** (1987) 26.
3. G. S. Atoyan, et. al., *Nucl. Instr. Meth.* **A320** (1992) 144.
4. W.R. Nelson, H. Hirayama and D.W.O. Rogers, SLAC-Report-165(1985).We used the EGS4 program with cut-off values of 100 keV for the kinetic energy of electrons and positrons and 10 keV for photons generated in the simulated shower development.
5. Kuraray Co. Ltd., Tokyo, Japan.
6. Doerun Co., Suite 400, 11885 Lackland Rd., St. Louis, MO 63146-4236.
7. D. Acosta, *et. al.*, *Nucl. Instr. Meth.* **A294** (1990) 193.

The STAR EM CALORIMETER
DESIGN
and
SMALL PROTOTYPE TEST RESULTS

David G. Underwood
High Energy Physics
Argonne National Laboratory
Argonne, Il, 60439

for the
STAR EMC Collaboration

ABSTRACT

The basis for several design features of The STAR Electromagnetic Calorimeter and Shower Maximum Detector is presented. This includes some of the tile-fiber optical design. We describe both the barrel and the end cap. Some preliminary analysis of electron acceptance vs pion rejection in test beam data is also discussed. Other information is given in the paper by G. Westfall in these proceedings.

INTRODUCTION

The EMC, Electromagnetic calorimeter in the STAR detector at RHIC, must function for a number of different physics goals.[1,2,3,4] The total em. energy and the distribution of em. energy must be measured for Heavy Ion events with thousands of soft pi-zeros. Jets must be reconstructed by using the EMC for neutral energy, and the TPC for charged track momentum and position. High energy electrons from W and Z must be measured. Direct photons must be separated from pi-zeros. The EMC also provides the high Pt trigger.[5]

In searching for the quark-gluon plasma, many possible signatures must be considered. The STAR collaboration has taken the approach of using large solid angle so that parameters can be measured within one event, rather than as an average over many events with different impact parameters and flucuations. One parameter is the temperature as measured from pion numbers and energies. Another approach is to distinguish the quenching of jets by a quark gluon plasma from effects of the partons passing through large nuclei. In searching for disordered chiral condensate, the ratio of charged to neutral energy may be measured over some substantial solid angle.

It is important to understand typical nucleus-nucleus collisions so that the import of unusual ones can be understood. A measurement of the gluon structure function in nuclei compared to that in nucleons is therefore needed. This measurement can be done with direct gamma plus jet measurements.

The design of the EMC is driven by covering large solid angle at mimimum cost with capability to do the above mentioned physics. This will be done by adapting what has become standard technology for other detectors such as CDF and SDC, along with some design features specific to STAR and recent advances in scintillator technology and possibly electronics.

DESIGN

The present design is to have a barrel calorimeter and one end cap calorimeter. The barrel will cover rapidity from -1 to +1 and 2pi in phi. The end cap will cover rapidity 1.05 to 2.0. The calorimeters are lead-scintillator sampling type. They are segmented in a projective tower structure and the initial cell size will be 0.1 in eta by 0.1 in phi. There are 1200 towers in the barrel and 540 in the end cap. There are provisions for segmentation in depth, possibly as an upgrade. The barrel will be about 18 radiation lengths deep and the end cap about 25. The resolution is expected to be about 16%/root(E), as measured in similar calorimeters and in our prototype. The scintillator tiles are read out by waveshifting fibers coupled to the edges. A fine structure shower maximum detector with crossed strips of 1.2 cm width will be used at a depth of 5 radiation lengths. The technology for this may be either wire/strip or scintillator with fiber readout. The wire/strip option is much more likely at the moment due to cost considerations.

These features of the design have been evaluated by Monte Carlo studies.
Monte Carlo studies show that using a shower maximum detector for event-by-event reconstruction of pi-zero to eliminate fake direct gammas gives much smaller error bars on a spin asymmetry measurement than use of a pre-shower detector.[6] The measurement involves the change in cross section for direct gamma production with beam spins aligned vs anti-aligned with both beams longitudinally polarized. The same general argument indicates that this is the most efficient way to do the direct gamma measurment to find the gluon distribution in nuclei.

For the charged to neutral ratio in nucleus-nucleus collisions, the studies show that the charged energy to neutral energy can be well measured at moderate to high A without reconstructing pi-zeros. For example, in Gold Gold collisions there will be on average about three gammas in each of the 1200 barrel towers, so that statistics will be adaquate in subsets of the 3.5 steradian covered.
The studies also show that the energy deposited in the EMC by charged pions can be subtracted out on average by use of counting the charged pions. This counting can be done in the TPC, or more quickly in a large number of scintillation counters used for triggering and time of flight.

The rapidity coverage provided by one end cap is needed to cover the appropriate xbj range in measurements of gluon distributions by means of direct gamma plus jet. The physics actually proposed so far does not require both end caps for kinematic coverage. The solid angle and rates would be improved with two of course. There is a conceptal design for the end cap and some finite element analysis for the mechanical aspects.[7]

Fiber - Tile Optics

One constraint on the optical scheme is the attenuation length of clear fiber. For this reason, the fibers are routed outward radially between the coil packages of the conventional magnet. They are also routed between the 30 iron flux return bars which run axially along the outside of the coils. This is the shortest route to phototubes which must be outside the magnetic field. This path is 2.5 meters on average.[8]

The tile-fiber optical scheme we have chosen for the barrel is shown in figures 1 and 2. It was outlined previously at Calor 92 Tests have been done at Argonne of various tile-fiber configurations using Bicron scintillator, Kuraray scintillator, and Russian injection molded scintillator for STAR and ATLAS tiles. These confirm that this design gives the best uniformity and light output that can be obtained without machining surface grooves as is done in the SDC and the CDF upgrade. A response map of a prototype tile twice as long as the tiles to be used and without amy masking corrections is shown in figure 3. The attenuation along the length of the tile due to attenuation in the fibers would be smaller in the actual tiles. The rise of 8 % near the edge can be flattened with a masking stripe on the wrapping.

Some comparative studies of ways to optically couple the fibers to the tiles are given in table 1. In all these tests, the size of the tile was approximately 24 cm between the fibers. The same pair of waveshifter fibers were used. These shifters were aluminized on one end and spliced to 1.5 m of clear fiber on the other end. The fibers were BCF-91a , 1 mm diameter. The tiles were 4 mm thick.

Table 1

Scintillator type	Configuration	photoelectrons per mip
Bicron Styrene	groove in square edges	0.9 pe
Bicron Styrene	parallel grooves in top surface, 5 mm from edges	1.0 pe
SCSN38	fibers on edges, parabolic wrapping , scint. hand polished	1.75 pe
SCSN38	" , " , diamond milled edges	1.65 pe
Injection molded (produced 1993)	fibers in edge grooves, parabolic scintillator shape	0.5 pe

With our prefered geometry, we have also studied a few cases of different fiber diameter, etc. This is shown in table 2

Table 2

Configuration	result
Standard.. 4 mm tile of SCSN38, one 1 mm fiber each of 2 edges	reference
Same tile but two 1 mm fibers each of 2 edges	30% more light
3 mm tile, one 1.5 mm fiber each of 2 edges	8% more light
6 mm tile, one 1.5 mm fiber each of 2 edges	40% more light

The gains are far less than proportional to the increases in material. We also prefer a maximum of 1 mm diameter fibers because of the bend radii involved in installation of the calorimeter.

Test Beam Results

Some results have been presented in the previous talk by Gary Westfall. The pion rejection tests and shower max. tests are described here. At high energy good pi vs electron discrimination is needed for W events. At low energy good gamma to charged pion discrimination is needed to reconstruct the pi-zero background to direct gammas. In the test beam electrons and pions of 0.3, 0.5, 1, 2, 4, 6, and 8 GeV/c were available. The electrons were tagged by one Cherenkov counter.

As shown in figure...a, about 97 % of pions are rejected at 8 GeV/c by using the E vs P cut. Events with measured energy two sigma below the mean energy in the calorimeter are rejected. This kind of cut is much less useful below 1 GeV/c. Another kind of cut is to use the energy deposition in the first 10 layers of the calorimeter vs that in the last 10 layers. The pion rejection for the differrence over the sum less than -.5 is shown in figure ...b. This rejects 60 to 70% of the pions below 2 GeV/c, but rejects most of the electrons at high energy. The use of both methods combined is shown in figure ...c. An optimum algorithm might make a transition to the E vs P method at a few GeV/c.

We obtained good spatial resolution and low background with a shower maximum detector utilizing strips of scintillator. There were strips in both directions transverse to the beam. The strips were 1.2 cm wide and 5 mm thick and were read out by waveshifter fibers to PHEU 85 phototubes. In one particular event with an 8 GeV electron, a strip with a direct hit had a pulse height 16 times the adjacent strips and the noise above pedestal in another strip was about 3% of the peak. This data and data with a wire/strip shower max is still being analyzed.

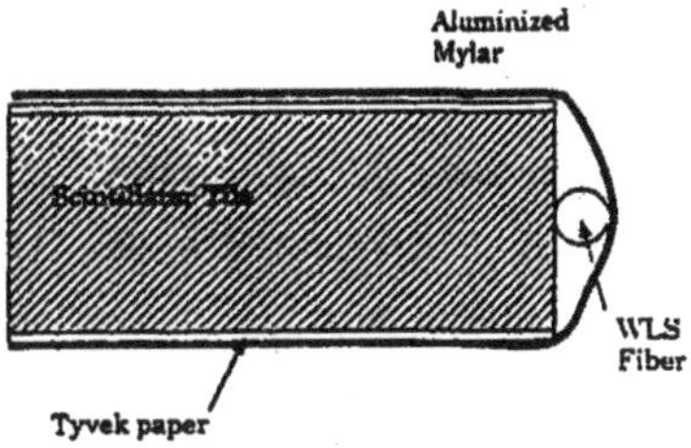

Figure 1 Configuration of a double-length tile from the small prototype

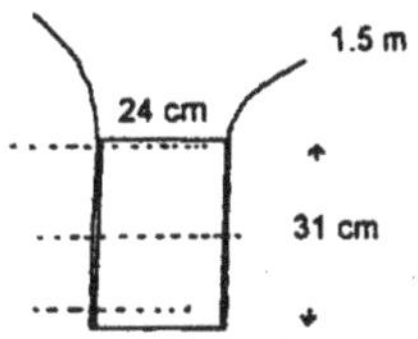

Figure 2 Response map of the double length tile without masking.

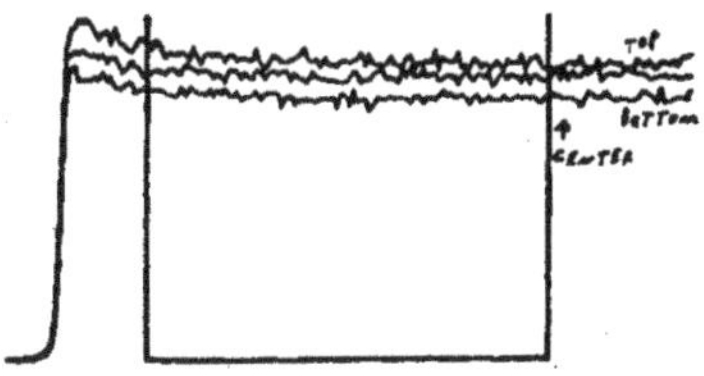

Figure 3 Light coupling to the waveshifter and wrapping of a tile.

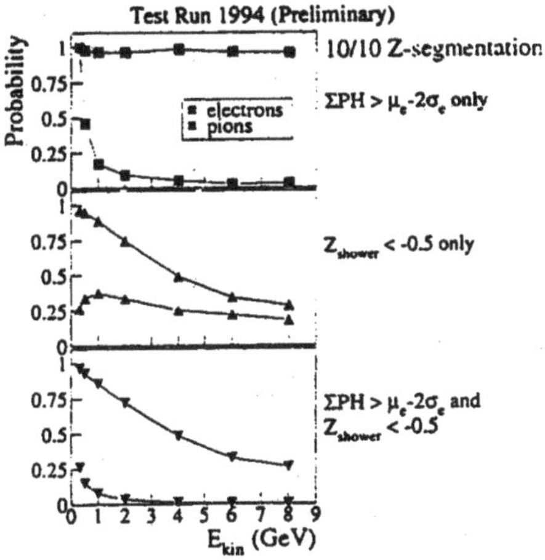

Figure 4 a) Pion rejection and electron acceptance using a) E/p, b) depth segmentation, c) both.

References

1) The Electromagnetic Calorimeter for the Solenoidal Tracker at RHIC, A Conceptual Design Report. The STAR EMC Collaboration, LBL Pub. 5380, (1993)

2) The STAR Electromagnetic Calorimeter, G. Westfall, Calorimetry in High Energy Physics, These proceedings.

3) The Electromagnetic Calorimeter for the STAR Experiment at RHIC, D. Underwood, Proceedings of the Third International Conference on Calorimetry in High Energy Physics, Corpus Christi, P. Hale and J Siegrist, editors, World Scientific, (1993), 673

4) Proposal on Spin Physics Using the RHIC Polarized Collider, update, RHIC Spin Collaboration, Sept, 1993

5) A Possible Level 0 Trigger Scheme for the STAR EMC, D. Underwood, Star Note 180, ANL-HEP-TR-94-4, http://rsgi01.rhic.bnl.gov/star/starlib/doc/www/sno/ice/sn0180.html

6) Estimates of Rates and Errors for Measurements of Direct gamma and Direct gamma plus Jet Production by Polarized Protons at RHIC, M. E. Beddo et al. Star Note 77, ANL-HEP-TR-92-59 http://rsgi01.rhic.bnl.gov/star/starlib/doc/www/sno/ice/sn0077.html

7) A Conceptual Design for the STAR Endcap Electromagnetic Calorimeter, E. Bielick et al. ANL-HEP-TR-93-62 and STAR Note 120 (1993)

8) Conceptual Design for the STAR Barrel Electromagnetic Calorimeter Support Rings, E. Bielick et al, ANL-HEP-TR-94-13, and STAR Note 154 (1994)

THE STAR ELECTROMAGNETIC CALORIMETER

GARY D. WESTFALL
National Superconducting Cyclotron Laboratory
and Department of Physics and Astronomy
Michigan State University
East Lansing, Michigan 48824-1321 USA

for the

STAR EMC COLLABORATION

ABSTRACT

The STAR electromagnetic calorimeter (EMC) will give STAR the ability to study the early, dense stages of Au+Au collisions at RHIC as well as providing access to spin physics studies in polarized p+p reactions. The STAR EMC is a lead-scintillator sampling calorimeter in the form of a barrel and an endcap. The light is collected using wavelength-shifting fibers spliced to clear fibers. A small prototype electromagnetic calorimeter was constructed and tested using beams of $e^{\pm}$, $\pi^{\pm}$, $\mu^{\pm}$ at momenta ranging from 0.3 to 8.0 GeV/c. Measurements were made to determine the energy calibration and resolution, the lateral and longitudinal profiles of electron and pion showers, the performance for non-normal particle incidence, and the performance of two different types of shower maximum detectors.

1. Overview

1.1. RHIC Physics Environment

The Relativistic Heavy Ion Collider (RHIC) presents unique challenges to calorimetry. RHIC will provide colliding beams of 100 GeV/nucleon Au + 100 GeV/nucleon Au at a luminosity of 2×10^{26} cm^{-2}s^{-1} which will provide an interaction rate of 10^3 collisions/s. The average transverse momentum, $< p_t >$, of the emitted particles is expected to be about 400 MeV/c. RHIC will also provide colliding beams of 250 GeV + 250 GeV polarized protons at a luminosity of 1.4×10^{31} cm^{-2}s^{-1} leading to an interaction rate of 6×10^5 collisions/s. The crossing time of the bunches will be 220 ns. A planned upgrade will lead to a luminosity of 2×10^{32} cm^{-2}s^{-1} with a crossing time of 110 ns. RHIC is scheduled for completion in mid-1999.

1.2. STAR

The Solenoidal Tracker at RHIC (STAR)[1] is a broad range detector composed of a large, room temperature solenoid with a field of 0.5 T, a silicon vertex tracker (SVT), a time projection chamber (TPC), a central trigger barrel/time-of-flight array (CTB/TOF), an electromagnetic calorimeter (EMC), and external TPCs (XTPCs) along with several beam definition and triggering counters. The magnet coils and return iron are designed to allow passage of the optical fibers of the EMC so that the phototubes can be located outside of the STAR magnet. A schematic drawing of STAR in cross section is shown in Fig. 1. The STAR EMC[2] is a lead-scintillator sampling calorimeter composed of a barrel and an endcap.

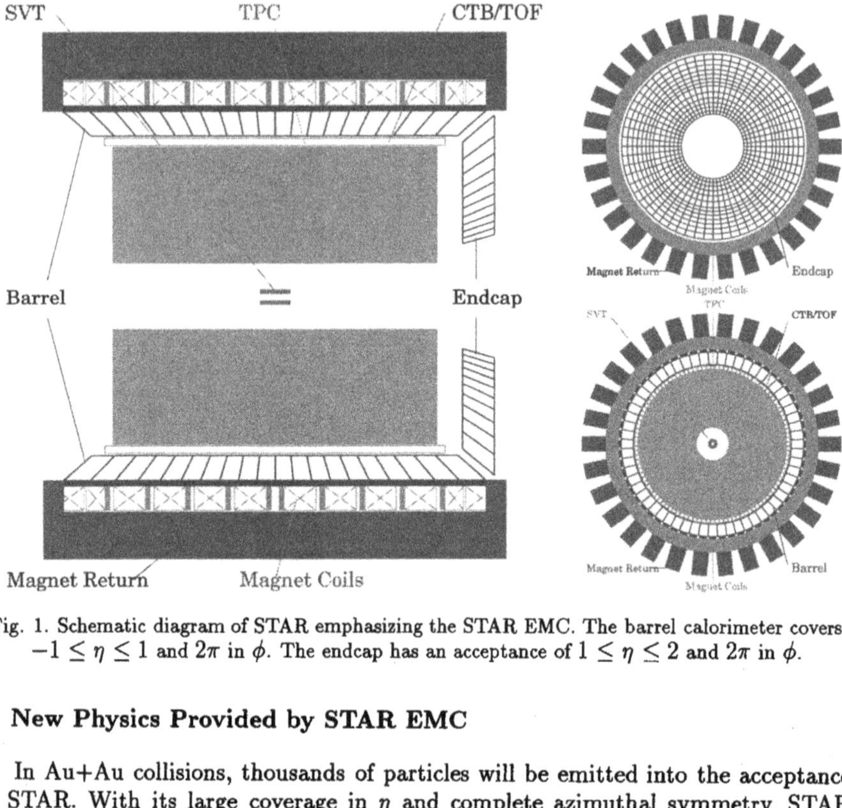

Fig. 1. Schematic diagram of STAR emphasizing the STAR EMC. The barrel calorimeter covers $-1 \leq \eta \leq 1$ and 2π in ϕ. The endcap has an acceptance of $1 \leq \eta \leq 2$ and 2π in ϕ.

2. New Physics Provided by STAR EMC

In Au+Au collisions, thousands of particles will be emitted into the acceptance of STAR. With its large coverage in η and complete azimuthal symmetry, STAR will characterize the hadronic final state of these collisions. The EMC will provide STAR a new dimension by allowing the study of the earliest, hottest, and most dense stage of the collisions. Another contribution of the STAR EMC is its measurement of parameters related to the hadronic final state such as the neutral energy which is a large fraction of the emitted energy and cannot be obtained in any other way. The STAR EMC will also make possible the study of polarized p+p reactions at $\sqrt{s}$ up to 500 GeV which will allow the study of spin as a fundamental observable in QCD.

2.1. Quark-Gluon Plasma Signatures

The EMC will enhance the ability of STAR to detect the quark-gluon plasma (QGP). One example involves the measurement of the total transverse energy versus the rapidity density of charged particles. In Fig. 2, the predicted effect of the QGP on these observables is shown.[3]

One observable that could be related to the QGP is the ratio of the observed electromagnetic energy to the charged particle multiplicity, E_t^{EM}/N_{ch}. A deviation of this ratio from its "normal" value (defined either as that of the majority of events, or as that from events at some value of A and $\sqrt{s}$ thought to be below the threshold for QGP formation) could also represent a change in the isospin ratio among the pions. Such unusual ratios between charged particles and neutral energy have been observed in high energy p+A events.[4] These events may signal phenomena such as disoriented chiral condensates which have also been suggested as unique signatures of the possible transition to a QGP.[5]

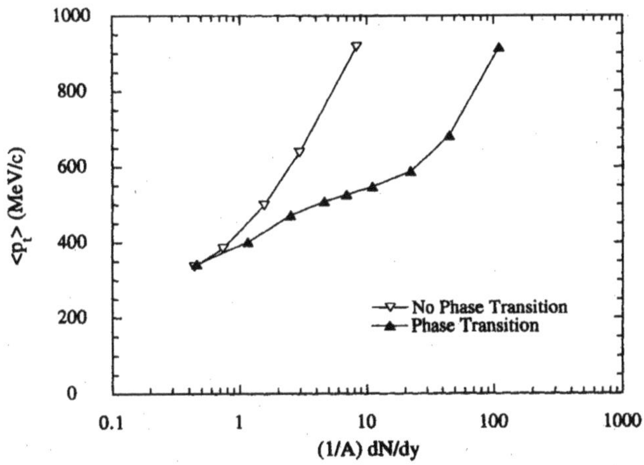

Fig. 2. The predictions for average transverse momentum, $< p_t >$, versus the normalized multiplicity, (1/A)dN/dy, with and without a first order phase transition.[3]

2.2. Spin Physics Program

The RHIC Spin Collaboration has planned an extensive physics program using the polarized proton beams. One class of experiments includes the study of the gluon spin distribution of the proton which will be studied through single jet production, di-jet production, direct-γ production, and direct-γ production with away-side jets detection. Other studies will include asymmetries in inclusive $W^{\pm}$ and Z^0 production and transverse quark structure functions of the proton. These experiments will investigate spin as a fundamental observable in QCD.

2.3. Trigger Capabilities of the EMC for STAR

The EMC will provide vital triggering for STAR including event topology, high p_t phenomena, and rare phenomena. The EMC will give STAR the ability to trigger on high p_t π^os, high p_t direct-γs, and the neutral component of jets. In p+Au and Au+Au collisions these triggers will allow access to observables such as the gluon structure function which leads to a more complete understanding of the results for the hadronic final state. In polarized p+p reactions, these triggers are vital to selecting the hard parton scattering events describable by perturbative QCD.

3. Basic Parameters of the STAR EMC

The STAR EMC consists of a barrel calorimeter and an endcap calorimeter. The barrel calorimeter covers $-1 \leq \eta \leq 1$ while the endcap has an acceptance of $1 \leq \eta \leq 2$. Both systems cover 2π in azimuth. The segmentation of the barrel calorimeter is $(\Delta\eta, \Delta\phi) = (0.1, 0.1)$ but can be upgraded to $(0.05, 0.1)$ simply by adding more phototubes and electronics. The endcap has a segmentation of $(\Delta\eta, \Delta\phi) = (0.05, 0.1)$ for $1 \leq \eta \leq 1.4$ and $(0.1, 0.1)$ for $1.4 \leq \eta \leq 2$.

The barrel calorimeter is composed of 19 layers of 0.5 mm thick lead plates interleaved with 20 layers of 0.4 mm thick scintillator tiles which corresponds to 18 X_0. There are 1200 pseudoprojective towers arranged in 120 modules each covering $0 \leq |\eta| \leq 1$ and $\Delta\phi = 6°$.

Each module can be installed separately allowing the EMC to be implemented in stages. In a given tower, each of the 20 layers of scintillator is made-up of 2 separate tiles read-out with wavelength shifting fibers spliced to clear fibers. The wavelength shifting fibers are aluminized on one end and are coupled to the edge of the tile with aluminized mylar. This tile/fiber technique allows future upgrades such as η and depth segmentation. A schematic drawing of a side-view of one of the barrel calorimeter modules is shown in Fig. 3.

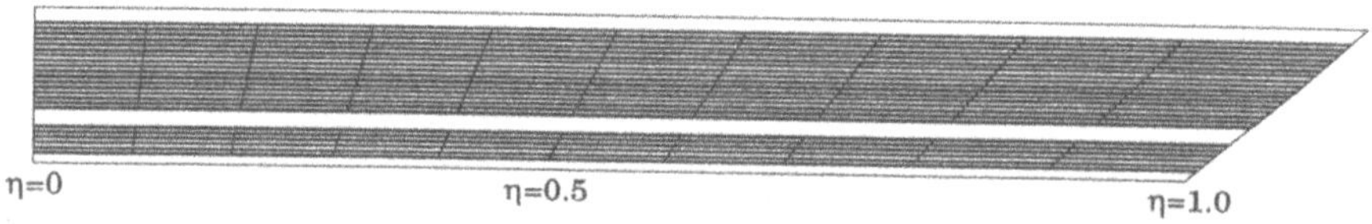

Fig. 3. Schematic side-view of one of the 120 modules of the barrel calorimeter showing the 20 layers of lead and scintillator and space for the shower maximum detectors.

The endcap calorimeter is constructed of 25 layers lead and 26 layers of scintillator of the same thickness as the barrel and corresponds to 23 X_0. The endcap is composed of 540 pseudoprojective towers arranged in two semicircular modules each covering $1 \leq \eta \leq 2$ and $\Delta\phi = 180°$.

Shower maximum detectors will be installed at a depth of 5 X_0 to provide a finer spatial resolution for the separation of direct-γs and those from π^0 decays. Two designs are under consideration including scintillating strips read-out with wavelength shifting fibers and wire/strip counters. Both designs would incorporate strips of about 1.4 cm width. The scintillator strip design is more straight-forward but relatively expensive while the wire/strip counters are less expensive but have technical challenges, which includes the placement of inaccessible amplifiers inside the calorimeter modules.

4. Test Beam Results

To test the concepts of the STAR EMC, a small prototype electromagnetic calorimeter (SPEMC) composed of 6 towers was constructed. There were four towers similar to those at $\eta \approx 0$ and two towers similar to those at $\eta \approx 1$. Like the barrel calorimeter, there were 20 layers of 0.5 mm thick lead and 0.4 mm thick scintillator (Kuraray SCSN 38). Each tile was read-out with two 1 mm diameter green wavelength shifting fibers (Bicron BCF-91A) coupled to 1 mm diameter clear fibers. Both types of fibers were single clad. The ends of the wavelength-shifting fibers were aluminized to minimize light loss. The fiber routing was reconfigurable to allow testing of depth segmentation. Hamamatsu R-580-A and R-580-17 photomultiplier tubes were used to detect the light. An assembly drawing and photograph of the SPEMC are shown in Fig. 4. Visible in this figure is the space for the two types of prototype shower maximum detectors.

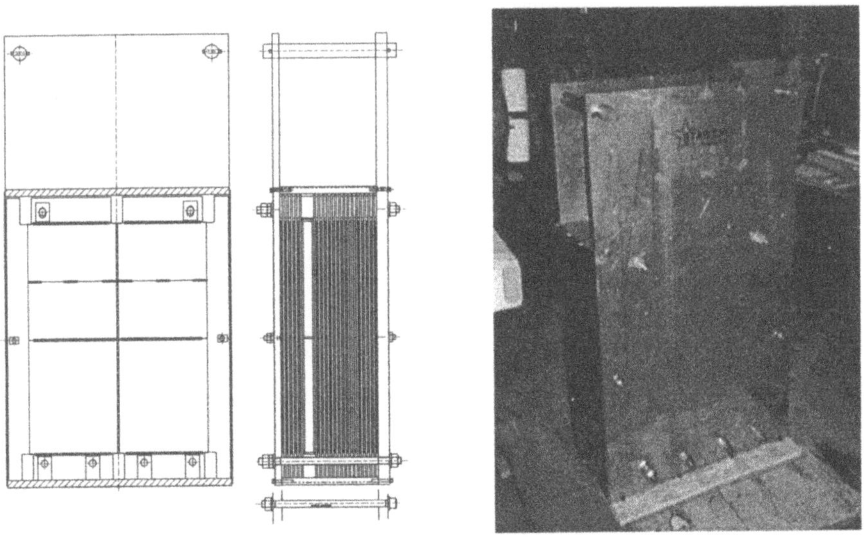

Fig. 4. Assembly drawing of the SPEMC showing front and side views. The photograph shows the SPEMC at the AGS test beam facility.

The AGS Test Beam Facility provided beams of $e^{\pm}, \pi^{\pm}, \mu^{\pm}$ at momenta of 0.3, 0.5, 1.0, 1.5, 2.0, 3.0, 4.0, 6.0, and 8.0 GeV/c. The measurements were carried out in July, 1994. Five different configurations of the SPEMC were extensively studied: normal incidence and no depth segmentation; 15° rotation and no depth segmentation; 15° rotation and a 10 layer/10 layer depth segmentation; 15° rotation and a 5 layer/15 layer depth segmentation; and normal incidence with a 5 layer/15 layer depth segmentation. Position scans were taken across the face of the calorimeter in all five configurations and at all available momenta. Data-taking benefitted greatly from the high speed acquisition system consisting of a five node transputer farm capable of recording 40,000 events/second to 8 mm tape. Beam definition consisted of two 32 element beam hodoscopes, two Cerenkov counters, two timing scintillators, and a muon tag.

The response of the calorimeter to electrons and pions+muons is shown in Fig. 5 for four incident momenta. The centroids of the these peaks are shown in Fig. 5 for all seven incident energies. The response of the SPEMC is very linear. The ratio of the widths of these peaks to the centroids is also shown in Fig. 5 scaled by $\sqrt{E}$ where E is the electron energy. The results for $E > 1$ GeV are approximately 16% $/\sqrt{E}$ which is above the sampling limit of 14% but below the required 20% which is needed to fulfill the physics goals of the STAR EMC. Other test beam results from the SPEMC will be presented in the paper by David Underwood.

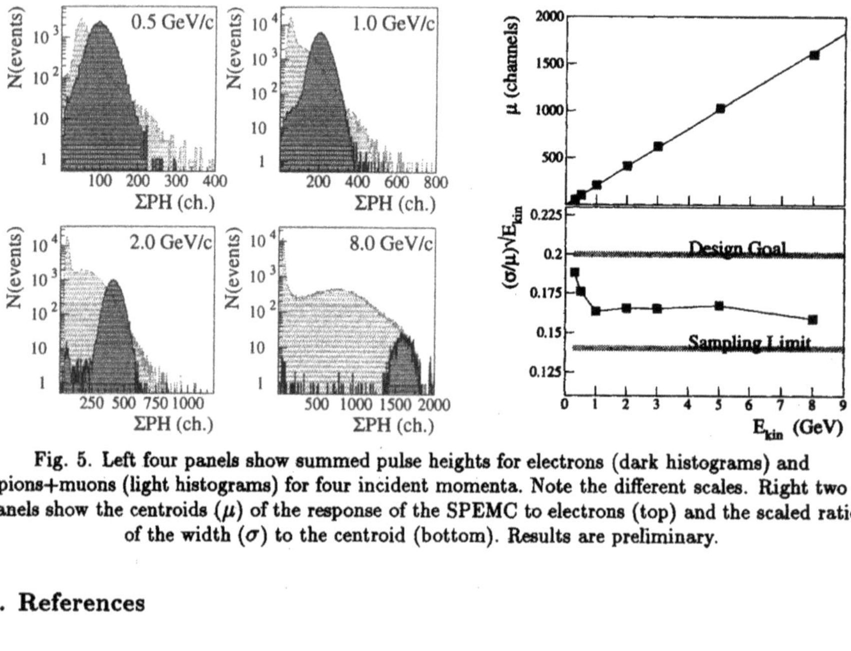

Fig. 5. Left four panels show summed pulse heights for electrons (dark histograms) and pions+muons (light histograms) for four incident momenta. Note the different scales. Right two panels show the centroids (μ) of the response of the SPEMC to electrons (top) and the scaled ratio of the width (σ) to the centroid (bottom). Results are preliminary.

5. References

1. Conceptual Design Report for the Solenoidal Tracker at RHIC (STAR), PUB-5347, (1992).
2. The Electromagnetic Calorimeter for the Solenoidal Tracker at RHIC, A Conceptual Design Report, PUB-5380, (1993).
3. M. Kataja, P.V. Ruuskanen, L.D. McLerran, H. von Gersdorff, Phys. Rev. **D34**, (1986) 2755.
4. Y. Takahashi and S. Duke, Nucl. Phys. **A461** (1987) 263c.
5. J.D. Bjorken, Acta Physica Polonica **B23** (1992) 637.

PHENIX PbSc Electromagnetic Calorimeter - results of test beam studies

E. Kistenev, G. David, S. White, C. Woody
Brookhaven National Laboratory, Upton NY 11973, USA

A. Basilevsky, A. Denisov, A. Durum, V. Kochetkov, Yu. Mikhailov, V. Semenov
Institute for High Energy Physics, Protvino, 142284, Russia

Abstract

Design and construction details are presented of a lead-scintillator electromagnetic calorimeter with penetrating wave-shifting fibers readout. Results are reported from a recent tests of a 256 channel calorimeter prototype in the electron and hadron beams from AGS at BNL. The energy linearity, energy, position and timing resolution are studied in the monentum range from 0.3 GeV/c to 4 GeV/c.

1. Introduction

The role of precision calorimetry in identifying particles and measuring their energies in high energy particle physics is well understood. A new field of applications for calorimetric devices opened recently following the growth of interest to the study of thermalized states in head-on collisions of heavy ions. Enormous multiplicities of the secondary particles expected to be a distinct feature of those processes create a new experimental environment and set a very stringent criteria which should be applied choosing adequate experimental technique. The search for a solutions which is able to provide the necessary performance, be easy to industrialize and still satisfy the tight budget requirements became a major issue for the calorimetry related R&D projects in recent years. This paper presents our answer to the challenge. It addresses the problem of designing the calorimeter able to provide a large area coverage at a competitive cost with a performance suitable for measuring electromagnetic energies in GeV range and an excellent multi hit handling capability.

2. PHENIX experiment at RHIC

Fig. 1 shows a three-dimensional cutaway view of the PHENIX detector, showing the locations of the various subsystems [1].

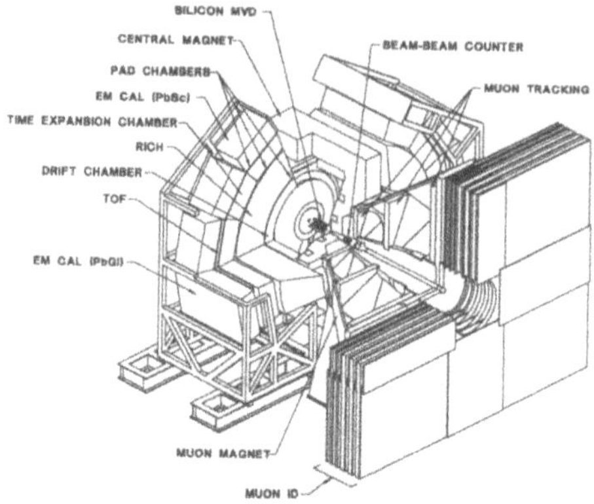

Figure 1: PHENIX Detector at RHIC

The calorimetry coverage is limited to the central detector arms which are designed to provide primarily for the photons and leptons measurement. The calorimeter acceptance measured in units of rapidity and azimuthal angle is equal to $\Delta\eta = \pm 0.35$, $\delta\phi = 180°$. It covers a total surface area of about 64 m^2 at a distance of 5 m from the crossing point. It employs two different technology: Sandwiched Lead-Scintillator plates for most of the coverage (75%) and Lead - Glass cherenkov counters for the rest of the acceptance. A cell size of $52.5 \times 52.5mm^2$ corresponding to the granularity of $\Delta\eta \times \delta\phi = 0.01 \times 0.01$ at $\eta = 0$ was chosen for PbSc calorimeter, finer ($40 \times 40mm^2$) PbGlass detector from WA98 [2] experiment is used everywhere else. This conservative choice of granularity result in low occupancy to carry out the desired measurements, even in the highest multiplicity Au + Au events at RHIC.

3. Design and prototyping

The EMCal design is based on planar sectors of individual modules that are first assembled into supermodules. Each of the two central arms of PHENIX comprises four sectors (octants) of EMCal modules (PbSc counters in 6 sectors, PbGlass counters in 2 sectors). The lead-scintillator modules are based penetrating wave-shifter fiber readout.

Standard calorimeter modules, shown schematically in Fig. 2, have transverse dimensions of 10.5×10.5 cm^2 and a depth of $18X_o$. Modules are constructed from alternating layers of 1.5-mm thick lead, white reflecting TYVEK paper, and 4.0-mm thick injection molded polystyrene-based

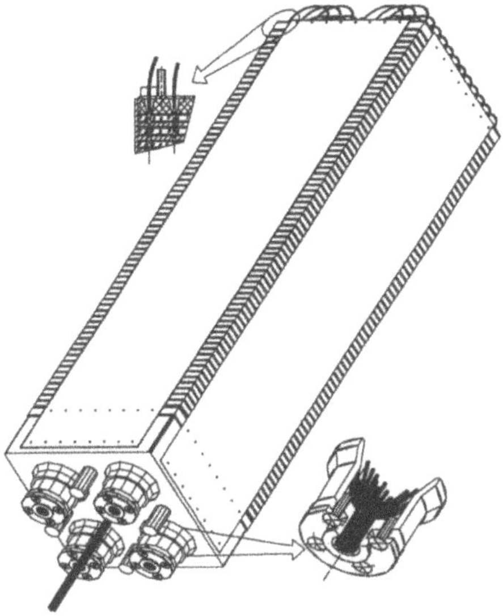

Figure 2: PHENIX PbSc EMCal module

scintillator tiles (1.5% PT / 0.01% POPOP). The injection molding device and lead-and-paper stamping devices incorporate a 1.2 mm diameter hole pattern on a 8.75-mm grid. The holes allow the WLS readout fibers (Bicron BCF-99-29a with 1.0-mm diameter, 2.8-ns decay time, and 500-nm peak emission) to be inserted into the assembled modules.

An important feature of this light collection technique is a photon yield which rivals more conventional techniques (such as WLS plates and edge-mounted WLS fibers). Due to the large number of fiber-scintillator couplings per volume occupied by the shower, it is rather insensitive to the variations in the fiber-scintillator coupling efficiency and to the nonuniformity of the light yield in the scintillator plates. The short emission center to WLS fiber light path eases the requirements on UV light attenuation in the scintillator and allows the consideration of alternative scintillator production technologies.

A wide range of tests have been performed with individual modules both in the laboratory, and in the beams from AGS, TRIUMF and U70 accelerator at IHEP (Protvino, Russian Federation). A recently constructed 256 channel supermodule was tested at the AGS in the spring of 1994. Results of the AGS tests are presented below.

4. Energy measurements

The prototype calorimeter was exposed to particle beams with mo-

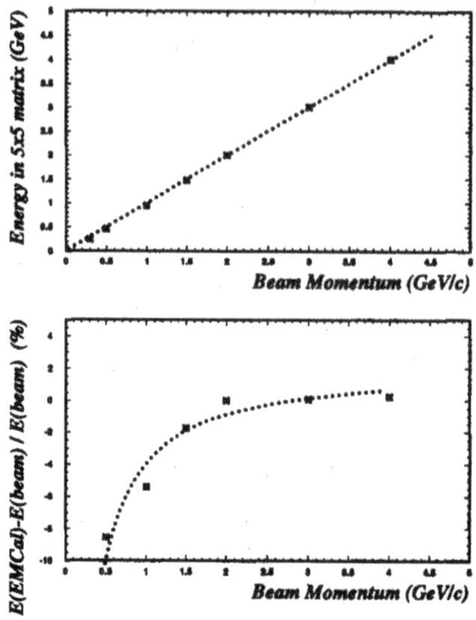

Figure 3: Energy measured in PbSc calorimeter (matrix of 5x5 counters) as function of the beam momentum

menta in the range 0.3 - 4 GeV/c. The calorimeter was moved across the beam in steps, the position of the impact point (when required) was deduced from the known stage coordinates and hit point measured in the beam wire chamber. The correlation plot between beam energy and the energy measured in the calorimeter is presented in Fig. 3, top part. The normalization was made at 3 GeV. The bottom part of the figure details observed deviations from linearity. They are negligible above 2 GeV/c while at lower energies the calorimeter certainly sees less energy then expected.

The finite light attenuation length (the average effective attenuation length of 200 cm measured exciting individual tiles with UV light) in the WS fibers is responsible for approximately 5% out of the the 8.5% missing energy at 0.5 GeV/c. The rest is likely due to: (a) too coarse a sampling for measuring low energy showers [3]; (b) energy loss in the scatterers downstream of momentum slit (10% of Lrad); (c) energy leakage through the calorimeter front face (albedo) [4].

The calorimeter energy resolution (corrected for the noise contribution) follows well the 1/sqrt(E) function form (see Fig. 4). The 8% value for the stochastic term is close to expected resolution from sampling as predicted by GEANT, the origin of a 1.4% constant term is discussed below. Major factors contributing to the constant term are the tower boundaries, hot spots at fiber positions and shower depth fluctuations. Simulation shows

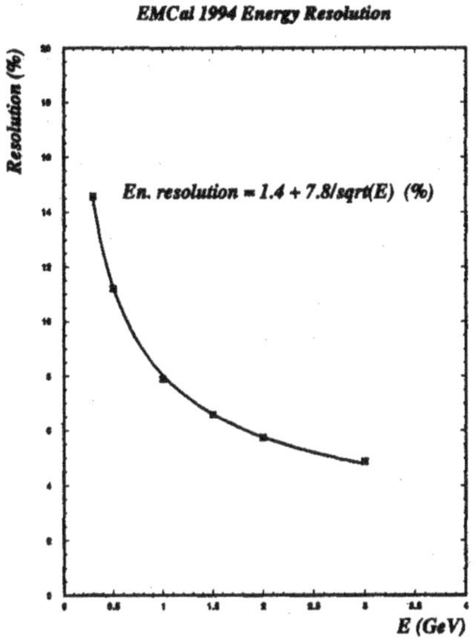

Figure 4: Dependence of the calorimeter energy resolution on the electron momentum

that 200 cm light attenuation length of our WLS fibers will result in a 0.5% contribution to the constant term. To study the importance of the lateral response nonuniformity we exposed different areas of the calorimeter to 2 GeV/c electrons measuring impact point coordinates in a single plane WC (given the beam divergence the estimated position resolution was about 3 mm). Fig. 5 illustrates the results of scanning the calorimeter along vertical line. Data for two adjacent towers from two geometricaly identical adjacent modules built of tiles differing by the type of edge treatment are shown separately. The response variation is ±2.5% for tiles plated with Al on the edges (technologically simple and cost effective solution) and ±5% for the tower wrapped with aluminized mylar. Large 5% nonuniformity in the "standard" mylar wrapped towers which were used for most of the coverage is a major effect responsible for 1.8% measured constant term.

The picture reveals the presence of a local nonuniformities associated with the hot spots at the points where fibers penetrate calorimeter depth. Unfortunately limited precision of the impact position measurements in our setup allows a qualitative estimate only: the induced nonuniformity is of the order of 2%.

A smooth polynomial approximation of the response map is shown in the right part of Fig. 6 ("standard" towers, 90° and 80° angles of incidence). Measured energy reaches maximum value in the center of the tower and drops down by nearly 8% at the corners. Tilting the calorimeter by 10° helps to extend the flat region in the center of the tower towards

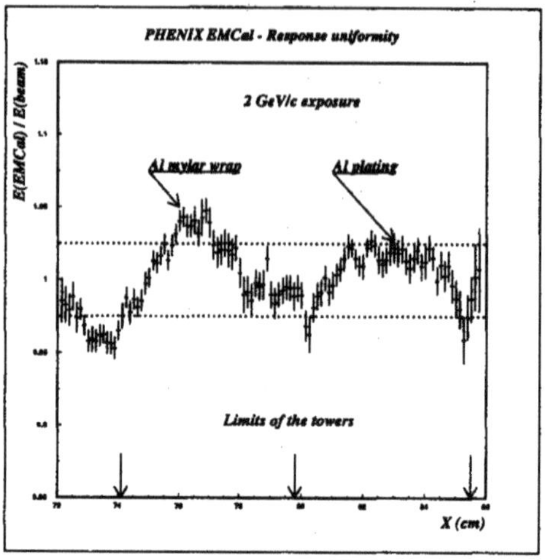

Figure 5: Average energy seen by the calorimeter exposed to 2 GeV/c electrons as function of the impact coordinate

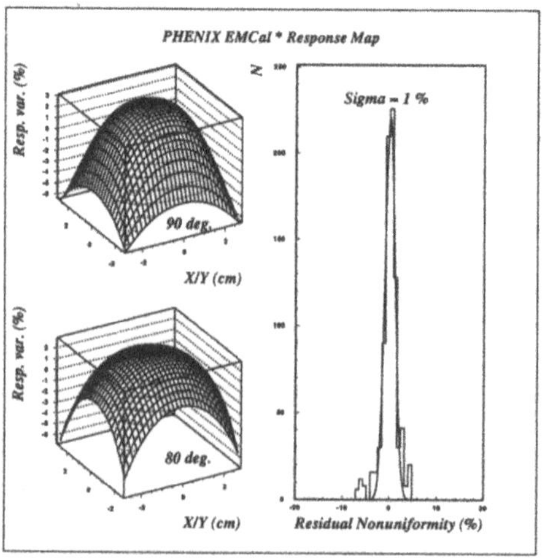

Figure 6: EMCal response map (left). Residual deviations of the corrected energy in the calorimeter from beam energy (right).

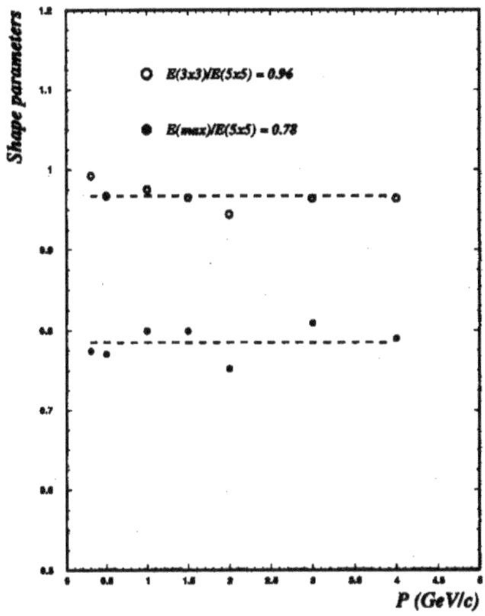

Figure 7: Dependence of the shower shape parameters $R1 = E_{max}/E_{5\times5}$ and $R2 = E_{3x3}/E_{5\times5}$ on the beam momentum

tower edges but solves the problem at the corners only partially.

A simple procedure is sufficient to reduce the constant term down to respectable 1% value. The histogram on the right in Fig. 6 shows the deviations of the data points from a beam energy (orthogonal impact) when weighted using inversed profile function. It has a $\sigma = 1\%$ which is due to the convolution of statistical uncertainties of individual measurements ($\sim 0.5\%$) and residual nonuniformities around fibers. Estimated residual contribution to the constant term value $\sim 0.5\%$.

A subset of showers with impact point coordinates within 0.5 cm from the tower center was used to study the energy sharing between counters. Momentum dependence of the ratios between energies measured in the central tower and summed over the 3x3 and 5x5 tower matrices around the central tower are shown in Fig. 7. Data are consistent with shower shape being energy independent (on a scale set by our tower size). The centrally hit tower contains 77% of shower energy corresponding to a Molier radius of $\sim 3cm$.

Using the measured energy resolution and line shape for incident hadrons (primarily $\pi^{\pm}$), hadron rejection was calculated for energy deposit cuts corresponding to 95% and 90% electron efficiency (Fig. 8). The hadron rejection ranges from at least 20:1 to 100:1 for p_T of 0.5–2.0 GeV/c. Further discrimination may be obtained by using shower-shape information (primarily useful for higher energies) or by time-of-flight information (for

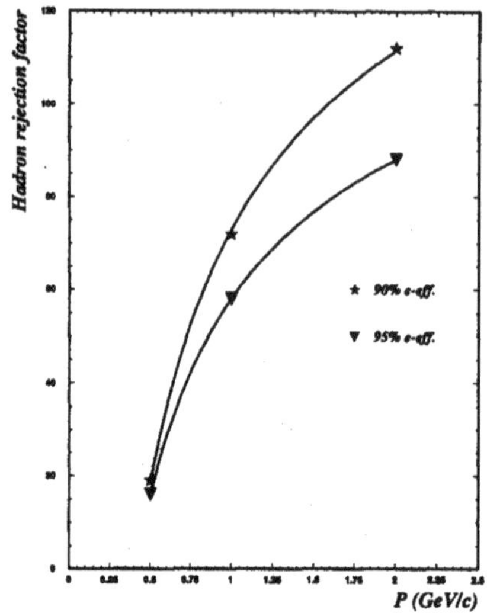

Figure 8: EMCal hadron rejection factor (via E/P matching) as function of the particle momentum

lower energies).

5. Position measurements

It is common to use the shower center of gravity computed weighting the coordinates of the tower centers by the corresponding energies [5]. A drawback of this approach is a classical S-shape of the 'measured' vs 'true' coordinates correlation plot. In our work we followed the approach of [6] and replaced the linear in the energy weights with the weights computed using the following expression:

$$w_i = max\left\{0, \left[W_0 + ln\left(\frac{E_i}{E_{tot}}\right)\right]\right\}, \qquad (1)$$

where E_i is the energy deposited in the tower i, E_{tot} is the total energy of the shower and W_0 is a free dimensionless parameter. For the purposes of current analyses we used $W_0 = 4.0$ similar to /3/ without attempting further optimization. Fig. 9 (upper right corner) shows the correlation plot between the average position calculated using this formulae and the measuments done in the wire chamber.

Points on the x-axis with coordinates -5.7 cm and 0 cm correspond to the centers of the two adjacent towers. The dependence is nearly perfect straight line with unity slope. The histogram of the differences between

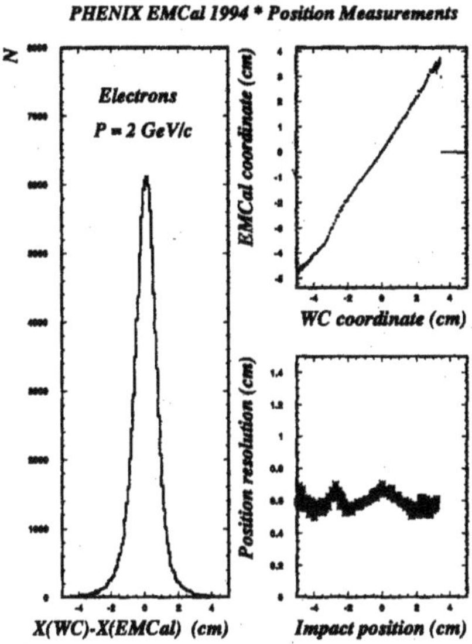

Figure 9: EMCal measurements of the impact point position when exposed to 2 GeV/c electrons

two coordinates (left) can be nicely fitted by the gaussian, the corresponding rms values (right, bottom) only weekly depend on the position of the impact point with respect to tower center.

The energy dependence of the position resolution measured with incident electrons is presented in Fig. 10.

To account for the WC resolution and beam divergence 0.3 cm were subtracted from the data in quadrature. The residual constant term of 0.37 cm accounts both for the systematical uncertainties in the data (nonzero impact angle, residual nonlinearities in the computations) and for the noise in the system.

6. Timing measurements

Earlier the timing response of the PbSc electromagnetic calorimeter was studied extensively exposing the calorimeter to electron and hadron beams with momenta greater than or equal to 1 GeV/c [7]. Measurements were done using larger size tower ($114 \times 114\ mm^2$) and fast timing XP2081 phototube with concave quartz window. One of the conclusions from those earlier measurements was the recognition of the fact that the cherenkov light produced by the charged particles crossing PMT entrance window can have an important and undesirable effect on the calorimeter measured

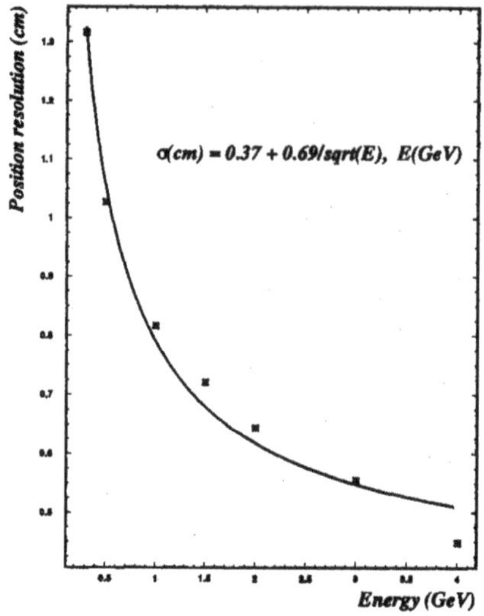

Figure 10: EMCal position resolution as function of the electron momentum

arrival times. Due to the differences in the light and shower propagation speeds in the calorimeter the presence of a precursor to the pulse can shift the arrival time for as much as 1 ns.

The PHENIX calorimeter will have a tower size of $52.5 \times 52.5\ mm^2$ and will employ specially designed 30 mm in diameter russian made PM115M phototubes with flat thin glass window. We used leading edge discriminators for timing measurements setting the thresholds to the instrumental limit of about 30 mV. At our gain setting the corresponding energy threshold was close to 50 MeV.

Resulting improvement can easily be judged from Fig. 11 showing the scatterplots of the arrival times measured in calorimeter versus the energy deposited in the central tower.

The points measured exposing the calorimeter to the particles in the 0.3 - 4 GeV/c momentum range are plotted together. Both scatterplots demonstrate comparable behavior - widening of the populated band at low values of deposited energies. The effect is appreciably more pronounced in case of hadrons. The same data projected onto the time axis are shown at the top of Fig. 12.

Projected distributions have a perfectly gaussian shape and a very few events in the tails what is consistent with systematics free measurements. Thiner phototube window, increased light yield (approximately 1500 photoelectrons per GeV of deposited energy) and correct threshold setting contributed to this improvement in the calorimeter timing performance.

The calorimeter timing resolution is plotted in the bottom part of the

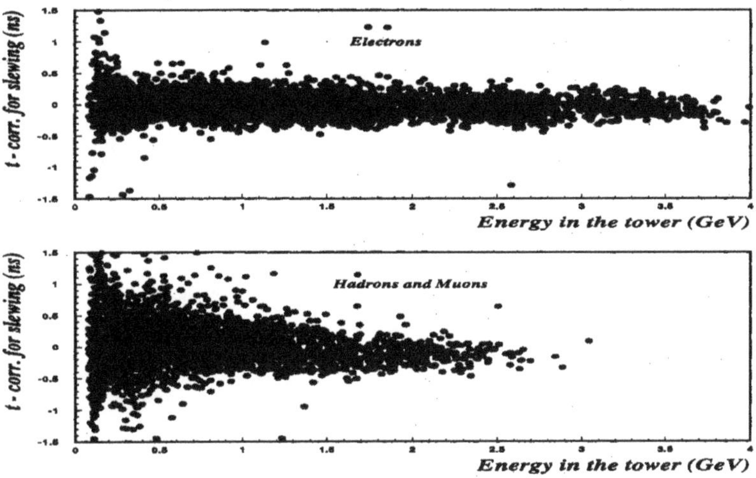

Figure 11: Slewing corrected arrival times measured exposing the EMCal to the beams of electrons (top) and hadrons (bottom)

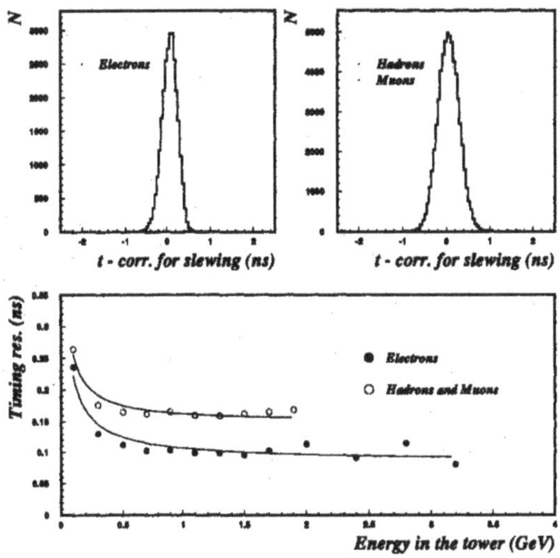

Figure 12: Lead-scintillator EMCal Time-of-Flight Resolution

Fig. 12 as a function of the energy deposited in the central tower. Above minimum ionizing peak it stays nearly constant (∼ 100 *ps for electrons*, ∼ 160 *ps for hadrons*) what corresponds to shower fluctuations being major contributor to measured values. Below minimum ionizing peak the calorimeter timing resolution deteriorates very rapidly consistent with photoelectron statistics being a dominant source of errors. The timing resolution can be well described by the functional form

$$\delta t = \delta t_0 + \delta t_1 / \sqrt{E_{Dep}}, \tag{2}$$

where δt_0 is an intrinsic timing resolution limit discussed above. The constant term t_0 resolution is approximately equal to 90 ps for electrons, and 150 ps for showering hadrons. It includes also the contribution from trigger counters which had a typical timing resolution of about 60 ps.

The second term in Eq. 2 involves the effective signal risetime and the number of photo-electrons, N_{pe}, which is proportional to deposited energy, E_{dep}. The data in Fig. 12 are consistent with it being equal to 65 ps both for electrons and hadrons.

7. Summary

Beam test results demonstrated an excellent energy and position resolution of the lead-scintillator sandwich calorimeter with penetrating wavelength shifting fiber readout. When tiles are properly treated around edges the lateral response nonuniformity stays within ±2.5%, further improvement is provided by the ever present angle between particle trajectory and a normal to the calorimeter front surface.

Our results are consistent with electromagnetic shower lateral shape being approximately independent of the energy. Mapping calorimeter response at one energy should be sufficient to correct for position dependence of the energy measurements for electrons and photons with energies above 1 GeV. Correcting energies below 1 GeV is probably of no importance - its effect can be neglected when compared to the stochastic contributions to energy resolution.

Our recent data show substantial improvement in the calorimeter timing performance when compared to earlier results. When energy seen in the tower is in excess of 200 MeV arrival times are measured with essentially energy independent resolution of ∼ 100 *ps* for electromagnetic and ∼ 160 *ps* for hadronic showers and minimum ionizing particles. At lower energies the effect of the the photon statistics becomes dominant and the timing resolution degrades inversely proportional to the square root of energy in the tower.

8. Acknowledgments

We would like to thank all our collaborators on PHENIX for their interest, discussions and support, in particular, S.Aronson, W.Gurin and D.Whitehouse. We would like also to express our gratitude to S.McCorkle for the help with building the Motorola based data acquisition system for our test beam studies, and Bo Yu for the help with beamline instrumentation.

The authors also acknowledge the support by US Department of Energy under contract DEAC02 - 76CH00016

References

1. PHENIX Conceptual Design Report, An Experiment to be Performed at the Brookhaven National Laboratory Relativistic Heavy Ion Collider, 29 January, 1993
2. "Proposal for a Large Acceptance Hadron and Photon Spectrometer", CERN SP-SLC/P260, May 1991, and revisions.
3. I.Stumer, private communication.
4. R.Wigmans, private communication.
5. G.A.Akopjanov et al., Nucl. Instr. and Meth. A242 (1986) 228.
6. T.C.Awes et al., Nucl. Instr. and Meth. A311 (1992) 130.
7. S.White et al., Proc. of the Third Int. Conf. on Calorimetry in High Energy Physics, Corpus Christi, Texas, (1992) 663.

AN ELECTROMAGNETIC SHOWER CALORIMETER WITH STEREO FIBER-OPTIC READOUT FOR A LARGE ACCEPTANCE SPECTROMETER AT CEBAF[1]

WILL BROOKS
Physics Division, CEBAF
12000 Jefferson Ave., Newport News VA 23606, USA

ABSTRACT

A lead-scintillator sandwich calorimeter is under construction for the CEBAF Large Acceptance Spectrometer (CLAS) in Hall B. It consists of six triangular submodules measuring approximately 4 meters on edge which are assembled together into a hexagonal endcap. 1296 "pixels" point at the target in each of the six sectors; each is segmented into two stages longitudinally. The scintillator strips are read out on one edge via semi-rigid fiber bundle assemblies fabricated in a flight-path-compensating projective geometry for optimal timing performance. Timing, amplitude, and uniformity characteristics have been measured for components of the system, and the results have been incorporated into simulations of the overall detector. Reconstruction algorithms have been developed and tested against data from a prototype and from the simulations.

1. Introduction

The Continuous Electron Beam Accelerator Facility (CEBAF) is a 4 GeV electron accelerator facility beginning operation in Newport News, Virginia. The maximum beam energy of 4 GeV will be gradually upgraded to 6 GeV over the next few years.

The advent of CW electron accelerators with 100% duty factors brings a new era of electromagnetic subatomic physics experiments into being. With the high luminosities becoming available at CEBAF many measurements which have been limited by statistical errors in the past will now be performed to much greater accuracy; additionally the study of a number of low rate processes and other small sized effects has become feasible. With the planned capability of polarization degrees of freedom, whole new classes of measurements have become possible. As of 1994 there were in total 67 experimental proposals approved by the program advisory committee.

The first experimental area to become operational was Hall C in July 1994; the second will be Hall A in early 1996. The third experimental area, Hall B, begins operation in late 1996. The experimental program in this hall is based on the CEBAF Large Acceptance Spectrometer (CLAS), described in the next section.

2. The CEBAF Large Acceptance Spectrometer – CLAS

2.1. Description

The CLAS is a "4π"-type spectrometer based on a toroidal magnetic field[2]. The field is generated by six superconducting coils which naturally segment the detector into six identical sectors. Fig. 1 shows a slice through the midplane of two sectors; conceptually, the full detector is approximately reproduced by rotating this figure around the beam axis. The structure shown is over 8 meters in diameter. There are four major detector subsystems in the CLAS: the drift chambers, Čerenkov counter, time-of-flight scintillators, and the electromagnetic shower calorimeter. The forward electromagnetic shower calorimeter provides a significant measure of electron-pion discrimination as well as serving several other critical functions such as measurement of high energy photons and neutrons, and identification of electrons at the first trigger level.

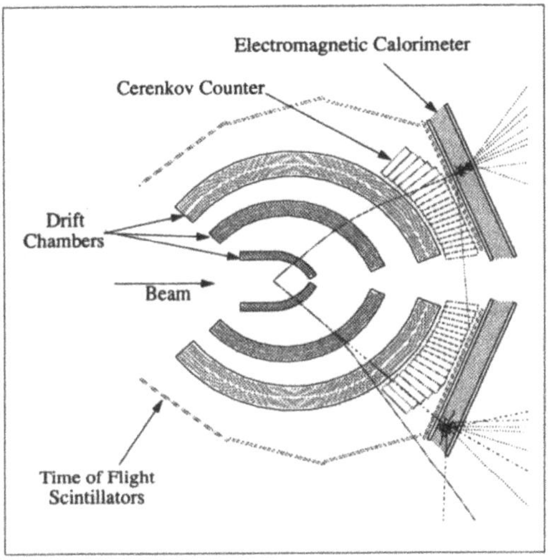

Fig.1 The CEBAF Large Acceptance Spectrometer.

2.2. Physics Program

Large acceptance spectrometers offer a number of experimental advantages. These include multiparticle detection, control of systematic errors, and compensation for experiments requiring low luminosity such as tagged photon beams (background limited) and polarized target experiments (beam intensity limited). The CLAS is ideally suited to address a wide range of physics topics by exploiting these advantages. The approved physics program includes a broad-based study of electroproduction of N^* resonances, hyperon production, multi-nucleon emission in electron-induced processes, the Drell-Hearn-Gerasimov sum rule, and precise measurements of nucleon form factors over a wide range in Q^2.

An example of an experiment which takes advantage of these features is CEBAF experiment 93-006, which proposes to measure two-pion decays of electroproduced light quark baryon resonances. This experiment requires the multiparticle capability of the CLAS, and in addition takes advantage of the capability of the electromagnetic shower calorimeter for neutral detection such as the neutrons and photons (from π^0 decay) in the reaction $ep \rightarrow en\pi^+\pi^0$. A second example is CEBAF experiment 94-017, a precise measurement of the neutron magnetic form factor. This experiment employs the reaction $ep \rightarrow en\pi^+$ to tag neutrons thereby measuring the neutron detection efficiency of the electromagnetic calorimeter simultaneous to the measurement of the ratio of e-n scattering to e-p scattering in a separate deuterium target. In addition to neutron detection and the multiparticle acceptance of the CLAS, this experiment relies on the large detector acceptance to study and minimize systematic errors.

2.3. Requirements for the Electromagnetic Shower Calorimeter

The versatility of the measurements accessible to the CLAS imposes a number of different requirements on the design of the electromagnetic calorimeter: to identify electrons of 0.4-6.0 GeV with essentially 100% efficiency; to be fast and have an energy measurement uncertainty $< 10\%/\sqrt{E}$ at the first trigger level; to detect photons with energies as low as 20-30 MeV; to have a good pion rejection ratio ($\leq 100:1$) over the 50 m^2 area; to allow good missing mass reconstruction for neutrals ($\delta M_{\pi^0} < 20$MeV, $\delta M_{\eta} < 80$MeV), and to be able to separate the two photons from π^0 decay; to allow multiple hit reconstruction in each sector; to detect neutrons with high efficiency and good timing; to present no mechanical obstruction; and to have good uniformity. The solution to these design challenges is described in the following section.

3. The CLAS Electromagnetic Shower Calorimeter

3.1. Description

The forward electromagnetic shower calorimeter covers scattering angles from $\theta = 8°$ to $45°$ with a ϕ coverage matched to that of the drift chambers. The detector is segmented into six triangular sectors, each of which is longitudinally segmented into an inner and outer component; additional modules will be located at larger θ but with more limited ϕ coverage. The forward calorimeter consists of 39 layers of plastic scintillator and lead positioned in a pointing geometry for a total of 16 radiation lengths. The lead contains a 5% admixture of antimony for increased stiffness. The front and back surface of the detector are composites fabricated from thin stainless steel skins epoxied to a layer of structural foam; the side walls of each triangular sector are made from aluminum plate. The nominal scintillator (BC412) thickness is 1 cm, and the lead has a nominal thickness of 2.3 mm. Each scintillator layer is further segmented into 36 strips approximately 10 cm wide with single-sided readout into one of three views. This provides good granularity and redundant position measurement for multiple-hit reconstruction. The three views are oriented approximately in the directions 0°, 120°, and 240° around the normal to the target.

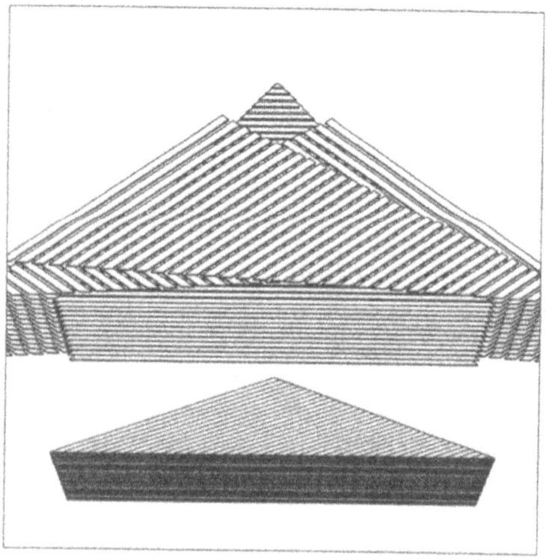
Fig.2 One sector in normal and exploded views.

Fig. 2 shows a single sector in a normal and an exploded view. Stereo readout is accomplished from each of the three readout views via custom fabricated cladded plastic fiber assemblies. These are arranged in a flight-path-compensating geometry to optimize timing resolution while preserving light collection efficiency. Approximately 20 fibers 3 mm in diameter (BCF98) view each strip. The fibers transport the light to a mixing lightguide which is attached to a phototube. This arrangement is shown in Fig. 3. Particles impinge on the detector from below in this drawing. As can be seen, for tracks which are parallel to the long dimension of the fibers the flight path timing compensation is optimal. The longitudinal segmentation can also be seen clearly in this diagram.

The fiber assemblies are fabricated in three steps: straight fiber sections are bent to the correct angles under tension in an oven; the fibers are glued into bundles in a compound angle geometry (which is different for each bundle in a given sector); finally the ends of the fiber bundles are diamond milled.

3.2. Fiber Optic Readout

There are several advantages to the fiber optic readout. The first is that light attenuation lengths are longer for this style of readout than for many other methods, such as a wavelength shifter readout. This is because the cladded fibers only accept light over a limited range of internal angles in the plastic, thus limiting the path lengths through which the light travels. Attenuation lengths for this readout system are on average 15% greater than for a system where the phototube is directly coupled to the scintillator. Attenuation lengths are being measured for each of the 8424 scintillator strips. For a sample of over 1400 scintillators ranging from 2 to 4.5 meters the attenuation lengths approximately follow a Gaussian distribution with a mean of 366 cm and a standard deviation of 45 cm.

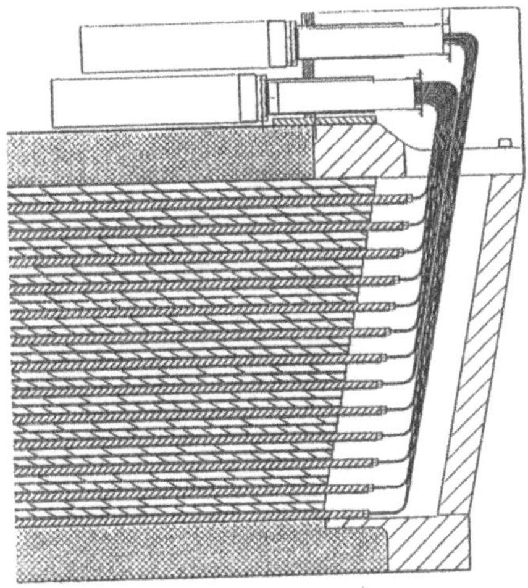

Fig.3 Cross sectional view of calorimeter readout.

The effect of limited light pathlengths is responsible for a second advantage of the fiber readout, which is that timing dispersion of light pulses travelling through the long scintillator is relatively small. The timing dispersion has been measured with a UV laser[3] and with cosmic rays[4] and approximately obeys the relation $\sigma^t{}_x = (1 + cx)\sigma^t{}_o$, with c≈0.1/m for the laser measurements and c≈0.12/m for the cosmic ray measurements. An additional timing advantage of this type of readout is that there is no additional time constant involved as in a waveshifter system, where the absorption and re-emission processes may degrade the timing performance. Measurements with a UV laser for a single strip[3] obtained timing uncertainties ranging from 1 ns at 10 photoelectrons at a distance of four meters to less than 100 ps for 1000 photoelectrons close to the phototube. In the full calorimeter the timing will be somewhat degraded from the latter values since there will be some flight path dispersion, however, detailed GEANT-based simulations indicate that showering events may still attain timing of a few hundred ps.

In addition to these advantages, it is possible to "compress" the fiber light using a light "funnel" which is typically impossible with conventional lightguide systems. It is possible here because the phase space of the fibers is smaller than that of the light funnel. As an example of this, the mixing lightguide in this system has an output area less than half of the input area, but over 90% of the fiber light is transmitted.

The efficiency of the light collection system has been measured to be 4-6 photoelectrons per MeV deposited in the plastic near the phototube[3]. The non-uniformity of the strip-to-strip light collection is <10%.

The optical properties of each of the four elements in the light transport and detection system have been measured as a function of wavelength. The results have been incorporated into a simple model of light propagation through the system. The model predictions for the shape of the attenuation curves is in good agreement with the measured data for long strips[5]. The light absorption characteristics of the system are dominated by the attenuation characteristics of the scintillator material and the photocathode, rather than the fiber lightguide or mixing lightguide.

3.3. Performance of Complete System

A small prototype of the detector was built and tested in an electron beam[6]. This prototype had the conventional waveshifter readout and was not constructed in pointing geometry, however all other features were similar to the final detector. The prototype attained a position measurement uncertainty of 2-3 cm at 1 GeV, and an energy uncertainty of $9.3\%/\sqrt{E}$, with a pion rejection efficiency greater than 98.5%. A GEANT simulation of the large calorimeter using the measured attenuation length for each scintillator (which are superior in quality to those in the prototype) predicts an uncertainty in the energy measurement of $8\%/\sqrt{E}$, of which 6.8% is due to the intrinsic sampling fluctuations. This study also showed a worsening in resolution due to longitudinal leakage of $0.5\%/\sqrt{E}$ over the range of 0.5 - 4.5 GeV incident electron energy. A subsequent study determined that a ± 10% variation in the fiber light couplings contributed no more than 0.25% to the energy resolution at 1 GeV.

A simulation has shown that the uniformity of the detector with constant 300 cm attenuation lengths is 4.4% at the trigger level, averaged over the detector face; the maximum deviation is 20%, which occurs in the corners of the triangle. This has been shown to be reduced to much less than 1% (maximum deviation ±2% using a scheme where signals provided to the trigger are weighted in the energy summing circuitry. Because the actual strips have a distribution of attenuation lengths the nonuniformity will be somewhat greater, however the expectation is still that it will be at the percent level. Cosmic ray testing prior to installation will be performed to obtain initial information on uniformity and calibration characteristics, including the use of a UV laser calibration system.

The response of the detector to neutrons has been studied using GEANT. An efficiency of over 60% with realistic thresholds is predicted for neutrons of momentum greater than 1.5 GeV/c. Neutron timing of about 0.5 ns or better is also predicted; an optimal reconstruction scheme is still under development. An angular measurement uncertainty of 0.5° is obtained for neutron momenta greater than 1 GeV/c.

A number of reconstruction algorithms have been developed[7]. Several levels of reconstruction are required; fast online versions, and offline versions with the ultimate accuracy which may depend on particle type. The elements which differentiate between the algorithms include the use of energy weighting, correction for average attenuation lengths, and degree of complexity of fitting functions for the energy profile. The most straightforward method locates energy centroids in each view and finds the intersection point on the detector subject to geometric constraints. The most robust method used is a Metropolis algorithm approach. While this method gives excellent results for multiple hit reconstruction, it is cpu intensive. The Metropolis algorithm may be most useful in offline analysis for disentangling events with overlapping multiple hits.

4. Acknowledgments

The author would like to thank Volker D. Burkert and B. A. Mecking for a critical reading of this manuscript.

5. References

1. The collaboration contributing to the design and construction project includes: V. Burkert, W. Brooks, J. O'Meara, W. Tuzel, CEBAF; K. Egiyan, S. Stepanyan, Y. Sharabian, R. Demirchyan, Y. Efremenko, G. Asryan, H. Egiyan, M. Amaryan, Yerevan Physics Institute, Armenia; R. Minehart, C. Smith, University of Virginia; K. Giovanetti, James Madison University; H. Funsten, T. Tung, A. Coleman, College of William and Mary; K. Beard, Hampton University.
2. CEBAF Conceptual Design Report, 1990.
3. V. Burkert et al., Light Readout System for the Electromagnetic Shower Calorimeter of the CEBAF Large Acceptance Spectometer, CLAS-Note 92-008, 1992.
4. W. Brooks, Time Dispersion in Long Scintillator Strips with Fiber-Optic Readout, in preparation.
5. W. Brooks, A Simple Model for the Propagation of Internally Reflected Light Through a Scintillator and Light Guide System, CLAS-Note 94-017, 1994.
6. R. Minehart et al., Lead-Scintillator Electromagnetic Calorimeter with Stereo Readout, CEBAF PR-90-026, 1990.
7. K. B. Beard, Forward Electromagnetic Calorimeter Reconstruction Software, CLAS-Note 93-012, 1993.

The response of the CHORUS calorimeter to π and e induced showers

S.Buontempo[b], A.Capone[c], A.G. Cocco[b], D.De Pedis[c], E.Di Capua[a], U.Dore[c], A.Ereditato[b], M.Ferroni[a], G.Fiorillo[b], P.F. Loverre[c], C.Luppi[a], D.Macina[c], F.Marchetti-Stasi[b], M.A.Mazzoni[c], P.Migliozzi[b], V.Palladino[b], G.Piredda[c], F.Riccardi[b], S.Ricciardi[a], P.Righini[c], B.Saitta[a], R.Santacesaria[c], P.Strolin[b] and P.Zucchelli[a]

[a] *Università di Ferrara and INFN Sezione di Ferrara, Ferrara, Italy*
[b] *Università "Federico II" and INFN Sezione di Napoli, Napoli, Italy*
[c] *Università "La Sapienza " di Roma and INFN Sezione di Roma, Roma, Italy*

presented by Antonio Capone

Abstract

The CHORUS Collaboration is operating at the CERN SPS-neutrino beam a hybrid detector, made of an emulsion target and electronic counters, with the aim to detect $\nu_\tau N \rightarrow \tau^- X$ in a background of ν_μ induced charged and neutral current interactions. The reconstruction of the event kinematics with electronics detectors permits to reduce the sample of events to be scanned and to locate the interaction vertex in the emulsions. Test beam results reported here show that the calorimeter allows to measure the hadronic showers energy with a resolution $\sim 30\%/\sqrt{E(GeV)}$ well in agreement with the design requirements

1. Introduction

Big effort was spent in the last decades to investigate the puzzle of a possible neutrino non-zero mass. If massive neutrinos (ν_i) exist, then they can be described as a linear combination of neutrino eigentstates of weak interactions (ν^l):

$$\begin{pmatrix} \nu_1 \\ \nu_2 \end{pmatrix} = \begin{pmatrix} \cos\theta & \sin\theta \\ -\sin\theta & \cos\theta \end{pmatrix} \begin{pmatrix} \nu^l \\ \nu^{l'} \end{pmatrix}$$

Neutrino flavours then mix and "oscillations" between different weak eigenstates are possible. The CHORUS experiment [1] searches for C.C. ν_τ interactions with a detector exposed to the CERN (400 GeV/c SPS proton) wide band neutrino beam. In this beam, the content of ν_τ directly produced at the proton target is negligible ($N_{\nu_\tau}/N_{\nu_\mu} < 10^{-7}$) so the observation of a ν_τ interaction would be an evidence for $\nu_\mu \longleftrightarrow \nu_\tau$ oscillation. The CHORUS detector [2] has been designed to detect different decay channels of the τ lepton produced in ν_τ charged current interactions:

$$\nu_\tau N \rightarrow X + \begin{cases} \tau^- \rightarrow \mu^- \nu_\tau \overline{\nu}_\mu & BR \sim 18\% \\ \tau^- \rightarrow h^-(n\pi^o)\nu_\tau & BR \sim 50\% \\ \tau^- \rightarrow \pi^-\pi^+\pi^-\nu_\tau & BR \sim 14\% \end{cases}$$

To enhance the discovery potential, allowing the identification of the τ^- lepton decay, CHORUS uses $800Kg$ of nuclear emulsions as target for ν interactions. The oscillation probability can be expressed in terms of the difference between mass squares of neutrino families (Δm^2_ν) and the mixing angle ($sin^2 2\theta_{\nu_\mu,\nu_\tau}$). Minimal supersymmetric extensions of the Standard Model of electro-weak interactions and the *see-saw mechanism* seem to favour a solution with $\Delta m^2_{\nu_\mu,\nu_\tau} \sim eV^2/c^4$. Present limits on the mixing angle in this region of Δm^2_ν is at $sin^2 2\theta_{\nu_\mu,\nu_\tau} \leq 2 \cdot 10^{-3}$. CHORUS intends to explore the $\Delta m^2_{\nu_\mu,\nu_\tau}$, $sin^2 2\theta_{\nu_\mu,\nu_\tau}$ parameter space in the region $\Delta m^2_{\nu_\mu,\nu_\tau} \geq 1eV^2/c^4$ down to $sin^2 2\theta_{\nu_\mu,\nu_\tau} \sim 10^{-4}$ (figure 1). This goal implies, given

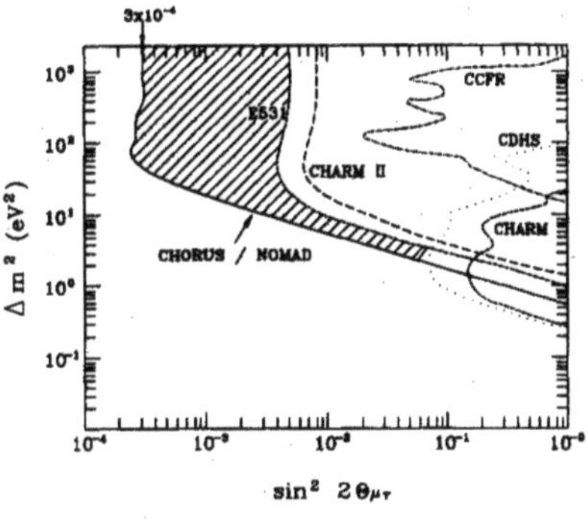

Figure 1. 90% confidence level limits and sensitivity of the CHORUS experiment to $(\nu_\mu \longleftrightarrow \nu_\tau)$ oscillations, if no signature for oscillations is found the present limit for $sin^2 2\theta_{\nu_\mu,\nu_\tau}$, at large $\Delta m^2_{\nu_\mu,\nu_\tau}$, will be improved by a factor of 16

the detection efficiency, that at least $5 \cdot 10^5 \nu_\mu$-induced charged current interactions have to be accumulated in the emulsion target. The Collaboration laboratories can scan ~ 50000 *events/year*. This implies that only a fraction of the collected sample of events can be analysed in a reasonable time. The electronic detector, located behind the emulsion region, along the beam axis, permits the reconstruction of the event kinematics. It allows therefore, a substantial background rejection, and provides useful informations (i.e. the interaction vertex region) for the scanning. It mainly consists of: 1) a high resolution tracker, placed in between the emulsions, used to locate the interaction vertex in the emulsions with a precision ($\Delta x \sim 300\mu m$, $\Delta\theta \sim 3mrad$); 2) a high resolution calorimeter to measure the hadronic shower energy and, with the help of the interaction vertex knowledge, the shower angle; 3) a muon spectrometer that completes the measurement of the event kinematics. The muon spectrometer consists of two parts: an air core magnet equipped with high resolution tracking planes, placed in front of the calorimeter, very useful for low momentum muons ($p_\mu \leq 3$ GeV/c), and a magnetised iron spectrometer equipped with drift chamber, streamer tubes and scintillators, placed behind the calorimeter. The kinematical cuts can reduce the number of events to be scanned down to ~ 50000. These results are reached if the hadronic shower energy can be measured with $\sim 30\%/\sqrt{E(GeV)}$ resolution.

2. The CHORUS calorimeter

The CHORUS calorimeter is made of lead (99% Pb and 1% Sb) and plastic scintillator. It consists of long modules, orthogonal to the neutrino beam line, equipped at the two sides with photomultipliers. There are three different types of modules (figure 2). The first part of the calorimeter, equivalent for a particle travelling in the beam direction to ~ 22 X_o or ~ 0.8 λ_{int}, is made of lead and scintillating fibers (1mm diameter KURARAY SCSF81), with a volume ratio lead-scintillator equal to $4:1$. The elementary unit of this calorimeter sector has dimensions $(40 \cdot 40 \cdot 2620) mm^3$ and is equipped with a pair of 1" photomultipliers (HAMAMATSU R 1355/SM). To reduce the light component with shorter attenuation length, a yellow filter (KODAK Wratten n. 3) is placed in front of each PMT. We will refer to this sector as EM since it performs the measurement of the "prompt" electromagnetic component of neutrino induced showers. Each EM module contains 2 basic elements. Also the second part of the calorimeter (HAD1), meant for the measurement of the hadronic shower component, is made of lead (99% Pb and 1% Sb) and 1 mm diameter scintillating fibers (BICRON BCF-12). It corresponds to ~ 55 X_o or ~ 2 λ_{int}. Each module, which dimensions are $(80 \cdot 80 \cdot 3350) mm^3$, is equipped at the two ends with yellow filters and the 2" PMT's EMI 9839 A. EM and HAD1 sectors amount in total to 2.8 λ_{int}, and therefore cannot fully contain hadronic showers at the experiment energies. A third sector, HAD2, is then placed behind the first two. It is also made of lead (99% Pb and 1% Sb) and scintillator strips, and improves the hadronic

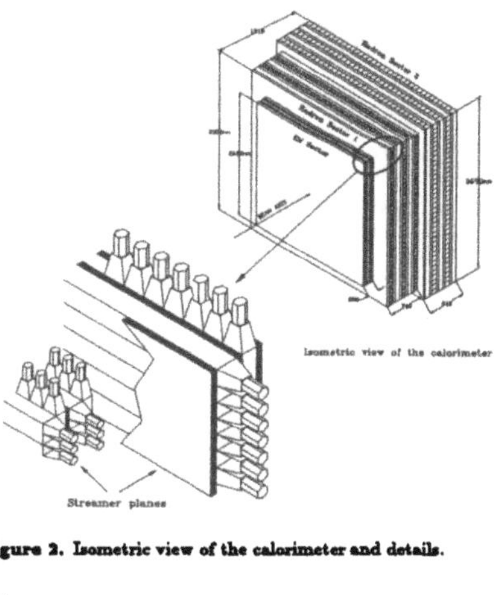

Figure 2. Isometric view of the calorimeter and details.

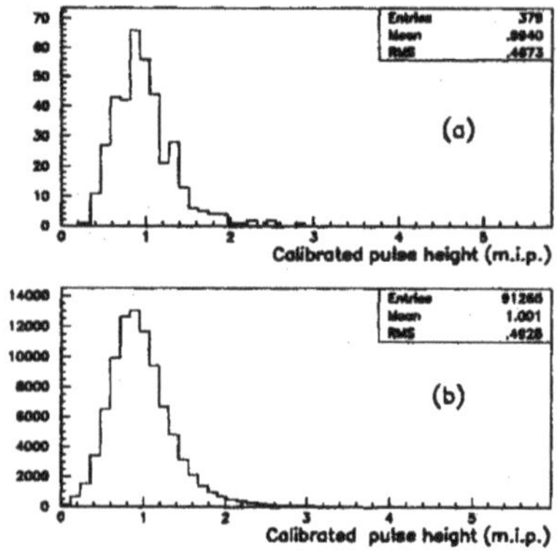

Figure 3. Pulse height distribution for cosmic rays in a sigle HAD1 module (a) and in the whole HAD1 sector (b)

shower containment with additional 2.4 λ_{int}. In each HAD2 module, with dimensions of $(100 \cdot 200 \cdot 3700) mm^3$, there are two elements with square cross section $(100 \cdot 100) mm^2$. Each element consists of five layers of lead bars, 16 mm thick, alternated with five scintillator strips (BICRON BC-408), 4 mm thick. A detailed description of the three module type construction and of a first test performed with electron beams has been already published [2]. Here we report on the response of the whole calorimeter to pions $(3 - 20 GeV/c)$ and electrons $(2.5 - 20 GeV/c)$ of the X9 test beam at the CERN-SPS.

3. Intercalibration of the EM, HAD1 and HAD2 module response

To ensure omogeneity between the response of different modules, belonging to the same part of the calorimeter (EM or HAD1 or HAD2), we compare their response to minimum ionising particles (cosmic ray muons). We select events with a cosmic track crossing the detector and collect the signals of each photomultiplier. For each PMT we consider the pulse height distribution for cosmics crossing the module in a region of $\pm$ 20cm around it's center (figure 3). A correction is applied to non orthogonal tracks. For each PMT we then calculate the *truncated mean* $= \mu_i$ of the pulse eight distribution (excluding from the average pulse heights bigger than five times the distribution peak). We can then express any PMT signal in terms of the pulse height produced by a m.i.p. crossing perpendicularly its central region:

$$N_j^{m.i.p.} = PH_j/\mu_i$$

For a calorimeter made by modules of the same size (and type), this "intercalibration" should be sufficient. Since EM, HAD1 and HAD2 differ in counter dimensions and, in the case of HAD2 also in the technology, we need to define intercalibration constants. The total measured energy is related to the energy deposited in the active material of different sectors through the expression :

$$E_{meas} = \alpha_{EM} E_{EM} + \alpha_{HAD1} E_{HAD1} + \alpha_{HAD2} E_{HAD2}$$

$$= \alpha_{EM}(E_{EM} + \beta_1 E_{HAD1} + \beta_2 E_{HAD2})$$

The intercalibration constants

$$\beta_1 = \frac{\alpha_{HAD1}}{\alpha_{EM}} \quad , \quad \beta_2 = \frac{\alpha_{HAD2}}{\alpha_{EM}} = \beta_1 \cdot \frac{\alpha_{HAD2}}{\alpha_{HAD1}}$$

can be obtained experimentally comparing the energy deposited in different sectors of the calorimeter by pion originated showers which interact at different depths. A pion interacting in HAD1, after having lost by ionisation part of its energy, will produce a shower which develops in HAD1 and HAD2 yielding the measured energy

$$E_{meas} = E^{\pi} = \alpha_{HAD1} E_{HAD1} + \alpha_{HAD2} E_{HAD2}$$

We selected events with pions interacting at different depth inside the calorimeter. Being the total measured energy of these events only function of the pion energy

$$E^{\pi} = E^{\pi}_{beam} - \int \frac{dE^{\pi}}{dx}$$

from a linear fit to the data one obtains the $\alpha_{HAD2}/\alpha_{HAD1}$. In similar way we determine $\alpha_{HAD1}/\alpha_{EM}$. The intercalibration constants calculated with this method:

$$\beta_1 = 2.34 \pm 0.06 \quad , \quad \beta_2 = 3.28 \pm 0.10$$

are, within the errors, independent of the pion energy.

4. Energy calibration

One can then relate the total pulse height of the calorimeter to the particle energy by:

$$E_{meas} = \sum_k \sum_{j=1,n} \alpha_k N_j^{m.i.p.} \quad , k = EM, HAD1, HAD2$$

where n is, for each sector, the number of active elements. As described above, each calorimeter unit is equipped with two PMT's. The signal of each PMT is a function of the particle impact point along the module. Thanks to the yellow filters, placed in front

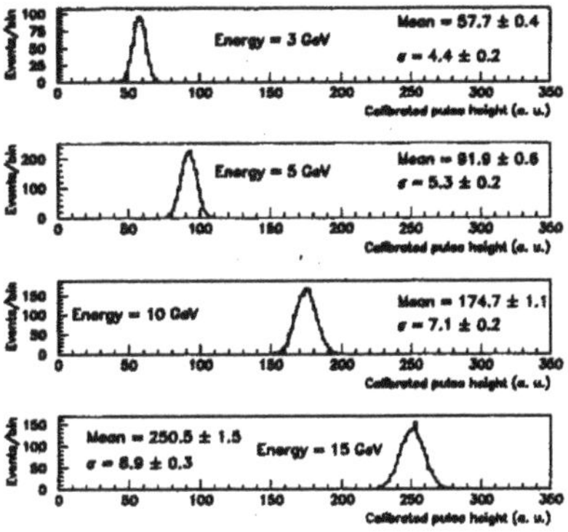

Figure 4. Calorimeter total pulse height for incoming electrons

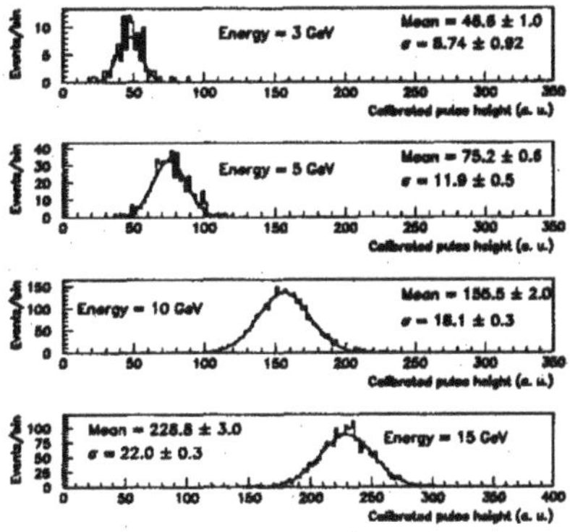

Figure 5. Calorimeter total pulse height for incoming pions

of each PMT, this function can be well approximated by a single exponential. The geometric average of the signals of the two PMT's is then, in good approximation, independent on the impact point, unlike the arithmetic mean. The following results are obtained adopting the geometric mean method. The energy resolution, for particle interacting at a fixed point along the detector, is the same for the two methods. In figure 4 and figure 5 the total calibrated pulse height distributions for incoming electrons and pions with energy between 3 and 15 GeV are shown. For pions, in order not to be affected by longitudinal leakage, we require them to interact in the EM sector. Figure 6 shows the total calibrated pulse height distributions for 10 GeV electrons and pions. The gaussian fit is very good and the absence of tails is evident.

5. Linearity and energy resolution

The calorimeter response to pion induced showers is linear (figure 7a) in the energy range $3 \div 20$ GeV; deviation from linearity are less then 2%. The hadronic showers energy resolution, shown as a function of the energy in (figure 7b), is well fitted by:

$$\frac{\sigma(E)}{E} = \frac{(30.3 \pm 2.3)\%}{\sqrt{E(GeV)}} + (1.8 \pm 0.7)\%$$

This experimental value is in very good agreement with expectations. It confirms the calorimeter design capabilities in reducing the number of events to

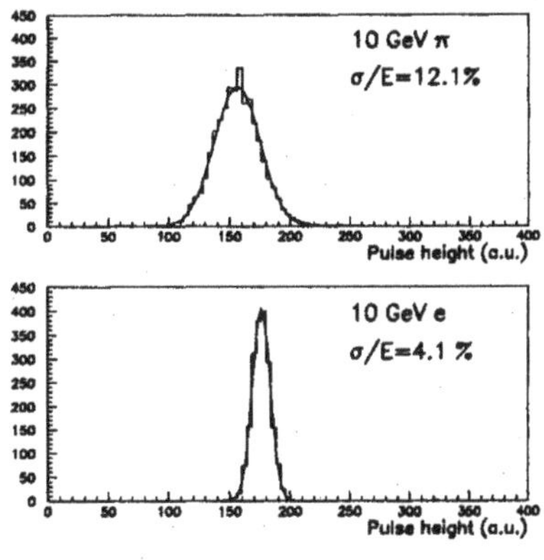

Figure 6. Total pulse eight in the calorimeter for 10 GeV pions and electrons

be scanned in the emulsions, as mentioned in the introduction. The response of the calorimeter to electromagnetic showers with energy lower than 5 *GeV* is linear: deviations from linearity are smaller than 1% (figure 8a), while they increase for $E \geq 5$ *GeV*.

This is, indeed, acceptable for this experiment, since $\sim$ 90% of the π^o produced in neutrino interactions

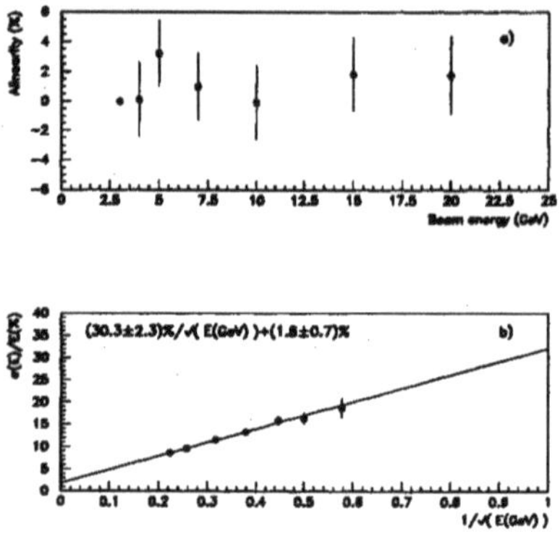

Figure 7. Response of the calorimeter to pion induced showers: a) deviation from linearity; b) energy resolution.

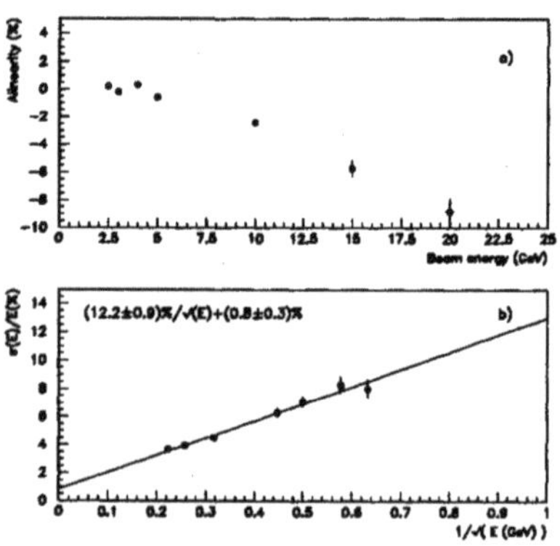

Figure 8. Response of the calorimeter to electromagnetic showers: a) deviation from a linear fit to $2.5 \div 5$ GeV data; b) energy resolution

have momentum lower than 5 GeV/c [2]. The reason of this measured deviation from linearity has not yet been clarified: we are still investigating whether is only caused by PMT and/or electronics non linearity, or also by light emission saturation in the active material. These effects may be due to the peculiar high energy

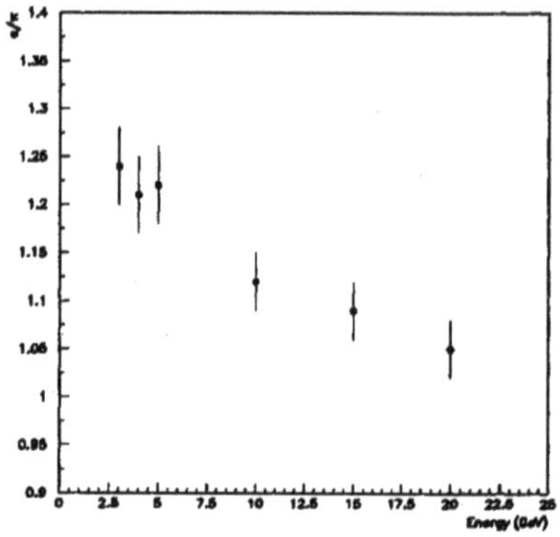

Figure 9. Energy dependence of e/π ratio

density deposition of the electromagnetic showers. The function

$$E_{meas} = aE^{(0.89\pm0.01)} + b$$

where $a = 22.5 \pm 1.1$ and $b = (1.96 \pm 1.96)GeV$, is therefore used to parametrise the response of the calorimeter to electromagnetic showers; the energy resolution, shown as function of energy in figure 8b, after correction for the non linear response, can be parametrized by:

$$\frac{\sigma(E)}{E} = \frac{(12.2 \pm 0.9)\%}{\sqrt{E(GeV)}} + (0.8 \pm 0.3)\%$$

This result agrees with previous measurements [3] and with the prediction of Montecarlo simulations based on the EGS4 package ($\sim 13\%/\sqrt{E(GeV)}$).

In a perfectly compensating calorimeter, the ratio between the responses of electromagnetic and hadronic showers, e/π, must be equal to 1. The LAA-SPACAL Collaboration [3] quoted $e/\pi \sim 1.1$ at 10 GeV, with a similar prototype calorimeter, but with fibers parallel to the beam direction. Our result, shown in figure 9 as a function of energy, is well compatible to the LAA-SPACAL one.

6. Conclusions

The CHORUS experiment has been taking data since April 1994. Its calorimeter makes use for the first time at large scale of the "spaghetti" technique. It has been calibrated with electron ($2.5 \div 20.0 GeV$) and pion ($3.0 \div 20.0 GeV$) test beams. The results show, in the energy range of interest for this experiment, linear response for both electromagnetic and hadronic showers. The energy resolutions are in agreement with expectations and confirm the potentiality of the CHORUS calorimeter to fulfill its task in the experiment.

References

[1] N. Armenise et al., CHORUS Collaboration, CERN-SPSC/90-42, M. de Jong et al., CHORUS Collaboration, CERN-PPE/93-134

[2] S. Buontempo et al., Nucl. Inst. and Meth. A 349 (1994) 70

[3] D. Acosta et al., Nucl. Instr. and Meth. A 294 (1990) 193; D. Acosta et al., Nucl. Instr. and Meth. A 308 (1991) 481

PERFORMANCE OF A SCINTILLATING FIBRES SEMIPROJECTIVE ELECTROMAGNETIC CALORIMETER

M. Bertino, C. Bini, G. De Zorzi, G. Diambrini Palazzi, G. Di Cosimo, A. Di Domenico, F. Garufi, P. Gauzzi and D. Zanello

Dipartimento di Fisica, Universita' "La Sapienza", Roma, and INFN Sezione di Roma, P.le A.Moro 2,00185 Roma, Italy

(Presented by G. De Zorzi)

ABSTRACT

A highly segmented scintillating fibres / lead electromagnetic calorimeter has been tested. Each calorimeter module has semiprojective geometry and is shaped as a wedge with an angle of (0.82)°. The fibres are however parallel to the wedge axis and the two small lateral regions are not fibre-instrumented. This simple and cheap approach to a projective geometry allows to achieve still good energy and space resolution. Results with electrons in the range 10-100 GeV are presented.

1. The Calorimeter

We have studied a lead-scintillating fibres segmented electromagnetic calorimeter with a semiprojective geometry. The modules have the shape of a truncated pyramid, in such a way that the sides are pointing towards an axis at a distance of 1.43 m. Figure 1a shows the geometry of a module.

The main feature of this calorimeter is that the fibres fill only the parallelepiped part of a module, so that the two small lateral regions are not fibre-instrumented. The purpose of our test has been to verify how much the calorimeter performance, in terms of energy and space resolution, is affected by the loss of energy deposited in the wedge-shaped regions, which are not sampled.

The calorimeter is segmented in 25 modules, 35 cm long, of 20.5 x 20.5 mm^2 (front) and 25.5 x 20.5 mm^2 (back) cross area. The main features of the calorimeter are summarized in table 1.

Table 1

Scintillating fibres	Kuraray SCSF81, 1 mm diameter
Absorber alloy (% in weight)	AF17 (52.5% Bi, 32% Pb, 15.5% Sn)
Volume ratio fibres : absorber	1 : 3.17
Average density	7.1 g/cm^3
Average radiation length	1.0 cm
Average Moliere radius	2.3 cm
Module cross area	20.5 x 20.5 mm^2 (front)
(horizontal x vertical)	25.5 x 20.5 mm^2 (back)
Calorimeter total cross area	10 x 10 cm^2 (front)
Calorimeter total length	35 cm

2. Test and Results

The test was performed at the CERN-SPS. The calorimeter was placed on a platform movable with respect to the electron beam both in the horizontal and vertical directions. A telescope of two delay wire chambers in front of the calorimeter was used to measure the electron impact point, with a space accuracy of about 0.2 mm. The calorimeter was operated with a tilt angle of 2.5° in both horizontal and vertical directions.

Figure 1b shows the detected energy as a function of the electron impact position for three calorimeter modules along horizontal and vertical directions, at an electron energy of 50 GeV. In the x-coordinate there is a systematic difference of about 20% between the measurement of the energy at the centre and the border of the modules, but as the impact point is measured by the calorimeter itself a correction can be applied to the detected energy.

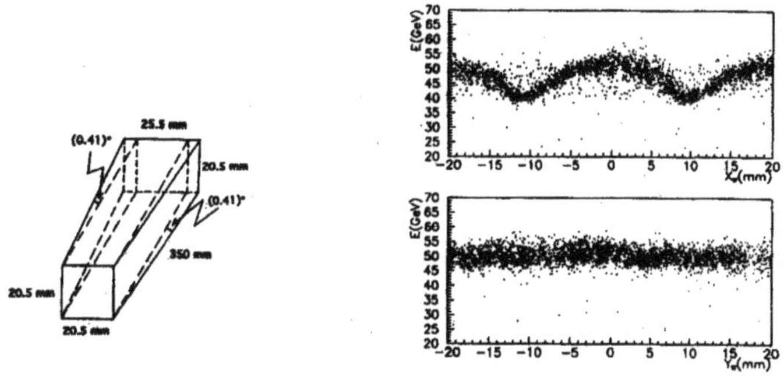

Fig. 1: a) Schematic view of a semiprojective module. The two lateral wedge-shaped regions with (0.41)° angular aperture are not filled with scintillating fibres. b) The calorimeter response uniformity in the horizontal and vertical directions. Energy detected by the calorimeter as a function of the x and y impact coordinates of 50 GeV electrons. A beam scan is performed from the centre of a module (-20 cm) through a second module, up to the centre of a third module (20 cm). Two dips are seen in the horizontal scan, corresponding to the two non-instrumented wedge-shaped regions.

The calorimeter response turns out to be linear within 2% in the explored energy range 10-100 GeV, as shown in fig. 2a. The linearity does not depend on the electron impact position on the module.

The energy resolution σ/E is a linear function of $1/\sqrt{(E(GeV))}$ as shown in fig.2b, but there is some dependence on the impact position (see fig.1b) :

$\sigma/E = (13.9 \pm 0.9)\%/\sqrt{(E(GeV))} + (1.4 \pm 0.2)\%$, at the module centre;

$\sigma/E = (11.0 \pm 1.0)\%/\sqrt{(E(GeV))} + (0.6 \pm 0.4)\%$, at the middle of the module x-border;

$\sigma/E = (14.0 \pm 1.0)\%/\sqrt{(E(GeV))} + (2.0 \pm 0.5)\%$, at the middle of the module y-border.

The x-border corresponds to an electron impinging between two modules, where only a fraction of the e.m. shower energy is sampled because the fibres are missing in the wedge-shaped volume. The y-border corresponds to the separation between two modules with full energy sampling.

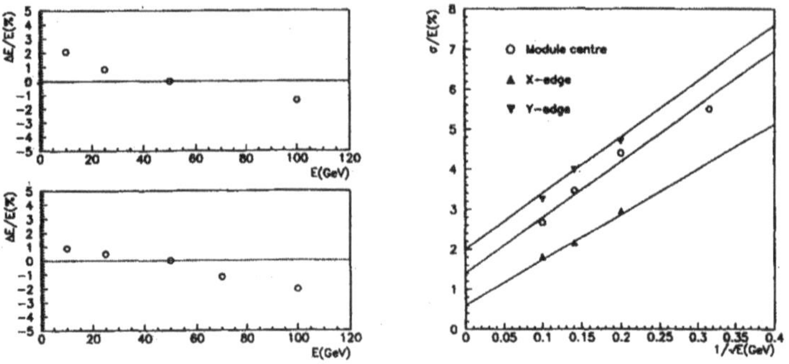

Fig. 2: a) Percentage non-linearity versus energy from 10 to 100 GeV. The horizontal and the vertical tilt angles are 2.5° (up); the horizontal tilt angle is 5° (down). b) Energy resolution as a funtion of $1/\sqrt{(E(GeV))}$. The electron beam hits the module centre (circles), the x-border (upward triangles), and the y-border (downward triangles). The horizontal and the vertical tilt angle are 2.5°.

The energy resolution of this calorimeter is in agreement with the typical performance of a 'spaghetti calorimeter' [1,2]. The present result is also comparable to the performance of the semiprojective [3] and fully projective [4] 'spaghetti' calorimeters having full fibre instrumented modules.

Moreover it should be noticed that the energy resolution turns out to be slightly better near the non-instrumented region (i.e. at the x-border) respect to the whole detector. Although the signal in this region is reduced by 20% with respect to the central region (see fig. 1b), the signal fluctuation is reduced even more. We do not have a simple explanation for this phenomenon.

Horizontal and vertical scans of the beam were done, in order to explore the calorimeter response, along the x-coordinate and the y-coordinate, for the three central modules. The electron energy was 25, 50, and 100 GeV.

The (x,y) position is directly measured by the calorimeter by the energy deposited in the modules surrounding the e.m. shower vertex. This can be done by a simple 'centre of gravity' algorithm but, as usual for segmented detectors, a better impact point determination can be achieved by more sophisticated algorithms.

We have parametrized the dependence of the 'centre of gravity' coordinate on the 'true' coordinate with the following formula [5] :

$$x_g = a + b \arctan [c (x_w + d)] \qquad (1)$$

where : x_w is the x-coordinate measured with the chambers; x_g is the x-coordinate measured with the calorimeter utilizing the simple 'centre of gravity' algorithm; a, b, c, and d are four parameters which are determined by fitting the experimental points (x_w, x_g) with the function (1). The same formula holds for the y-coordinate.

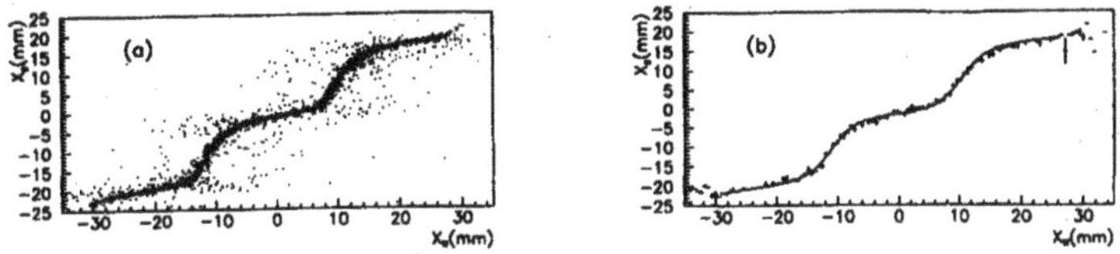

Fig. 3: a) Comparison between the x-coordinate measured by the beam chambers, and the same coordinate reconstructed by the calorimeter with a simple 'centre of gravity' algorithm. The 50 GeV electron beam scans 3 calorimeter modules, i.e. from -30 cm to 30 cm. b) The same result is shown with superimposed the fit of equation (1).

Figure 3a shows the x_g coordinate versus the x_w coordinate for 50 GeV electrons detected in the three central modules: the waving behaviour is typical of any modular calorimeter. Figure 3b shows the x-coordinate data with the fit of equation (1) superimposed. The fit is carried on from a module centre to the adjacent module centre. The values of the four fitted parameters do not depend on the electron energy in our energy interval.

The parametrization (1) can be inverted as follows:

$$x_c = -d + (1/c)\tan[(x_g - a)/b] \qquad (2)$$

and applied event by event in order to obtain an unbiased x_c coordinate from x_g.

The waving behaviour in both horizontal and vertical coordinates is clearly reduced to an almost straight line by applying the equation (2). Figure 4a shows x_c (y_c) versus x_w (y_w) at 50 GeV. A small residual waving is still present, but it does not deteriorates the space resolution which is basically determined by the width of the (x_w, x_c) distribution, and (y_w, y_c) distribution, for the horizontal and the vertical coordinate respectively.

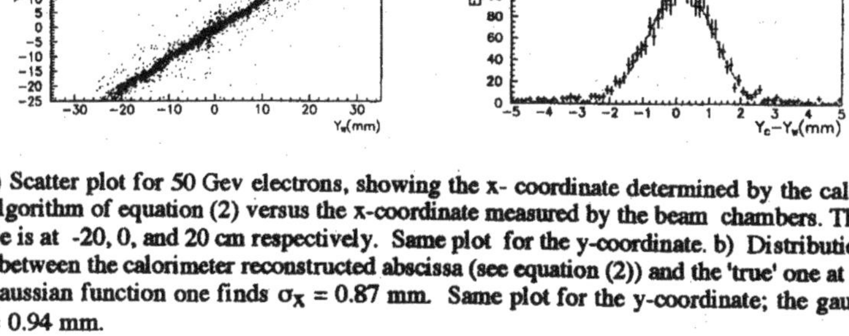

Fig. 4: (a) Scatter plot for 50 Gev electrons, showing the x- coordinate determined by the calorimeter with the algorithm of equation (2) versus the x-coordinate measured by the beam chambers. The centre of a module is at -20, 0, and 20 cm respectively. Same plot for the y-coordinate. b) Distribution of the difference between the calorimeter reconstructed abscissa (see equation (2)) and the 'true' one at 50 GeV. Fitting a gaussian function one finds $\sigma_x = 0.87$ mm. Same plot for the y-coordinate; the gaussian fit gives $\sigma_y = 0.94$ mm.

An example of impact points distribution as a function of the difference ($x_W - x_C$) and ($y_W - y_C$), is given in the fig. 4b at 50 GeV. The impact point resolution turns out to be slightly dependent on the impact position , as shown in fig. 5a at 50 GeV for both coordinates.

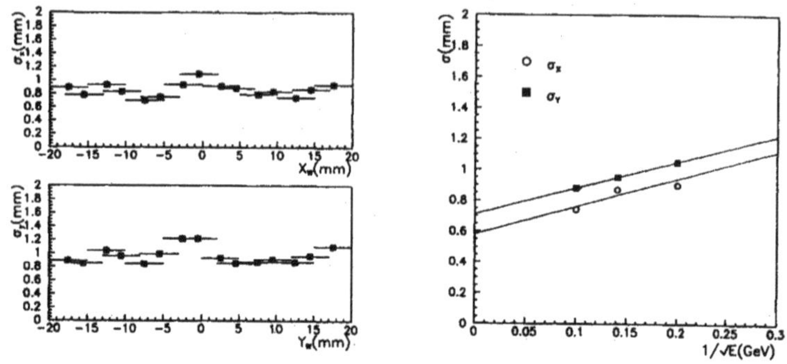

Fig. 5: a) Impact position resolution in x-coordinate and in y-coordinate, as a function of the impact position x and y respectively. The module centre is at about 0 cm, and its edges at -10 cm and 10 cm. The electron energy is 50 GeV. b) Impact point resolution of the electron shower as a function of the electron energy for the horizontal (circles) and the vertical (squares) coordinate respectively. The linear fit in $1/\sqrt{(E(GeV))}$ gives the result quoted in section 2.

After all corrections the space resolution is a linear function of $1/\sqrt{(E(GeV))}$:

$$\sigma_X = (1.8 \pm 0.1) \text{ mm} /\sqrt{(E(GeV))} + (0.6 \pm 0.1) \text{ mm}$$
$$\sigma_Y = (1.7 \pm 0.1) \text{ mm} /\sqrt{(E(GeV))} + (0.7 \pm 0.1) \text{ mm}.$$

This result is obtained by averaging σ over the full module size and is plotted in fig. 5b.

It is worth noting that the resolution is the same in both x and y coordinates. The different energy sampling at the two module borders in the horizontal coordinate does not affect the impact point reconstruction accuracy of this calorimeter.

The present result should be compared with the resolution obtained with highly segmented [2] and less segmented calorimeters [1,6] , and furthermore having different fibre : lead volume ratio.

3. Conclusions

A simple approach to the design of a segmented semiprojective e. m. calorimeter has been carried out. The main feature of this calorimeter is that the wedge-shaped lateral region, at both the sides of a module, is not filled with fibres.

The energy and the space resolution have been measured on an electron beam in an energy range from 10 to 100 GeV.

Our results show that both these features are uniform over the whole calorimeter, i.e. no worsening effect is found when the electron hits the wedge-shaped regions. The energy resolution compares with the common values of the so called spaghetti calorimeters: $\sigma/E = 14.0\% /\sqrt{(E(GeV))} + 1.5\%$. The space resolution, in both the coordinates, is as good as that obtained with highly segmented, but non-projective, fibres/lead calorimeters: $\sigma = 1.7 \text{ mm} /\sqrt{(E(GeV))} + 0.6 \text{ mm}$.

4. Acknowledgements

We warmly thank R. De Salvo for many discussions and for helping us in the preparation of the test.

5. References

[1] R. Wigmans, *Nucl. Instr. and Meth.* **A 315** (1992) 299.
[2] A. Asmone et al., *Nucl. Instr. and Meth.* **A 326** (1993) 477.
[3] B. Bencheikh et al., *Nucl. Instr. and Meth.* **A 315** (1992)354.
[4] J. Badier et al., CERN-PPE/93-20 (1993).
[5] D. Acosta et al., *Nucl. Instr. and Meth.* **A 305** (1991) 55.
[6] S. Dagoret et al., CERN-PPE/93-147 (1993).

EM Preshower

Session Chair: Giorgio Apollinari (Rockefeller)

CONSTRUCTION AND TEST OF A FINE-GRAINED LIQUID ARGON PRESHOWER PROTOTYPE

RD3 Collaboration
presented by Johann COLLOT
Institut des Sciences Nucléaires
CNRS-IN2P3-Joseph Fourier University
53,Av. des Martyrs 38026 Grenoble France
e-mail:collot@frcpn11.in2p3.fr

ABSTRACT

Over the last two years, the RD3 collaboration has been studying a separate LAr preshower consisting of two layers featuring a fine granularity of $2.5\ 10^{-3}$. A prototype covering approximately 0.8 in pseudo-rapidity and 9° in azimuth was built and tested at CERN in July 94. It has a modular structure which is appropriate for a much larger scale construction. Its modules were produced using a moulding technique developed for this application. CMOS and GaAs VLSI preamplifiers were also designed and tested for this occasion. Preliminary test beam results are presented.

1. Introduction

The idea of equipping the accordion LAr electromagnetic calorimeter, designed for the future large hadron collider of CERN, with a fine-grained preshower detector is dictated by several arguments. A device is needed to independently measure the energy lost in the inactive material upstream of the calorimeter, thereby preserving the energy resolution of electron and photon signals. By combining the position measurements of the first calorimeter compartment and of the preshower, a precise determination of the particle direction can be derived which helps in keeping the two photon invariant mass resolution within a tolerable range to observe the $H\rightarrow \gamma\gamma$ decay mode. Furthermore it was demonstrated in previous analyses that a jet rejection factor of 10^4 is mandatory to reduce the huge background contribution stemming from γ-jet QCD events below the $H\rightarrow \gamma\gamma$ signal. This implies a rejection factor better than 3 of the preshower-calorimeter system against isolated π_0's arising from parton fragmentation fluctuations.

The potential of a fine-grained preshower with at least 3 X_0 of material to convert photons in front of its second layer measuring the longitudinal position was explored with a small prototype (6×6 cm^2) at CERN a few years ago [1] . The obtained performance showed the adequacy of the concept regarding the LHC requirements . Here we present the study of a modular system that constitutes a step forward in the integration of such a preshower into the ATLAS calorimetry system. Some other details may be found in [2].

2. Detector geometry

At the scale of ATLAS such a detector would exhibit a polygonal structure formed by assembling 32 identical sectors of 2.7 m each, covering half a barrel (pseudo-rapidity spread: 1.4). These sectors would be suspended on mini-rails anchored in front of the electromagnetic calorimeter. Such an assembly concept would allow one to slide in and out the preshower sectors during the initial mounting phase or any exceptional maintenance operations without moving the electromagnetic calorimeter out of its cryostat.

The transverse cut of a preshower sector prototype featuring an azimuthal angle of 9° is presented in figure 1.

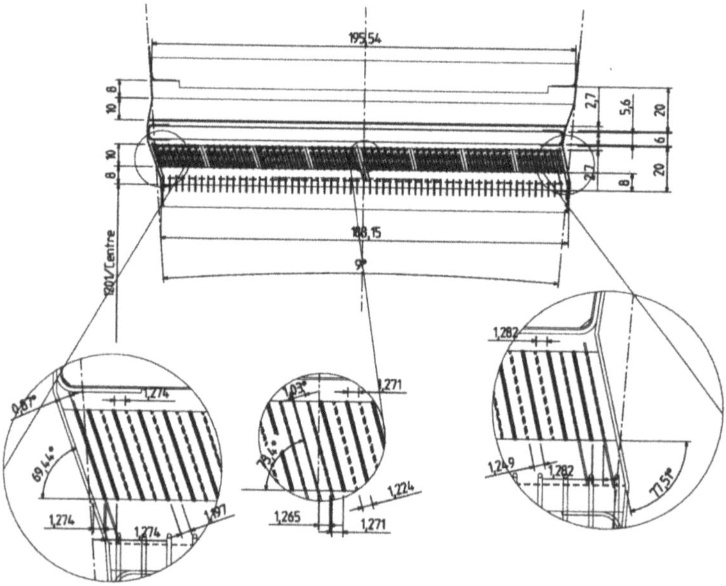

Figure 1: Transverse cut of the preshower prototype

Ten modules arranged in two layers reading the ϕ and η coordinates and sandwiching an internal lead converter plate, are supported by two lateral thin skirts made with stainless steel. The granularity of the active layers in the fine-grained dimension is $\Delta\phi$ or $\Delta\eta$=2.5 10^{-3} and is kept constant over the entire structure. The length of the ϕ strips is constant and fixed to 110 mm. In the η layer , this length is 94 mm representing half of the sector opening. The liquid argon gap varies from 1.2 mm in the ϕ cells to 1.6 mm for the last η module.The strips are slanted so that they offer a 100% geometric efficiency to muons and improve the space resolution. The internal lead layer adds one more radiation length in front of the second active sampling, while an external lead plate fixed on the skirts complements the amount of material to a constant thickness of 2 X_0 before the ϕ active layer. They are both tapered in the longitudinal direction to keep the amount of material constant in front of their respective sampling.

A preshower module reads 64 elementary intervals in the fine-grained direction and 2 in the coarse one; therefore there are 128 individual cells on each module. Seven modules - 4 in the ϕ direction and 3 in the η one - were mounted and tested at CERN in front of the accordion 2m prototype [5] in July 94. They approximately covered a pseudo-rapidity range of 0.2 to 0.8 .

3. Module production technique

A preshower module is produced by moulding a layer of epoxy on top of the strips stacked in a hermetic mould and precisely positioned by spacing shims. The cathodes are made with Cu-Be and are 0.2 mm thick. The anodes are multi-layer kapton electrodes with Cu-clad HV pads on their external faces. They are .25 mm thick on average. Numerous tests were performed to select the epoxy system and the composite structure in order to obtain a correct cryogenic behavior of the modules. The epoxy resin finally selected (CIBA - GEIGY system MY745, HY905, DY072) is loaded using micro glass fibers (50% in weight). Its relative shrinkage coefficient from room temperature to LN_2 temperature after loading is $\Delta L/L = 0.7$ %. Moreover, to obtain an object remaining flat when cooled down to LN_2 temperature, we had to insert glass-epoxy reinforcement bars buried in the epoxy resin. Some of them are placed along the lateral edges, the others are laid down in the x and y directions; they further reduce the shrinkage coefficient of the composite object down to a value comparable to the glass epoxy one. The resin is injected with the mould placed in a vacuum-oven. A differential pressure of 0.3 bar is applied during the injection. The polymerization cycle lasts 16 h and reaches 130 °C. After unmoulding and ethanol cleaning, NOMEX honeycomb strips are slid into the cells to safely maintain the liquid argon gap.

4. Electronics

Given the high channel density of such a detector, VLSI techniques have to be employed to design the front-end electronics. This work has started by devising and producing octal preamplifier chips in CMOS and GaAs monolithic processes.

4.1. CMOS preamplifiers

ICON is a current conveyer preamplifier based on a layout that was originally considered for the readout of silicon detectors [3]. To meet the particular needs of the LAr calorimetry, some improvements had to be made regarding the current gain and the noise. This study was carried out using CMOS monolithic processes of three different firms: MIETEC, ES2 and AMS. So far the best performance with respect to the speed was measured at 77 °K on the AMS 1.2 μm version (t_p(260 ns triangle)=50 ns and 60 ns when filtered by a CR-RC^2 filter of 50 ns shaping time), while the equivalent noise current was found at 19 nA for a CR-RC^2 shaping time of 50 ns. Because of the lack of time, 512 channels of the MIETEC version, which has a lower performance in terms of speed (t_p(260 ns triangle)=80 ns and equivalent noise current=9 nA), were utilized to equip 4 preshower modules in July 94. Their outputs were integrated by 40 ns gated integrators and subsequently digitized by 12-bits ADC's.

4.2. GaAs preamplifiers

The second readout device under study is a monolithic charge-sensitive preamplifier made with a GaAs ion-implanted MESFET process [4] . The chip developed contains dominant pole preamplifiers with a gain-bandwidth product of about 1 GHz. The preamplifier input capacitance is 12 pF. The series noise at 87°K is 0.54 $nV/\sqrt{Hz}$ and the corner frequency of the 1/f noise does not reach 1 MHz. Its power dissipation is 7.5 mW. With CR^2-RC^2 bipolar shaping an ENC of 436 + 21 e^-/pF at 28 ns was measured at 87°K. 384 channels of that type were mounted on 3 preshower modules for the July 94 test beam period. Their outputs were filtered through 50 ns CR^2-RC^2 shapers, and then sent to 12-bit ADC's.

5. Performance in a test beam

The results presented in this paragraph are from the analysis of the data collected at CERN in July and August 94 in a test beam of electrons and photons. Because of limited time, no effort has so far been devoted to ameliorate the precision of the energy calibration coefficients of the preshower channels that were determined on-line with an accuracy of about ±15%. However as the average fraction of the electron energy deposited upstream of the calorimeter does not exceed 6% at high energy (≥287 GeV), this has limited effect on the energy resolution of the preshower-calorimeter system. As an example, the energy deposited by 287 GeV electrons is presented in figure 2. The electron signal was cumulated over a calorimeter cell nonet and 100 (50/layer) cells in the preshower centered on the most energetic channel. The represented quantity is the total energy obtained as $E_{tot} = a(\alpha E_{psh}+\beta E_1+ E_2+E_3)$, where E_{psh},E_1,E_2,E_3 refer to the energy collected in the preshower and the three calorimeter radial compartments; α,β are energy dependent weights introduced to minimize the width of the energy distribution; a is an overall calibration coefficient. The standard position-dependent energy corrections were applied to correct for the shower containment in η and the modulation of the transverse LAr thickness in ϕ [5].

The signal from 150 GeV muons as reconstructed in 4 (2/layer) preshower strips equipped with CMOS preamplifiers, was measured with a signal/noise ratio of 5.6, that is 9 nA of current noise per channel in very good agreement with the value quoted in section 4.1.

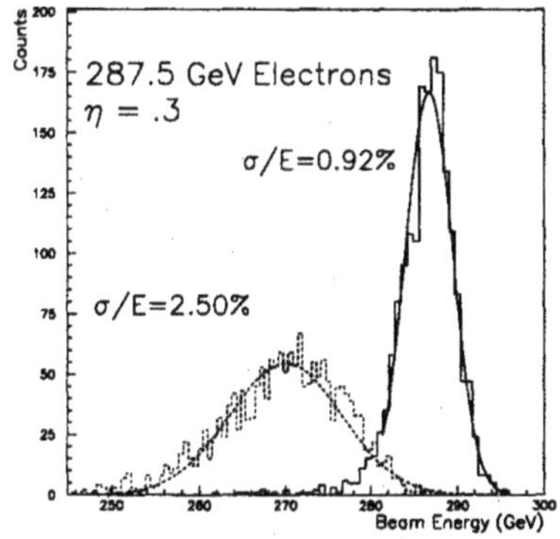

Figure 2: Energy reconstructed in the preshower-calorimeter system from 287 GeV e$^-$ before and after adding the preshower contribution

The electron shower position in each view is obtained by calculating the energy-weighted barycentre of a triplet of cells centered on the channel with the largest signal. From the comparison of this reconstructed position to the one extrapolated from 3 beam chambers located upstream of the cryostat, a typical s-shape correction was fitted and used to correct the barycentre for a bias in the vicinity of the cell center. Figure 3 shows the position resolution obtained in both views as a function of the incident electron energy after s-shape correction and unfolding of the beam chamber contribution (σ = 250-300 μm).

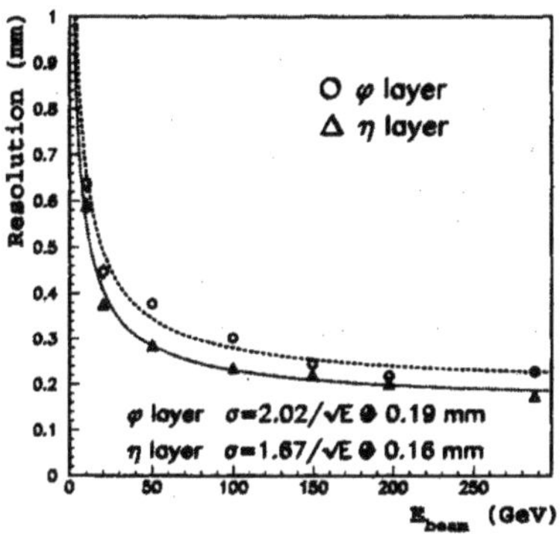

Figure 3: Space resolution of the preshower after s-shape correction and unfolding of the beam chamber contribution

The curves which are superimposed on the data points agree very well with the predictions of some Monte-Carlo simulations: σ_ϕ=1.83 mm/ $\sqrt{E(\mathrm{GeV})}\oplus$ 0.24 mm ; σ_η=1.72 mm/ $\sqrt{E(\mathrm{GeV})}\oplus$ 0.19 mm. The space resolution is slightly better in the η layer where the signal is bigger owing to the additional lead placed in front. Above 200 GeV, the position resolution is better than 220 μm in both views.

6. Conclusion

The moulding process developed over the last two years to produce preshower modules is technologically a good candidate for larger scale construction. Its industrialization is now under way. The performance of the prototype measured in an electron test beam shows that this concept meets the LHC requirements especially with respect to the space resolution. Although GaAs cold analog electronics seem more promising for this application, CMOS preamplifiers may constitute a good alternative solution if their resistance to ionizing radiation can be achieved. The difficulty of the integration of such a device in a detector such as ATLAS comes from the huge number of electronic channels (120,000 for $|\eta| \leq 1.4$) to bring out of the cryostat. Analog optical links [6] that allow the reduction of the overall diameter of the cryostat feedthroughs, may be envisaged.

7. Acknowledgements

The author wishes to thank the technical staff of the participating laboratories who devoted a substantial part of their precious time to the construction of this device.

8. References

1. RD3 Collaboration, Nucl. Inst. and Meth. A330(1993) 405.
2. RD3 Collaboration, CERN/DRDC/92-40(1992).
3. D. Dzahini et al., A CMOS preamplifier and shaper with 50Ω line driver for a LAr preshower detector, to be presented at the Nuclear Science Symposium, Norfolk October 1994.
4. D. Camin, N. Fedyakin and G. Pessina, Monolithic GaAs readout for a LAr preshower detector, proceedings of the fourth international conference in high energy physics, La Biodola, September 1993.
   D. Camin et al., Nucl. Inst. and Meth. A315(1992)385.
5. RD3 Collaboration, Performance of a large scale prototype of the ATLAS accordion electromagnetic calorimeter, submitted to Nucl. Inst. and Meth. A.
6. S. Tisserant, Analog optical link in liquid argon, in these proceedings.

An Integrated UV preshower in the RD3 Liquid Argon Accordion Calorimeter: First Test Beam Results

Isabelle Wingerter-Seez
LAPP
BP 110, 74941 Annecy-le-vieux, Cedex, France
for the RD3 collaboration *

ABSTRACT

In this paper we report on the construction and the tests of an integrated UV preshower within the RD3 Liquid Argon (LAr) accordion electromagnetic calorimeter (EM). Both design and construction followed the same techniques as for the EM calorimeter: no problem was encountered. Beam tests were performed during summer 1994 and good performances of the integrated preshower have been measured: no energy resolution loss, position resolution better than 300 μm and angular resolution better than 6 mrad for electrons with energy above 100 GeV .

1. Introduction

In the framework of the ATLAS EM calorimeter elaboration and following the idea of the GEM collaboration [1] to integrate the preshower in the first sampling of the EM calorimeter, the RD3 collaboration decided to build and test an *integrated stereo UV preshower*. The RD3 collaboration has also tested a *separate preshower* and results were presented at this conference [2]. So were results of an *integrated* η *preshower* in an accordion Liquid Krypton calorimeter [3].

1.1. Physics requirements

Most of the constraints on the EM calorimeter requirements come from the necessity for the ATLAS experiment to be able to search for the Higgs boson in the mass region $M_H \geq 80\ GeV$ [4]. For $M_H \leq 130\ GeV$ the search of the Higgs decaying into 2 γ seems to be the only possible way to extract the signal over the QCD background. The $\gamma\gamma$ mass resolution has two contributions: one from the energy and one from angular measurements; in ATLAS, both benefit from the use of a preshower:

- due to the amount of dead material in front of the calorimeter(tracker, cryostat, coil, clearance), an early sample of the EM shower allows to correct for the energy loss in this material ($\simeq 1.2\ X_0$ at $\eta = 0$ to $\simeq 2.5\ X_0$ at $\eta = 1.4$).

- the angular resolution benefits from a highly segmented first sampling, where the EM shower is still narrow.

- a good position resolution improves the π^0 rejection, important when fighting the background to single photons.

*Alberta, Alma-Ata, Annecy, Barcelona, Brookhaven, CERN, Saclay, Grenoble, Madrid, Marseille, Milano, Montreal,Nevis, Orsay, Paris, Stockolm and Victoria

1.2. Construction constraints

When building a large detector as ATLAS, ease of construction is an important aspect; the main advantage of an integrated preshower is that it does not involve extra mechanical structure as it is part of the EM calorimeter; it is therefore intrinsincly aligned with it. Having two views (U and V) is also an apriori advantage as it should improve π^0 rejection.

In the next section, the construction of the two sectors of the RD3 EM prototype, equiped with UV electrodes, will be described and the noise figures summarized. In the third section, the test-beam setup and the analysis will be described leading to the position and angular resolution measurements. In the last section conlusions will be drawn and prospects presented.

2. Contruction

2.1. ideas

The RD3 EM prototype is segmented in three samplings in depth; in this test, the depth of the first sampling was reduced and the transverse segmentation modified to draw the strips of the integrated preshower.

In the GEM design [1], the integrated preshower had a fine segmentation in the η direction but a very coarse one in ϕ. The RD3 collaboration has developped the idea of a stereo segmentation in the U and V directions: making use of the accordion geometry ($\phi \neq cte$), two sets of strips with borders following the equations of Fig. 1 were drawn on the kapton electrodes; then, when stacking the module, U and V electrodes were interleaved. The rest of the construction follows the procedure described in [5]. To limit the number of electronic channels, six U (V) stips, with the same $\eta - (+)\phi$, were connected together. The total number of channels in the preshower sampling is twice the number of the middle sampling.

The strip width is $\simeq 5\ mm$. The segmentation is $\Delta\eta \times \Delta\phi = 0.00441 \times 0.0784$ to be compared to $\Delta\eta \times \Delta\phi = 0.00176 \times 0.0196$ for the middle sampling.

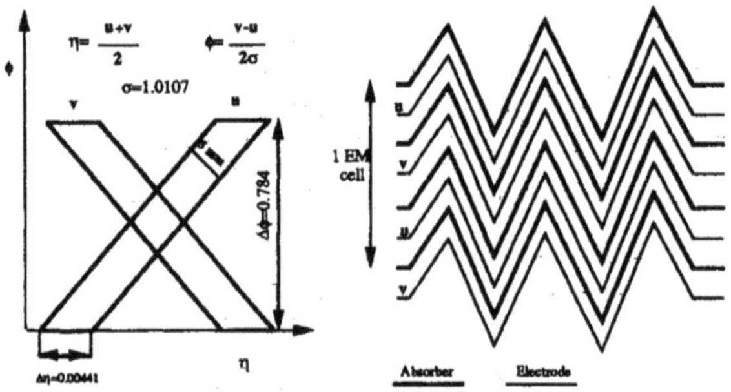

Fig. 1. *a) Schematic representation of the U and V strips in the η, ϕ plane. The value of $\sigma = 1.0107$ comes from the geometrical constraints of the RD3 prototype. b) Azimuthal view of the absorbers and electrodes.*

2.2. High voltage

In the RD3 EM prototype, the high voltage (HV) was distributed from the front face of the calorimeter for the front sampling, and from the back face for the middle and the back samplings. In order to reduce the number of connections on the front face, the HV for the preshower sampling was also brought from the back using resistive ink. The aim was to have a resistance of $\simeq 100\ k\Omega$ for these connections; only $\simeq 10\ k\Omega$ were achieved; this low resistance probably explains the cross-talk between the middle and the preshower sampling that was measured at the level of 1% per cell.

2.3. Electronics and read-out

Standard motherboards [5] equiped with hybrid GaAs preamplifiers were used for the middle and the back samplings. Multi-layers (10) cards, equiped with monolithic GaAs preamplifiers [7] were developped for this test; the total thickness of the electronics on the front face (summation+preamps+calibration) was $\simeq 27\ mm$. The standard acquisition chain was then used (shapers with $\tau_\delta \simeq 20\ ns$, track and hold and ADC); standard calibration was used.

2.4. Electronics noise and muon signal

Using the test-pulse system, the gains were measured to be 10 MeV/ADC for the preshower strips and 40 MeV/ADC for the middle and back samplings. The noise figures are respectively 15, 60 and 40 MeV (differences with [5] come from oscillations in the preamplifiers of the middle sampling). The cross-talk between adjacent U(V) strips was measured at the level of 2.5%.

Using muon data, the ratio of a mip signal to noise could be measured; summing the signal in two U and two V strips (chosen according to the beam impact), the ratio was found to be 3.3 (for a shaping time $\tau_\delta = 20ns$).

2.5. Conclusions on the construction

In summary, both design, construction and performances of the integrated UV preshower followed expectations: no major problem arose building the two sectors and the noise figures are as naively predicted; the operation of such a device is identical as of the calorimeter.

The design of the device could not be fully optimized as there were existing constraints from the already existing protoype: there was no massless gap, the overlap between cell definition and the kapton electrodes was odd (cf Fig. 1 and 3), the cross-talk was not studied in details.

Improvements are forseen for the HV connections between the middle and the preshower samplings and also a thinner summation system is under study as dead material before the calorimetr is critical for both energy and angular resolutions.

3. test-beam setup and analysis

3.1. Test-beam setup

The prototype was installed in the H8 beam at the CERN SPS; details about the installation can be found in [5]. The integrated preshower covers the region 0.42 to 0.85 in rapidity and 0.31 to 0.47 in azimuth. Energy scans with electrons of energies

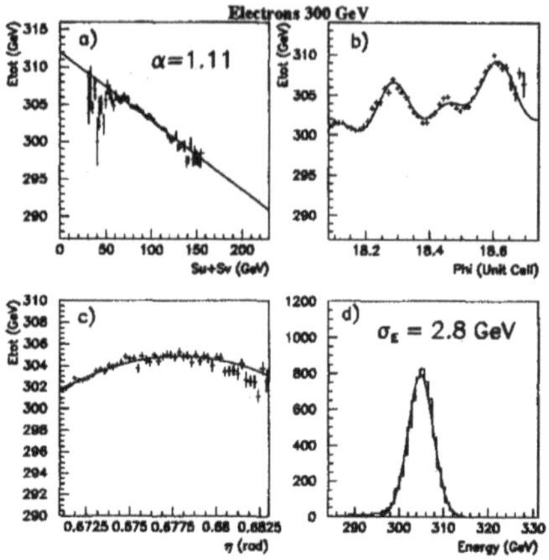

Fig. 2. *Energy reconstruction: a)* $E^{total}_{cluster}$ *vs* $E^{U+V}_{cluster}$*; the slope is fitted to* 1.1. *b)* ϕ *modulation across a cell of the middle sampling (18.0 corresponds to the middle of the cell); the fitted curve is a sum of sinusoides. c) lateral leakage across a cell in* η*; the fitted curve is a parabola. d)* $E^{corrected}_{cluster}$*; the beam energy resolution has not been unfolded.*

10, 20, 50, 100, 150, 200 and 300 GeV were performed at $\eta = 0.53$ and $\eta = 0.61$. The analysis presented here concentrates on data at $\eta = 0.53$ where the amount of dead material before the active part of the calorimeter amounts to $1.4X_0$ (beam line, cryostat, dead LAr), the depth of the three samplings being 5.6, 14.6 and 9.1 X_0 from front to back.

3.2. Energy reconstruction and shower profile

The preshower sampling is thin enough to be used to correct for the shower energy loss in the dead material (cf Fig. 2). The EM shower is reconstruted, in the preshower sampling, as 11 strips around the strip with maximum energy for U and V strips independantly; in the middle sampling the cluster is 3×3 cells and 2×3 in the back sampling. The total energy is calculted as $E_{total} = \alpha(S_U + S_V) + S_{middle} + S_{back}$. where S is the signal in the cluster for the given sampling, calibrated using the test-pulse system and the expected ionisation electrons yield in LAr. After corrections for the ϕ modulation (absorber vs electrodes) and the lateral leakage on the side of the cluster, the energy resolution for 300 *GeV* electrons is measured at $\frac{\Delta E}{E} = 0.9\%$ (cf Fig. 2), before unfolding the beam energy resolution contribution. This is comparable to what will be published in [5] without a preshower.

The fine granularity can be used to measure the shower width after 7 X_0. On Fig. 3 is shown the shower profile measured with the V strips, for 200 GeV electrons, the beam spread unfolded. The fit of the distribution gives a measurement of the shower width of 4 *mm*.

3.3. position resolution

The beam direction is measured with three MWPC [8] which resolution, both in η and ϕ directions is between 250 and 300 μm. The shower position measured with the preshower is calculated as the energy weighted barycenter on three strips (one strip on each side of the strip with maximum energy) for both U and V; then $\eta = \frac{U+V}{2}$ and $\phi = \frac{U-V}{2\sigma}$ are calculated. Some geometrical corrections, illustrated by Fig. 3, are applied independantly of the energy and relying only on the calorimeter. The beam chambers resolution is unfolded. The energy dependance of the position resolution in the η direction (the relevant one for ATLAS) has been fitted between 10 and 200 GeV by the function:

$$\sigma_\eta = \frac{1.55\pm0.16}{E} \oplus \frac{6.7\pm0.4}{\sqrt{E}} \oplus 0.24 \pm 0.01 \ mm.$$

The value for the constant term is not yet understood; the cross-talk is probably one explanation. Work is going on to try to understand these numbers, in particular the difference between the measurement at 100 GeV of $\sigma_\eta = 360\ \mu m$ and the prediction from a full monte-carlo simulation of 180 μm.

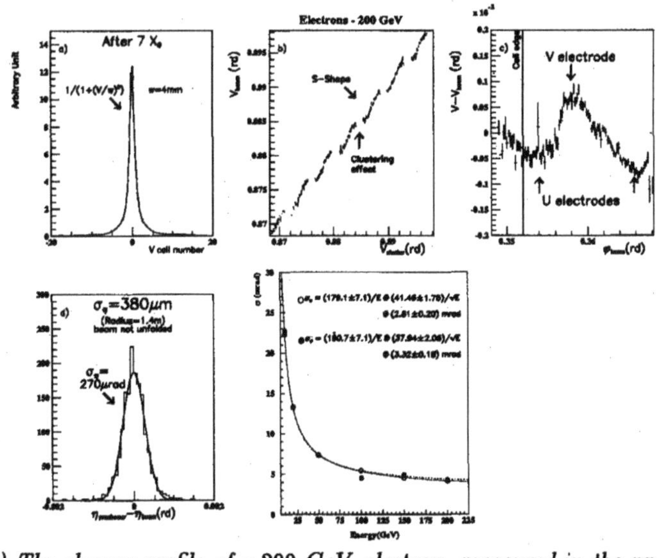

Fig. 3. *a) The shower profile of a 200 GeV electron, measured in the preshower sampling using the V strips. The curve is the result of the fit of a function of the form* $\frac{1}{1+(\frac{V}{w})^2}$ *where w is the width of the shower. w is found to be 0.8 which gives a width of 4 mm. b) Beam impact vs cluster position in* η *(radians). Two effects are visible: the clusterisation (only 3 strips are used) and barycenter definition "S-shape". c)* ϕ *modulation; d)* $\eta_{strip} - \eta_{beam}$ *in radians; e) the energy dependance of the angular resolution measured with the preshower and the middle sampling and a lever arm of 10 cm.*

3.4. angular resolution

Using the preshower and the middle sampling, and knowing the level arm between the 2 samplings, one can compute the shower direction. The position in the middle sampling is computed also as the weighted baycenter in the 3×3 cluster. The level arm has been estimated by monte-carlo to be 10 *cm*. On Fig. 3 the energy dependance of the angular resolution is shown, together with the fitted function. The angular resolution in the η direction is 8 *mrad* at 50 *GeV* and better than 6 *mrad* above 100 *GeV*.

4. Conclusions

The RD3 collaboration has built and operated with succes an integrated UV preshower: no fundamental problem was encountered. It has been tested in july 94 with electrons of energies between 10 and 300 GeV and good performances were measured:

- The energy resolution is not degraded by the presence of the preshower.
- The position resolution in the η direction is 380 μm for 50 GeV electrons, and better than 300 μm for energies above 100 GeV.
- The angular resolution, measured with the preshower and the middle sampling, is 8 mrad for 50 GeV and better than 6 mrad above 100 GeV.

Work is going on to understand better these numbers and to complete the analysis; the linearity, uniformity will be studied. Events with one γ (with energy between 10 and 100 GeV) and one electron were taken in the UV preshower; their analysis should allow to measure the $\gamma\gamma$ separation and to estimate the π^0 rejection.

5. References

1. The GEM collaboration, GEM technical design report,GEM-TN-93-262, SSCL-SR-1219.
2. J. Collot for the RD3 collaboration, Performance of a high granularity preshower, contribution to this conference.
3. D. Lissauer, Precision Electromagnetic Calorimetry with Liquid Krypton, contribution to this conference.
4. Z. Kunst and F. Zwirner,Testing the Higgs sector of the MSSM at large hadron colliders, *Nuclear Physics* **B385** (1992) 3-75.
5. D.M. Gingrich et al., The RD3 collaboration, Performance of a Large Scale Prototype of the ATLAS Accordion Electromagnetic Calorimeter, to be published in NIM.
6. J. Colas, S. Jézéquel and B. Mansoulié, Principles of a stereo preshower integrated in the accordion calorimeter, RD3 Note 55/ ATLAS-CAL-NO-40.
7. D. Camin for the RD3 collaboration, Monolithic front-end electronic for the accordion LAr calorimeter, contribution to this conference.
8. A. Manarin and G. Vismara, The Delay Wire (DWC) decription, LEP/BI-TA/Note 85-3.

BEAM TEST OF A NOBLE LIQUID PRESHOWER DETECTOR FOR NEUTRAL PION IDENTIFICATION

Jon K. Eisenberg and Paul M. Mockett
Physics Department, FM-15, Univ. of Washington
Seattle, WA 98195 USA

ABSTRACT

We present results from a beam test of a liquid argon π^0 detector placed near shower maximum in a sampling calorimeter. It employs 1 mm pitch strips with every sixth strip ganged together at the electrode dramatically reducing the number of electronics channels required but still allowing excellent π^0 rejection and good position determination. The position resolution is (0.67 ± 0.03) mm for 15 GeV/c electrons. Monte Carlo studies show that these strips would give π^0 rejection by a factor of 3 with better than 95% efficiency up to 100 GeV at 1 m from the interaction point.

1. Detector Design

To reduce the π^0 background for the $H \to \gamma\gamma$ channel in a hadron collider experiment, one must discriminate π^0 decay photons from single photons. In the first 3-4 X_0 of a shower the showers are only about one millimeter wide, requiring millimeter-level granularity. The transverse scale over which to discriminate is set by the photon opening angle (5.6 mrad minimum opening angle for 50 GeV π^0s).

We developed a liquid argon (LAr) π^0 detector with fine granularity incorporating a high degree of spatial multiplexing. The multiplexing and granularity were chosen to reduce the electronics channel count while providing good π^0 rejection over a large energy range.

The detector is comprised of 15 layers (fig. 1). The first 13 form a sampling calorimeter with 10 cm $\times$ 10 cm lateral segmentation. Two millimeter LAr gaps are read out by copper-clad polyimide circuit board electrodes. Fast striplines with four channels each bring charge from towers to charge-sensitive

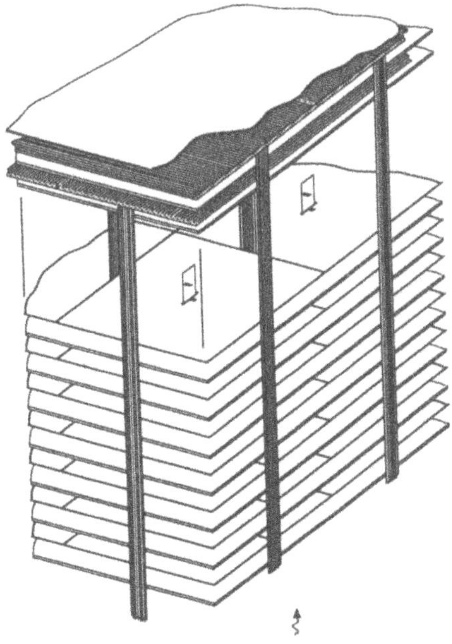

Fig. 1. Overall schematic of the detector. Particles are incident from below.

preamplifiers mounted upstream. Both signal and ground connections are made at each layer with soldered tabs so as to keep the inductance low. The striplines pass through 1 mm × 10 mm slots in the electrodes and absorbers. They make low-inductance signal and ground connections to a decoupling capacitor made from a multi-layer polyimide circuit board. The capacitors are then connected in a similar fashion to the preamplifier board. The absorber plates are 1 mm lead clad with with copper foils by means of a carbon fiber composite. The composite allows matching the thermal contraction coefficients of the absorber and electrode.

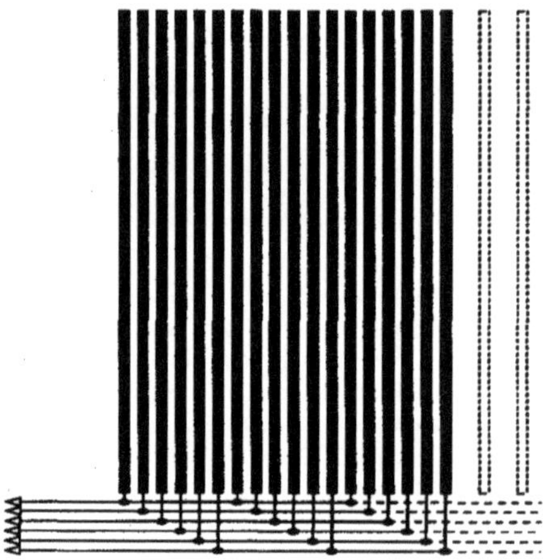

Fig. 2. Strip ganging schematic.

The last two layers, at about 3-4 X_0 depth, have electrodes with 1 mm pitch strips with every sixth strip tied together (fig. 2). Each layer reads out one coordinate. Six channels read out each 10 cm × 10 cm transverse region in 17 cycles. Thus there are 102 strips, each 10 cm long. One channel reads strips 1, 7, 13, ..., a second reads strips 2, 8, 14, ...and so on. The ganging is done using plated-through holes and a small bus region on the edge of the electrode, occupying 5% of the area. These channels are fed to the preamplifiers by six-channel microstriplines which are etched as an integral part of the electrodes so that no solder connections need be made in the gaps.

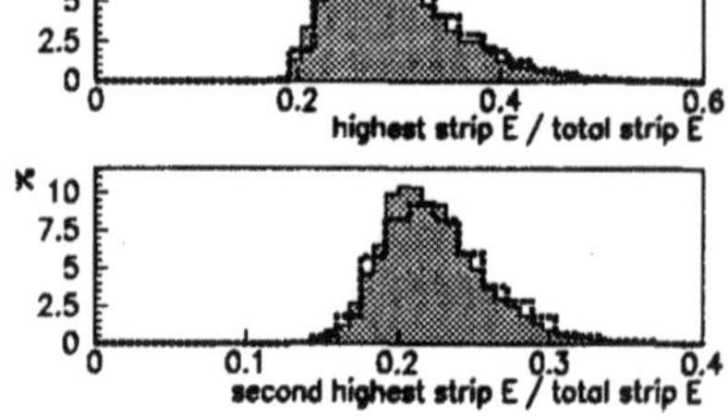

Fig. 3. Ratios of highest and second-highest x strip energies to the total energy deposited in the x strips. Filled region is data; dashed line is Monte Carlo.

The preamplifiers are read with an electronics system using a 40 ns shaping time. In this test, the electronic noise is 5 MeV. The energy scale here is the energy deposited in the LAr, assuming a long shaping time where all charge is collected. Much of this noise was due to extra capacitance in interfacing required for this prototype run. An optimized design could have 1 MeV noise.

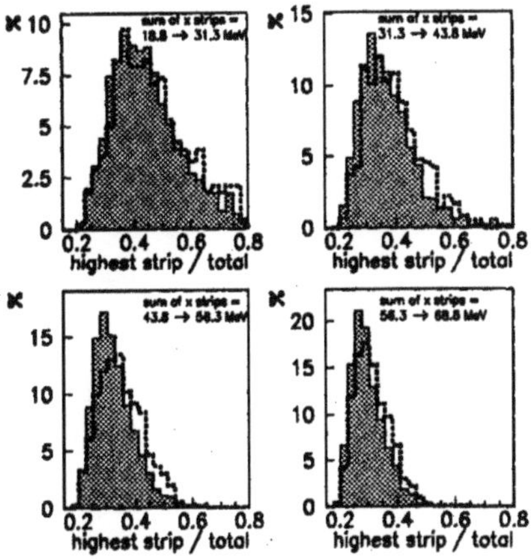

Fig. 4. Ratios of highest energy strip to total strip energy of total energy for subsequent bins of total energy deposited in the x strips. Filled region is data; dashed line is Monte Carlo.

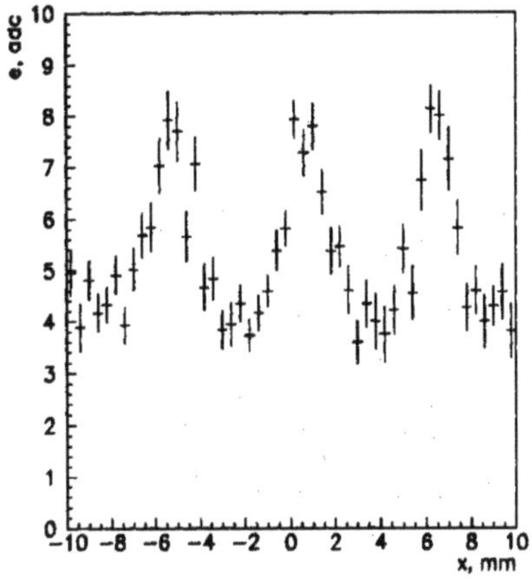

Fig. 5. Strip energy modulation in x for a single strip.

2. Results

The detector was tested with a 15 GeV/c momentum beam with an electron trigger at the AGS A3 beamline at Brookhaven National Laboratory. Agreement with Monte Carlo simulation using GEANT 3.16 was tested by comparing the ratios of energies deposited in the strips. Figure 3 shows the very good agreement between measurements and Monte Carlo. Comparing the ratios differentially in bins of total energy deposited in the strips also yields good agreement (fig. 4).

The position resolution of the strips is calculated by comparing the shower position measured in the strips with the electron position measured by a 150 μm accuracy drift chamber system. Given that the test was conducted only with 15 GeV electrons, events depositing a total energy above 50 MeV in the x strips are selected to show how the detector would perform at higher energies.

Plotting the energy in an x-measuring strip as a function of the x electron track position in the drift chamber, x_{DC}, yields the expected 6 mm period modulation (fig. 5). The position measured by the strips, x_{strip}, was calculated as the center of gravity of the highest energy strip and its highest neighbor. This information has an ambiguity of a multiple of 6 mm so that the correlation between track and strip positions, $x_{strip} - x_{DC}$, will give a peak every 6 mm corresponding to each of the cycles the beam strikes (fig. 6).

A Gaussian fit to the peaks gives a width of (0.81 $\pm$ 0.04) mm (fig. 7). Fitting to a Gaussian plus the background (from an as-of-yet unidentified source), the width is (0.67 $\pm$ .03) mm (fig. 8). Monte Carlo simulation including the drift chamber resolution and measured electronic noise level but neglecting any upstream interactions predicts a width of

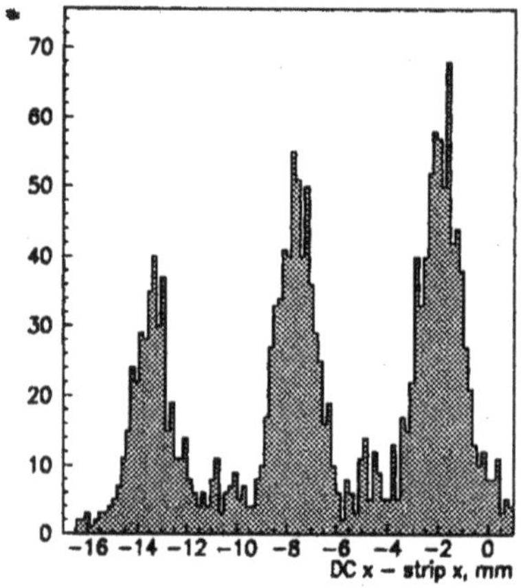

Fig. 6. Cycles in $x_{DC} - x_{strip}$. Peak heights reflect the beam intensity profile.

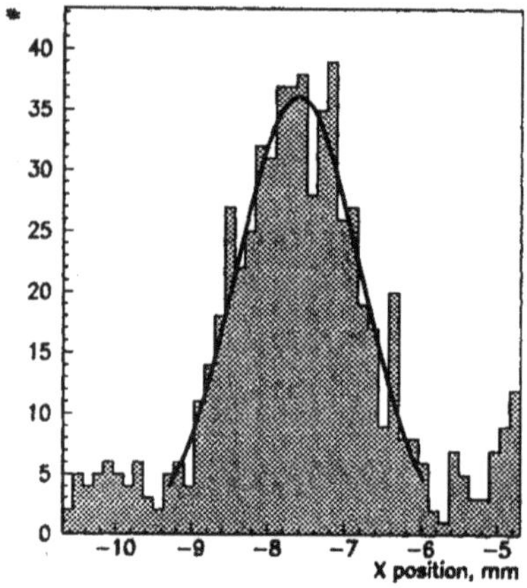

Fig. 7. Strip position resolution.

(0.62 ± .04) mm. If the intrinsic detector resolution is 0.67 mm, Monte Carlo π^0 rejection results for a distance of 850 mm from the interaction point (IP) will yield the same performance at a distance of 850mm × 0.67/0.62 = 920 mm.

3. Monte Carlo of π^0 Rejection

The detector configuration is simulated with GEANT 3.16. Isolated photons and π^0s are generated at a distance of 850 mm from the IP. A full shower simulation is used to generate detector signals for the strips. Electronics noise of 1 MeV is included. Two pattern recognition cuts selecting π^0s are applied to each coordinate separately:

- one strip above 15 MeV and a second with more than 50% of the height of the first and separated by one or more strips with less energy than the second
- for an incident energy of more than 85 GeV, more than 2 consecutive strips with greater than 60% of the highest energy strip

A π^0 is likely to give a skewed center of gravity when measured in an early longitudinal segment, The distance between the mean position of the shower as measured in the strips, x_{strip} and y_{strip}, and that measured in a calorimeter behind the π^0 detector, x_{cal} and y_{cal}, is compared using

$$r > \begin{cases} 6.75\text{mm}/\sqrt{E} & E < 85\text{GeV} \\ 0.65\text{mm} & E > 85\text{GeV} \end{cases} \quad (1)$$

where

$$r = ((x_{cal} - x_{strip})^2 + (y_{cal} - y_{strip})^2)^{1/2} \quad (2)$$

Corresponding to $7.9\text{mrad}/\sqrt{E}$, the cut is well within the capabilities of sampling LAr calorimeters. These cuts result in a π^0 rejection of a factor of 3 with a efficiency above 95% from 25 to 100 GeV (fig. 9).

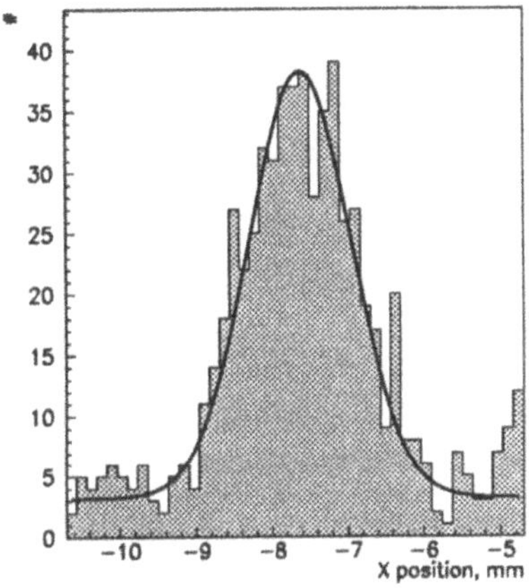

Fig. 8. Strip position resolution with background.

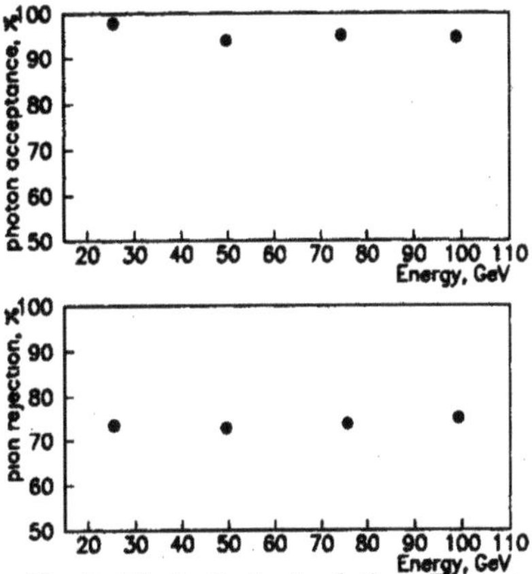

Fig. 9. Monte Carlo simulation. Top: γ efficiency. Bottom: π^0 rejection.

4. Summary

We have constructed and tested a π^0 detector that provides excellent rejection with high efficiency. A ganging scheme yields a low channel count while preserving good rejection performance. The Monte Carlo simulation using simple pattern-recognition cuts gives a π^0 rejection factor of 3 with better than 95% efficiency for these 1 mm strips. With the observed position resolution, this performance would be realized at a distance of approximately 1 m.

Acknowledgments

This work was supported in part by the Department of Energy under grant number DE-FG06-91ER4014 and the Texas National Laboratory Commission under grant number RGFY92-113.

THE NEW SMALL ANGLE CALORIMETER IN DELPHI

(Presented by Vincent Hedberg)

S.J. Alvsvaag, O.A. Maeland, A. Klovning
Department of Physics, University of Bergen, Allegaten 55, N-5007 Bergen, Norway

A.C. Benvenuti, V. Giordano, M. Guerzoni, F.L. Navarria, M.G. Verardi
Dipartimento di Fisica, Università di Bologna and INFN, Via Irnerio 46, I-40126 Bologna, Italy

T. Camporesi, E. Vallazza
CERN, European Organization for Nuclear Research, CH-1211 Geneva 23, Switzerland

M. Bozzo, R. Cereseto
Dipartimento di Fisica, Università di Genova and INFN, Via Dodecaneso 33, I-16146 Genoa, Italy

G. Barreira, M.C. Espirito Santo, A. Maio, A. Onofre, M. Pimenta, B. Tome
LIP and IST-FCUL Coimbra and Lisbon, Av. Elias Garcia 14-1, P-1000 Lisbon Codex, Portugal

H. Carling, E. Falk, V. Hedberg, G. Jarlskog, I. Kronkvist
Department of Physics, University of Lund, Sölvegaten 14, S-22363 Lund, Sweden

M. Bonesini, F. Chignoli, P. Ferrari, S. Gumenyuk, R. Leoni, R. Mazza, P. Negri, M. Paganoni, L. Petrovykh
Dipartimento di Fisica, Università di Milano and INFN, Via Celoria 16, I-20133 Milan, Italy

D.R. Dharmasiri, B. Nossum, A. L. Read, B. Skaali
Physics Department, University of Oslo, Blindern, N-1000 Oslo 3, Norway

L. Castellani, M.Pegoraro
Dipartimento di Fisica, Università di Padova and INFN, Italy, Via Marzolo 8, I-35131 Padua, Italy

A. Fenyuk, Yu. Gouz, Yu. Ivanyushenkov, A. Karyukhin, V. Obraztsov, E. Vlasov, A. Zaitsev
Institute for High Energy Physics, Serpukov P.O. Box 35, Protvino (Moscow Region), Russian Federation

M. Bigi, V. Cassio, D. Gamba, E. Migliore, A. Romero, L. Simonetti, E. Torassa, P.P. Trapani
Dipartimento di Fisica Sperimentale, Università di Torino and INFN, Via P. Giuria 1, I-10125 Turin, Italy

M. Bari, G. Della Ricca, L. Lanceri, P. Poropat, M. Prest
Dipartimento di Fisica, Università di Trieste and INFN, Via A. Valerio 2, I-34127 Trieste, Italy

ABSTRACT

Two lead-scintillator sampling calorimeters have been built for the DELPHI experiment at LEP. Wave length shifting optical fibers are used for light collection. The energy resolution of the detector is $13.5\%/\sqrt{E} \oplus 1.5\%$ and the radial shower position of 45 GeV electrons is measured with a resolution which varies between 250 μm at the edge between two towers to 1.2 mm at the center of a tower. In order to measure the angle of the shower axis, two planes of Si strip detectors have been installed inside the calorimeter at a depth of 4.0 and 7.4 radiation lengths. This shower maximum detector has an angular resolution of $\sim$13 mrad for 45 GeV electrons.

1. Introduction

The DELPHI collaboration installed two new Small angle scintillator TIle Calorimeters (STIC)[1 2] before the 1994 LEP run. The main purpose of the calorimeters is to measure the luminosity with a systematic accuracy of the order of 0.1%. This requires very hermetic calorimeters with a uniform energy response over the calorimeter surface at the level of a few per cent and an understanding of the internal geometry to better than 50 μm. With an energy resolution of 2-3% for 45 GeV electrons, the radiative Bhabha events can be separated from the background which is also important for high-precision luminosity measurements.

Luminosity is measured at LEP by counting the number of Bhabha scattered electron-positron pairs in a well-known acceptance region and since the measured cross section $\propto 1/\theta^3$ (where θ is the angle of the scattered electrons with respect to the beamline) it is the lower edge of the acceptance that is important to understand. The angular dependence of the cross section also explains why the luminosity detectors surround the beampipe at minimum angles.

DELPHI has previously relied on a tungsten ring around the beampipe to define the inner acceptance[3]. With this method an energy cut translates into a very sharp radial cut and one can obtain a radial resolution that is ~25 μm. A tungsten ring is used also with STIC but in addition radial position measurements with this new detector makes it possible to define other smaller acceptance regions and determine the luminosity without using the tungsten ring, thus providing an (almost) independent luminosity measurement.

The new STIC detectors should moreover improve the hermeticity of DELPHI in the forward region in view of the search for new particles when LEP will run at energies above the Z^0. It will be an important tool for tagging the photon from radiative events at LEP200. These events can be used for several physics studies e.g. α_s and the hadronic to muon cross section ratios for $m_Z \leq \sqrt{s} \leq m_{ww}$. STIC can furthermore tag the scattered small angle electrons in collisions between virtual photons (so-called two photon physics).

To improve the rejection of off momentum electrons the detector should be able to measure the direction of a shower with a ~10 mrad accuracy. For this reason the calorimeter has been equipped with two planes of Si strip detectors which act as a shower maximum device. These detectors will also provide a limited hadron identification.

2. Mechanical design

2.1. The calorimeter

The two cylindrical calorimeters surround the beampipe on each side of the interaction region in DELPHI. The first scintillator plane in each calorimeter is at a distance of 2200 mm from the interaction point and extends radially from 65 mm to 417.5 mm which translates into an angular coverage of 29 mrad $\leq \theta \leq$ 188 mrad. The calorimeters consist of 47 lead/scintillator layers and two planes of Si strip detectors. Both the lead and the scintillator layers have a thickness of 3 mm and the total thickness of the calorimeters corresponds to 27 radiation lengths.

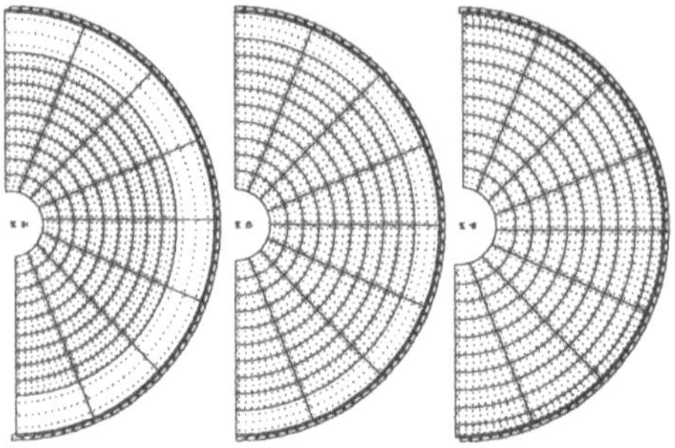

Plane 1 *Plane 25* *Plane 49*

Figure 1: The first, middle and last scintillator plane in one half of one calorimeter. The sector borders are slightly displaced in the different layers. Due to the projective geometry the width of the tower rings increases from layer to layer with exception for the outer ring which decreases in width.

The scintillator layers are made of injection molded polystyrene plates doped with 1.5% paraterphenyl and 0.05% POPOP. These plates have been cut into a projective tower structure with 10 rings and 16 sectors as depicted in figure 1. The rings in the first scintillator layer are 30 mm±25 μm wide (with exception for the 35 mm wide inner ring and the 78 mm outer ring). The sectors in all scintillator planes cover an angle of 22.5°. The sector borders, however, are displaced from plane to plane by 375 μm i.e by a total of 18 mm from first to last plane. The reason for this displacement is to obtain a perfect hermeticity and to avoid the possibility that particles traverse the calorimeter in the holes for the optical fibers or in the crack between the two half-cylinders that make up one calorimeter. The converter plates are continuous half-circular steel laminated lead plates (with a total thickness of 3.4 mm). The scintillator tiles are mounted with an accuracy better than 50 μm on the lead plates by using ⊘1 mm steel alignment pins. The tiles are optically isolated from each other by white 120 μm thick Tyvek® diffuser paper.

The calorimeter is of the so-called Shashlik type i.e. the blue light produced in the scintillators is read out by ⊘1 mm optical plastic fibers (Y7 Kuraray) doped with green wave length shifter. The fiber density is 0.79 fibers/cm^2 except for the outer ring where it is 0.30 fibers/cm^2. The green light in the fibers is carried to the back of the calorimeter and the fiber bundle coming from each tower is viewed by a Hamamatsu 1" R2149 phototetrode, positioned 5 mm from the polished end of the fibers.

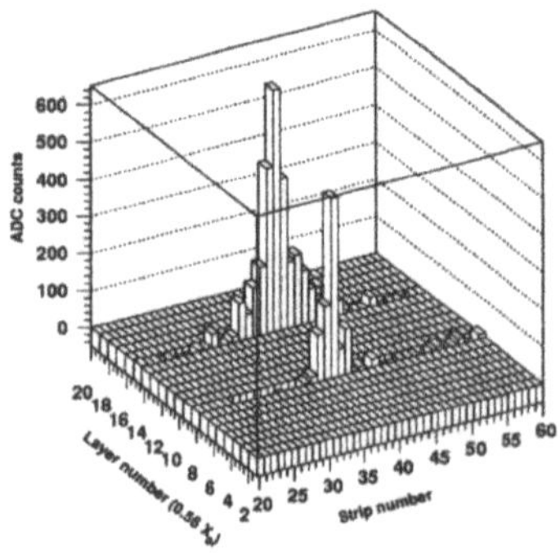

Figure 2: The online pulse height distribution for different strips after pedestal and noise subtraction.

The tetrodes have an average gain of 16 in the 1.2 Tesla field of DELPHI when they are mounted with a 15° tilt with respect to the magnetic field direction (the gain is 30 without magnetic field). The quantum efficiency is ~10% and the photo electron yield is estimated to be 1000 p.e. per GeV of incident particle energy.

The electronic readout chain consists of the preamplifiers, Fastbus shapers and Fastbus ADC cards developed for the triode readout of the forward lead glass electromagnetic calorimeter[4] (FEMC) in DELPHI.

2.2. The shower maximum detector

The calorimeter planes no. 8 and no. 15 (at a depth of 4.0 and 7.4 rad. lengths) consist of Si strip detectors[2]. These 300 μm thick detectors have circular strips which are 1.712 and 1.754 mm wide in the radial direction and match the calorimeter's projective geometry and 22.5° sector segmentation. They cover the angular region 32 mrad $\leq \theta \leq$ 77 mrad i.e. the inner three and a half rings of towers. An unusual feature of these detectors is given by the holes (laser-cut with a precision of about 10 μm) which are needed to let the wavelength shifting fibers go through the strips.

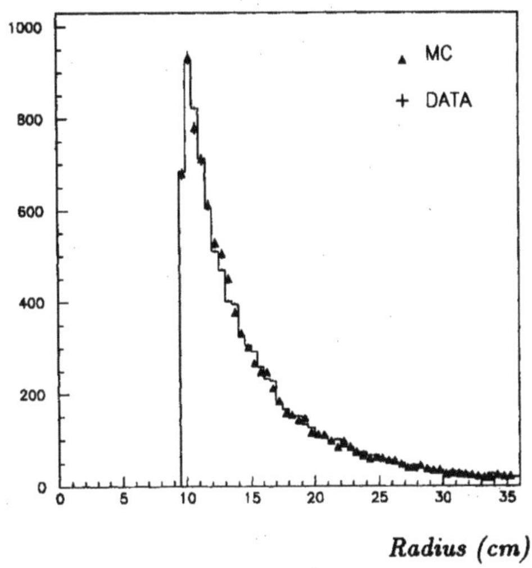

Figure 4: The radial distribution of Bhabha electrons measured by STIC. The histogram represents data and the triangles a Monte Carlo simulation (GEANT).

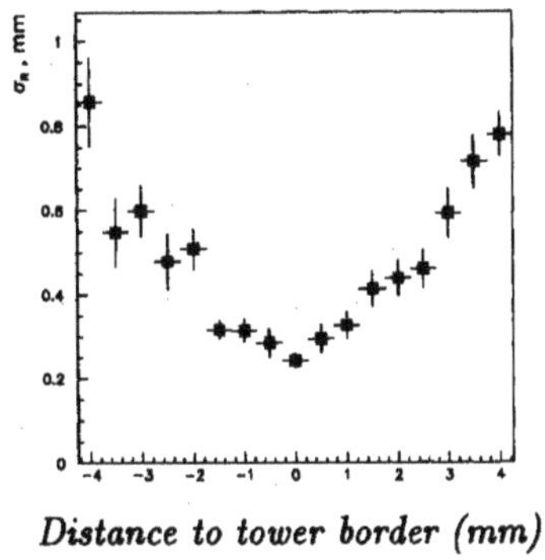

Figure 3: The resolution of the measurement of the radial shower position in a region around the border of the first and second tower rings.

The signals from the detectors are taken to the outside of the calorimeter by 35 cm long Kapton cables. These cables increase the detector capacitance from 25 pF to 45 pF and are connected to Microplex 4-type[5] 128 channel charge amplifiers with multiplexed output. The signals from the Microplex cards are amplified ~20 times by Amplifier cards and these are in turn readout sequentially by Multiplexer cards which send the data to the control room via a 8-to-1 analog multiplexer. The data from each calorimeter is clocked into a Sirocco IV Fastbus module[6] with a frequency of 1 MHz and the data is then digitized by a flash ADC in the Sirocco.

Figure 2 shows an example of an online pulse height distribution caused by one 45 GeV electron after pedestal and coherent noise subtraction.

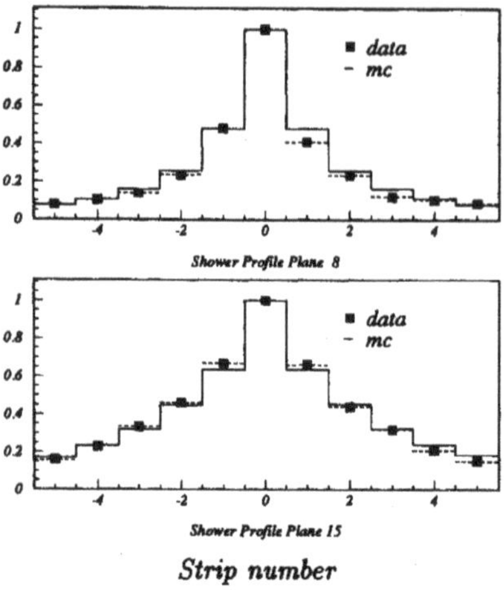

Figure 5: The lateral shower profiles measured by the two silicon strip planes compared to GEANT Monte Carlo simulations.

3. Position measurements

In a luminosity detector the accuracy of the shower position measurement is more important than the energy resolution. Thus a large effort was devoted to build STIC with a projective geometry and to a very high mechanical precision. When the tungsten ring is not used, the border between tower ring one and two defines the inner edge of the acceptance for Bhabha electrons and consequently a high radial resolution at this ring border is required.

A STIC-prototype was studied in the X5 testbeam at CERN in 1992 and a radial resolution better than 200 μm was obtained[1] for 45 GeV electrons close to a tower border.

During 1993, two different STIC-modules (one module is one half of a calorimeter) were studied in the X3, X5 and X7 testbeams. A silicon microstrip telescope was used in the X3 beamline to measure the beam particles' trajectory with a resolution of the impact point of $\sim$40 μm.

The ratio of energy below and above a tower ring was measured as a function of the impact point and this measurement was then parametrized to translate an energy measurement into a radial position. The resolution of this position measurement for 45 GeV electrons is shown in figure 3 in a region $\pm$4 mm from a tower ring border. At the border the resolution is 250 μm and it increases to a maximum of 1.2 mm at the center of a tower. The azimuthal angle can also be measured by the calorimeter albeit with a poorer resolution due to the coarser sector segmentation and because the sector borders are not well defined due to the displacement of the scintillator layers. A resolution of $\sim$1 mm at the sector borders increasing to $\sim$4 mm at the center of a tower has been obtained for 45 GeV electrons in ring two. The fits to testbeam data can be used to reconstruct the radius of Bhabha scattered electrons and positrons in LEP. Figure 4 shows such a distribution and a comparison with a GEANT Monte Carlo simulation. The distribution follows the characteristic Radius^{-3} dependence expected for Bhabha

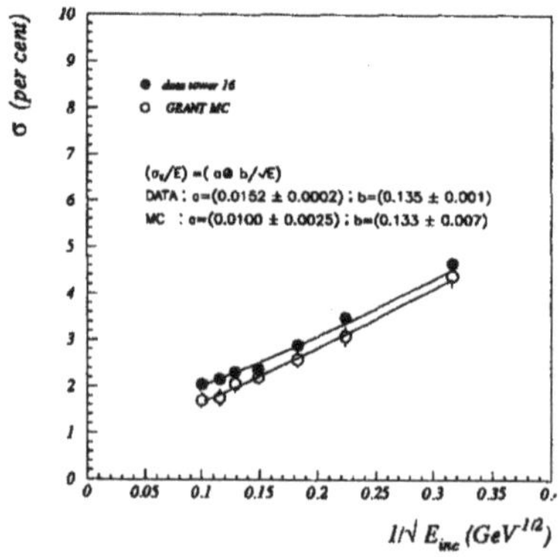

Figure 6: The measured and GEANT simulated energy resolution for electrons going into the center of one tower.

electrons and the overall agreement between data and Monte Carlo is good. This work, however, is still preliminary and future improvements are expected.

In 1994 a testbeam measurement was carried out in the X7 beamline to study the performance of the Si strip detectors in STIC. The data agreed very well with GEANT Monte Carlo simulations as can be seen in figure 5 which shows the shower profiles for 45 GeV electrons measured by the two planes of Si strip detectors. The Monte Carlo predicts a radial resolution of 440 μm and 610 μm for the two planes. The distribution of the difference of the measured radius in the two planes has been used to calculate an angular resolution of $\sim$13 mrad (rms).

4. Energy measurements

An energy scan was carried out in 1993 at the X5 testbeam with electron beams of 10,20,45,60,70 and 100 GeV of energy. The beam had a size (rms) of 3.5 mm vertically and 6.0 mm horizontally and hit the center of a tower in ring 6. The result of these measurements is depicted in figure 6 which shows the energy resolution as a function of $1/\sqrt{E}$. The data can be fitted with $\sigma/E = 13.5\%/\sqrt{E} \oplus 1.52\%$ and a negligible noise term of 90 MeV. A GEANT Monte Carlo simulation of the detector predicted a resolution of $\sigma/E = 13.3\%/\sqrt{E} \oplus 1.00\%$ i.e. the stochastic term is well described by the Monte Carlo while the constant term is not perfectly reproduced. This is probably due to the fact that fluctuations and imperfections in the light collection process are not yet incorporated in the Monte Carlo. The energy response of the calorimeter was found to deviate from linearity by less than $\pm1\%$. The spatial uniformity has been studied with both testbeam data and LEP data. With a 45 GeV electron beam the variation of the measured energy in radial scans was found to be less than $\pm2\%$. Figure 8 shows a measurement of the energy of Bhabha scattered electrons at a fixed radius and as a function of the azimuthal angle (where both radius and angle were measured by the calorimeter). The calorimeter azimuthal angle resolution is sufficient to show the energy modulations due to the fiber structure in the detector. The spatial non-uniformities are at the order of $\pm3\%$.

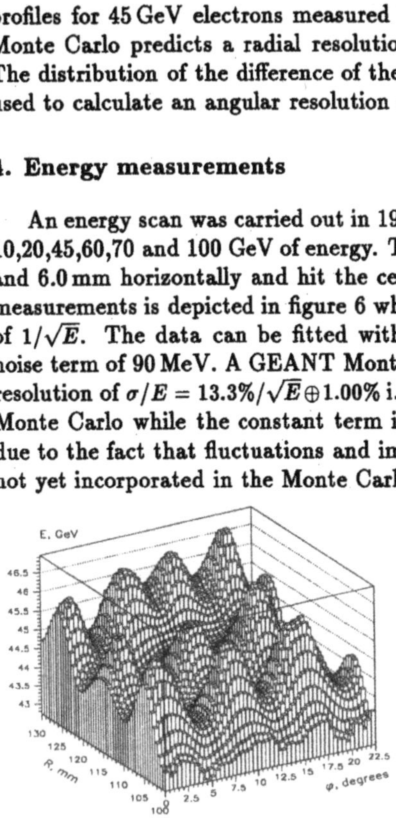

Figure 7: Map of the measured energy as a function of the radius and azimuthal angle measured by the calorimeter (for towers in ring two).

Using the LEP data it is possible to map the energy response of the calorimeter as a function of the measured shower position (by the calorimeter). Such an energy response map for a tower in the second ring is shown in figure 7. The map has been produced by fitting a smoothing function to the data. These maps can be used to correct the energy measured by the calorimeter after the position has been estimated.

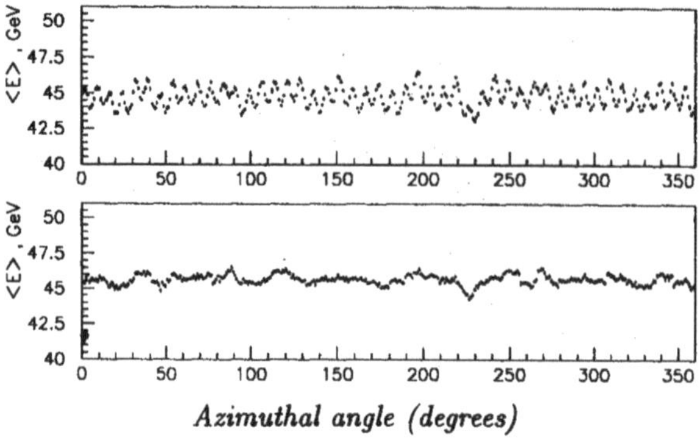

Figure 8: The measured energy at a radius of 10.5 cm as a function of azimuthal angle for 45 GeV electrons at LEP. Both angle and radius were measured by the calorimeter itself. The top plot is before and the bottom plot after energy correction.

An azimuthal angle scan after such a correction was applied is shown in figure 8. Much of the modulations due to the fibers have disappeared. The energy resolution is also improved from a resolution for 45 GeV electrons at LEP of 3.2% to 2.7% after the energy map correction. This is very close to the 2.5% resolution obtained under ideal conditions in a testbeam with the beam going into the center of a tower.

5. Conclusions

The small angle scintillating tile calorimeter in DELPHI satisfies the design requirements with a radial position resolution between 250 μm and 1.2 mm and an energy resolution of 2.7% for 45 GeV electrons at LEP. The shower maximum detector has an angular resolution of ~13 mrad.

The present 1994 LEP run has demonstrated that the detector can provide a reliable on-line luminosity measurement as well as a high precision offline estimate of the luminosity.

6. References

1. A. Benvenuti et al., *IEEE Trans. Nucl. Sci.* **40** (1993) 537;
   A. Benvenuti et al., to be published in the proceedings of *The International Europhysics Conference on high energy physics*, Marseilles, July 1993;
   A. Benvenuti et al., to be published in the proceedings of *The 4th International Conference on calorimetry in high energy physics*, La Biodola, Elba, September 1993.
2. S.J. Alvsvaag et al., to be published in the proceedings of *The 6th Pisa Meeting on Advanced Detectors*, La Biodola, Elba, May 1994.
3. P. Abreu et al., *Nucl. Phys.* **B367** (1991) 511.
4. I. Lippi et al., *Nucl. Instr. and Meth.* **A286** (1990) 243.
5. J.T. Walker et al., *Nucl. Instr. and Meth.* **A226** (1984) 200.
   P.P. Allport et al., *Nucl. Instr. and Meth.* **A273** (1988) 630.
6. M. Burns et al., *Nucl. Instr. and Meth.* **A277** (1989) 154.

Test of a ZEUS BCAL Module Including Presampler and Shower Max Detector in a Low Energy Beam

DAIVA MIKUNAS
High Energy Physics Division
Argonne National Laboratory
Argonne, IL 60439, U.S.A.

and

Department of Physics
Illinois Institute of Technology
Chicago, IL 60616, U.S.A.

Abstract

Two possible upgrades, a shower max detector and a presampler, designed to improve the low energy electron/hadron separation capabilities of the ZEUS barrel calorimeter are described. Testbeam results are reported.

I. Introduction

ZEUS, one of two detectors at the storage ring HERA at DESY in Hamburg, intercepts the products of collision between 27 GeV electrons and 820 GeV protons. A major component of the ZEUS detector is the compensating, uranium/scintillator calorimeter which is divided into three parts, the forward, rear and barrel calorimeters. The barrel calorimeter, or BCAL, is itself composed of 32 modules. Each module is divided into an electromagnetic section (EMC), one interaction length deep (21 Xo), and two hadronic sections (HAC1 and HAC2), each two interaction lengths deep. [1]

Besides providing energy measurements of electrons and jets, the BCAL can also be used for electron/hadron separation. The standard way to do this, is to exploit differences in electron and hadron shower shapes. The electron shower occurs earlier in the calorimeter, is narrower, and is contained in the EMC section of the BCAL, while the hadron shower is later, broader, and deposits most of its energy into the HAC's. This method is not efficient at lower energies where the hadrons also often deposit a sizeable fraction of their energy into the EMC section.

II. Two Possible Solutions

For 2 GeV electrons, their shower maximum occurs after they have traversed about four radiation lengths of matter. Something in the detector sensitive to shower activity, placed after four radiation lengths, would give a second handle on identifying these low energy electrons.

In the EMC section of the BCAL, between the third and fourth layers of uranium, (one layer of uranium is 1 Xo thick) a space was provided for just such a shower max detector. For ZEUS, the shower max detector is called a Barrel Hadron Electron Separator, or BHES for short. The type of shower max detector proposed as an upgrade to the BCAL is a multi-wire proportional chamber, with wire and pad read-out. [2]

The mechanical support for the BHES is a three sided, aluminum, box with rectangular grooves in which the wires are strung. The wires run parallel to the beam axis providing measurement of the phi coordinate. The separation between adjacent wires is 0.68 cm. The pads are on the circuit boards which close the aluminum box. The pads are perpendicular to the wires and so provide information on the theta coordinate. The separation between centers of adjacent pads is 1.74 cm.

Another possible handle on identifying lower energy electrons is provided by a presampler (PRE). Between the ZEUS interaction point and the BCAL modules, there is a superconducting coil that provides the magnetic field for the detector. This coil is about one radiation length thick, and so the low energy electrons are much more likely to start showering in this coil than are the low energy hadrons. A layer of scintillator tiles between the coil and BCAL would show a multi-mip signal for the showered electrons and a single mip signal for the unshowered hadrons, thus providing discrimination between the two.

III. Testbeam Set-up

Beam tests of the two possible upgrades to the BCAL, the BHES and the presampler, were done at the Brookhaven National Laboratory AGS testbeam using a prototype module that is the same in essential characteristics to the production version. The BHES was inserted into the prototype BCAL module and the presampler was placed between a mock aluminum coil and the module. Data was taken with electrons and pions in the energy range of 1 to 6 GeV. A schematic of the trigger is shown in Figure 1. Scintillator counters S2, S3, Sa and Sb were used to define the beam, and C1 and C2 were Čerenkov counters used to select electrons or pions. The prototype BCAL module was mounted on a stand (the muon stand) that allowed the module to be positioned either in the horizontal or vertical direction with an accuracy of 0.5 mm.

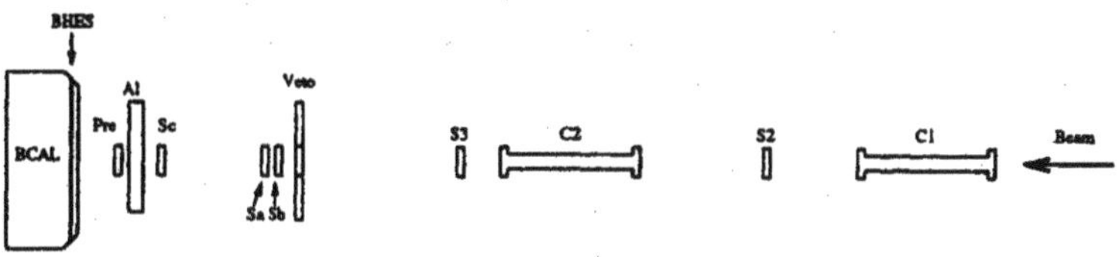

Figure 1 : Schematic layout of the testbeam trigger counters.

IV. Data Analysis

The BHES measures shower activity by means of the ionization produced by the particles in the shower traversing the volume of gas flowing through the BHES. This ionization charge is collected on the wires and pads. To be less sensitive to the noise in the BHES detector coming from the uranium radioactivity, a minimum charge cut Qmin was used. For a Qmin from 0.01 to 0.1 picocoulombs, the signal from the showering particles was larger than these values.

Two quantities were used in the analysis. The first was the number of wires per event with charge above Qmin. A histogram of this quantity for a central energy of 3 GeV is shown in Figure 2(a) for electrons and in Figure 2(b) for pions. If all events with more than one wire with charge above Qmin are called electrons, then this cut keeps 97.1% electrons for a pion contamination of 26.9%.

The second quantity is the total charge deposited per event. This is shown for both 3GeV electrons and pions in Figure 2(c). If the cut is placed at 0.35 Picocoulombs, then for an electron efficiency of 91.5% there is a pion contamination of 10.0%. The two cuts combined give an electron efficiency of 90.4% for a pion contamination of 8.4%.

The last step in the BHES analysis is to include data from the BCAL itself. To discriminate between electrons and pions using the BCAL, a cut was made on the fraction of the total energy deposited in the EMC sections. The numbers from combining the cuts, as well as the numbers for the BHES and BCAL separately, are shown in Figure 4 for 90% electron efficiency and in Figure 5 for 75% electron efficiency for beam momenta between 1 and 6 GeV. For example, at 3 GeV, combining the two cuts reduces the pion contamination at 90% electron efficiency from 6.2% without the BHES, to 3.2% with the BHES. At 75% electron efficiency, the pion contamination of 3.2% with the BCAL only, is reduced to 0.9% when the BHES is included.

In particular, at 1 GeV for an electron efficiency of 75%, the pion contamination is cut from 15.8% to 6.4% when the BHES is combined with the BCAL.

A similar analysis was done using the presampler data. The presampler signal for 3 GeV electrons and pions is shown in Figure 3. The final numbers for 75% electron efficiency are shown in Figure 6, for PRE alone, BCAL alone, and PRE and BCAL combined. The improvement in electron identification with the presampler over with the BCAL alone is a factor of about two to three.

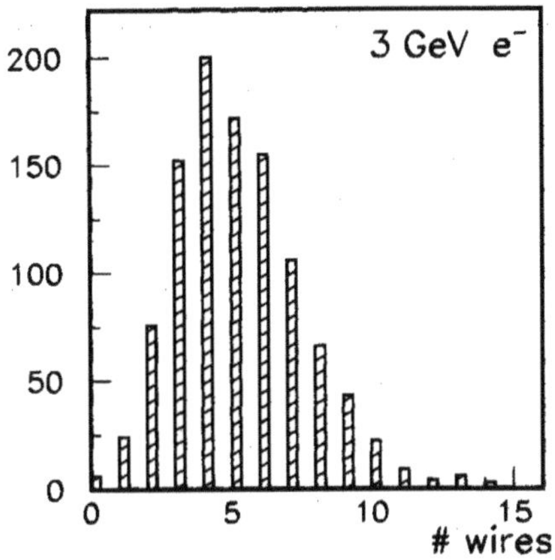

Figure 2a. Number of wires per event with charge above Q_{MIN} for 3 GeV e^-

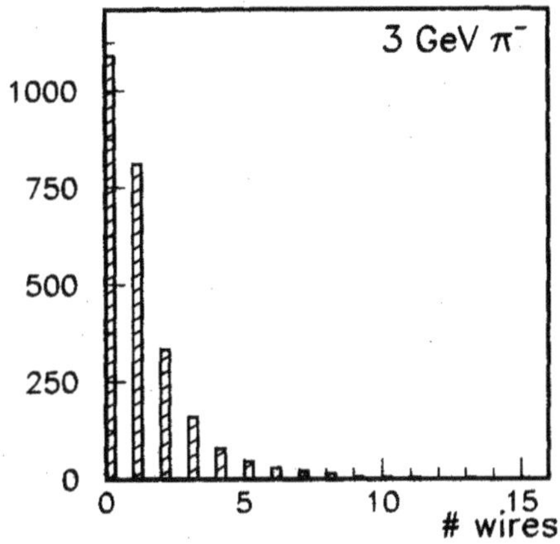

Figure 2b. Number of wires per event with charge above Q_{MIN} for 3 GeV π^-

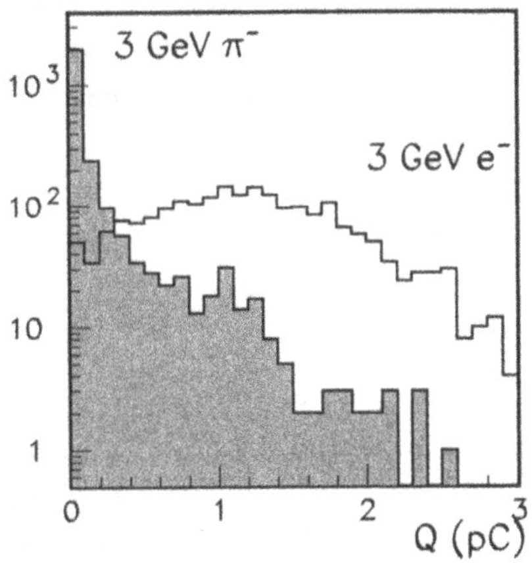

Figure 2c. Total charge deposited on the wires for 3 GeV e^- (open) and 3 GeV π^- (shaded histogram)

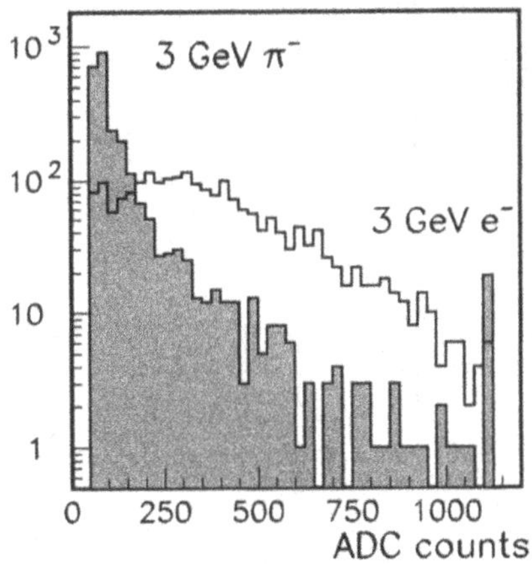

Figure 3. PREsampler pulse height for 3 GeV e^- (open) and 3 GeV π^- (shaded histogram)

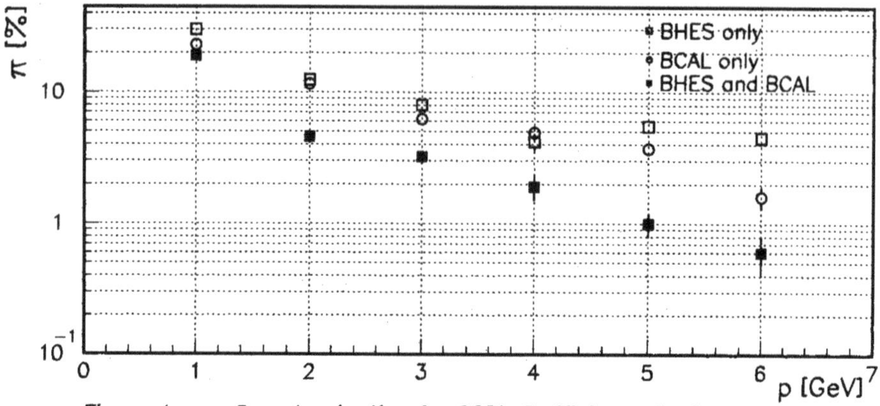

Figure 4. π^- contamination for 90% e^- efficiency for beam momenta between 1 and 6 GeV using the BHES and the BCAL.

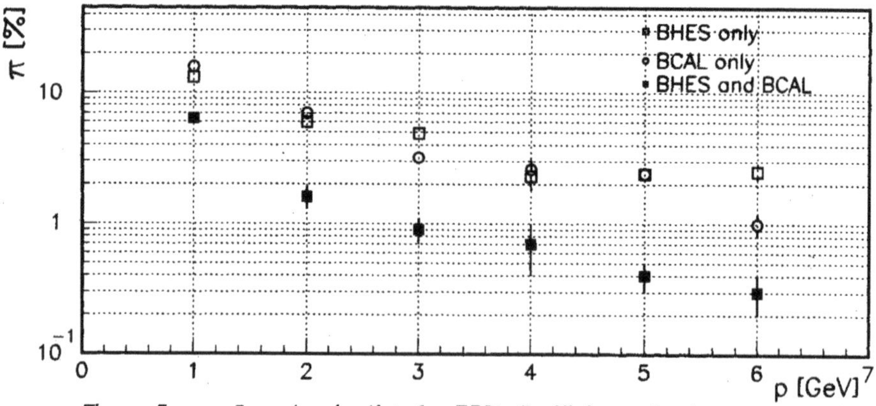

Figure 5. π^- contamination for 75% e^- efficiency for beam momenta between 1 and 6 GeV using the BHES and the BCAL.

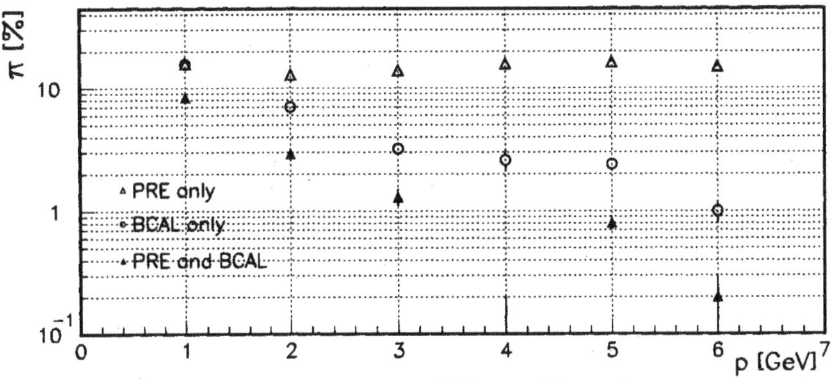

Figure 6. π^- contamination for 75% e^- efficiency for beam momenta between 1 and 6 GeV using the PRE and the BCAL.

V. Position Resolution

The BHES also improves the position resolution of electromagnetic showers in the BCAL. The front face segmentation of the BCAL module is (4.9cm) x (23.3cm). In the 4.9 cm dimension, energy sharing between towers is used to find the position (the theta coordinate), and in the 23.3 cm dimension, the position (the phi coordinate) is estimated from energy sharing between phototubes that read out on opposite sides of the same tower. For the BHES, the position is obtained from the charge weighted average for both the wires and pads. The finer segmentation of the BHES, described in section II, produces a position resolution improved by about a factor of three, as seen in figures 7(a) and 7(b). For 1 and 2 GeV, the electron showers were confined to one tower, and so energy sharing between towers could not be used to find the theta coordinate.

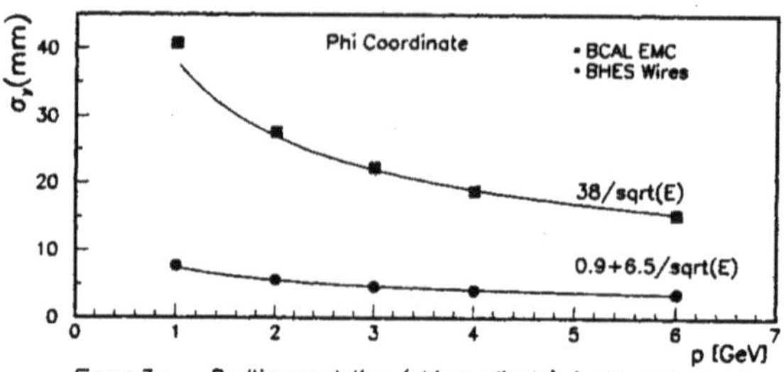

Figure 7a. Position resolution (phi coordinate) for beam momenta between 1 and 6 GeV using the BHES and the BCAL.

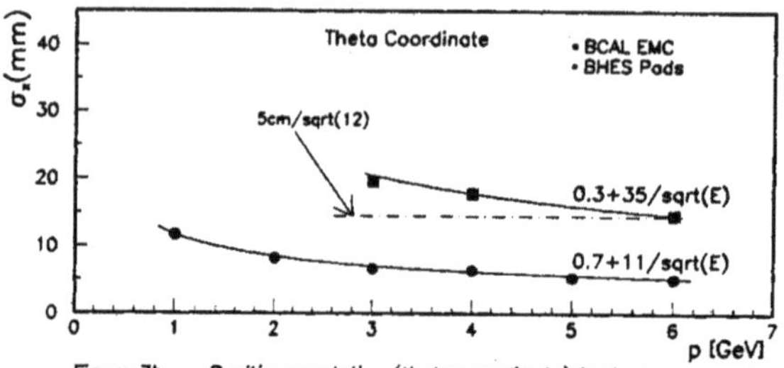

Figure 7b. Position resolution (theta coordinate) for beam momenta between 1 and 6 GeV using the BHES and the BCAL.

VI. Conclusion

The primary purpose of the testbeam was to see how and by how much the electron identification capabilities of the BCAL could be improved. The results are:

****The BHES improves the electron identification capabilities of the BCAL by a factor of 2 to 3, which is especially important at 1 and 2 GeV, where the pion contamination is large.

****If 75% electron efficiency is acceptable, then the PREsampler is also a good tool for separating electrons and hadrons.

****Lastly, the position resolution of electromagnetic showers in the BCAL is significantly improved by the BHES.

References

1. M.Derrick et al., *NIM* **A309**, 77(1991).

2. L.Balka et al., *NIM* **A267**, 272(1988).

SHOWER MAXIMUM DETECTOR FOR SDC CALORIMETRY

J. ERNWEIN
CEA, DAPNIA/SPP, CE-Saclay
91191 Gif-sur-Yvette, France

ABSTRACT

A prototype for the SDC endcap (EM) calorimeter complete with a preshower and a shower maximum detector was tested in beams of electrons and π's at CERN by an SDC subsystem group [1]. The prototype was manufactured from scintillator tiles and strips read out with 1 mm diameter wave-length shifting fibers. The design and construction of the shower maximum detector is described, and results of laboratory tests on light yield and performance of the scintillator-fiber system are given. Preliminary results on energy and position measurements with the shower max detector in the test beam are shown.

1. Introduction

The SDC was to build a general purpose detector to study proton-proton interactions at 40 TeV at the Superconducting Super Collider (SSC) [2].

The detector was designed to measure electrons, gammas, muons, jets, and missing energy. Good particle identification is of great importance for achieving this goal. The shower maximum detector (SMD), located at $\sim 5X_0$ in the EM central calorimeter, provides a fine segmentation of each tower and makes it possible to measure the position of showers to a few mm by using energy sharing in the strips. By matching the tracks with the shower position, it is possible to distinguish charged pions accompanied by γ's or π^0's from electrons. In addition, π^0's can be distinguished from single γ's on an event by event basis up to energies of $\sim$ 60 GeV above which the two γ's are too close together.

A 128-strip SMD built at Saclay was inserted in a 16-tower prototype of the endcap EM calorimeter built at LBL. The requirements which must be met by the detector to allow for additional particle identification beyond the standard techniques such as energy sharing between calorimeter towers, electromagenetic to hadronic energy ratio, and E/p cuts, are the following: *i)* a segmentation of 8 strips per EM tower is needed to ensure adequate position resolution, *ii)* the strip length should not exceed 4 EM towers to keep the occupancy at a low level, *iii)* the light yield must exceed 1 $p.e./MIP$ to maintain good resolution at low energy, and *iv)* the detector must be radiation resistant.

2. Design and Construction

2.1. Shower Maximum Module Design

The SMD is made of a single mechanical module which is inserted in the EM calorimeter after lead plate #5 and covers 16 towers as shown in figure 1. The transverse dimensions of the EM tower are 11 cm x 11 cm. There are two shower max (SM) planes in which strips run in two orthogonal directions. The width of a SM

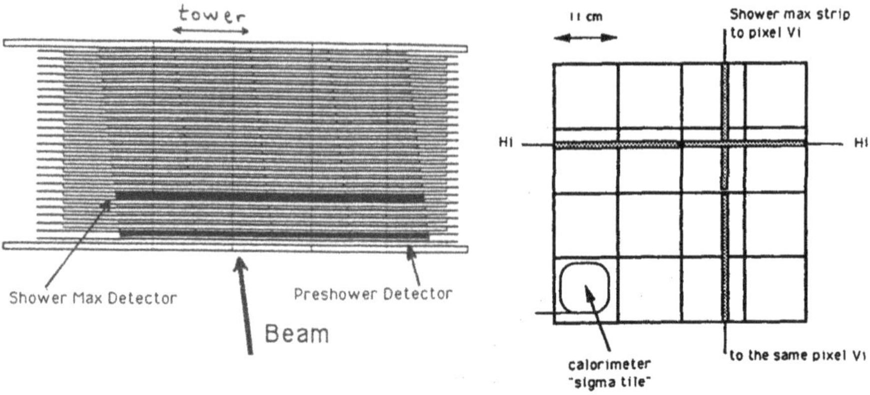

Figure 1: **Lay-out of the 16-tower ECEM calorimeter test module with the Preshower and Shower Maximum Detectors.**

scintillator strip covers 1/8 of a tower and the length covers two towers. The strips are read out by wave length shifting fibers (WLS) spliced to clear fibers. Two strips have their fibers plugged into the same photodetector pixel in such a way that a logical strip spans 4 towers (figure 1).

2.2 Scintillator-Fiber System

Plates of 4 mm thick SCSN-38 scintillator from KURURAY are machined into rectangular SM strips, 22 cm long and 1.3 cm wide. The edges are polished and a groove shaped like a keyhole is made along the center of the strip (figure 2). The strips are then wrapped in sheets of white reflecting material (TYVEK [3]) and glued onto an aluminum support plate to form a 32-strip tile. Four such tiles are needed to build the two-layer (H,V) shower maximum module.

Figure 2 shows the optical path of photons from the scintillator strip to the photodetector. The strip is read out by a WLS fiber inserted into the groove. The WLS fibers are doped with 200 ppm of Y11. One end of the WLS fiber is polished and covered with a reflective aluminum coating obtained by sputtering. The long fiber path described in figure 2 is needed in the SDC calorimeter to bring the light signal to the photodetectors located several meters away from the SM detector. The optical disconnects make it easier to link the fibers to the photodetectors.

3. Test Bench Performance

3.1. Light Yield and Response Uniformity

The light yield of the system is measured with a β-source (Ru^{106}) placed on top of the strip and triggered by scintillation counters. The measurement is made with a RCA 8850 Quantacon which has an estimated quantum efficiency of 10% at 510 nm,

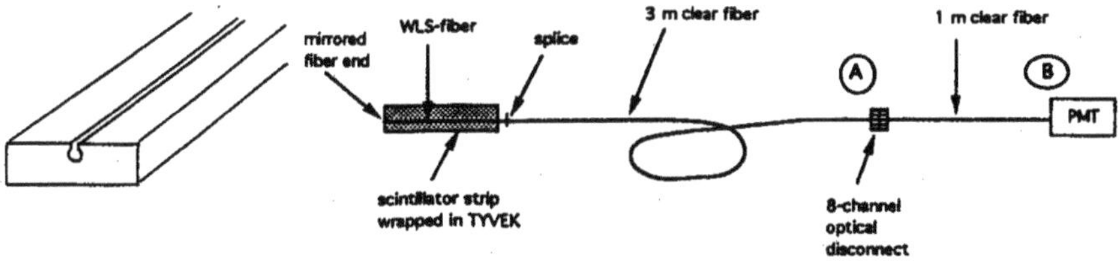

Figure 2: Optics of a shower max strip and fiber. The strips are made of 4 mm thick SCSN-38 plastic scintillator; the fibers are multi-clad, 1 mm in diameter; the Y11 WLS-fibers are spliced to 3 m long clear fibers. The link to the photodetector is through an 8-channel optical disconnect and 1 m of clear fiber. Both scintillator and fibers are from KURURAY.

the peak of the Y11 emission spectrum. Figure 3a shows the pulse height spectrum obtained with one strip at the optical disconnect (point A in figure 2).

The global performance of the SM module can be seen in Figure 3b. The r.m.s. spread of the light yield of the 128 strips is 5.5%. This spread is remarkably small and includes the following contributions: scintillator, polishing, wrapping, fiber mirroring, fiber to fiber variations, splices, and optical disconnects. The average value corresponds to a light yield at the PMT entrance of 3.8 $p.e./MIP$.

3.2 Technical Developments

The use of double clad fibers increases the light yield by 30 % because of their increased numerical aperture. In order to further improve the performances, it was necessary to develop special techniques such as fiber mirroring, fiber splicing, and manufacturing of multichannel fiber disconnects.

The end of the fiber is coated with aluminum by using a sputtering technique. This reflector increases the light yield at the photodetector by 40 %.

An automated splicing machine was built for the SDC shower maximum detector fabrication. It is based on a technique developed for the UA2 experiment [4]. The splices take no extra space (no sleeves) and are very reproducible. Tests in which fibers were repeatedly bent and pulled showed that the mechanical strength is excellent, despite the absence of a sleeve; furthermore, with this splicing technique, spliced multi-clad fibers are stronger than single-clad fibers. For a large number of splices, the average optical transmission is 80% and the r.m.s. spread is $< 3\%$.

Optical disconnects are needed for easy installation and for interchanging photodetectors. Eight-channel plastic connectors which are very thin (7 mm) have been developed. Fibers are inserted and glued in holes drilled in the two pieces which are then polished and assembled with screws exerting the necessary pressure. With this dry fiber-to-fiber contact, the optical transmission is 84% with an r.m.s. spread of 4%.

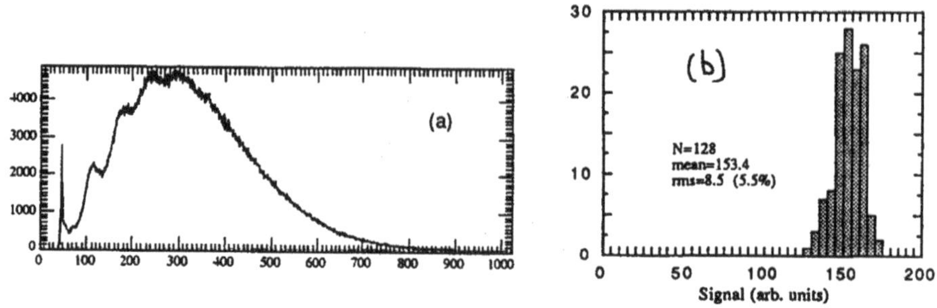

Figure 3: a) Pulse height distribution measured with a Shower Max strip at the optical disconnect (point A), as described in fig. 2. The various peaks correspond to 1, 2, 3, ...$p.e./MIP$. The PMT used is a RCA 8850 "Quantacon" with an estimated quantum efficiency of 10% at 510 nm, the peak of the Y11 emission spectrum. b) Light yield distribution of the 128 strips; the r.m.s. spread is 5.5% and the average light yield 3.8 $p.e/MIP$.

4. Test Beam Results

4.1. Experimental Set-up

The tests were conducted at CERN during fall 1993. Data were collected with incident particle momenta ranging from 3 to 120GeV. The detector was placed on a table capable of remote controlled horizontal and vertical movements. This made it possible to direct the beam anywhere in the 16 towers. A drift chamber was placed on the same table in front of the detector to measure the position of the incident particle to better than 0.5 mm. The trigger required the beam particle to pass through several scintillation counters in coincidence. Electrons were tagged by the signals from Cerenkov counters.

Starting from upstream, the following detector elements are encountered: *i)* the drift chamber; *ii)* the preshower detector, consisting of a 6 mm thick 11x11 cm^2 scintillator tile placed after the second lead absorber sheet in the EM calorimeter; this tile is read out separately; *iii)* the shower max detector placed after absorber sheet #5 (see section 2.1 and figure 1); *iv)* the EM calorimeter in which the two preceeding detectors are imbedded; it consists of 22 layers, each made of 6.35 mm lead sheets and 4 mm thick 11x11 cm^2 scintillator tiles read out by WLS fibers.

The SDC standard portable acquisition system UNIDAQ, based on a UNIX system running on a SUN workstation, was used.

4.2 Light Yield

The light yield obtained in one SMD strip was measured in the test beam by reducing the light yield by a factor 4 with the use of a neutral density filter. From

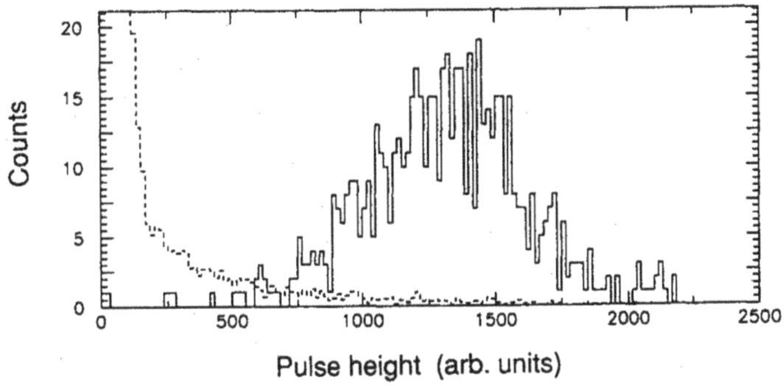

Figure 4: **Energy response (pulse height) of one layer of the shower maximum detector to 8 GeV electrons and pions.**

the widening of the pulse height distribution when the filter is used, one finds a light yield of the order of 2.5 $p.e./MIP$. This value is lower than that obtained on the test bench because of the smaller quantum efficiency of the multichannel PMT.

4.3 Energy and Position Measurements

The light signals from the SMD strips are carried by clear fibers (2 fibers per pixel) to the photodetector , a prototype Philips XP 1723/D1 multichannel PMT. The 64 output signals from the PMT enter two 32-channel ADC modules gated by the beam signal.

The segmentation of the SMD is such that most of the energy of a EM shower is contained in 3 strips. The electron "energy" is obtained in each view by summing the pedestal-subtracted pulse height signals of the three adjacent strips with the highest energy deposit. Figure 4 shows the pulse height distributions for 8 GeV electrons and pions in one layer of the SMD. With the SMD alone, a charged pion rejection factor of 25 is obtained for 95 % electron efficiency.

The energy resolution for electrons is defined as the ratio r.m.s./mean of the pulse height distribution. Figure 5a shows the comparison of a GEANT Monte Carlo prediction including the effect of 2 $p.e./MIP$ photostatistics and the data. The SMD energy resolution is less than $\simeq$ 25% above 10 GeV.

The shower position in the SMD is obtained from the shower profile given by the pulse heights in the strips. The centroid of this distribution is calculated using the 3 adjacent strips with the largest signals. The position resolution is then defined as the r.m.s. of the distribution of the difference between the position of the shower centroid and the position of the electron track measured with the drift chamber. The position resolution of the SMD is shown as a function of energy in figure 5b; it is better than 2.5 mm at energies above 10 GeV.

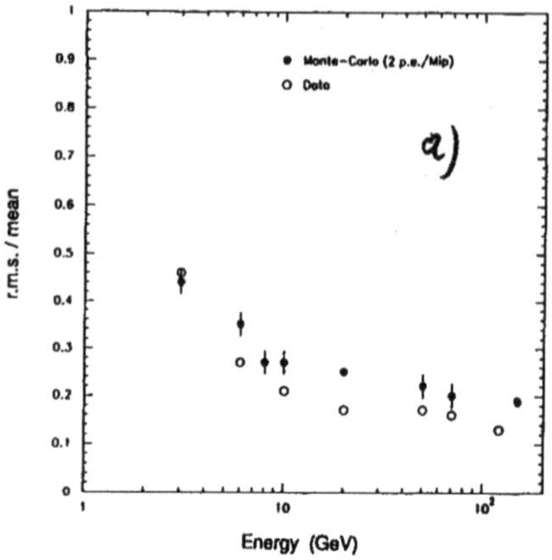

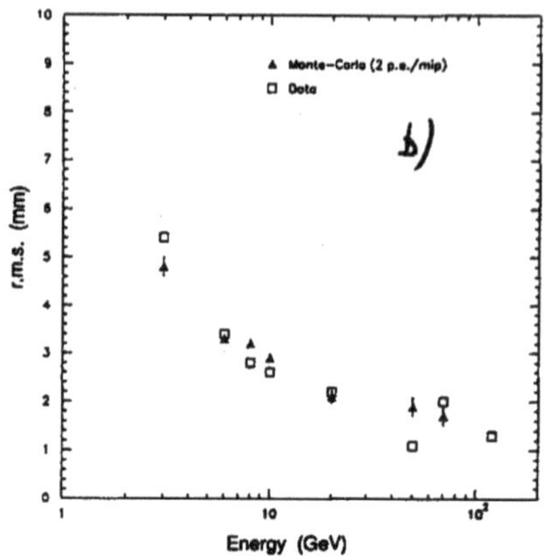

Figure 5: Energy resolution (a) and position resolution (b) of one layer of the shower maximum detector as a function of electron energy. Data are compared with a GEANT Monte Carlo calculation using a light yield of 2 $p.e./MIP$.

5. References

1. This work was performed within the Shower Maximum and Endcap Electromagnetic Calorimeter subsystems of the SDC: *Lawrence Berkeley Laboratory:* G. Abrams, W. Ashmanskas, A. Ciocio, C. Hearty, R. Kadel, J. Kadyk, M. Pripstein. *University of Minnesota:* P. Cushman, R. Rusack, V. Singh. *Northeastern University:* S. Reucroft, D. Ruuska. *The Rockefeller University:* A. Bhatti, P. Melese. *CE-Saclay:* A. Amadon, P. Bonamy, J. Bystricky, G. Cohen-Tannoudji, J. Ernwein, J.F. Glicenstein, J.R. Hubbard, P. Le Du, J.F. Picard, J.M. Reymond, F. Rondeaux, G. Villet. *SSC Laboratory:* J. Yarba. *UCLA:* J. Hauser, D. Joyce, M. Lindgren, T. Muller.
2. SDC Collaboration, *"Technical Design Report", Superconducting Super Collider Laboratory special report* **SSCL-SR-1215** (1992).
3. TYVEK is a registered trademark of the Dupont de Nemours Company.
4. J. Calvet et al., CE Saclay, *Process for connection of optical fibers and connecting devices,* United States Patent #4, 675, 136, June, 1987.

Hadronic Calorimeter

Session Chair: Konstantin Goulianos (Rockefeller)

THE ATLAS TILE CALORIMETER PROJECT

The Tilecal Collaboration
Argonne National Lab., Barcelona Autonomous Univ., Univ. of Chicago, Inst. for Atomic Physics Bucharest, Clermont-Ferrand Univ., CERN, Dubna JINR, Erevan Phys. Inst., Univ. of Illinois, Lisbon LIP, Univ. of Pisa, Charles University Prague, Academy of Science of the Czech Republic, Univ. Fed. Rio de Janeiro, Serpukhov IHEP, Stockohlm Univ., Valentia IFIC

Presented by Marzio Nessi
CERN, Division PPE , Geneva

ABSTRACT

The ATLAS detector at LHC has chosen a hadron barrel calorimeter design with iron as absorber and scintillating tiles placed in planes perpendicular to the colliding beams. After an initial period where the principle has been prouven in beam tests, the Tilecal collaboration is now facing the design and construction of the ATLAS calorimeter. Its mechanical layout, component tests, scintillator peoperties, radiation hardeness as well as the readout characteristics are presented.

1. Introduction

The bulk of the ATLAS hadronic calorimetry consists of a large scintillating tile calorimeter with steel as a showering material. The technology for this calorimeter is based on a sampling technique with scintillating plates readout by wavelength shifting (WLS) fibers. New in this approach is the orientation of the scintillating tiles, which are placed in planes perpendicular to the colliding beams and are staggered in depth. This, as predicted from Monte Carlo simulations [1], provides good sampling homogeneity (particularly when placed, as it is, behind an EM calorimeter and a coil which add up to about two interaction lengths of material).

The calorimeter consists of a cylindrical structure with an inner radius of 2280 mm and an outer radius of 4230 mm. It is subdivided into a 5640 mm long central barrel and two 2650 mm long external barrels. Each of them is further subdivided into 64 independent sector modules. The scintillator tiles lie in the $r - \phi$ plane and span the width of the sector in the ϕ direction. WLS fibers running radially read out the tiles at both of their open edges. Readout cells are then defined by grouping together a set of fibers onto a photomultiplier (PMT), to obtain both axial and transverse segmentation. This idea has been successfully tested in beams on large size prototypes as part of the RD34 project [2].

2. Mechanics

The steel structure of each sector module consists of a stack of repetitive elements (periods). As seen in Figure 1 each period is a stack of 4 layers. The first and third layer are formed by large, 5 mm thick trapezoidal steel plates (master plates), spanning the full radial dimension of the module. In the second and fourth layer, smaller trapezoidal steel plates (spacer plates) and scintillator tiles alternate along the radial direction. The spacer plates are 4 mm thick and of 11 different sizes. Scintillator tiles of 3 mm thickness are inserted into the modules after completion of the mechanical steel assembly. The iron to scintillator ratio is 4.67 to 1 in volume. Two holes are located on the radial axis of each spacer, in the corresponding master plates and in each scintillator tile. These are used to position the spacer plates and the scintillators and also for calibration purposes. A rigid girder at the outer radius and a 10 mm front plate at the inner radius provide structural integrity and allow modules to be joined into a self-supporting cylinder. Key slots at the inner and outer radii of the master plates are used to align the absorber structure to the girder and to the plate at the inner radius. The girder also houses the photomultiplier readout and in addition to the structural rigidity it also provides sufficient steel cross section for the solenoidal field flux return.

Each central barrel module will be stacked with 311 periods arranged along the barrel axis. Four distinct approaches to containment of the absorber structure have been evaluated, and after review an epoxied stack will be utilized. A module will be assembled from prefabricated 300 mm long submodules with two 2 cm thick endplates added to the extremities.

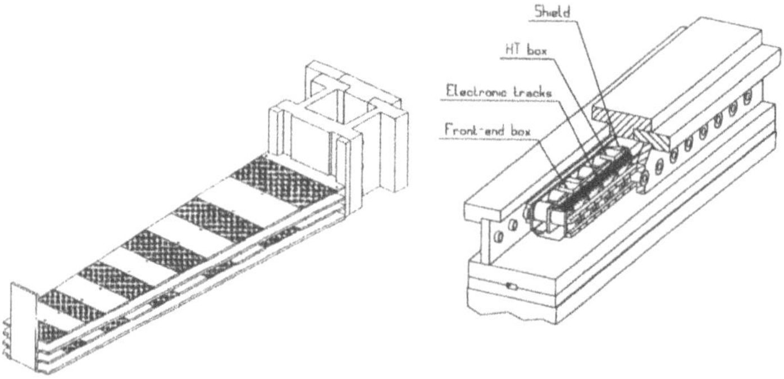

Figure 1: (a) Detail of a tilecal period (b) Layout design of a drawer inside the girder , which accomodates all readout electronics and PMT's.

The construction of the calorimeter will be realized in 5 steps:

Absorber procurement and cutting. About 2500 tons of laminated steel of 4 and 5 mm thickness will be cut into master and spacer plates.

Submodule assembly. About 2500 glued stacks of 30 cm thickness will be constructed by various institutions and then delivered to module assembly sites. The weight of each submodule is about 1 ton.

Module assembly. 19 submodules for the central barrel or 9 submodules for an external barrel will be joined to the individual girders. A central (external) barrel module weighs about 20 (10) tons and will be assembled in industry or in larger laboratories.

Module instrumentation and calibration. Wrapped scintillator tiles will be inserted into the mechanical structure, WLS fiberswill be coupled to the tiles and fiber bundles will be prepared. The drawers with readout components will be inserted and the readout cells will be intercalibrated using the radioactive source system. A few of these modules will be calibrated in test beams.

Final assembly. Individual modules will be preassembled into supermodules (6 to 8 modules, up to 180 tons) at CERN, interconnected, lowered into the ATLAS pit and then assembled into the final configuration.

The 300 mm thick submodules will be glued together starting from glued half–period preassemblies. For each half period, elastic dowel pins precisely position spacers on master plates. A layer of epoxy is applied on the spacer to master contact face, the half–periods are stacked to 30 cm height and compressed with appropriate tooling, keeping the outer faces parallel. The glue has the role of keeping the stack compact and at the same time of compensating for surface planarity defects. The final stack thickness will be held to a fixed value by the use of calibrated columns in the pressing tool. Half periods are precisely positioned to each other by means of 2 key ways placed at the inner and the outer radius. Referencing to these keys will minimize tolerance buildup. Once the glue is polymerized, four strips are welded at 4 corners of the submodule for handling and transportation.

Modules of 5.64 m length for the central barrel and 2.65 m long for the external barrels are built by bolting submodules to the girder and welding two end plates and a front plate. These operations are performed on a dedicated assembly jig, placing the submodules vertically on its horizontal carrying girder. Particular care is taken to the levelness of the girder surface ($\pm 0.1\ mm$). Alignment of the submodules to the girder is by the same key slots used in the 300 mm assembly. The submodule positioning accuracy of $\pm 0.020''$ ($\pm 0.15\ mm$) is checked by a level gauge at each step, allowing corrections for deflections from the vertical axis. With a cross rail and brackets bolted to the inner radius surface, the $\sim 6m$ long module can well be lifted and transported in the vertical position.

All PMTs and associated electronics are housed in a movable drawer system inside the girder structure (shown schematically in Figure 1). Such a design allows rapid access to PMTs and electronics with minimal interference to the rest of the detector. Two 1.5 m drawers are sequentially inserted from each end of the barrel (or one end of the external barrel). Faulty readout elements can quickly be replaced by retracting the external barrel by 2 m and replacing the appropiate drawers. The total number of drawers will be 512 and each drawer will be of identical construction. The drawers will slide into position so that a small air gap between the fiber bundles fixed in the girders and the light mixers in the drawers is maintained. There are many constraints on the drawer specifications. These include the girder dimensions, the number of readout channels and the size of PMTs, the location of the electronics, the tolerances in the PMT positioning in front of the fibers, the simplicity for the handling and the costs.

A drawer is auto-centered in the girder by fiber rings and plate springs between the girder and supporting rails. The readout components (light mixer, PMT, base electronics) are assembled in a compact tube (PMT block) made of μmetal and soft iron. The PMT blocks are 120 mm long. Two shielding sheets protect the electronics and provide the cooling.

The relative position between the PMT blocks and the fiber bunches is controlled by indexers. Only one sensor per drawer is required for a longitudinal positioning precision of ± 1.5 *mm*. After gluing the fibers into the ring, a diamond tool works into the girder to polish the fiber bunches. Special tools will be developped to insert, position and transport drawers.

The central barrel and external barrel assembly will be done in the ATLAS pit, using preassembled supermodules (up to 8 single modules). Supermodules (maximum weight 180 tons) will be strengthened over their inner radius to allow better and safe manipulation. After the lower half of the calorimeter is assembled, the LAr cryostat and calorimeter are installed, including all cryogenic services and feedthroughs. Following this, the second half of the tile calorimeter is put in place. The cylinder is self-supporting, and links between modules are created externally on the girder. The contact between adjacent modules is made at the inner and at the outer radius, using localized bearing surfaces and a design gap between wedges of less than 1 mm. The resulting cylinder rests on feet placed every 50 cm and sitting on rails.

3. Optical components

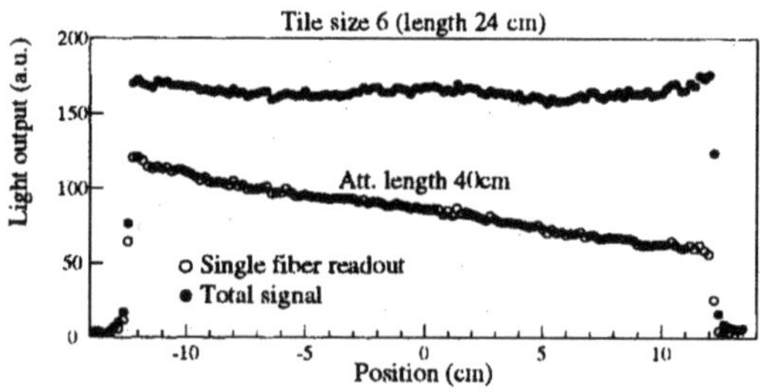

Figure 2: Average response to ionizing particles along the direction between the two fibres.

The light emitted by 3 mm thick scintillator tiles is transmitted to 1 mm diameter WLS fibers placed along both sides of the scintillator. Fibers are air-coupled to each tile and one fiber sees 1 or 2 tiles. The individual scintillators are wrapped with white paper, the optical contact with the fiber is fixed by an aluminized mylar reflector. Tile length and width vary with depth from approximately 220×100 mm^2 to 380×190 mm^2, for a total of 11 different sizes. All optical components are designed to give sufficient photoelectron statistics and good response uniformity across and along all the modules to minimize the energy resolution.

Tiles are produced by injection molding for reason of costs and high production rate. Optically transparent granulated polystyrene is used as a base material with the addition of two scintillation additives (1.5% of PTP and 0.04% of POPOP). In general no additional mechanical treatment (e.g. polishing) is needed for the final product.

Natural ageing and radiation hardness depend on the components used. Calorimeters based on molded scintillators have been used for about 10 years at IHEP, Protvino, and no significant performance deterioration has been reported.

1mm diameter WLS fibers incorporating a shifter with a decay time of ~10 ns will be used to transmit the scintillation light to the PMTs. The fiber ends opposite to the PMTs are aluminized by a sputtering technique which gives a reflectivity of ~85%. Attenuation lengths of about 2 m have been achieved. The exact choice of fiber is not yet specified, however a comparative investigation of commercially available fibers produced by Bicron, Kuraray and Pol.Hi.Tec has been made. Sample fibers have been investigated as a function of, amongst other things, cladding type, dopant concentration, UV absorber. Bicron BCF91A and Kuraray Y11(200)MS are probable choices. Multiclad fibers, as manufactured by Kuraray, have been shown to improve the light yield by a factor of 1.6.

Tile/fiber setups have been extensively tested inside the collaboration using collimated β-sources (^{106}Ru or ^{90}Sr) and scanning along the scintillator surface (Fig. 2). The achieved light uniformity is typically of 5% when adding the left and right readouts, with attenuation lengths in the scintillator between 30 and 40 cm. The nonuniformity caused by the holes in the tiles is negligible. The uniformity and the light yield improves by using a simple masking technique and a proper tile-fiber coupling (flat edges instead of grooves). The average response for the tiles coupled to Y11(200)MS fibres is ~40 pe/GeV. The response spread within a cell is reasonably small.

The tile to tile response fluctuation and fiber to fiber (including fiber/tile coupling) non–uniformity have been measured on a sample of about 4000 injection molded tiles produced and used in the prototype R&D program. For both cases the achieved uniformity of the response is about 5–6%. All these characteristics are more than adequate for an hadronic calorimeter.

Particular care has been given to the issue of radiation resistence. The maximum expected annual peak dose in the TILECAL calorimeter is ~38 Gy at η=1.2. In order to assess the scale of the problem, tile/fiber setups have been irradiated with varying doses from 650 Gy to 10 kGy, at rates of 16 Gy/h and 220 Gy/h, and the optical properties have been monitored. The results agree with previous systematic tests using a 2.5 GeV electron beam. A phenomenological model that takes into account the radiation damage evolution shows that for long term irradiation only the permanent damage needs to be considered. For 10 LHC years, i.e., an integrated luminosity of 10^6 pb^{-1}, a light loss of no more than 5% is expected in the tiles of the first longitudinal sampling. Using those results and assuming no radiation damage for the LAr EM calorimeter, the degradation on the energy resolution and jet signal is negligible.

4. Electronics and Calibration

The calorimeter readout is mounted in a system of drawers, placed inside the girder structure. This system has been designed as a very modular one. The smallest unit is a PMT block, which contains a light mixer, the phototube (for example Hamamatsu R5900 series, 10 stage metal channel dynodes), the base and the magnetic shielding. In a 3 in 1 concept the base contains on the same board a voltage divider, a shaping amplifier and a slow integrator for the current mode readout. The digitizing electronics is be resident on the same drawer, so that cabling and services for the system will be kept to a minimum. The proposed granularity (0.1×0.1), together with a 3 fold radial segmentation, requires approximately 10000 readout channels. The dynamic range of the system is designed to cover single muons identification and up to 1 TeV jet energies in a single hadronic cell, for a value of 16 bits equivalent.

The calibration and monitoring system is designed to achieve a 2% percent constant term in the energy resolution. Short term PMT gain drifts will be calibrated out by means of a laser driven light pulsing system. A YLF solid state laser pulse generator has already been successfuly tested on prototype modules; a continuous monitoring precision has been achieved at the level of 0.5%. A movable radioactive source system will allow to first equalize and then monitor the gains of all cells. The absolute energy calibration uses a test beam (for a fraction of the modules) and will be maintained in time using the movable source system. The concept of the source is simple; every scintillator tile is traversed by a rod and a hollow tube, into which a gamma source can be inserted. An integrator is used to transform into a DC voltage the source-induced current in the PMTs. The same technique is used for the minimum bias activity monitor.

All these electronics subsystems and calibration methods have been successfully prototyped and applied to the 1993 and 1994 RD34 test beam activities [3].

5. Future plans

The tilecal collaboration is now facing the construction of the ATLAS calorimeter, its detailed engineering and financement. The plans for 1995 are to construct and test in beams a module 0 of 6 meter length. All assembly and manipulation tools will be designed and constructed for the first time. This prototyping will include both mechanical and electronics aspects. The start-up of the ATLAS production phase is planned for 1996. Installation in the ATLAS pit is scheduled to take place beginning of 2001, leaving just 5 years for the full construction and test of this complex and large hadron calorimeter.

6. References

1. O. Gildemeister, F. Nessi-Tedaldi, M. Nessi, Proc. 2nd Int. Conf. on Calorimetry in HEP, Capri, 1991, page 189.
2. F. Ariztizabal et al., NIM A 349(1994)384 and M. Bosman et al. (RD34 Collaboration), CERN/DRDC/93-3 (1993).
3. A. Amorimi et at. (The TILECAL collaboration) *In These Proceedings.*

ATLAS TILECAL TEST BEAM RESULTS

The TILECAL Collaboration

Argonne national Labroratory, Institut de Fisica d'Altes Energies—U. A. Barcelona
Inst. for Atomic Physics—Bucharest, Institute of Atomic Physics—U. Chicago
LPC Clremont-Ferrand—U. Blaise Pascal / CNRS-IN2P3, CERN—Geneva
JINR—Dubna, Michigan State University—East Lansing, Erevan Physics Institute—Erevan,
LIP and CFNUL—Lisbon, Charles U. and Inst. of Physics—Prague
Univ. Fed. Rio de Janeiro, IHEP—Serpukhov, Stockolm University
University of Illinois—Urbana-Champaign, IFIC—Valentia

presented by A. Amorim
LIP and CFNUL, Lisbon, Portugal

ABSTRACT

Several prototype modules of the proposed ATLAS tile hadron calorimeter have been constructed and exposed to particle beams at the CERN SPS. Results have been obtained on energy resolution, linearity and compensation using muons, electrons and pions. The Monte Carlo predictions are compared with the test beam results. All readout cells have been systematically intercalibrated using a movable radioactive source. Gain and linearity of the electronics chain have been monitored with a Laser.

1. Introduction

This paper describes the test beam performance of a prototype of the proposed barrel hadronic calorimeter[1] for ATLAS[2], where it will be placed downstream of a Lead-Liquid Argon electromagnetic compartment equivalent to about 2 nuclear interaction lengths[3]. It is characterized by having scintillating tiles placed in planes perpendicular to the beams and staggered in depth. Three modules have been built and tested in 1993 and 2 additional have been included in the 1994 tests. The calorimeter modules were exposed to high energy pion (20-150 GeV), electron and muons beams at the CERN SPS without an electromagnetic compartment in front[4]. The test beam conditions, where the full particle shower is developed inside the tile calorimeter, allow the detailed comparison with the Monte Carlo predictions and help to pinpoint the detector specific properties. A test beam program of a combined electromagnetic plus hadronic calorimeters, in the ATLAS configuration, is in progress.

This novel technique is illustrated in Fig. 1 showing one of the modules used in the test beam run which are similar to an azimuthally modular sector of the ATLAS "barrel" or "external barrel" calorimeters[2]. They consist of a iron/scintillator sandwich, in which the tiles are placed inside the iron absorber structure in $R-\phi$ planes. WaveLengthShifter (WLS) fibers coupled to the radial edges of the tiles can thus be kept straight in the radial direction allowing for good hermeticity, easy assembly and low cost. Outside the absorber structure the fibers are grouped according to the desired

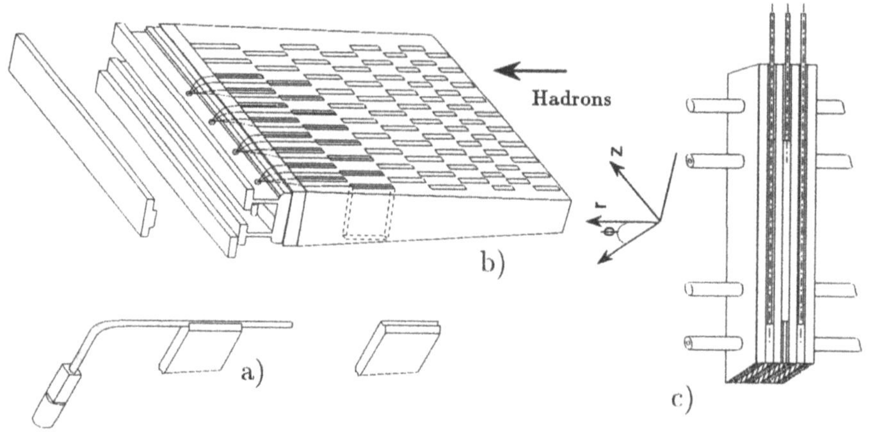

Figure 1: Principle of the TILECAL calorimeter. a) Scintillator coupling to fiber and PMTs. b) Tilecal module. c) Calorimeter layer structure.

axial (along z) segmentation of the calorimeter and coupled to the photomultiplier tubes (PMTs).

2. The Prototype Calorimeter

The prototype calorimeter consists of five sector modules (Fig. 1), each spanning $2\pi/64$ in azimuth and 9 interaction lengths in the radial direction from an inner radius of 200 cm to an outer radius of 380 cm. They are formed by 57 repeated "periods" in axial direction, each consisting of four layers formed by 5 mm thick trapezoidal steel plates spanning the full radial dimension (first and third layers) and 10 cm long steel plates and scintillator tiles alternating between the second and fourth layer. The smaller steel plates and the scintillator tiles are 4 mm and 3 mm thick respectively giving an iron to scintillator volume ratio of 4.67:1. Each spacer plate has two 8 mm diameter holes on the radial axis of the trapezoid that, together with the corresponding holes on the master plates and scintillators, allow to compress the stack of plates with a series of rods. Every second rod has an axial hole where a ^{137}Cs source can be inserted for calibration purposes.

The scintillator tiles were made using an injection molding technology at Protvino IHEP from a granulated polystyrene base doped with 1.5% PTP and 0.03% POPOP. They were wrapped with white paper and covered with aluminized mylar enclosing also two Bicron BCF91A fibers air-coupled to the tile at both ends. The fiber attenuation length is about 2 meters and the uniformity is improved by aluminizing the end opposite to the PMTs. As the result of optimization the fifth module uses Kururay Y11(200)MS WLS fibers. The prototype calorimeter was radially divided into 4 sections by reading radial groups of tiles using different fibers. The 3 inner tiles

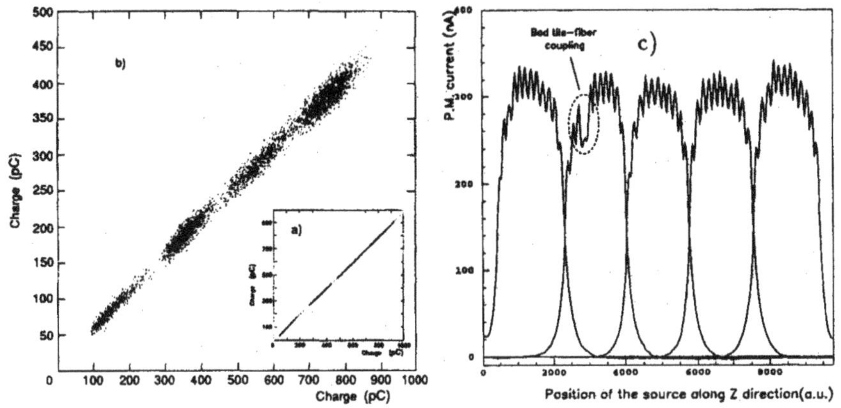

Figure 2: (a): Correlation of the charges measured by the two monitoring diodes; (b): PMT charge vs mean of diode monitor signal; (c): Current measured in the PMTs as a function of the position of the source along the rod.

formed section 1, the following 4 formed section 2, the next 5 section 3 and the last 6 section 4. The fibers in each module were also divided into 5 readout cells in the z direction of about 20cm each and connected to XP2012 5/4" PMTs via light mixers.

3. Monitoring and Intercalibration

The laser monitoring system consists of a triggerable laser in which the intensity of each pulse can be externally modulated. The light goes trough a splitter from which clear fibers go to each of the module PMTs and to two monitoring photodiodes. A precision better than 1% was achieved in monitoring the laser intensity as can be seen in the correlation of the response of the two photodiodes (Fig. 2a). The response of the photomultiplier to four different pulse intensities is shown in Fig. 2b. From the dispersion in response along the vertical axis a photoelectron yield of 20 photoelectron/GeV(p.e.) was obtained in the 1993 tests, independently measured with an electron beam. As the result of further optimization, this value has improved to 40 p.e./GeV for the new module (see Sec. 6).

The cell to cell intercalibration has been handled by performing space average of the DC current induced by the ^{137}Cs source for each PMT (Fig. 2c). The reproducibility of the ^{137}Cs calibration method has been measured to be better than 1%[4]. In the particular case of the pion response, a simulation was performed by miscalibrating the signal in each cell according to randomly picked distributions with a given r.m.s. The induced constant term due to miscalibration is 0.3 times the r.m.s./mean of the distribution of cell calibration errors. In particular, if this errors are of the order of 4% they contribute only 1.2% to the constant term of the pion resolution.

4. Experimental Setup

The calorimeter was installed on a platform that moves horizontally perpendicular to the beam line (z direction). The detector can also be rotated on its vertical (θ) and horizontal (ϕ) axis. Upstream of the calorimeter, a trigger counter telescope was installed including two delay-line wire chambers allowing to reconstruct the impact point of the beam particles to better than $\pm$ 1 mm.

5. Test Beam results

Data were taken with electron beams of 20, 40, 80 and 150 GeV, pion beams of 20, 40, 80, 150, 200 and 300 GeV and muons of 225 GeV at various θ and ϕ angles of incidence.

5.1. Muon Signal

To check the cell intercalibration, the calorimeter was exposed to muons impinging at 90° in the geometrical center of each of the 18 tiles. The correlation of the fitted muons peak values for each PMT with the corresponding source calibration constants gives an r.m.s. spread of 6%. It is expected that this value is dominated by the error in the position of the muon peak, because the small dynamic range of the muon signal makes the peak value relatively sensitive to small pedestal shifts.

The response to 225 GeV muons incident at small θ angles was studied showing an energy deposition in the calorimeter of about 2.8 GeV. This value can be compared to the expected pileup noise at LHC nominal luminosity that is expected to be 0.3 GeV or less, on one cell with 0.1×0.1 granularity, giving a comfortable margin to detect isolated muons at the LHC.

5.2. Electron Response

The calorimeter longitudinal segmentation (10 cm) is much larger than the radiation length in iron, making the response to electrons at small angles dependent on whether the initial impact point is in iron or in scintillator. This is of no consequence for ATLAS where the hadronic calorimeter will be preceeded by an electromagnetic calorimeter. Calculating the average response as the mean over one 18mm period, one sees that it is linear versus energy for $\theta = 10^o, 20^o$.

An electron beam of 20 GeV impinging sideways on the calorimeter was used to measure the photoelectron yield . Using the absorption filter technique, based on studying the contribution of the photoelectron statistics to the statistical term in the resolution, the number of p.e. per GeV obtained using the laser was confirmed (see Sec. 3)[4].

5.3. Pion Response

The uniformity along the z coordinate, energy resolution and linearity of the hadronic response were extensively studied. Fig. 3a shows the uniformity over an

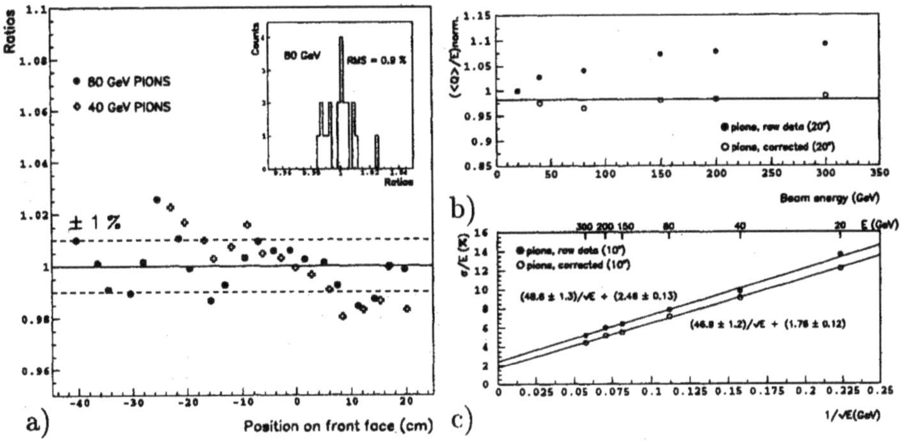

Figure 3: (a): Uniformity scan with 40 and 80 GeV pions at 10°; (b): Pion response for raw and corrected energy spectra; (c): The corresponding energy resolution.

extended region in the z direction, measured using 40 GeV and 80 GeV pions. The observed energy spectra in the full energy range are quite symmetric at all energies and display small or no tails on the high energy side[4]. The energy resolution, for the raw data, is given by $\sigma/E = 49\%/\sqrt{E} + 2.5\%$ at $\theta = 10^o$, as shown in Fig. 3b. At $\theta = 20^o$ the pion energy resolution is given by $\sigma/E = 47\%/\sqrt{E} + 2.2\%$. The e/π response has been measured to be 1.2 at 20 GeV.

Several corrections can be applied to compensate for the $e/\pi > 1$ by adjusting downwards the response of readout cells with large signals, due to electromagnetic clusters. The results, shown together with the raw data in Figs. 3(a,b), were obtained by replacing the signal in each readout cell Q_{ij} by $Q_{ij} - (P_j/Q_j)Q_{ij}^2$, where i, j are indexes for the transverse and longitudinal segmentation respectively. The variable Q_j is the sum of signals in each longitudinal sample. The constants P_j were obtained empirically to be $P_1 = 0.15$, $P_2 = P_3 = 0.25$ and $P_4 = 0$ and a small offset of about 1.7 GeV was applyed to obtain the best proportionality to beam energy. This procedure improves the resolution (Fig. 3c) and gives a linear response versus beam energy (Fig. 3b).

5.4. Comparison with the Monte-Carlo predictions

The comparison of test beam data with the simulation results is an important step to validate the predictions obtained in the ATLAS environment. An exhaustive simulation program has been carried out, using the GEANT package of the CERN program library, where the exact test beam conditions have been reproduced.

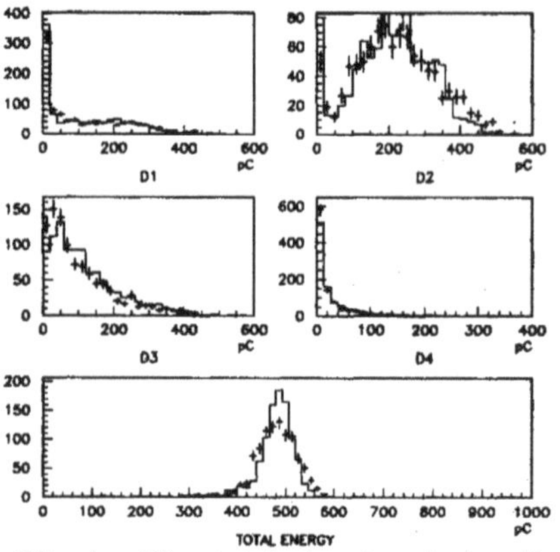

Fig. 4: The total signal and the distribution on depths D1,...,D4 predicted by the Monte Carlo are compared with data (dots) for 80 GeV pions at $\theta = 20^o$.

The Tilecal Monte Carlo uses a detailed geometrical description of the detector and includes the tile and fiber response variations and photoelectron statistics. An example of the comparison with test beam data is shown in Fig. 4. The normalization applyed to the Monte Carlo results was obtained by compraing its predictions to muon data. These results were obtained using the FLUKA hadronic package in GEANT version 3.21. The Monte Carlo predicts the correct statistical term in σ/E but underestimates the constant term. These results, together with simulation prediction of a more linear response versus energy in comparison with the test beam data, are being extensively studied.

6. Optimization Achievements and Conclusions

The test beam results for the Tilecal prototype, without an electromagnetic calorimeter in front, show a good hadronic resolution which is in agreement with the Monte Carlo predictions. The resolution can be further improved with a simple sampling correction that yields a very good linearity. It was shown that the Tilecal can be precisely and reliably intercalibrated and the readout could be accurately monitored.

Specific R&D was able to explore the very good performance of the calorimeter and gave new solutions that are now used in the prototypes. In particular, a uniform tile response along $r - \phi$ was achieved by replacing the tile groove and chamfer edges at 45° by a flat surface. This, together with the use of double cladding Kururay Y11(200)MS fibers, has led to an increase in light production by a factor of ≈ 2.

7. Acknowledgments

We are indebted to V. Barnes from Purdue University who lent us the ^{137}Cs source with its mechanical drive and associated electronics used in the test beam runs.

8. References

1. R&D 13 Proposal, *CERN/DRDC/93-3*, (1993)
2. The ATLAS Collaboration, Progress Report on ATLAS Milestones, *CERN/LHCC/93-51*, Oct. 1993
3. M.Nessi et at. (The TILECAL collaboration) *In These Proceedings.*
4. F. Ariztizabal et al. (The RD34 collaboration) *Nucl. Instr. and Meth.* **A349** (1994) 384.

PERFORMANCE OF A LIQUID ARGON ACCORDION HADRONIC CALORIMETER PROTOTYPE

Jérôme Schwindling
DAPNIA/SPP
CE Saclay
91191 Gif sur Yvette Cedex, France
for
the RD3 collaboration

ABSTRACT

A liquid argon hadronic calorimeter using the "accordion" geometry and the electrostatic transformer readout scheme has been tested at CERN, together with a liquid argon accordion electromagnetic prototype. The results obtained for pions on the linearity, the energy resolution and the uniformity of the calorimeter response are well within the requirements for operation at the LHC.

1. The Calorimeter Prototype

The hadronic prototype [1] is made of 10 mm thick ground and readout stainless steel plates interleaved with 3 mm liquid argon gaps. It is built in two separate modules in depth. Each module has two readout samplings, 1.2 and 1.6 absorption lengths deep respectively.

The converter plates were given an accordion shape using a cold profiling technique. After profiling, the readout plates were cut along projective η lines and into the two samplings in depth. Six (in the first module) or eight (in the second module) argon gaps are grouped to form projective towers in ϕ. The tower size is $\Delta\eta \times \Delta\phi = 0.045 \times 0.05$. The prototype is eight towers wide in ϕ and about ten towers long in η.

The gaps of a cell are connected, according to the EST readout scheme [2], in series rather than in parallel as usual. To first order, the cell capacitance is the gap capacitance divided by the number of gaps in series instead of being multiplied by it. Thus, the cell capacitance is much lower and the charge transfer time much shorter. It has been checked by a simulation of the EST that the crosstalk induced, by this readout scheme, between adjacent cells in η has no visible effect on the energy resolution or the shower profile.

2. Test Beam Results

2.1. Noise Measurements and Muon Data Analysis

The noise has been measured by looking at the energy seen in regions of various sizes in random trigger events. The noise is the quadratic sum of random and coherent noises. Typical random noise values are between 50 and 90 MeV per cell, for a shaping with a 90 ns peaking time response to a delta pulse.

Most of the coherent noise was traced to a faulty high voltage supply. Its value ranges from 4 to 13 MeV per cell.

In order to study the response of the calorimeter to muons, a precise alignment of the beam chambers and the calorimeters was performed by using electron and pion data. Muon events were selected from a 180 GeV tertiary beam by requiring no more than 10 GeV energy deposited in the calorimeters.

Because of the accordion shape of the calorimeter, two towers in ϕ were considered to measure the muon energies. Using the tracks reconstructed from the beam chambers, a fiducial volume within these two towers was defined and muons not in this volume were rejected. The electronic noise was measured for the same group of cells using random trigger events.

Figure 1 shows the energy deposited in the hadronic calorimeter for muons and the energy computed for random trigger events. To define the energy scale, the calibration factors for high energy densities, obtained by the weighting method described in the next section, were used. The signal over noise ratio is 7.9 when the full calorimeter is used and 6.3 when only the first readout sampling is considered.

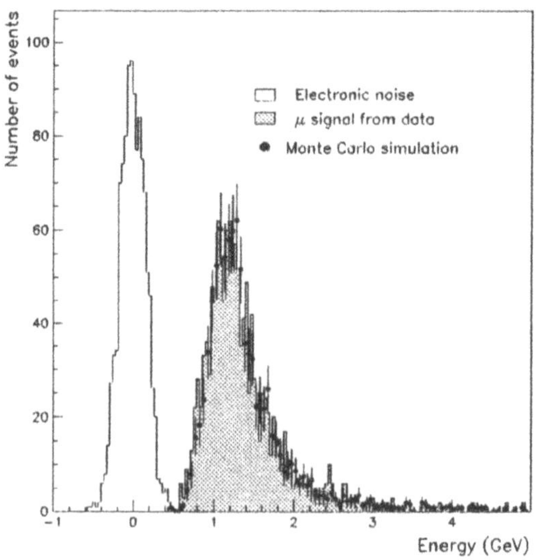

Figure 1: Energy distribution in the hadronic calorimeter for muons and for random trigger events.

2.2. Pion Data Analysis

The measurement of pion energies is performed in an area of 11 × 11 cells in the electromagnetic prototype and 7 × 7 cells in the hadronic modules, around the beam direction. Only cells with absolute signals greater than 1.3 times their measured electronic noise value are used.

In order to correct for the e/π ratio not being one, the cell energy density ρ = (cell energy) / (cell volume) is used to compute weights for each cell:

$$w_{EM}(\boldsymbol{\theta},\rho) = (1 - e^{-|\rho|/e_1}) \cdot (e_2 \cdot e^{-|\rho|/e_3} + e_4) \quad (1)$$

$$w_{HAD}(\boldsymbol{\theta},\rho) = (1 - e^{-|\rho|/h_1}) \cdot (h_2 \cdot e^{-|\rho|/h_3} + h_4) \quad (2)$$

where $\boldsymbol{\theta} = (e_1, e_2, e_3, e_4, h_1, h_2, h_3, h_4)$ is the vector of function parameters.

The particle energy is estimated as:

$$\tilde{E} = \sum_{k=1}^{3} c_k \sum_{j\epsilon(EM)_k} w_{EM}(\boldsymbol{\theta}, |\rho_j|) \cdot E_j + \sum_{k=4}^{7} c_k \sum_{j\epsilon(HAD)_k} w_{HAD}(\boldsymbol{\theta}, |\rho_j|) \cdot E_j \quad (3)$$

The weights $e_1, \ldots, e_4$, $h_1, \ldots, h_4$ and the calibration factors $c_1, \ldots, c_7$ are optimized on a separate set of events in order to get the best energy resolution. Their dependence with the beam energy is rather smooth and is parameterized by linear or exponential functions.

The energy of any pion event at any beam energy is then computed in the following way: A first estimate of the energy is computed as the sum of signals over the calorimeter cells and is used to calculate the weighting parameters. A new weighted energy is then estimated and the procedure is iterated four times, no improvement being seen with more than four iterations. Figure 2 shows the energy distribution obtained in the data and in a Monte Carlo simulation using GEANT with the GHEISHA hadronic shower package, for 100 GeV pions. For all beam energies between 20 and 400 GeV, the difference between the average reconstructed energy and the beam energy is less than 2%.

The energy resolution as a function of the beam energy is shown in figure 3 for the data and the GHEISHA and FLUKA simulations. The resolution of the data is well fitted by:

$$\sigma/E = \frac{(51.6 \pm 2.6)\%}{\sqrt{E}} + (3.3 \pm 0.1)\% + \frac{2.5 \pm 0.1}{E} \quad (E \text{ in GeV})$$

The fitted "noise" term (2.5 GeV) is in agreement with the quadratic sum of the electonic noise (1.8 GeV) and the RMS energy loss (1.5 GeV) due to the 1.3 sigma

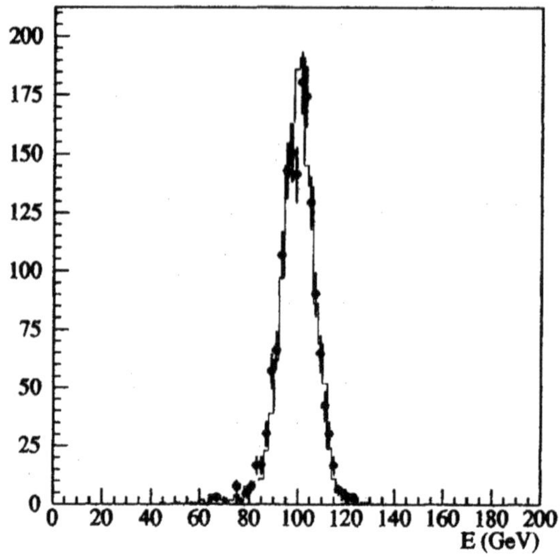

Figure 2: Distribution of the reconstructed energy for 100 GeV pions. The histogram represents the data, dots are the GHEISHA Monte-Carlo simulation results.

cut. However, as already discussed, part of the electronic noise is due to a faulty high voltage supply. A non negligible fraction of the constant term should come from longitudinal leakage.

The effect of the EST on the resolution seems to be small. In the GEANT-GHEISHA simulation, the resolution is (7.0 ± 0.2)% at 100 GeV if currents from all the gaps are just summed up, and (7.4 ± 0.2)% if EST readout scheme and all intercapacitances are invoked. If, rather than a weighting procedure, only two calibration constants are used (one for the EM calorimeter and one for the HAD calorimeter), the resolution is 12.2% instead of 6.5%, at 100 GeV in the data.

Figure 3 shows also the expected resolution from the Monte Carlo simulations. Although both GHEISHA and FLUKA are in reasonable agreement with the data for the shower profiles, FLUKA predicts a resolution better, by about 20%, than the resolution obtained in the data.

The calibration factors and weights determined at one incident beam position have been used on data taken at other positions. Over the size-limited hadronic calorimeter, the uniformity is better than 1%.

3. References

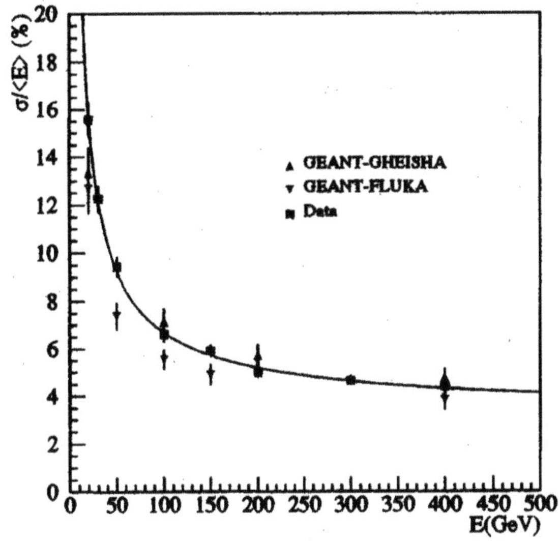

Figure 3: Energy resolution as a function of the beam energy (calibration + weighting). Error bars are 3 sigma errors and contain a systematic error evaluated by changing the weights within their errors.

1. The RD3 collaboration, CERN-PPE 94-127, to be published in Nucl. Instr. and Meth.
2. J. Colas, M. Pripstein and W.A. Wenzel: *The Electostatic Transformer* Nucl. Instr. and Meth. **A294** (1990) 583.

Hadron Calorimeter Production for the CDF Plug Upgrade

Qifeng Shen
Department of Physics
Purdue University
West Lafayette, IN 47907

Representing the CDF Plug Upgrade Group

ABSTRACT

To accommodate the running conditions of the Fermilab Collider Run II, the end-plug of CDF is being upgraded to scintillator based calorimeters. In this paper, we report the results obtained from the R&D and the production of the hadron portion of the calorimeter. The construction procedure of the hadron calorimeter is described here. The test results during the production indicate that the calorimeter meets our design specifications.

1. Introduction

In anticipating the bunch spacing of 132 ns and high luminosity during the Fermilab Collider Run II, the CDF collaboration is upgrading its present gas based end-plug calorimeters with scintillator based calorimeters.[1] Fig. 1 shows a cross sectional view of one quadrant of the upgraded plug.

The plug has an electromagnetic compartment (EM) in front and a hadronic compartment (HAD) behind, covering the pseudo-rapidity (η) region from 1.1 to 3.5. Scintillator plates of 4 mm thick are sandwiched between 5 mm thick lead plates for EM, while those of 6 mm thick are inserted into the gaps between 5 cm thick steel plates for the HAD. There are 23 layers in EM, with 23 X_0 and 1 λ_{int}, and 22 layers in HAD, with 7 λ_{int}. The energy resolutions are $1\% \oplus 17\%/\sqrt{E}$ for EM and $5\% \oplus 80$–$90\%/\sqrt{E}$ for HAD. A shower maximum detector made of scintillator strips for position measurement is located behind the 6th lead plate in the EM compartment[2].[3]

The scintillator plates in both the EM and the HAD are segmented into small tiles in the $\eta - \phi$ plane, forming projective towers. The tiles in large η regions cover 15° in ϕ (15° tiles), those in small η regions cover 7.5° in ϕ (7.5° tiles). A wave-length-shifting (WLS) fiber embedded in each tile is used to collect the light. Optical ribbon cables are used to transport the light to a decoder box in which the layer-based cables are sorted into tower-based groups. Each tower is read out by one phototube.

To calibrate the calorimeters, thin steel tubes are placed close to the scintillator plate, enabling radioactive sources to traverse the tiles. The source measurements are used to set the initial tower gains when the plug is first installed and to monitor and calibrate the plug after it is commissioned.

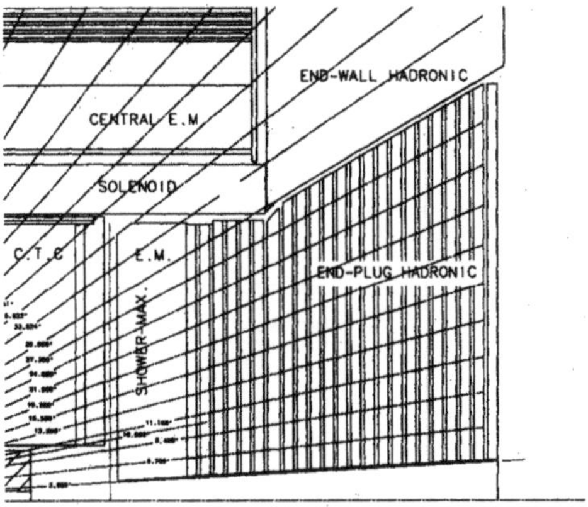

Fig. 1. A sectional view of one quadrant of the upgraded CDF plug calorimeters, fitting into the CDF detector structure.

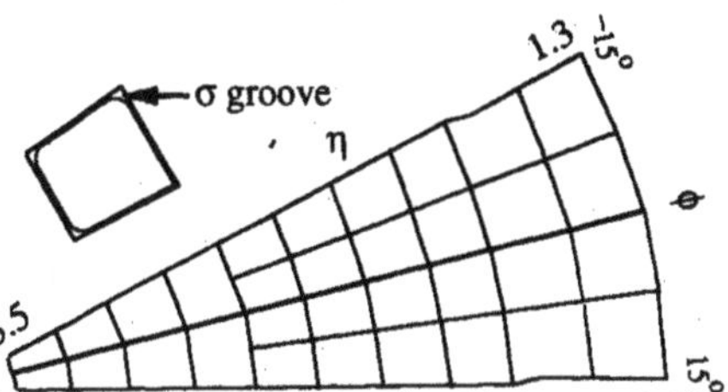

Fig. 2. A layer 6 megatile. Shown here are the tile segmentation and the fiber grooves.

Similar methods in production and test are used for both the HAD and the EM. In the following sections, we will concentrate on discussing the hadron calorimeter.

2. Hadron Megatile Production

The tiles in a 30° section of the HAD layer are cut from a large plate of Kuraray SCSN-38 polystyrene scintillator as shown in Fig. 2. About 0.2 mm of the 6 mm thick scintillator are left uncut at the bottom of the grooves separating the tiles. The grooves are filled with white epoxy for both structural support and light isolation.[4] A black line is painted along the groove at the bottom surface of the scintillator, to reduce the optical cross talk, which is measured to be 1% per edge of the tile. The mechanical unit of this 30° section, consisting of nominally 32 tiles, is called a megatile.

The fiber groove is σ shaped.[5] Its depth of 2.5 mm is optimized to give a uniform light yield across the tile. The WLS fiber is the 0.83 mm Kuraray multiclad Y11 fiber. It is mirrored at one end and spliced to a clear fiber at the other end.[6] The clear fiber

Fig. 3. Top side of the white plastic with fiber routine grooves, terminated at four connectors.

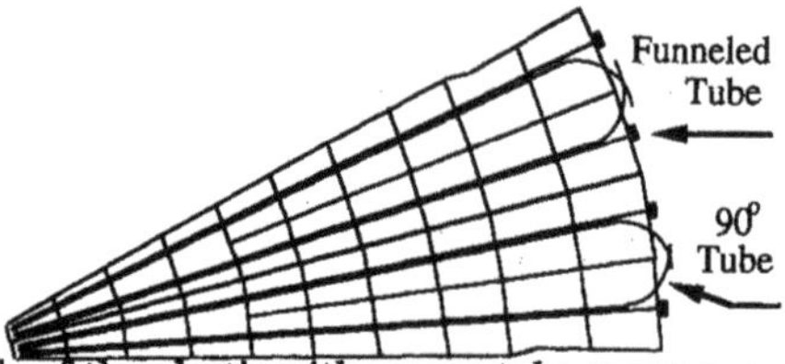

Fig. 4. Bottom side of the plastic with source tube grooves, overlayed on the scintillator.

is terminated at a mass connector mounted at the outer edge of the megatile. The light yield of the tiles, with a 4 m long clear fiber cable, is measured to be 4 pe/MIP, which is more than the 1 pe/MIP estimated to obtain the designed calorimeter resolution.[7]

The scintillator plate is wrapped with highly reflective white Tyvek. Then a white polystyrene sheet is placed on top of the scintillator. Grooves are milled on the top side of the plastic to route the clear fiber and a portion of the WLS fiber. The WLS fiber is inserted from the white plastic into the fiber groove in the tile. As show in Fig. 3, the fibers from the 30° section are terminated at four connectors. The relative light yields, normalized to the average for each fiber-connector configuration obtained with a UV light test setup before the fibers are inserted, is plotted in Fig. 5. The RMS of that distribution is 3.2%, which includes contributions from the mirroring, splicing, attenuation, and the connector polishing.

Calibration source tubes are positioned inside the grooves cut in the bottom side of the plastic, as shown in Fig. 4. The tubes are stainless steel with OD of 1.3 mm and ID of 1.0 mm. The grooves are positioned such that the radioactive source can traverse near the geometric center of the 7.5° tiles, and about one quarter in ϕ direction away from the edges of the 15° tiles. The tile response is very sensitive to the distance between the steel tube and the tile surface.

There are two types of tubes in terms of accessibility. The first type, which is instrumented for all layers, is terminated at a funnel at the edge of the megatile. These tubes are only accessible when the plug is extracted. Source runs after the plug is assembled, along with the results from the production tests and from the test beam, can be used to set the initial gains of the towers. The second type is installed for only four selected layers. The tubes are bent 90° at the outer edge of the megatile as shown in Fig. 4 and are coupled to tubes which are routed all the way to the back of the plug, such that the source can be driven into those layers from time to time during the experiment to monitor the stability of the calorimeter. This calibration in time takes all the factors from the scintillator, to fiber, to the phototube into account.

The scintillator/white plastic assembly is covered with black Tedlar for light-tightness. It is then protected by top and bottom aluminum plates. The edge of this assembly is taped for further light-tightness. The total thickness of the assembly is 1.5 cm. Twenty-seven megatiles are produced for each layer.

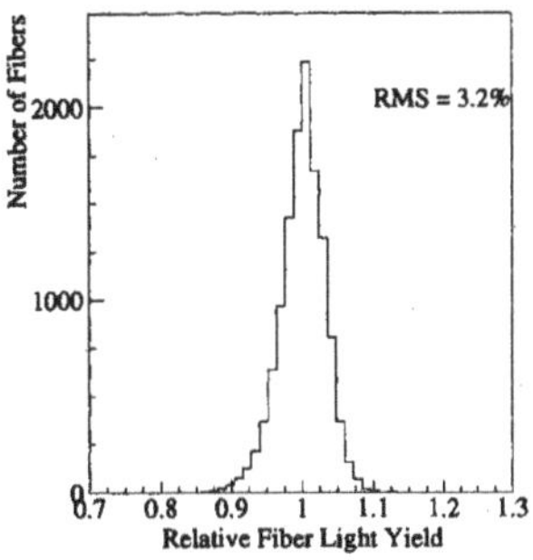

Fig. 5. Relative light yields of fibers normalized to the average of each fiber-connector type.

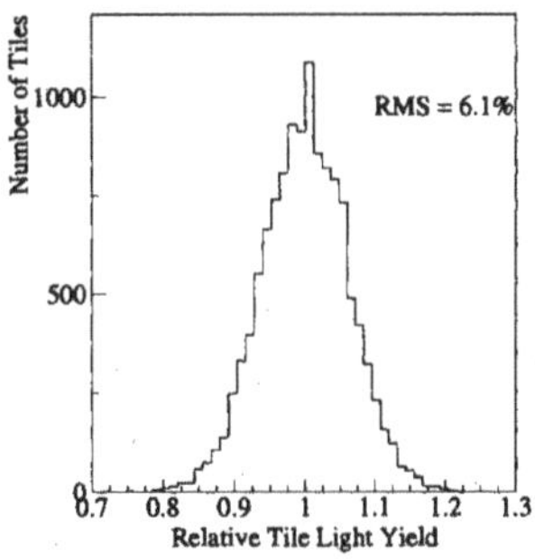

Fig. 6. Relative light yields of tiles normalized to the average of each tile size.

3. Quality Control Tests

After being assembled, each megatile is tested in a scanner box with two radioactive sources for quality control.[8] Four optical cables are used to transport the light from the megatile to a decoder box, which also houses four phototubes. The decoder is designed in a double checker-board pattern such that no neighboring tiles are read out by the same phototube. The gains of the phototubes are monitored and corrected by eight control tiles.

Both radioactive sources used in the QC tests are ^{137}Cs. One is collimated with a lead cone 4.4 cm deep having 22.7° half angle. The other is un-collimated and is located on the tip of a long flexible wire. The collimated source is positioned above the geometric center of each tile. The light yield from the tile is read out in terms of a DC current from the phototube. The typical current is 50 nA.

Fig. 6 shows the relative light yield of the tiles normalized to the average for each tile size (there are 27 identical tiles of each size). If the light yield of one tile falls more than 22% below the average, that tile is checked for possible defects in its optical readout chain. In most cases, a single WLS fiber is replaced due to damage that occurred during the insertion. In rare cases, the defect is in the clear fiber or the connector itself in which case the whole fiber-connector assembly has to be replaced. As shown in the figure, the RMS of the relative light yield is 6%. This variation includes contributions from the whole tile/fiber system of the megatile: scintillator thickness, groove quality, optical coupling between the scintillator and WLS fiber, and the fiber-connector assembly.

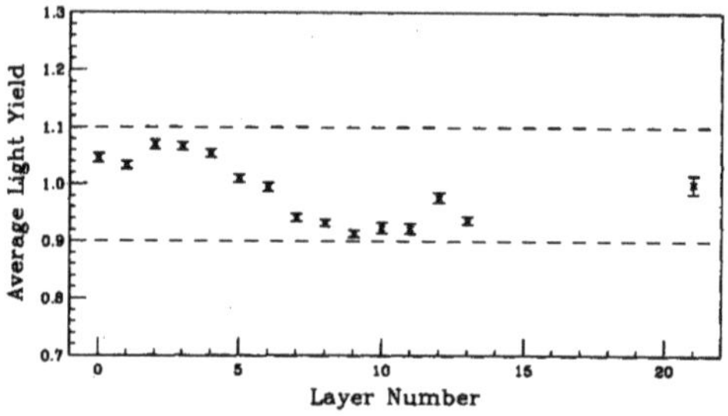

Fig. 7. Average tile light yield as a function of layer number, normalized to that of layer 21. Layer 0 is at front, layer 21 is at back. The data are corrected for light attenuation in the fiber ribbon cable, the estimated light yield increase in magnetic field, and the effect of the finite size of the collimator.

Our design criteria require that the light yield variation in a tower from the front layer to the back layer be within 10%, which contributes about 3% to the 5% constant term in resolution. Fig. 7 shows the average light yield versus layer number. The points are all within $\pm$ 10%, satisfying out requirement.

After correcting for the effect of the finite size of the collimator based on a calculation of volume integral of the radiation field in the tiles, the collimated source scan result becomes an absolute measure of the tile response to a given excitation source. This should be proportional to the tile response to muons. If the longitudinal shower profile is known, this result can be used to predict the tower response to hadron showers.

But it is not possible to have the collimated source scan the tiles after the megatiles are installed into the steel structure. The wire source is used to bridge the collimated source results and the tile response to the hadron showers.

The wire source scan is performed right after the collimated source scan. The maximum current recorded when the wire source moves across a tile is used to represent that tile's response to the source. The ratio of two measurements for each tile, one during the wire insertion, the other during the extraction, is plotted in Fig. 8. The RMS of this distribution is 0.6%. This repeatability is adequate for long term monitoring and calibration.

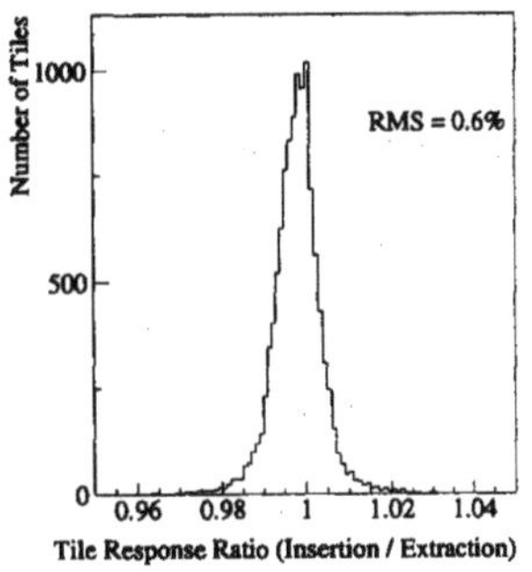

Fig. 8. Repeatability of the wire source measurements, represented by the ratio of tile responses to the source during wire insertion and wire extraction.

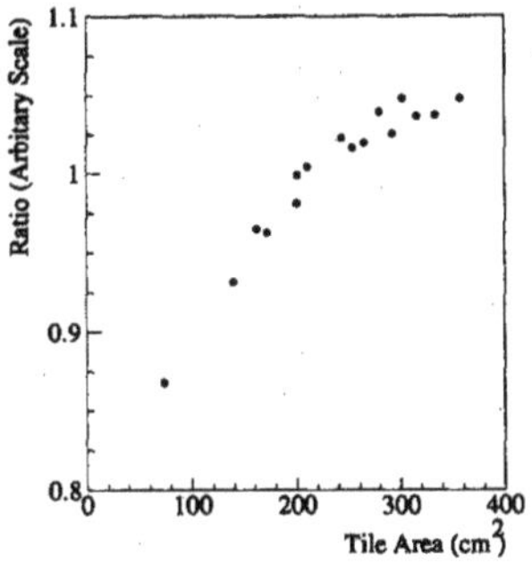

Fig. 9. Ratio of tile responses to the wire and the collimated sources as a function of tile size. Data from layer 13 is shown.

The tile response ratio of wire to collimated source is established for each tile. This ratio as a function of the tile area is shown in Fig. 9. The ratio normalized to each tile size is plotted in Fig. 10. The RMS of 1.2% indicates accurate placement of the source tubes. The correlations between the source scan results and the tower responses to MIPs and to hadron showers should be better than this 1.2%, since the particles are sampled in many layers. A 60° test beam module will be built and tested. The results from the test beam muons and the pions will be compared to the predictions based on the source scan results obtained at production time and at the test beam.

The tile size dependence of the ratio shown in Fig. 9 comes mostly from the geometric effect of the un-collimated wire source. The predicted ratios from the volume integral calculation match very well with the measured ratios in most cases. The exceptions are the two highest η tiles, where the leakages of the radiation field become important. The ratio of the measured ratio to the predicted ratio is plotted in Fig. 11. Each entry is for one tile type. The two highest η tiles from all layers are excluded from this plot. The width of this distribution is 1%, which indicates the level of our understanding of the tile response to the sources for most tiles. In addition, it indicates that for future construction of tile calorimeters a similar calculation can be used to predict the tile response to sources.

4. Conclusion

Our test results show that the light yields from the tiles are 4 pe/MIP, the tile-to-tile light yield variation is 6%, and the relative light yields between the front layer and back layer are within 10%, satisfying our design requirements. The wire source measurements are repeatable to better than 1%, insuring our ability to monitor and calibrate the calorimeter over the long term. The tile responses to collimated and wire sources correlate at the level of 1.2%. These results, along with our test beam results,

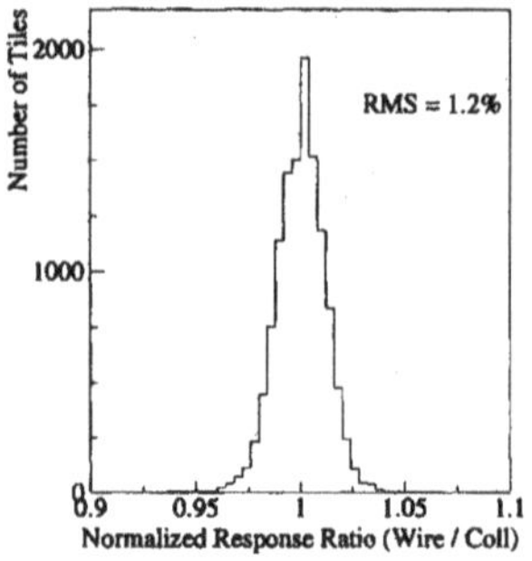

Fig. 10. Ratio of tile response to the wire source and the collimated source normalized to each tile size. RMS: 1.2%

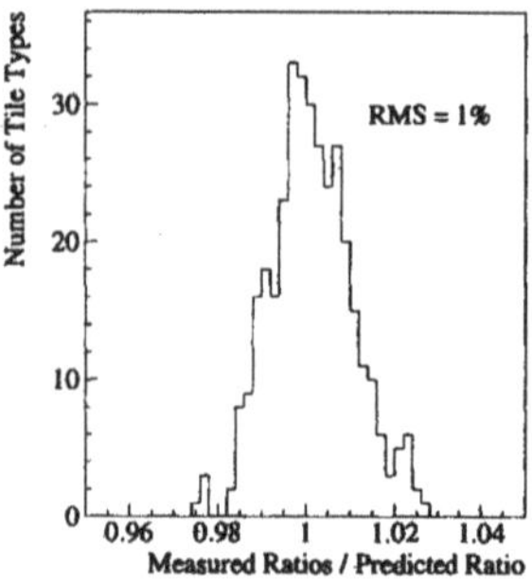

Fig. 11. Comparison between the measured ratio and the calculated ratio. They agree at the level of 1%

will enable us to predict tower response to muons and hadron showers.

Acknowledgements

We wish to thank the technical staff members of our institutions and Fermilab Technical Support Section for their help in the construction and test of the hadron calorimeter. This work is in part supported by the U.S Department of Energy.

References

1. The CDF Collaboration, "Proposal for an Upgraded CDF Detector", *CDF Note* 1172, (1990).
2. M. Lindgren, *Proceedings of the Third International Conference on Calorimetry in High Energy Physics*, ed. P. Hale and J. Siegrist, World Scientific, (1993) 61.
3. G. Apollinari *et al.*, *Proceedings of the Fourth International Conference on Calorimetry in High Energy Physics*, ed. A. Menzione and A. Scribano, World Scientific, (1994) 200-225.
4. M. Olsson *et al.*, "Techniques for Optical Isolation and Construction of Multi-tile Assemblies in Scintillator Tile-Fiber Calorimeters Using White Epoxy", *CDF Note* 2582, (1994).
5. P. de Barbaro *et al.*, *Nucl. Instr. and Meth. in Phys. Res.* **A 315** (1992) 317-321.
6. J. P. Mansour *et al.*, "A Semi-automated Splicer for Plastic Optical Fibers", presented at *the 1993 Scintillator Fiber Workshop*, Notre Dame, IN, Nov 5-12 1993.
7. P. de Barbaro, *et al.*, "CDF End Plug Upgrade Hadron Calorimeter Design", *CDF Note* 2545, (1994).
8. P. de Barbaro *al.*, "Quality Control Studies of Scintillating File/Fiber Megatile Production for the CDF End Plug Upgrade Hadron Calorimeter", *CDF Note* 2778 (1994).

Simulation of the ZEUS calorimeter

Yoshihisa Iga*
DESY, Notkestrasse 85
22603 Hamburg, Germany
for ZEUS Collaboration

ABSTRACT

The simulation of the ZEUS uranium-scintillator calorimeter is presented. It is based on the GEANT and GHEISHA Monte Carlo packages. To simulate the high resolution calorimeter with adequate precision and feasible execution time, two fast, tunable shower terminators have been developed. One of them (electromagnetic shower terminator) is applied for electromagnetic particles and the other (neutron shower terminator) for hadronic interactions inside the uranium calorimeter. The calorimeter simulation was tuned and verified with extensive data, obtained with prototype and production modules in test-beams at CERN and Fermilab. The simulation reproduces the energy resolution and e/h ratio for momenta down to $0.5\,GeV/c$. The simulated longitudinal and transverse shower profiles also agree reasonably well with data. Effects caused by the boundary between read-out cells and by extra material in front of the calorimeter are also reproduced well by the simulation. The shower terminators reduce the execution time by a factor of about 15.

1. Introduction

The ZEUS detector has been taking data at the electron-proton collider, HERA, at DESY since summer 1992. Details about the ZEUS detector are described elsewhere [1,2]. One of the main components of the ZEUS detector is the hermetic, high resolution calorimeter. The sampling structure consists of $1X_0$ depleted uranium plates interleaved with plastic scintillator tiles. The ratio of the thickness of the scintillator to the absorber was chosen to yield $e/h = 1$ (compensation), hence achieving excellent energy resolution of hadronic showers [3].

The ZEUS collaboration uses the GEANT [4] and GHEISHA [5] packages to simulate the ZEUS detector. We started with GEANT 3.11 and GHEISHA7, and afterwards updated to GEANT 3.13.

Although GEANT simulated most features of the ZEUS detector well, it did a poor job with the ZEUS uranium-scintillator calorimeter response. For example, test-beam data with the prototype at CERN gave a resolution, excluding the contribution of photostatistics, of $15\%/\sqrt{E}$ for electron beams [3,6]. On the other hand, the Monte Carlo, with electrons and photons tracked down to $1\,MeV$, gave $\sim 17.5\%/\sqrt{E}$ [7]. For hadronic showers, the Monte Carlo, with charged hadrons tracked down to $10\,MeV$

*e-mail address:iga@desy.de

and neutrons down to $10\,keV$, gave the resolution $\sim 45\%/\sqrt{E}$ with $e/\pi \sim 1.2$ for 100 ns gate width [7], to be compared to the measured values of $35\%/\sqrt{E}$ with $e/\pi \sim 1.0$ [3,6].

To overcome the discrepancies between test-beam data and the Monte Carlo simulation, we have developed two fast, tunable shower terminators [7,8]. In this approach the general features of GEANT, like the cross sections, dynamics and so on, are kept, and GEANT simulates most components as before. The first terminator, called the *E/M shower terminator*, operates on e^-, e^+ and γ with kinetic energy less than $200\,MeV$. The *E/M shower terminator* is of general use both inside and outside the calorimeter, although it is not used in the ZEUS inner tracking region. The second terminator, or the *neutron shower terminator*, operates on the evaporation energy [†] of excited nuclei at nuclear interactions simulated with GHEISHA and on neutrons with energy below $50\,MeV$. The *neutron shower terminator* acts only in the interior of the ZEUS calorimeter.

The ZEUS calorimeter simulation with shower terminators gives reasonable agreements with test-beam data for e/h, for energy resolution of both electromagnetic and hadronic showers, and for shower profiles in the momentum range of $0.5 - 100\,GeV/c$. The shower terminators also make the program execution faster by a factor ~ 15.

Precision measurements from the prototype with CERN test-beam enabled us to tune the shower terminators, and to include some geometrical effects, e.g. the inefficiency near boundaries. Correct simulation of such detailed but important effects reduces the errors due to calorimeter simulation systematics in the physics analysis.

The ZEUS calorimeter is described briefly in Section 2, and the shower terminators are described in Sections 3 and 4. The comparisons between data and Monte Carlo with shower terminators are shown in Section 5. Finally we summarise the ZEUS calorimeter simulation.

2. ZEUS Calorimeter

The ZEUS calorimeter is divided into three parts: the forward (FCAL), barrel (BCAL), and rear (RCAL), which cover $2.2° < \theta < 39.9°$, $36.7° < \theta < 129.1°$, and $128.1° < \theta < 176.5°$ respectively. (The downstream proton direction defines $\theta = 0$.) The calorimeter covers 99.7% of the total solid angle.

The basic read-out unit in the ZEUS calorimeter is a tower, whose cross-section is approximately $20\,\text{cm} \times 20\,\text{cm}$. For read-out, each tower is subdivided longitudinally into an electromagnetic (EMC) and hadronic (HAC) sections. FCAL and BCAL have two HAC sections and RCAL has one. Each EMC section is subdivided transversely into 4 cells $5\,\text{cm} \times 20\,\text{cm}$ ($\sim 5\,\text{cm} \times 25\,\text{cm}$) for FCAL(BCAL), and 2 cells $10\,\text{cm} \times 20\,\text{cm}$ for RCAL. HAC sections are not subdivided transversely and so the transverse cell size is $\sim 20\,\text{cm} \times 20\,\text{cm}$. The scintillators of each cell are read out on each side through a wave-length shifter (WLS) coupled by a light guide (LG) to a photomultipliers (PMT).

The sampling structure consists of $3.3\,mm$ thick uranium plates and $2.6\,mm$ thick

[†]Evaporation energy makes very low energy ($K.E \leq 20\,MeV$) neutrons or protons.

scintillator tiles for each cell. An EMC section has a depth of $25X_0$, and a HAC section has 3λ for FCAL and RCAL and 2λ for BCAL. The innermost EMC section in RCAL also has silicon diode planes to assist hadron-electron separation (HES). The unit size of HES, at $3\,\text{cm} \times 3\,\text{cm}$, is much smaller than the EMC cell, thus providing more accurate position information. Equivalent devices will also be installed in FCAL, and possibly also in BCAL, at a future date. Further details of the ZEUS calorimeter can be found elsewhere [10,11,12,13].

In order to investigate the performance of the ZEUS calorimeter, an FCAL prototype was built and thoroughly tested in CERN test-beams in momentum range between 0.5 and $120\,GeV/c$. The prototype size is 4×4 towers, with the same structure as FCAL. The measured energy resolution is $35\%/\sqrt{E(GeV)}$ for hadrons and $18\%/\sqrt{E(GeV)}$ for electrons, and e/h=1 [3,6]. Similar tests were made at FNAL with BCAL production modules.

3. E/M Shower Terminator

The calorimeter provides information on energy, position, timing, and particle identification. The energy resolution depends on the sampling fraction. The position resolution depends upon the longitudinal and transverse shower profiles, which also give information useful for particle identification. It is important for the simulation code to reproduce accurately the sampling fraction and shower profiles.

The essential problem of shower simulation is to reproduce the measured shower characteristics without prohibitive use of CPU time in tracking the myriad shower particles and their interactions. Electromagnetic (*em*) showers have a remarkable feature, namely tight longitudinal collimation due to the characteristics of electromagnetic physics processes. For simulation purposes, the highly collimated nature suggests that a low energy shower below Δ (e.g. $\Delta = 200\,MeV$), called *showerlet*, could be described by one representative since all the shower particles are almost parallel. while a lower threshold δ (e.g. $\delta = 1\,MeV$) might be necessary to reproduce the longitudinal shower profile. This characteristic of the *em* showers allows us to use the following prescription for the *E/M shower terminator* [7].

For the *em* interactions at energy below Δ, the *E/M shower terminator* keeps one secondary and discards the others, instead of tracing every shower particle. The *E/M shower terminator* also keeps account of energy left, called undistributed energy, of the original *showerlet* to take care of energy conservation. The retained particle is tracked by GEANT as usual. Each energy deposit during GEANT tracking is scaled up by the ratio of undistributed energy of the original *showerlet* to that of the survivor, so that the *E/M shower terminator* also deposits contributions from discarded particles, and then the undistributed energy is updated.

The *E/M shower terminator* continues to apply this discard algorithm for each *em* interaction, until the retained secondary particle's energy is less than δ.

This basic prescription of the *E/M shower terminator* describes only very crudely the correct shower profile. In order to generate energy deposits by GEANT, the survivor is chosen from among the e^- and e^+, excluding the γ. The resulting energy losses by charged particles, instead of by photons, makes longitudinal penetration too shallow.

In order to compensate for this problem, a "stretched" path length phase was introduced [8]. The origin of the *showerlet* is replaced by a photon with an empirically longer mean free path of pair-production, even when it was actually an electron or a positron. After this "stretched" photon has an interaction, the original discard algorithm takes it over.

On the other hand, the basic *E/M shower terminator* also makes the transverse shower profile narrower due to discarding half of the binary products. To recover the correct transverse shower profile, a post *E/M shower terminator* procedure is carried out. When the retained particle's energy falls below δ, the undistributed energy of the parent *showerlet* is still relatively large, typically 0.5Δ. This undistributed energy is spread along the *showerlet* axis using a GEANTINO (a fictitious particle provided by GEANT) according to an empirical formula which parameterises the transverse shower profile as a function of radiation length.

Inside the ZEUS calorimeter, an additional procedure is necessary to reproduce the measured energy resolution. The basic *E/M shower terminator* prescription gives too wide energy distribution, since in effect the signal from low energy showers fluctuates as though showers consisted of one particle rather than many. To correct this sampling problem, the visible energy deposited in scintillator is compared to the expected energy based on the sampling fraction, which is 0.037 from an EGS study [9]. Deviations in the energy resolution for electrons are then scaled by a factor of 15/17.5 (see section 1) [7].

4. Neutron Shower Terminator

The *neutron shower terminator* was introduced to overcome the discrepancies in hadronic showers between the test-beam data and Monte Carlo simulation. The main uncertainty, which plays a major role in e/h and the energy resolution of hadronic showers, comes from the complex nuclear processes.

The ratio e/h was measured as a function of the incident kinetic energy with the FCAL prototype at CERN. At high energies $e/h = 1$. As the incident kinetic energy decreases below $4\,GeV$, e/h also decreases, eventually approaching e/mip, where mip stands for minimum ionising particles [6]. When GEANT generates a hadronic or nuclear interaction inside the ZEUS calorimeter using GHEISHA, the *neutron shower terminator* intercepts

- the evaporation energy, and
- the energy of any spallation neutrons below $50\,MeV$.

To simulate the measured e/h behaviour at low kinetic energy, an energy dependent scale factor is applied to the evaporation energy in GHEISHA interactions. Most of the evaporation energy is distributed isotropically by one or more GEANTINOs, according to the form $e^{-\Delta x/\sigma}$ where Δx is the distance in x from the interaction point and σ is set to $7\,cm$, and similarly in y and z. It is also distributed in time, t, according to the "lethargy plot" based on results of very detail shower simulation [7,14]. To represent the penetrating component in the evaporation energy, the rest is thrown as a proton or neutron randomly chosen according to the target's nuclear composition, whose direction is the same as that of the incident hadron.

Spallation neutron energy is also thrown with the same prescription, except in the spatial distribution, which is biased toward the neutron's direction, and the time distribution, which is prompt.

The energy distributed as GEANTINOs by the *neutron shower terminator* has a sampling fraction empirically chosen to yield $e/h = 1$ for $10\,GeV/c$ π^-. The GEANTINO energy deposits are also smeared to reproduce the overall hadronic energy resolution of $\sim 35\%/\sqrt{E}$ at this energy.

The *neutron shower terminator* is used only inside the ZEUS calorimeter, and is limited to the above mentioned components of nuclear interactions. GHEISHA takes care of the energy dependence and many other features in hadronic showers, including fluctuations and correlations.

5. Comparisons to Data

5.1. *Electromagnetic showers*

First, we compare Monte Carlo simulations with and without the *E/M shower terminator*. In fig. 1, the longitudinal distribution simulated with the *E/M shower terminator* for $5\,GeV/c$ electrons is compared to that from normal GEANT (i.e. without the *E/M shower terminator*), where tracking is done to $100\,keV$ for γ, and to $1\,MeV$ for e^- and e^+. They agree very well.

In order to study the transverse shower profile, data were collected in three adjacent electromagnetic cells (EMC1, EMC2, and EMC3), each measuring $5\,\mathrm{cm} \times 20\,\mathrm{cm}$. The ratio of energy deposited in EMC1 to the sum of the three, as the electron beam normal to the EMC's face is scanned across the central cell (EMC2), characterises the transverse shower profile. This ratio is shown, for both data and Monte Carlo simulation, for $15\,GeV/c$ e^- in fig. 2, as a function of beam position. The Monte Carlo results are in reasonable agreement with the data. We should note that the *E/M shower terminator* was already tuned to the test-beam for some of the light propagation effects.

5.2. *Hadronic showers*

An important feature in the ZEUS calorimeter is the excellent energy resolution for hadronic showers due to the compensation mechanism, giving $e/h = 1$. Fig. 3

shows e/h as a function of kinetic energy. Data were taken from the FCAL prototype with CERN test-beam for $h = \pi$ and p. The corresponding Monte Carlo results are also shown in fig. 3 for both of pions and protons. The energy dependence of e/h is reproduced very well, even in the energy region below $4\,GeV$.

The energy resolution of hadronic showers is plotted as a function of incident particle momentum in fig. 4. Proton and pion data are presented separately, along with Monte Carlo results. The CERN test-beam data shows a resolution of $\sim 35\%/\sqrt{E}$ in the high energy region. The Monte Carlo reproduces this behaviour, even in the low energy region, for both protons and pions.

To study the longitudinal shower development, fig. 5 shows the fraction of the energy deposited by incident π^+s in the EMC, HAC1 and HAC2 sections, of the FCAL prototype, as a function of incident beam momentum. An EMC section is $25X_0$ thick, and HAC1 and HAC2 are each 3λ thick. Monte Carlo simulations are in fair agreements with data for each section in the momentum range of $0.5 - 100\,GeV/c$. The ZEUS barrel calorimeter (BCAL) has thinner HAC sections: HAC1 and HAC2 are each 2λ thick. In fig. 6 the same quantity is shown for BCAL as a function of the incident π^+ momentum. Simulation results are shown to agree well with data across the entire momentum range, from 20 to $100\,GeV/c$, where BCAL test-beam results are available.

The lateral spread of $30\,GeV$ hadronic showers is presented in fig. 7a and 7b for data and Monte Carlo, respectively. The FCAL prototype consists of 4×4 towers, with the beam incident on the tower positioned in the third column and the second row. Each box shows the ratio of energy deposited in that tower to the sum from all 16 towers. The data and Monte Carlo agree with each other at the level of $\pm 0.5\%$.

5.3. Calorimeter response

In this section we describe the simulation of various more detailed aspects of the ZEUS calorimeter response.

The mechanical structure includes cracks between towers, and between cells, consisting of air gaps, wrapping papers, and spacers. Such cracks reduce signal output near the boundaries. CERN tests with production modules measured this crack effect precisely. In fig. 8 the signal reduction in RCAL for $10\,GeV/c$ e^- is shown as a function of beam position. The solid curve represents the CERN test result, while dots are the simulation results. The dip in the right side corresponds to the crack between EMC cells, and the left dip is from the crack between neighboring towers. In this worst case, the beam is perpendicular to the calorimeter surface. The signal reduction at the tower boundaries is $\sim 10\%$. At the cell boundaries the effect is $\sim 7\%$, because there are small spacers between towers which protect scintillator tiles. The crack effect is very well duplicated by Monte Carlo. In practice, this effect is not as serious for the ZEUS physics data, because the incident particles are not generally normal to the FCAL and RCAL surface.

There is a similar effect at the boundary between modules, where WLSs and LGs transport scintillation light to the PMTs. When charged particles deposit energy in WLS and LG, there is no scintillation light, but there can be Čerenkov light under the appropriate circumstances. In order to compensate the reduced scintillation light by Čerenkov light in the WLS, there is a single lead sheet of 2.5 *mm* thickness between each pair of modules in the ZEUS calorimeter. This thickness was chosen on the basis of a CERN test [3,15]. In the simulation code the factor for Čerenkov light in WLS was adjusted to reproduce this test-beam data. Fig. 9 shows Monte Carlo results with this configuration. The contribution from WLSs and scintillators are individually shown in fig. 9a and 9b respectively, and the total is in fig. 9c as a function of the impact position of electrons. The position of $x = 30$ corresponds to an intermodule boundary, with the e^- shot from the ZEUS interaction point. Based on the test-beam data, the predicted fluctuation is less than 5%.

There is also the problem of extra material in front of the calorimeter, e.g. other components, signal cables and beampipes. In the CERN test with the prototype, this "dead material effect" was investigated [16]. The results are compared to the simulation in fig.10 for 1, 5 and 30 GeV/c electrons, where the signal reduction ratio is plotted against the radiation length of the extra material in front of calorimeter. Monte Carlo and data agree reasonably well at each momentum.

6. Conclusions

The ZEUS calorimeter simulation is based on GEANT and GHEISHA, and two fast, tunable adaptations, *E/M shower terminator* and *neutron shower terminator*. The *neutron shower terminator* is necessary to reproduce $e/h = 1$ and the excellent energy resolution of the calorimeter. The *E/M shower terminator* is needed for the correct shower profile. Using these shower terminators also makes the CPU execution faster by a factor of ~ 15.

Data and Monte Carlo were compared for several distributions, the ratio e/h, energy resolutions for both hadronic and electromagnetic showers, and shower profiles. The Monte Carlo simulations agree well with the measured data in the momentum range, $0.5 - 100\,GeV/c$, where the test-beam data is available.

An important application of the extensive test-beam data was for fine tuning the shower terminators. Geometrical effects due to the boundary between cells, towers, and modules are nicely reproduced, as well as the effect of extra material in front of the calorimeter.

This calorimeter simulation is currently used to generate useful Monte Carlo data samples for physics analysis in the ZEUS collaboration.

7. References

1. The ZEUS detector, Status Report 1993, DESY (1993).
2. ZEUS collab., M.Derrick et al, *Phys. Lett.* **B293** (1992) 465.

3. ZEUS Calorimeter Group, U.Behrens et al, *Nucl. Inst. and Meth.*, **A289** (1990) 115.
4. R.Brun et al, GEANT 3, CERN/DD/EE/84-1 (1987).
5. H.C.Fesefeldt, *Simulation of Hadronic Showers*, PITHA 85-02, RWTH Aachen (1985).
6. ZEUS Calorimeter Group, A.Andresen et al, *Nucl. Inst. and Meth.*, **A290** (1990) 95.
7. G.F.Hartner, ZEUS-Note 88-49 (1988).
8. G.F.Hartner and Y.Iga, ZEUS-Note 90-84 (1990).
9. B.J.Stacey, G.F.Hartner and J.F.Martin, ZEUS-Note 88-29 (1988).
10. A.Andresen et al, *Nucl. Inst. and Meth.* **A309** (1991) 101.
11. A.Bernstein et al, *Nucl. Inst. and Meth.* **A336** (1993) 23.
12. A.Caldwell et al, *Nucl. Inst. and Meth.* **A321** (1992) 356.
13. J.A.Crittenden, *The Performance and Understanding of the ZEUS Calorimeter*, in this proceedings.
14. H.Brückmann et al, *On the Theoretical Understanding and Calculation of Sampling Calorimeters*, DESY 87-170 (1987).
15. J.A.Crittenden and R.Yoshida, Proceedings of the Workshop 'Physics at HERA' vol.1, DESY (1992) 121.
16. J.Krüger, *Length of Calorimeter and Effect of Absorbers in Front of Calorimeters*, DESY 90-163 (1990).

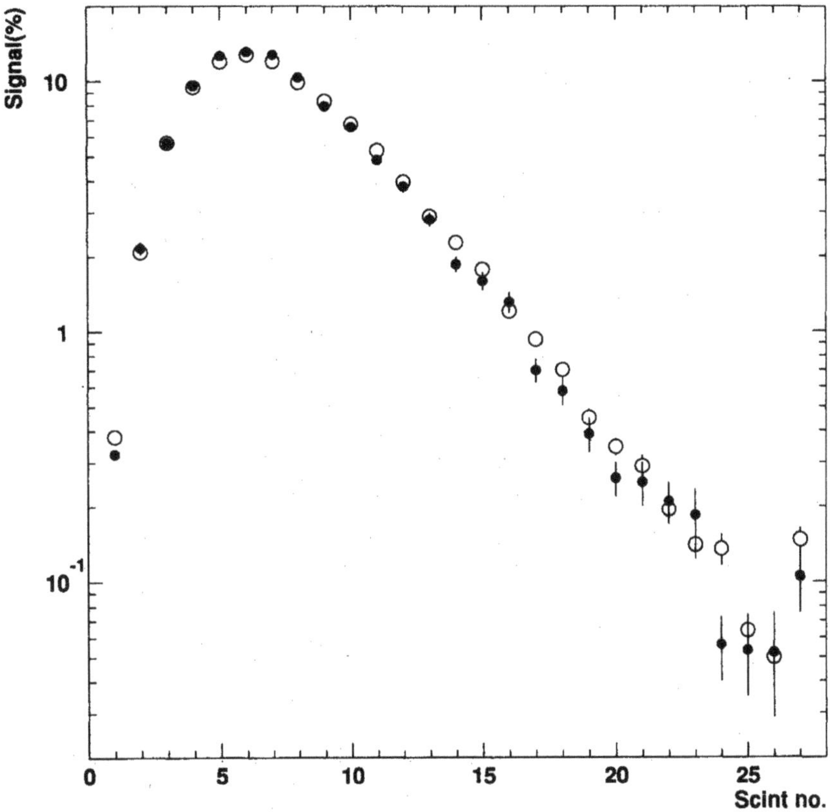

Figure 1: *Mean percentage of FCAL signal in each EMC layer for 5 GeV/c electron. The summed HACs is in the last point.* ○ : *with E/M shower terminator and* ● : *without E/M shower terminator*

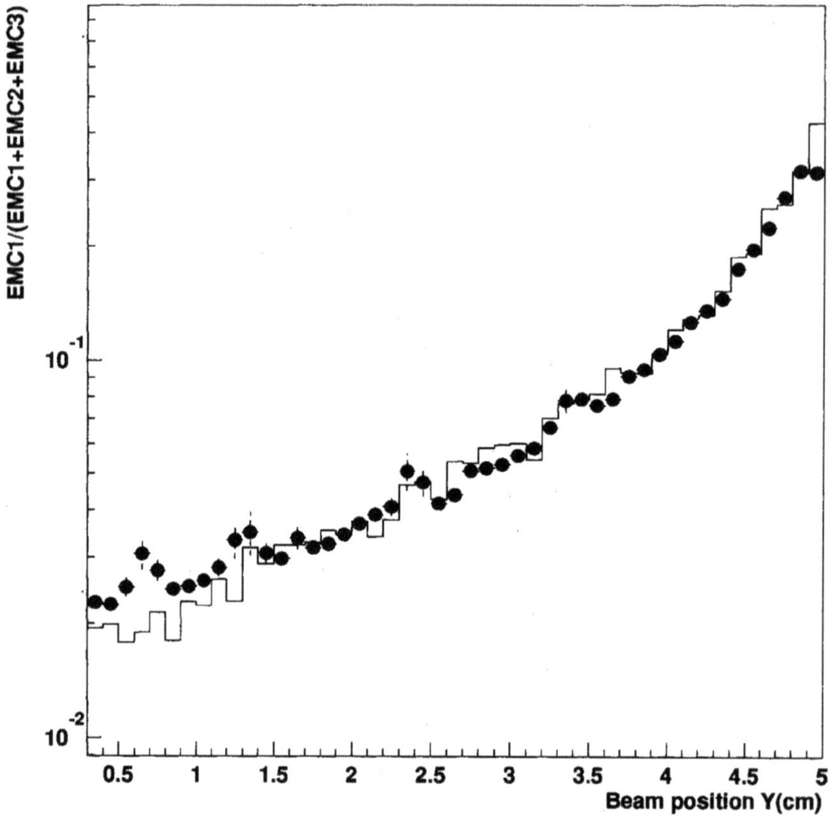

Figure 2: *Transverse shower profile for 15 GeV/c e−. Ratio of energy deposited in the right side (EMC1) to the sum of three adjacent EMC cells as a function of the beam position. Y=0 to 5 is corresponding to the central cell (EMC2). Histogram is for MC result and dots for data.*

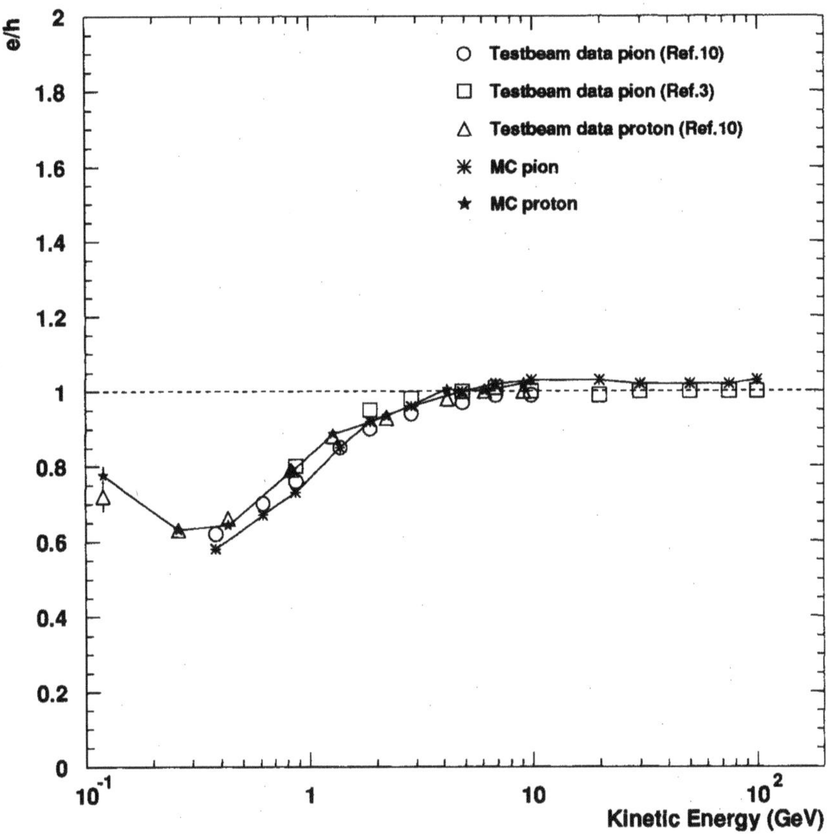

Figure 3: *e/h ratio as a function of kinetic energy for pions and protons. Testbeam data are shown with open symbols and MC points are connected by solid lines.*

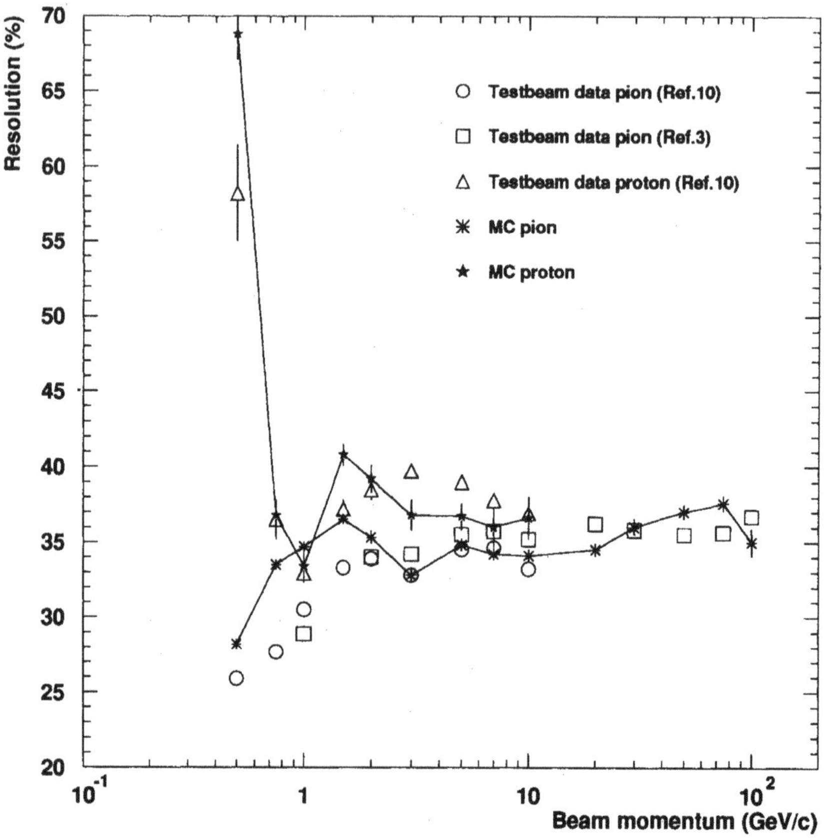

Figure 4: *Energy resolution of hadronic showers as a function of beam momentum. Testbeam data are shown with open symbols and MC points are connected by solid lines.*

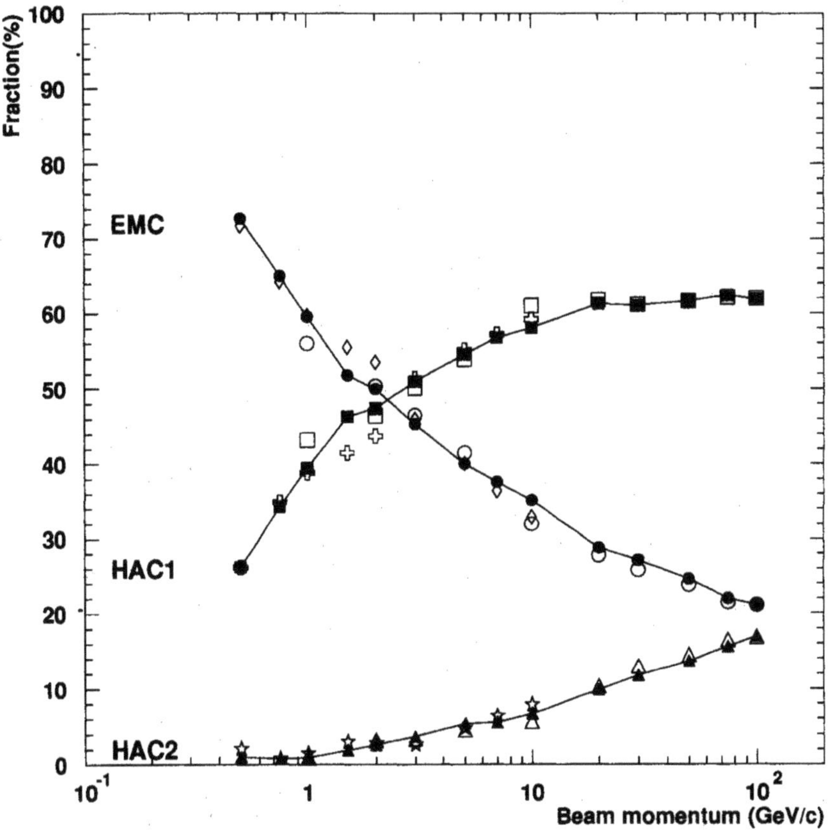

Figure 5: *Longitudinal shower profile in FCAL for π^+. The fraction of energy deposited in each section to total deposited energy are shown as a function of beam momentum. Open symbols are for testbeam data (Refs.3 and 10) and closed symbols for MC results with solid lines.*

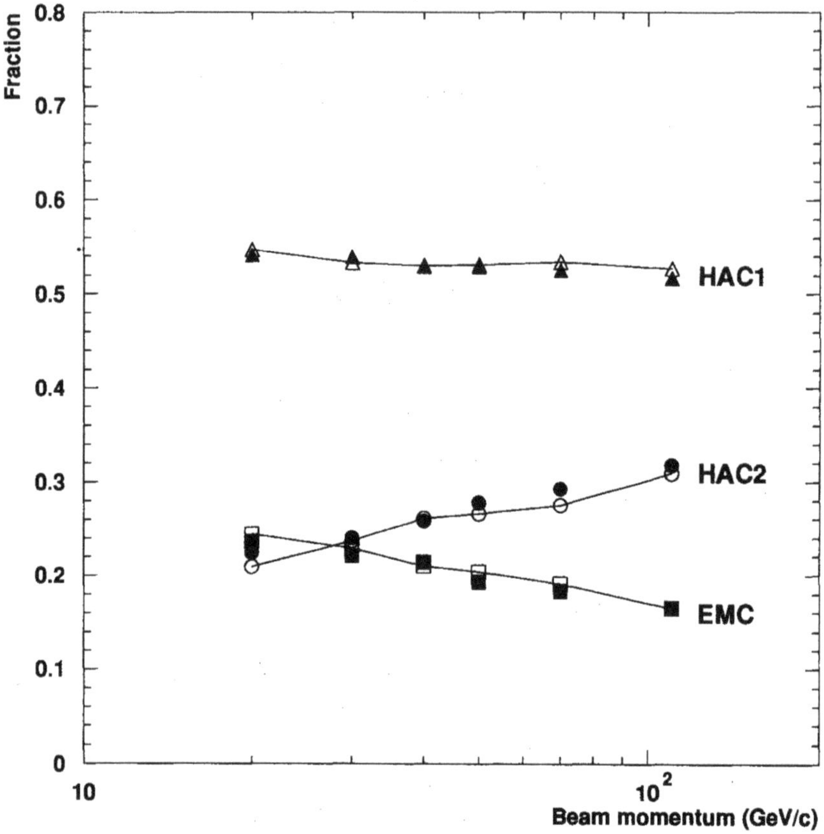

Figure 6: *Longitudinal shower profile in BCAL for π^+. The fraction of energy deposited in each section to total deposited energy are shown as a function of beam momentum. Open symbols are for testbeam data and closed symbols for MC results with solid lines.*

0.26	1.3	3.8	1.3
0.31	3.8	77.7	3.8
0.26	1.3	3.8	1.3
0.09	0.26	0.31	0.26

(a)

0.15	1.2	4.5	1.2
0.34	4.3	76.4	4.3
0.16	1.2	4.5	1.2
0.05	0.16	0.36	0.16

(b)

Figure 7: *Lateral spread of 30 GeV hadronic shower. (a) Testbeam data and (b) Monte Carlo result. Beam position is in the third column and the second row. Each box shows the ratio of energy deposited in that tower to the sum of all 16 towers.*

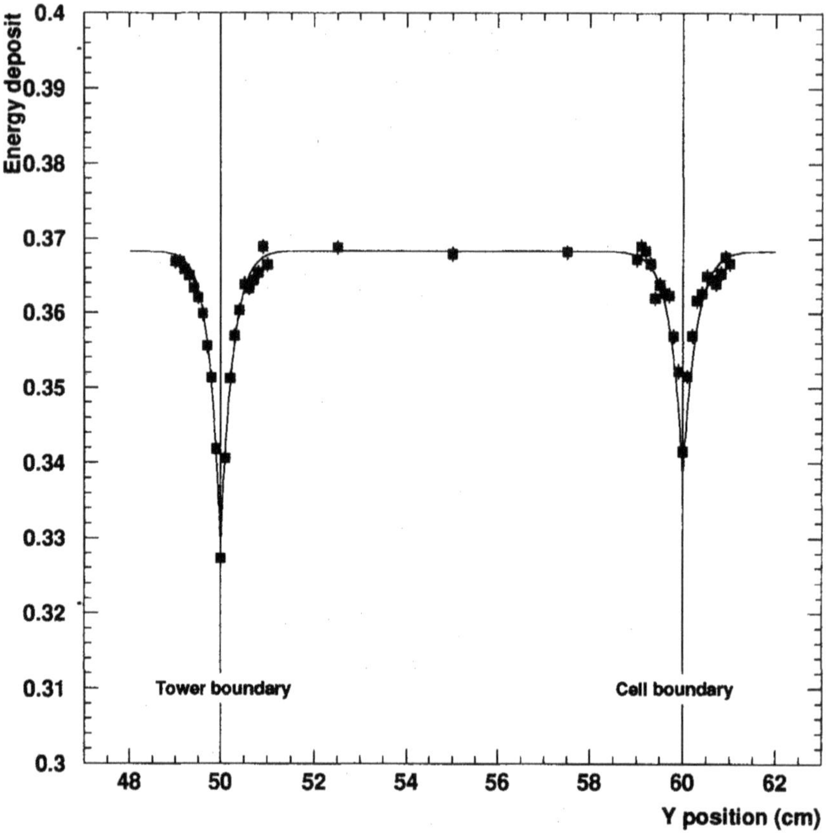

Figure 8: *Signal reduction due to cracks in RCAL for 10 GeV/c* e^- *as a function of beam position. Y=50 corresponds to the tower boundary and Y=60 to the cell boundary. Solid line shows testbeam data and square symbols present Monte Carlo result.*

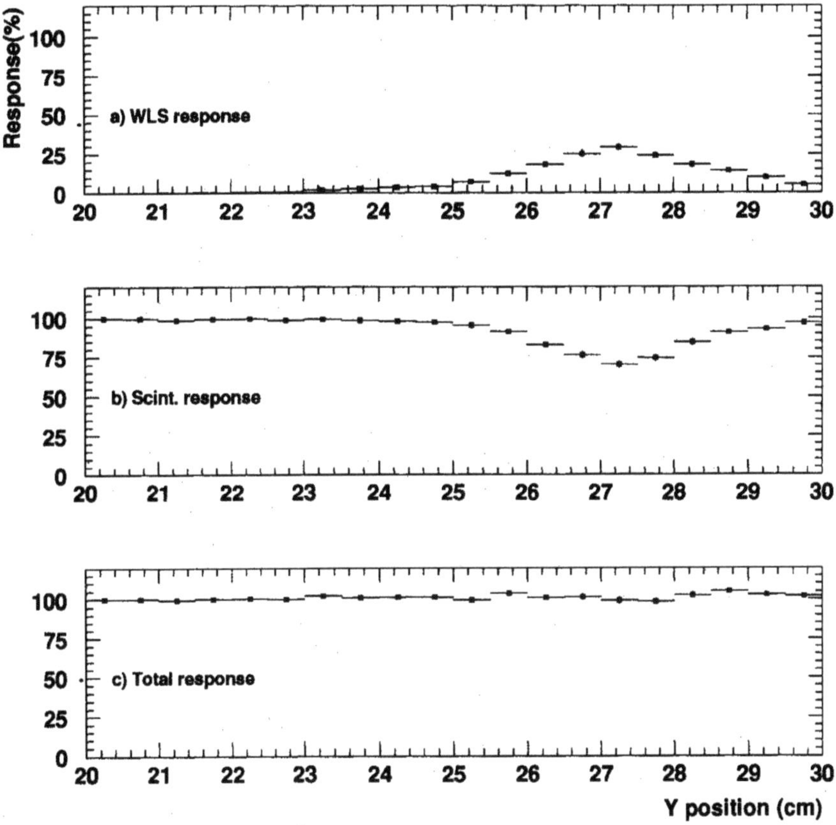

Figure 9: *Calorimeter response in the module boundary for electrons as a function of beam position from Monte Carlo, normalised to incident beam energy. a). WLS, b). Scintillator and c). Total response.*

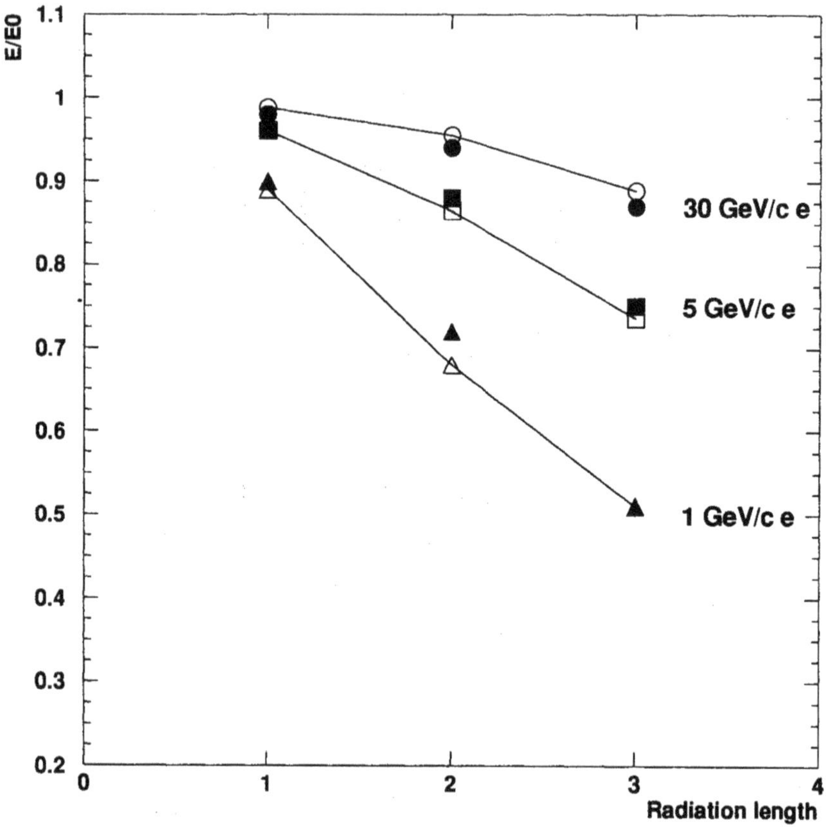

Figure 10: *Signal reduction ratio as a function of the radiation length of extra material in front of the FCAL prototype. Closed symbols show the testbeam data in ref.16 and open symbols present Monte Carlo results and they are connected by solid lines.*

A CALORIMETER FOR COSMIC RAY HADRONS UP TO 10 TEV

H.H. Mielke, H. Kornmayer, M. Locher, J. Engler, and J. Knapp*†
Kernforschungszentrum und Universität Karlsruhe,
P.O. Box 3640, D-76021 Karlsruhe, Germany

ABSTRACT

A 320 m^2 iron calorimeter for cosmic ray hadrons to be used in the extensive air shower experiment KASCADE is described. The active elements are 10^4 liquid ionisation chambers filled with the room temperature liquid tetramethylsilane (TMS). A prototype of 6 m^2 and 9 λ_0 thickness has been operative for more than 2 years and records cosmic ray hadrons with energies up to 10 TeV. Experiences with this first large scale warm liquid calorimeter and its performance are presented. The energy flux of cosmic ray hadrons has been monitored with this prototype and was used to survey the signal stability of the ionisation chambers. No signal decrease could be observed during this period. For energies of 0.1 to 10 TeV shower profiles and transition curves have been measured. The results are reproduced by Monte Carlo calculations using the Gheisha code in the energy regime below 500 GeV, and the Fluka program for the higher energy particles.

1. The KASCADE Experiment and its Hadron Calorimeter

KASCADE[1] is an extensive air shower experiment presently under construction at Karlsruhe, Germany. The basic configuration of the experiment consists of a large array of detector stations with e/γ and μ detectors and a large central hadron calorimeter with muon detectors underneath. This set-up is well suited to measure the e/γ-, the μ- and the hadronic component of individual air showers simultaneously. By performing a multidimensional measurement including all correlations between the observables we aim to determine the elemental composition and the flux of the cosmic radiation in the energy range $3 \cdot 10^{14}$ eV to $5 \cdot 10^{16}$ eV.

The hadron calorimeter has an area of 20×16 m^2 and is 11 λ_0 deep. It consists of iron slabs of 12 to 36 cm thickness interspersed with layers of a new type of liquid ionization chamber[2] as active elements, and a layer of plastic scintillators for fast timing and trigger purposes. The deepest active layer is positioned in the basement below the iron tower under a 70 cm concrete ceiling. The calorimeter is thick enough to absorb even 10 TeV hadrons completely.

The scintillator trigger is placed below the third absorber layer at a depth of approximately 2.5 λ_0, where on average the maximum of the hadronic shower development is reached. In total 228 scintillators of 1 m^2, viewed by photomultipliers are spread over the calorimeter. Their time resolution is better than 2 ns allowing a good arrival time determination of muons and hadrons.

*now at Enrico Fermi Institute, The University of Chicago, Chicago, IL 60637, USA
†Feodor Lynen-Fellow of the Alexander von Humboldt-Foundation

2. Liquid Ionization Chambers as Active Elements

A liquid ionization chamber is basically a plane capacitor filled with a dielectric liquid. The electrons and ions that are released by a penetrating ionizing particle are separated in the electrical field of the capacitor and drift to the anodes with velocities proportional to their mobilities. Suitable liquids consist usually of nonpolar molecules that exhibit good conductivity for free electrons leading to a signal rise time in the 0.1 to 0.5 μs range. The signal of the positive ions is not collected, as the ion's drift velocity is typically smaller by a factor of 10^4 to 10^5.

Liquid ionization chambers offer several advantages that suggest their usage in a cosmic ray experiment. The signals are directly proportional to the energy deposited in the chamber and do not exhibit saturation. This implies a large dynamic range of the detectors, which is important for measuring cosmic rays with particle energies varying over many orders of magnitude. In addition, a large dynamic range is essential for good energy resolution. The signals do not depend on the location of the particle's track through the chamber and no edge effects are observed. Feed-back amplification allows stable operation over a long period of time, especially important for a multi-year experiment located on the lawn without any temperature stabilization. The absolute calibration, which gives the charge output for a given energy deposit in the ionization chamber, can be calculated from the detector geometry, the electric field applied, and the lifetime of free electrons in the liquid. The latter depends on the concentration of electronegative impurities in the liquid, and can easily be obtained by recording the rise in gain of a well known signal (e.g. minimum ionizing particle, radioactive source) with increasing high voltage. A further advantage of liquid ionization chambers is the low price per channel, permitting the construction of large, fine grain calorimeters.

Up to now, liquified noble gases have generally been used as ionization liquids. However, several nonpolar room temperature liquids[3] have shown free electron conduction as well and are, therefore, suitable as ionization liquids. Liquid noble gases as dielectric media require thermal insulation and cooling devices. Warm liquid ionization chambers are operated at room temperature and, therefore, do not need a cryostat and are much easier to handle.

Because of this we have chosen liquid ionization chambers filled with the room temperature liquid tetramethylsilane (TMS) for the KASCADE calorimeter. TMS consists of a silicon atom surrounded by four methyl groups and forms a very spherical molecule. The mobility of free electrons in TMS is 98 cm^2/Vs and the charge yield is about 0.8 e^- per 100 eV energy deposit in an external field of 5 kV/cm.

A KASCADE ionization chamber consists of a $50 \times 50 \times 1$ cm^3 stainless steel box in which 4 pads of 25×25 cm^2 are kept in position by ceramics spacers. They are supplied with high voltage of 3 to 5 kV and read out via a 4-fold ceramics feedthrough. Each pad is connected to a separate electronic channel. A schematic view of a chamber is shown in fig. 1. The fine spatial segmentation permits a detailed measurement of the energy deposit in the cores of extensive air showers and to further identify single hadrons and muons rather than energy sums only. The gap width of 2×3.5 mm is

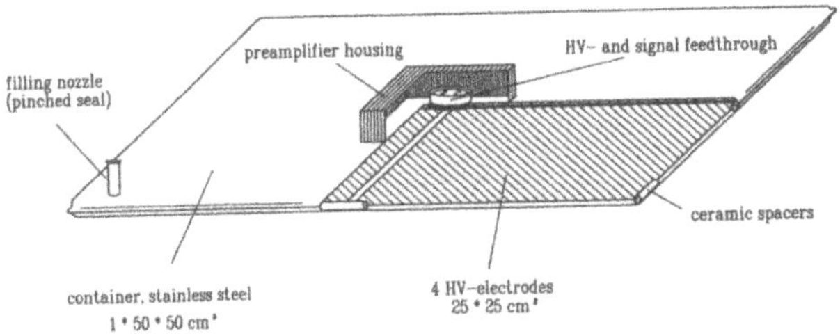

Figure 1: **Schematic view of a TMS ionization chamber.**

large enough to record single minimum ionizing particles, which deposit $\approx 1\ MeV$ (corresponding to 5000 e^-), well above the level of the preamplifier noise of 1700 e^-. Moreover, in the maximum of a 10 TeV hadron shower several GeV are deposited in a single channel. Thus, we require a dynamic range of the electronics chain of about 10^4. The dynamic range of the detectors is sufficiently large to measure 10 TeV hadrons without any saturation, the dynamic range being entirely limited by the electronics.

About 10^4 chambers are needed to equip the KASCADE calorimeter completely.

3. Purification of Liquid and Chamber Parts

As mentioned above, polar impurities in the liquid capture free electrons and reduce the fast signal. Therefore, the liquid and all parts which are in contact with it need to be cleaned carefully.

The liquid is purified by multiple vacuum distillation and passage over oxysorb and molecular sieves of various pore sizes to absorb oxygen and impurities. Typically a residual free electron life time of $\tau_e > 50\ \mu s$, corresponding to a rest impurity of less than 4 ppb O_2 equivalent is obtained.

Before assembling the chamber, all the chamber parts are cleaned in an alkaline bath, then in an ultrasonic bath with distilled water and are finally rinsed with ultrapure water of more than 18 $M\Omega cm$ volume resistivity. After being dryed at $250°C$ the chamber is mounted in a clean bench and eventually welded helium tight all around.

The assembled chamber is further cleaned by a glow discharge before being filled. The cleaning is performed with 200 Pa Argon gas and at 350 V with a limited current of 40 $\mu A/cm^2$. During the glow discharge the cathodes are struck by Ar-ions, which efficiently remove H_2O, CO, CO_2 and O_2. After 15 seconds the gas is removed, the polarity is changed, and the cycle starts again. After 60 cycles and a total dose of 10^{17} to 10^{18} ions/cm^2 the cleaning is sufficiently advanced and the chambers are ready for

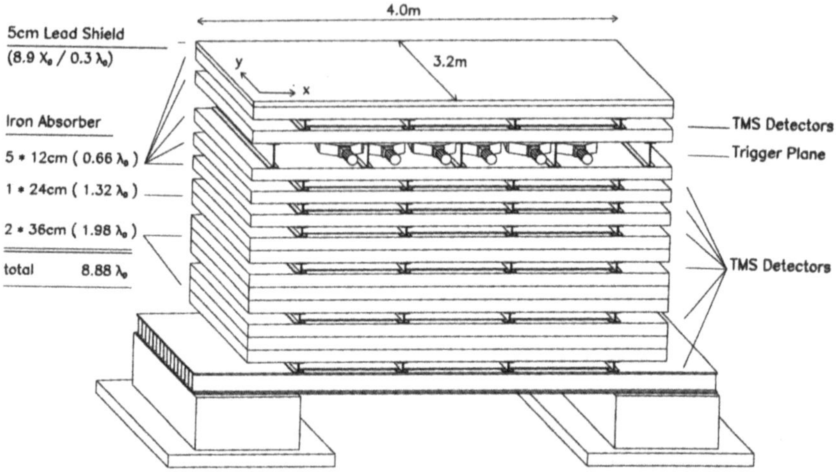

Figure 2: The prototype of the KASCADE calorimeter.

filling. The chambers are finally flushed two times with clean TMS, then filled and closed by cold welding of the Ni filling nozzle.

All the chambers are tested with cosmic ray muons and the free electron lifetime is determined individually. Chambers with lifetimes above 1 μs are accepted for use in the calorimeter. A lifetime of 1 μs and a maximal electron drift time through the detector gap of $\approx$ 300 ns lead to a signal loss due to impurities of less than 10%. Only 4% of the chambers do not fulfill this condition, and need to be filled again. For the completed chambers the typical lifetimes are 10 μs, corresponding to impurities of less than 20 ppb O_2 equivalent.

4. The Prototype Calorimeter

A prototype was built which differs only slightly from the KASCADE hadron calorimeter design. It has almost the same longitudinal absorber structure but the last layer is missing and the triggerplane is placed in the second slot. The prototype is 8.9 λ_0 thick. Thus, leakage becomes important at lower energies as in the KASCADE calorimeter. With a first set of chambers, 7 layers were equipped with 6 m^2 active area each. For the readout we have used existing electronics with a limited dynamic range of only 10^3, entailing saturation effects in the electronics at higher energies. The trigger is positioned at a depth of 1.62 λ_0 and consists of scintillator paddles which are viewed from both sides. Fig. 2 shows the prototype set-up in a schematic view.

5. Data Taking and Event Reconstruction

Data were taken with this prototype for more than 2 years, recording about 10^7 events. An event is defined by coincident signals at both ends of the scintillator, both larger than $30\times$ that of a minimum ionizing particle. The trigger efficiency for protons is 50% at 50 GeV and reaches its plateau of 80% at 200 GeV. The missing 20% are due to hadrons that penetrate the triggerplane without undergoing a hadronic reaction in the material above. Several cuts and a track finding algorithm are used to select cosmic ray hadrons and to reject muons and electrons. Specifically, in at least 5 of the 7 layers, clusters must be found with energies above 5 MeV, the energy clusters and the trigger signal must line up on a straight line within one pad width, and the total energy must not be deposited predominantely in one layer. These requirements are very restrictive, especially for low energy hadrons, leading to an overall acceptance of 10% at 100 GeV and 45% at 700 GeV. The low values are due to the inclined hadrons that leave the small calorimeter area before passing 5 layers of detectors. When using vertical showers only, the acceptance reaches 80%.

6. The Energy Estimation

To reconstruct the hadron energy from the measured charges, we need to know the fraction of the hadron's energy that is deposited in the active medium. Besides the charge to energy calibration, this fraction is usually evaluated by a beam test with particles of a known energy. For hadrons in the TeV range, however, no testbeam is presently available. We may use Monte Carlo simulations to evaluate this fraction at higher energies, extrapolating from lower energies where the simulations have been shown to reproduce the experimental results. The extrapolation to higher energies, however, is not experimentally verified as no test beam in the TeV region exists. In this situation we only can perform consistency checks, showing that the simulation, including its extrapolation of the interaction model to higher energies, reproduces the measured distributions. The simulations were performed with the GEANT package[4], using GHEISHA[5] and FLUKA[6] for the hadronic interactions.

Hadron showers of unknown energies are compared to simulations that produce the same energy sum E_S in the active medium. The energy sum of an event is a weighted sum, the weights being the absorber thickness in front of the active layer.

For a totally absorbing calorimeter with a sufficiently fine and homogenous sampling, the fraction of energy in the active medium $c_a = E_S/E_h$ should not depend on energy. Thus, the energy deposit E_S in the active layers should be proportional to the primary energy of the particle. A variation of c_a would require a change of the energy fraction escaping detection (neutrinos, binding energy) or, in case of a non compensating calorimeter, a systematic change of the hadronic and electromagnetic energy fraction. In the Monte Carlo calculation the energy in the active medium is indeed proportional to the hadron's energy from 100 GeV up to 20 TeV, as shown in fig. 3. The solid line shows the energy deposit obtained by the Monte Carlo versus the hadron energy. The dashed line gives the reconstructed energy deposit after

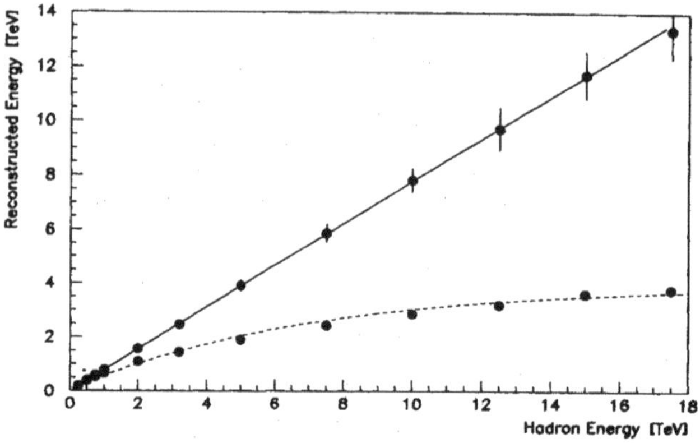

Figure 3: The simulated energy sum in TMS versus the hadron energy. The upper curve represents the pure simulation. The lower curve gives the energy sum when detector effects, electronics saturation and reconstruction are taken into account.

applying a realistic detector simulation including side and rear leakage, the electronics saturation and a subsequent trackfinding algorithm to the simulated events. The latter curve is used to attribute the energy to the experimental showers. The figure shows that this energy estimate is a linear extension of the region below 300 GeV, where the simulation programs have been checked[7] and are reliable.

7. Comparison of Experimental Data with Monte Carlo Simulation

When attributing an energy to a hadron on the basis of its energy sum according to the simulation program, we cannot check whether the simulation predicts the correct total energy deposit or not. If the Monte Carlo program is wrong we might compare hadrons with simulations of a different energy. We can, however, check whether the pattern of energy deposition in the detector is correctly reproduced by the simulation.

For this purpose we compare the longitudinal and lateral distributions of showers with the same reconstructed energy deposit E_S^*. In fig. 4 the average energy deposit versus depth in the calorimeter is shown for hadrons of various energies. Fig. 4a gives the pure MC simulation of the energy deposit versus the depth. The energy deposit scales with the hadron energy, the maximum of the shower development shifts logarithmically to bigger depths, and the attenuation in the rear part of the showers is independent of the energy, as expected from a simple linear extrapolation of the interaction model to higher energies. From the experiment we find the shower curves shown as full dots in fig. 4b. Clearly the effects of electronics saturation and acceptance distort the form of the shower curves. The maxima are no longer so well pronounced, their heights do not scale at higher energy and the attenuation length

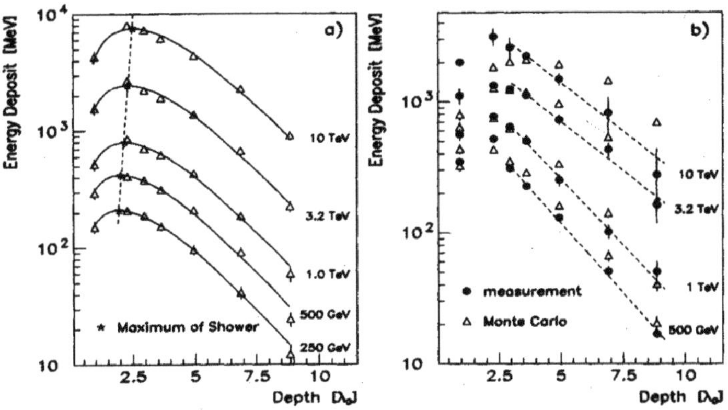

Figure 4: **The longitudinal shower development. a) Pure simulations of energy deposit versus depth. b) Experimental results (full circles) of the reconstructed energy deposit versus the depth and Monte Carlo simulations (open triangles) including leakage, electronics saturation and track finding.**

seems to become smaller with increasing energy. The Monte Carlo events, after the full simulation, including all the detector and reconstruction effects, show, however, almost exactly the same behavior as the data at energies up to 3 TeV. We therefore can conclude that the experimental longitudinal energy deposit in hadron showers behaves very much like the simulations. The distortions of the ideal shower curves can be attributed to detector effects and reconstruction inefficiencies.

The discrepancy between data and simulation that occur at 3.2 TeV and 10 TeV are due to the fact that very high energy hadrons are likely to be accompanied by low energy electromagnetic radiation being created somewhere in the air above the detector. Those particles may penetrate the lead and iron on top of the calorimeter with a small but finite probability and deposit part of their energy in the first active layers. Thus, we compare lower energy hadrons accompanied by electromagnetic radiation with a simulation of unaccompanied hadrons of higher energy.

A comparison of the lateral energy distribution in the third layer between data and simulation is shown in fig. 5. The energy density as a function of the radius exhibits an exponential shape with a change in the slope at 35 cm. The flatter part at larger radii is attributed to secondary low energy neutrons in the shower. Low energy neutrons penetrate the calorimeter setup better than charged particles due to their low interaction cross-section. The TMS with its large fraction of hydrogen, however, is sensitive to it and detects efficiently the recoil protons. The width of the lateral distributions does not vary significantly with energy.

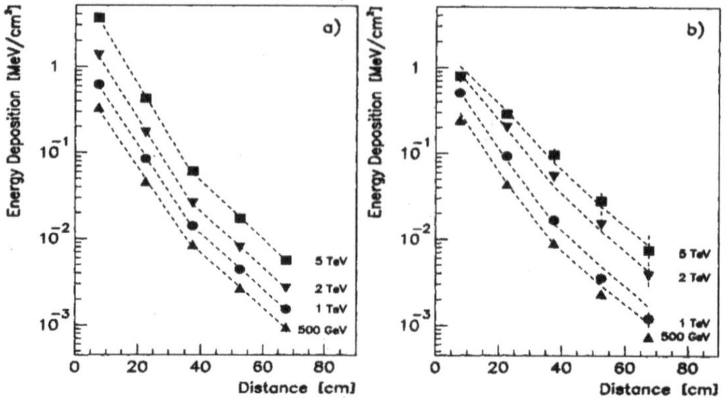

Figure 5: The lateral shower development in the third layer. a) Pure simulations of the energy density versus distance of the shower axis. b) Experimental results (symbols) and simulations (dashed lines) of energy density versus radius including leakage, electronics saturation and track finding.

The experimental findings are shown as symbols in fig. 5 b) and are well reproduced by the simulations that are represented by the dashed lines. One recognizes the effect of electronics saturation and reconstruction on the form of the distributions. For higher energy hadrons the energy densities in the region of small radii are diminished and consequently the slope of the distribution is changed.

The lateral distributions become slightly wider with increasing depth in the calorimeter in agreement with the Monte Carlo predictions.

The cylinder around the shower axis that contains 95% of the hadron energy has a radius of $\approx 30\ cm$ in good agreement with the effective absorption length of $\lambda_{eff} = 29\ cm$ calculated for the calorimeter structure.

8. Comparison of Flux Measurements

We performed an independent check of the hadron energy reconstruction by evaluating the energy spectrum of the cosmic hadrons using the energy assignment described above. Our flux values are shown in fig. 6 together with several other measurements[8]. The agreement is reasonably good. Some of these experiments used magnet spectrometers, where the energy assignment did not rely on simulation of hadronic interactions, while others used sampling or single layer calorimeters with similar problems in calibrating th energy scale. The coincidence, especially at the higher energies, gives us confidence that the Monte Carlo simulations with the FLUKA program are reliable.

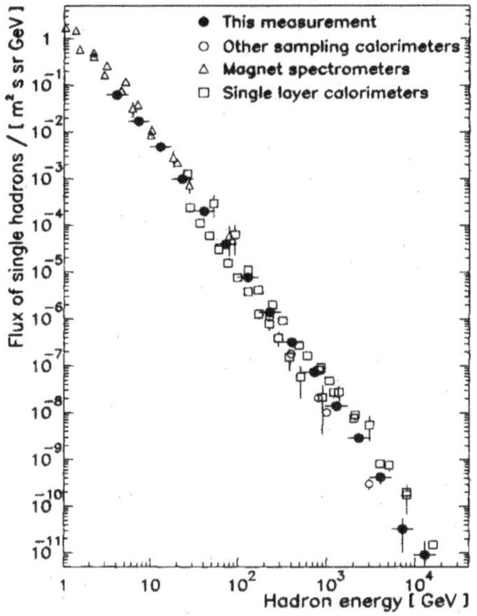

Figure 6: The cosmic hadron energy spectrum. The fluxes obtained with our energy reconstruction compare well with those of other experiments.

9. The Signal Stability

We can survey the signal stability of the TMS chambers versus time using the cosmic ray hadron flux assuming that the flux of cosmic hadrons does not vary with time. If we then plot the rate of hadrons above a given energy threshold we will see a decrease with time if the signal of the TMS chambers deteriorates, e.g. due to outgasing of polar impurities from the chamber body into the liquid. Due to the steep flux decrease with energy, this test is very sensitive. In fig. 7 the cosmic hadron rates for 4 different energy thresholds are shown versus time. No significant deterioration is observed in the data recorded over more than two years.

10. Conclusion and Outlook

The prototype of the KASCADE calorimeter is the first large scale calorimeter successfully using warm liquid ionization chambers. The TMS ionization chambers have been proven to operate stably for more than two years. No signal deterioration with time has been observed.

With a set-up of 6 m^2 active area and 8.9 λ_0 total thickness cosmic ray hadrons have been recorded up to 10 TeV. Simulations with the GHEISHA and FLUKA

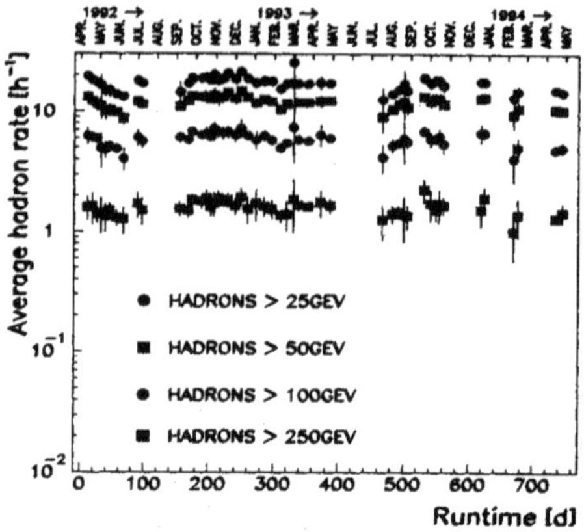

Figure 7: The average rate of hadrons above four different energy thresholds recorded over more than two years.

program describe well the longitudinal and lateral energy distribution of hadrons in the calorimeter. At energies above 3 TeV radiation accompanying the hadron has to be taken into account.

The KASCADE calorimeter has 320 m^2 active area and is 11 λ_0 thick. It has 8 layers of liquid ionization chambers which are read out via 4×10^4 electronic channels. It absorbs up to 10 TeV hadrons without leakage and measures the energy deposit without any saturation.

The production of TMS chambers for the KASCADE calorimeter is well advanced. All chambers are assembled, 40% are filled with TMS and tested with cosmic rays and 20% are inserted in the calorimeter. By the end of 1994, data taking will begin with 100 m^2 operational calorimeter. The assembly of all 320 m^2 will be completed in 1996.

For the fully equipped KASCADE calorimeter we predict an energy resolution of better than 25%, a spatial resolution of better than 5 cm, and an angular resolution of better than 2° for $E_h > 100\ GeV$.

11. References

1. P. Doll et al., The Karlsruhe Cosmic Ray Project KASCADE, *KfK Report 4686* (1990)
2. H.H. Mielke et al., A calorimeter for cosmic ray hadrons up to 10 TeV, *submitted to Nucl. Instr. and Meth.* (1994)
3. J. Engler et al., *Nucl. Instr. and Meth.*, **A223** (1984) 47
   R. Apsimon et al., *Nucl. Inst. and Meth.*, **A305** (1991) 331
   R.A. Holroyd, D.F. Anderson, *Nucl. Instr. and Meth.*, **A236** (1985) 294
4. GEANT 3.15, Detector Description and Simulation Tool, CERN Program Library Office, CERN
5. H. Fesefeldt, The Simulation of Hadronic Showers -Physics and Applications-, *Technical Report PITHA 85/02* III Physikalisches Institut, RWTH Aachen, Germany (1985)
6. P.A. Aarnio et al., FLUKA user's guide. Technical Report, TIS-RP-190, CERN, 1987, 1990
   P.A. Aarnio et al., FLUKA: hadronic benchmarks and applications, Proceedings MC 93 Int. Conf. on Monte Carlo Simulation in High-Energy and Nuclear Physics, Tallahassee, Florida, 22-26 Feb. 1993
7. A. Fassò et al., *Nucl. Instr. and Meth.*, **A332** (1993) 459.
8. H.H. Mielke et al., *J. Phys. G: Nucl. Part. Phys.*, **20** (1994) 637.

Construction, Calibration and Characterization of an SDC Hadron Calorimeter Wedge Using a Moving Radioactive Source

Virgil E. Barnes
Department of Physics
Purdue University
West Lafayette, IN 47907

Representing the SDC Hadron Calorimeter Calibration Group*

ABSTRACT

One SDC Hadron Calorimeter steel wedge has been partially instrumented and is being studied to determine the accuracy of the construction techniques and of moving radioactive gamma source and cosmic ray calibration techniques. The radioactive source system and its goals, use and operation are described. We report measurements of the uniformity of response of the tiles within towers (including all optical fiber and coupling effects) and the systematic accuracy and reproducibility of the tile measurements, as seen with the source moving transversely and longitudinally to the tower axes. Good accuracy and excellent reproducibility of certain source measurement are seen, and the tile uniformity within towers is generally good. Cosmic ray data are being taken but are not yet available for key comparisons with the source measurements.

1. Introduction

One SDC HCAL wedge, 1/32 of one half-barrel, has had twenty towers instrumented with scintillating tiles with wavelength shifting fiber readout. This calorimeter module has been installed in a cosmic ray test stand and has also been instrumented with a set of fine longitudinal and transverse steel tubes into which a wire carrying a radioactive gamma source can be inserted to access every tile for calibration and other purposes.

The general purposes of such a moving radioactive source system are: 1) to provide Quality Control/Quality Assurance via measurement of installed tile-to-tile response variations; 2) to set initial phototube gains; 3) to predict the relative responses of all towers to testbeams and cosmic ray muons, based on the "fingerprints", or profiles, of tile responses in each tower; 4) to monitor phototube gains and tower responses; 5) to measure possible long-term changes in the tower fingerprints due to radiation

***J. Nasiatka R. Stanek, Argonne National Laboratory; S. Gourlay, M. Lamm, P. Mantsch, J. Ozelis, D. Torretta, Fermilab; V. Hagopian, S. Jones, H. Whitaker, Florida State University; V. Barnes, A. Bujak, M. Fahling, L. Gutay, P. Higley, J. Horvath, A. Laasanen, J. Ross, Q. Shen, Purdue University; P. Draper, University of Texas at Arlington; J. Kochocke, L. Mo, Virginia Polytechnic Institute.**

damage, aging, or disturbance of the optical couplings to the phototransducers; and 6) to measure directly the effects of any magnetic fields in brightening the scintillator responses.

For item 3), in practice a subset of typical towers would be calibrated in test beams and these absolute calibrations carried to the rest of the calorimeter using the source. The systematic accuracy of this procedure is best measured by comparing source predictions with testbeam results for a number of towers – a study which is not planned for the SDC Hadron wedge design. Nearly as good validation can be obtained by comparing source predictions to actual cosmic ray responses of towers. This will be done in the near future.

2. The Moving Source System

An early version of the moving source system has been described elsewhere[1].[2] The present upgraded source system consists of a source-wire driver and associated small- diameter tubes to access every tile in the calorimeter. The "wire" (actually a 0.71 mm diameter stainless steel hypodermic tube) contains a pointlike ~5 mCi ^{137}Cs source ~5 mm long. The source is inside and near the tip of the "wire", which can be up to ~11 m long. The wire is stored in a 25 cm diameter reel on the source driver, which has a footprint 33 cm x 66 cm and is 25 cm high. The driver can select any of up to 380 channels. Approximately 190 nylon tubes lead to the present calorimeter, where they are coupled to stainless 1.83 mm OD steel tubes inside the calorimeter (the tubes have 1.27 mm OD in the CDF Endplug Upgrade Calorimeter[3] and in the design of the CMS HCAL).

The driver uses reversible 12 VDC motors and is controllable by manual switches or by computer via CAMAC. The wire position is read out with 0.1 mm least count. The source speed is typically ~5 cm/s, but in special cases can be switched to < 1 cm/s; a typical round trip in one tube takes less than 5 minutes. DAQ and motor control are via CAMAC, controlled from a PC, a VAX, or a Sun/Solflower-VME-CAMAC system in the present case.

Phototube anode currents are typically 50 to 500 nA; in some "indirect" sourcing configurations where the source is remote from the tile, peak currents as low as ~1 nA are digitized. Source current DAQ can be done by the DSP2032 autoranging 32 channel DVM where the voltage is developed passively in an external circuit with precision 100 kΩ resistors and a time constant of 100 ms; or in the present case in the ANL-designed 48 channel MUX and op-amp signal conditioner with 1 or 20 $M\Omega$ current to voltage conversion (12 channels of the latter) and a time constant of 22 ms. CDF Rabbit system current DAQ has been used with another calorimeter, with manual motor control; Rabbit offers simultaneous fast pulse and DC current digitizations, a highly desirable feature for source calibrations.

The typical sampling rate in the present system is 5 to 10 Hz. We use the single largest current reading to be the peak response value of a tile or tower. Wire extension and retraction peaks are averaged.

A second, more sophisticated peak value algorithm is also studied: find the peak position from the 20 highest samples, then do a truncated mean of all readings within ± 2 cm of the peak, (cut at 2.5 σ or 6%); average the extension and retraction peaks.

3. Corrections Needed for Non-local Source Effects

Because the source is uncollimated and non-local, tile and tower responses depend on the source position and also on tile size and on radiation sharing to other tiles in the tower. All these effects can be calculated using the volume integral over the tile of the inverse square radiation field, attenuated by absorption. These calculations require detailed knowledge of the materials and geometry of the calorimeter.

For example, we calculate 20 - 25 %/mm tile response variation with source height when the source tube is in a groove in the steel absorber, whereas the variation is 5.6 - 7.9 %/mm (larger tiles are less sensitive) when the source tube is in air near the tile. These calculations have been a valuable design tool and aid to intuition.

Larger tiles intercept more of the uncollimated gamma radiation. We calculate a 7.5 % response variation over the full depth of the calorimeter.

Radiation sharing to other layers in the tower adds 32 % (6.2 %) to the directly-sourced tile's response in the front (rear) Had1(2) compartments. This sharing diminishes apparent tile variations within a tower, and must be unfolded to get individual tile responses, by solving the linear equations which relate the vector of raw source responses (all layers) to the vector of individual tile responses, using a calculated matrix of sharing fractions $f_{i,j}$ for the response of tile-i when the source is in layer-j.

Unfolded fingerprints are also corrected for the calculated tile -size dependence. Similar size- and sharing- calculations have been checked at CDF using wire and collimated gamma source measurements.[3] Similar SDC Hadron calorimeter calculation have been checked against measured responses of towers to sources in layers outside of Had1 or 2; also against measured responses where the source is separated from the nearest tile by one steel plate. Good agreement is seen.

4. Hadron Calorimeter Wedge Parameters and Construction

The wedge design and some parameters are given in the SDC Technical Design Report.[4] The steel wedge structure contains 13 towers in eta x 2 towers in phi, with 0.1 x 0.1 granularity. The steel structure was built in China, of stacked and welded steel plates with pockets continuous in ϕ for the scintillating tiles. Grooves to accomodate source tubes run the full length of the calorimeter in Z. At Fermilab 20 towers (5 eta x 2 phi x 2 depths) were instrumented. Had1 and 2 have 27 mm (54 mm) thick steel plates, and 27 to 23 (15 to 12) scintillator layers, respectively, as eta increases. The scintillating tiles are SCSN 81, 3.7 mm thick, ranging in size from $\sim$ 25 x 26 cm^2 to $\sim$ 44 x 44 cm^2 from front to back of the calorimeter. A sigma-shaped groove pattern is milled into the tile (see Fig. 1). The wavelength-shifting (WLS) fibers are Y-11 multiclad Kuraray, 0.83 mm in diameter, heat-fused to clear multiclad fibers. The ends of the WLS fibers are mirrored. These optical components were made in Japan.

Grooved tiles are wrapped with a single layer of white Tyvek and installed at Fermilab in the pockets in the steel. Then the WLS fibers are slid into the grooves in the tiles, and each tower's clear fibers are routed to a cookie, light mixer and phototube at the backplane of the wedge.

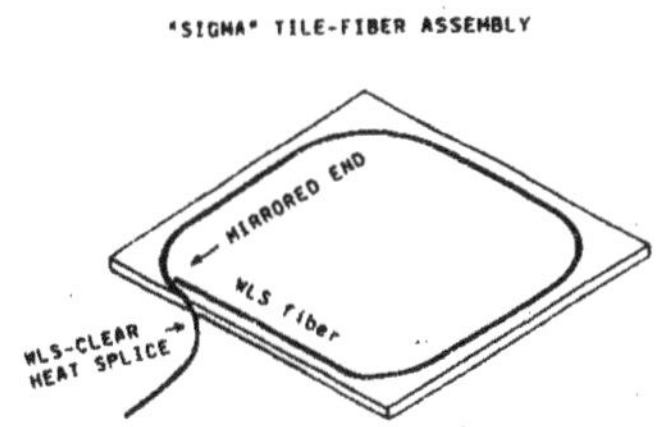

Fig. 1. Scintillating tile with "sigma"-pattern mirrored WLS fiber spliced to clear fiber.

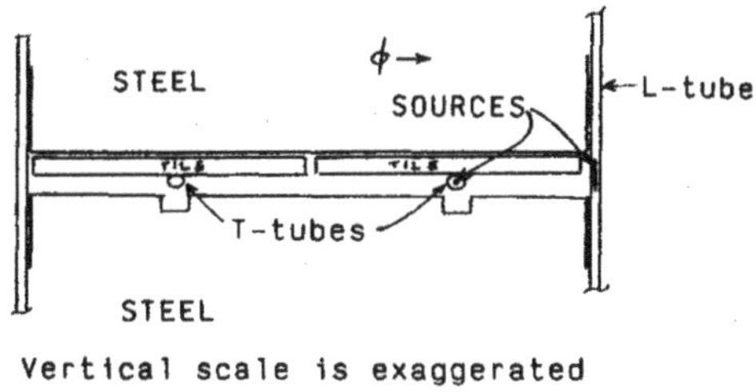

Fig. 2. Schematic of calorimeter showing one air pocket, source tubes, and typical source position when directly (indirectly) exciting right (left) tile.

Transverse (T-) source tubes cross near tile centers (Fig. 2). Tubes and tiles are held in air by venetian-blind springs (not shown). Longitudinal (L-) source tubes were installed outside of the thin light-tight skin of the wedge ("StaticShield" paper and black PVC tape). The L-tubes run over the midpoints of the tile edges, and parallel to the tower axes (Fig. 2), and are kept snug against the wedge by small permanent magnets, except in the region of a non-projective "dogleg" in the face of the wedge. The distances from the L-tubes to the tile edges are uncertain by several mm.

5. Source Measurements

A typical direct T-scan response peak for one layer of one tower is seen in Fig. 3. The "plateau", (defined as being within 1% of the peak reading), is $\sim$ 6cm wide. Noise is negligible – the vertical spread in regions of steep slope is due to (a) three current digitizations for each reel position reading and (b) a few cm backlash of the source wire in the curved nylon tubes leading to the calorimeter (both extension and retraction data are plotted).

An indirect T-scan response peak is seen in Fig. 4. These are typically 2% to 4% as strong as direct T-responses. Noise is $\sim \pm 3\%$; occasional single fluctuations of +8% or more are seen in some peaks (not shown), requiring the second, more robust, peak- finding algorithm. Collimation by the steel plates essentially eliminates response sharing between layers.

A direct L-scan of one tower is shown in Fig. 5; a comb structure is seen above a level caused by sharing of the source activity between layers. The drop in level at the two ends of the tower is due to non-containment effects. Reel counts have been interpolated between readings, assuming constant source speed. The source speed is slowed to $\sim$ 0.8 cm/s, using an alternate gear head; the data are too sparse for the second peak finding algorithm, but are suitable for the simple peak finder. The present analysis does not calculate or unfold the source sharing.

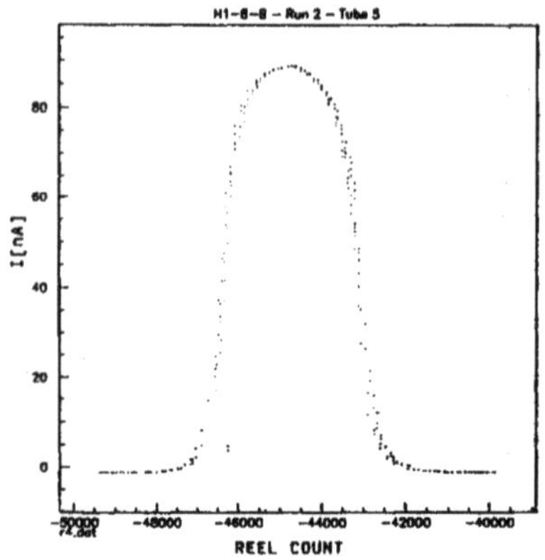

Fig. 3. Direct T-scan, one layer of one tower. Reel count = 0.1 mm.

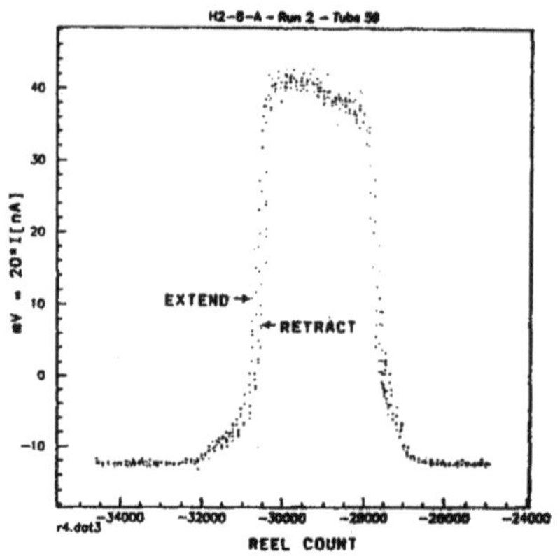

Fig. 4. Indirect T-scan, one layer.

Indirect L-scans (Fig. 6) have extremely narrow peaks, due to collimation by the steel plates; data points are clearly too sparse for good peak sampling, and noise is typically 10%. The peaks range from 0.3% to 1.3% of the direct T-peaks.

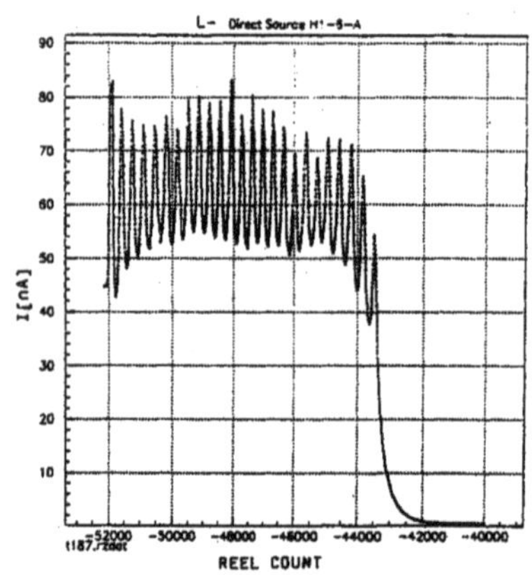

Fig. 5. Direct L-scan, one tower.

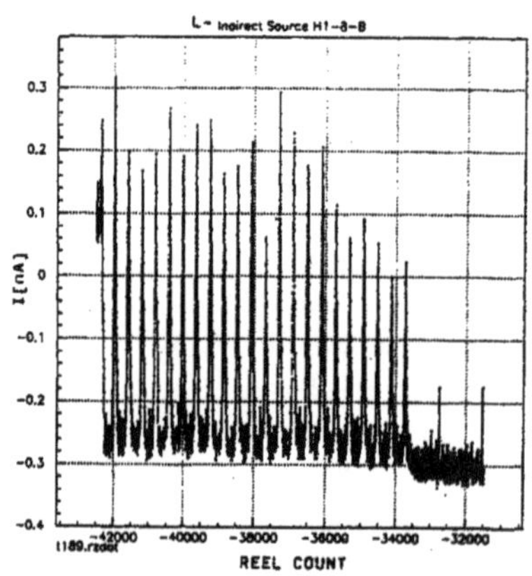

Fig. 6. Indirect L-scan, one tower.

An unfolded direct T-scan fingerprint of one tower is shown in Fig. 7. It is the worst tower, with a weak tile (layer 9) and an RMS/Mean of 18.3% (omitting layer 9 this becomes 7.5%). For each of the 20 towers, the RMS deviation within the fingerprint is calculated. It ranges from 3.8%, to 18.3%, and averages 8.0%. We can estimate the effect of repairing 2% of the tiles, (7 tiles out of a total of 382, affecting 5 towers) which deviate by more than 25% from the mean, by culling them from the data. The worst tower RMS is then 9.1%, and the average is 6.7%, corresponding to a ~3% contribution to the constant term in $\sigma(E)/E$.

The effects of the remaining tile-to-tile variations are partly correctable, on average, by convolving the measured tile responses with (energy-dependent) hadron shower depth profiles. This gives correction factors for calibration which (a) equalize tower responses to hadrons and (b) remove energy nonlinearities. Unweighted sums of the tile responses, scaled by $1/\sin(\theta)$ for muon path length, are expected to predict cosmic ray muon responses.

The source wire has 0.64 mm clearance inside of the source tube, leading to an estimated RMS systematic error of 1.3% (1.8%) for large (small) tiles in the present calorimeter. The CDF endplug upgrade[3] and the CMS HCAL use smaller source tubes, with 0.25 mm wire clearance. In either case, some enlargement of the wire tip would reduce the systematic error.

The measured reproducibility of direct T-scan tile measurements can be better than the estimated systematic accuracy – the wire acquires a permanent curvature which could cause the tip to travel at the same height above the tiles each time it enters a given tube. We have compared tower fingerprints taken two months apart, using August/June response ratios for each tile (simple peak finder, responses unfolded for sharing). Three towers are removed because the ratios show secular changes with layer number, indicating phototube gain drift during the eight-hour-long calibration runs. Two anomalous ratios of layer-1 tile responses are also removed. RMS/Mean within towers ranges from 0.22% to 0.51%, and averages 0.32%, indicating excellent repeatability.

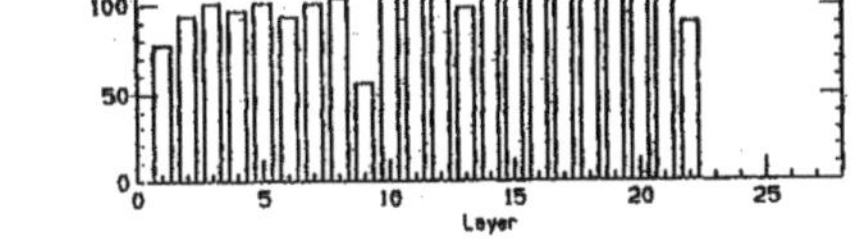

Fig. 7. Unfolded Direct T-scan "fingerprint" of a tower with one weak tile

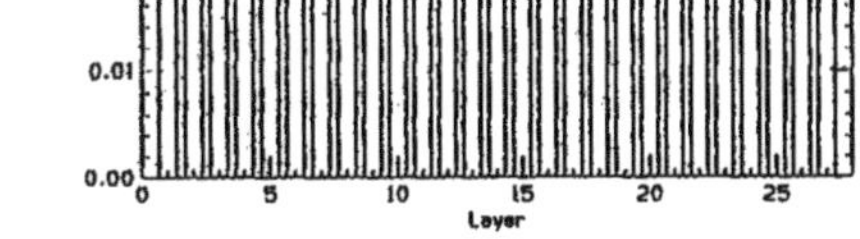

Fig. 8. Indirect T-/Direct T- ratios for one tower

Indirect T-scan responses, using the simple peak finder, are compared tile-by-tile with unfolded direct T-scan responses obtained during the same calibration run. Ratios for a typical tower are plotted in Fig. 8. The source-to-tile distance increases linearly from front to back of the tower. A linear fit in this tower to the tile ratios vs. layer number has an RMS/Midpoint of 11.1%, and for the 20 towers the RMS/Midpoint averages 9.5%. Omitting 6 towers with obvious outliers (probably due to noise and the simple peak finder) the average RMS/Midpoint is 8.4%. Using the second peak finder, which should be less sensitive to noise, the 20-tower average is 8.7%. These numbers are dominated by indirect T-scan systematic errors, which we believe arise from variations in tile placement and, in part, from layer-to-layer staggering of the source tubes in the phi-direction.

Reproducibility of the indirect T-scan measurements, from August/June comparisons, using the simple peak finder, ranges from 1% to 2% in different towers, and averages 1.6% in 14 towers. The more robust peak finder gives reproducibility of 1.3% and permits use of data from all 20 towers.

Systematic variations of each tile are easy to measure relative to the accurate direct T-scans, and can be kept in a permanent data base. Thus, while indirect T-scans are not suitable for initial fingerprinting of towers in this calorimeter, they could be used, at a $\sim$ 1.3% level of reproducibility, to (a) monitor towers adjacent to directly sourced towers and (b) measure changes in fingerprints with time (for faster monitoring runs, or when access to some tubes is restricted).

The systematic accuracy of direct L-scans is estimated by comparing to direct unfolded T-scans taken in the same source calibration run, again using tile-by-tile ratios in each tower. Sharing fractions have not been calculated for this geometry, and no unfolding has been done – this will contribute at some level to the deviations between L- and T- responses seen in the present analysis. The direct-L/direct-T tile ratios within towers have RMS/Mean ranging from 5% to 12.6% and averaging 7.5%. This scatter is dominated by the direct L-scan systematic errors, which can be due in part to variations in tile placement relative to the L-source tubes, and in part to not unfolding the L-response sharing in the present analysis. Reproducibility was checked in two Had1 and two Had2 towers, and is 0.56% (RMS) in scans taken a few days apart. Direct L-scans may be useful in ways similar to indirect T-scans (and with reproducibility approaching that of direct T-scans), and additionally L-tubes offer alternate routes into calorimeters which might offer engineering advantages in future designs.

For the indirect L-scans, the comparison to direct T-scans has an average scatter of 17%, reflecting the sparse and noisy signals seen in Fig. 6. In four towers, the reproducibility is 7.9% (RMS). Indirect L-scans are unlikely to be useful, at the phototube gains and source speed studied so far.

6. Conclusions

The SDC prototype Hadron Calorimeter wedge was constructed with generally good tile uniformity within each tower. Direct T-scan source measurements of individual tiles have an estimated systematic accuracy of 1.3% - 1.8%, and a measured reproducibility of 0.3%. The systematic accuracy should permit accurate source predictions of tower responses to hadrons and muons (only the latter will be studied); and the reproducibility is excellent for monitoring changes with time. Indirect T-scans and direct L-scans offer some potential operational or engineering advantages for monitoring changes in response, but with reduced accuracy.

Acknowledgements

We wish to thank the technical staff members of our institutions and Fermilab for their help in the construction of the calorimeter, source system, and cosmic ray test stand. The steel structure was built at the Xin He shipyard under the supervision of IHEP Beijing. The tiles and fiber assemblies were made at the Tsukuba University Physics Institute under the direction of K. Takikawa and K. Hara. This work is supported in part by the U.S. Department of Energy.

References

1. Virgil E. Barnes and Alvin T. Laasanen, "Proceedings of the First International Conference on Calorimetry in High Energy Physics", ed. D. F. Anderson, M. Derrick, H. E. Fisk, A. Para and C. Sazama, World Scientific, (1991), 189.
2. Virgil E. Barnes, Alvin T. Laasanen and James A. Ross, "Proceedings of the Second International Conference on Calorimetry in High Energy Physics", ed. Antonio Ereditato, World Scientific, (1992), 195.
3. Qifeng Shen, "Hadron Calorimeter Production for the CDF Plug Upgrade", proceedings of this conference.
4. Solenoidal Detector Collaboration, "Technical Design Report", SDC Note SDC-92-201, (April, 1992).

Hadronic Forward

Session Chair: Richard Wigmans (Texas Tech. Univ.)

TEST BEAM RESULTS FROM A VERY FORWARD LIQUID SCINTILLATOR CALORIMETER FOR THE LHC

A. Artamonov[1], V. Barassi[2], S. Buontempo[2], V. Epstein[1], A. Ereditato[2], G. Fiorillo[2], F. Garufi[2], P. Gorbunov[1], V. Jemanov[1], V. Khovansky[1], S. Kruchinin[1], A. Maslennikov[1], M. Rjabinin[1], V. Zaitsev[1], S. Zeldovich[1], I. Zuckerman[1]

[1] *ITEP, Moscow, Russia*

[2] *Università "Federico II" and Istituto Nazionale di Fisica Nucleare (INFN) Naples, Italy*

presented by

F. Garufi

Università "Federico II" and INFN Naples, Italy

ABSTRACT

We report on beam tests performed on a liquid scintillator calorimeter prototype designed to cover the very forward region in an experiment at the LHC. The liquid scintillator technique has known advantages related to high speed, low noise and radiation hardness. The proposed solution is a SPACAL-like calorimeter with liquid scintillator circulating in quartz tubes positioned in a lead matrix. Tests were performed at the CERN SPS with electron and pion beams in the range 20-120 GeV. Results on the energy response and resolution are presented.

1. Introduction

The liquid scintillator technique matches all the requirements for the very forward calorimetry at an LHC experiments i.e. lateral granularity in the range $\Delta\eta \times \Delta\phi \leq 0.2 \times 0.2$, energy resolution better than $100\%/\sqrt{E} \oplus 10\%$, high speed ($\leq 25ns$), radiation hardness and low noise [1] . The proposed solution is a SPACAL-like [2] calorimeter with liquid scintillator circulating in hollow light guides.

An *R&D* study is being carried on as a collaboration between the ITEP group and INFN Naples. In this paper we report on the design, construction and test on electron and pion beams of a four module prototype calorimeter. Results are compared with MC predictions and estrapolated to full containment.

2. Design and construction

The calorimeter has a modular structure [3,4,5]. Each module has a parallelepiped shape with 66×66 mm^2 cross section and 2380 mm length, including a 380 mm long

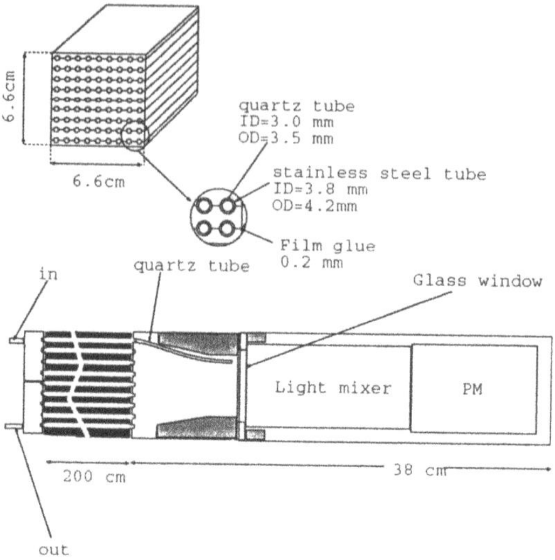

Fig. 1. Calorimeter module design

photodetector housing (see Fig. 1).

It consists of an absorber lead matrix containing 81 quartz tubes with inner diameter of 3 mm and wall thickness of 250 μm. The tubes run parallel to each other along the module to form a square transverse pattern and the modules are aligned parallel to the beam line. The spacing between the tubes (center-to-center) is 7.3 mm, providing a 6:1 ratio between lead and scintillator [3,4].

The modules were constructed from grooved lead plates produced by casting. Then, 2010 mm long stainless steel tubes with inner diameter of 3.8 mm and wall thickness of 200 μm were placed into the grooves. The lead plates were glued together by means of a special 100 μm thick tape placed under and above the tubes and by compression at high temperature. The compression was released after the glue hardening. The tubes coming out at both ends of the module were inserted into the holes drilled in stainless steel manifolds and then tin soldered. The front manifold has inlet and outlet pipes separated by a membrane, the back one is provided with a frame with a glass window to transmit the scintillation light. The module is followed by a 15 cm long plexiglass light mixer and a photomultiplier.

The modules were washed with a purified alcohol flow circulated by a pump during a few hours and then dried with air. The quartz tubes were inserted from the back of the modules. The last 10 cm of the tubes' far end were bent to form a bundle of 42×42 mm^2 cross section.

The modules were blown with Argon, then filled under Argon atmosphere. Finally the liquid was circulated by a pump to remove bubbles. The outlet pipe was then

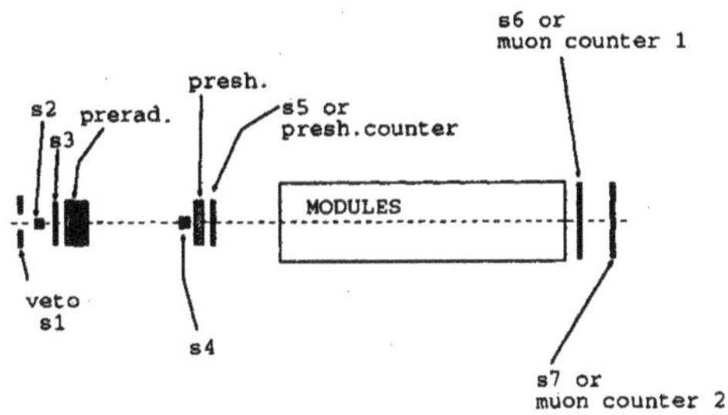

Fig. 2. Position of modules and counters along the beam line

connected to a buffer volume filled up to 1/3 with scintillator and 2/3 with Argon.

3. The beam tests

A prototype calorimeter made of 2×2 full scale modules was tested with pions and electrons beams at the CERN SPS in July 1994. The calorimeter was placed parallel to the beam direction on a movable table allowing displacement in the vertical and horizontal directions and tilt in the horizontal plane.

For three of the modules BC599-13G BICRON scintillator (green emitting) was used, the fourth was filled with methylnaphtalene. All the modules were equipped with filters to enhance the attenuation length of the light, and read by EMI 9839A photomultipliers.

A sketch of the experimental set-up is shown in Fig 2. A veto counter limited the lateral dimensions of the beam, then two scintillation counters (s2 and s3) were placed before a variable amount of material used as a preradiator, to define a beam spot size of 7×7 mm^2. Another 7×7 mm^2 scintillation counter (s5) preceded a $2X_o$ lead preshower and a scintillation counter placed in front of the modules. The modules were followed by two scintillator plates used to detect passing through muons. All counters were read out to allow an off line event selection while the trigger was given by the the coincidence of S2 and S3 only.

4. Experimental results

The calorimeter was exposed to electron and pion beams of 20, 40, 80, 120 GeV; the modules were calibrated by hitting 120 GeV pions on their center and equalizing the response.

Fig. 3 (a) shows the energy distributions for 120 GeV electrons with 1° tilt angle with and without a $3X_o$ preradiator. In the case of no preradiator a high energy tail is clearly visible, due to channelling effects, The use of preradiator strongly reduces this effect and the RMS of distributions is decreased by 50%.

Fig. 3 (b) shows distributions with the $3X_o$ preradiator for 1° and 3° tilt angles.

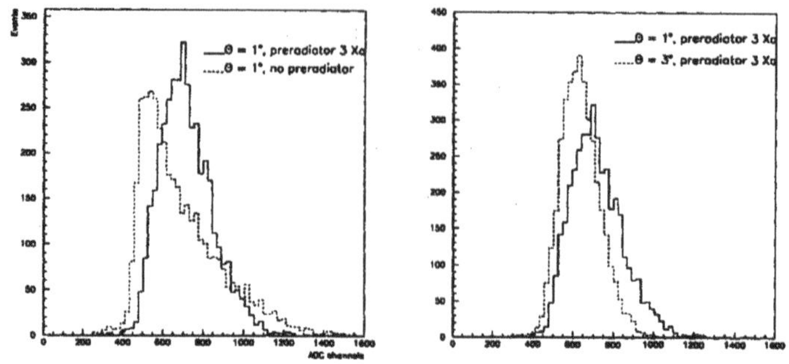

Fig. 3. (a) Energy distribution for 120 GeV electrons with and without preradiator; (b) distribution with preradiator at two different angles

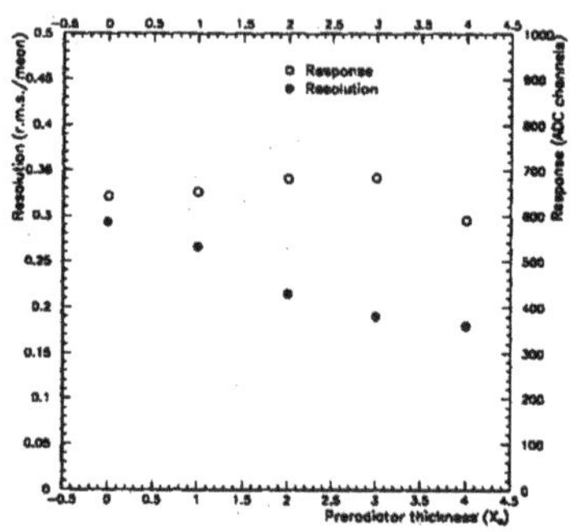

Fig. 4. Energy response and resolution for 120 GeV electrons as a function of the preradiator thickness

As expected, the increase of the tilt angle results in a better energy resolution.

Response and resolutions have been studied for 120 GeV electrons also as a function of the preradiator thickness. Fig. 4 shows that while the response is almost constant for the whole range explored, the resolution improves by the thickness.

The calorimeter was also tested with pions for two different values of the tilt angle (0° and 1°).

Fig 5 shows the response distribution for 120 GeV pions. The low energy tail is due to lateral leakage and well explained by a MC simulation. Energy respose and resolution are shown in fig 6. Unlike for the electon events. the resolution for pions is obtained from a gaussian fit of the distributions in a $\pm 2\sigma$ region around the peak.

It turns out to be $\sigma/E = 86\%/\sqrt{E(GeV)}+7\%$ corresponding to a $51\%/\sqrt{E(GeV)}+8\%$ resolution for the full calorimeter, well below the limits imposed by the ATLAS design [1].

Pions seem not to be affected by the material in front of the calorimeter in the thickness range explored above a threshold needed to avoid channeling effects [3].

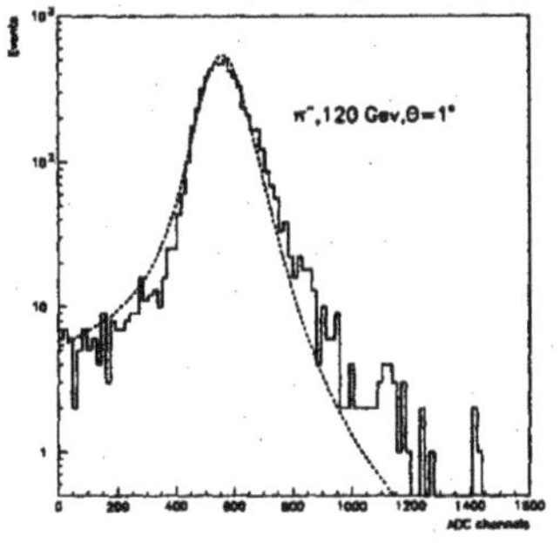

Fig. 5. Energy distribution for 120 GeV pions at 1°. The dashed line is the result of a MC simulation

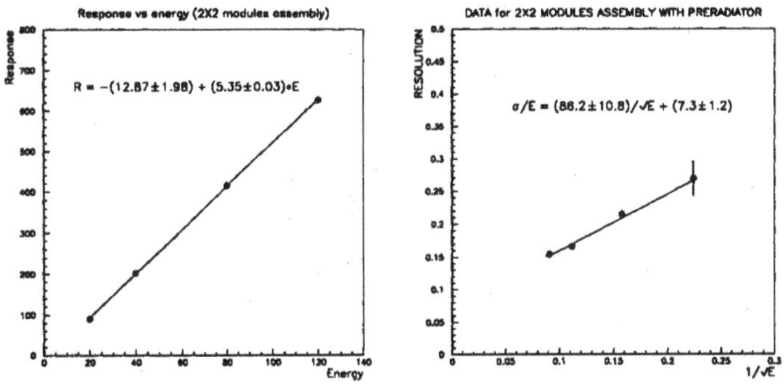

Fig. 6. Pion energy response and resolution as a function of $1/\sqrt{E(GeV)}$

5. Conclusions

We have constructed and tested a prototype lead-liquid scintillator calorimeter to be possibly used as a forward calorimeter in a LHC detector. Good energy resolution for electrons is obtained even at small incidence angles by using a passive preradiator in front of the modules. The response to pions is linear and the energy resolution is $\sigma/E = 86\%/\sqrt{E} + 7\%$ corresponding to $51\%/\sqrt{E} + 8\%$ for the full calorimeter.

References

1. ATLAS Letter of Intent for a General-Purpose p-p Experiment at the Large Hadron Collider at CERN, CERN/LHCC/92-4 LHCC/I 2
2. D. Acosta, et al., *Nucl. Instr and Meth* **A294** (1990) 193-210
3. A. Artamonov, et al., *Liquid scintillator Calorimetry for the LHC* in *Proc. Frontiers Detectors for Frontier Physics*, 6th Pisa meeting on Advanced detectors, 22-28 May, 1994 - La Biodola, Isola d'Elba. Italy.
4. A. Artamonov, et al., in *Proc. III Conference on Calorimetry in High Energy Physics*, 29 Sept. - 2 Oct. 1992 - Corpus Christi, Texas, U.S.A.
5. A. Artamonov, et al., in *Proc. II Conference on Calorimetry in High Energy Physics*, 14-18 Oct., 1991 - Capri, Italy.

A Spaghetti Calorimeter for the H1–Detector

Gerhard Müller for the H1–Collaboration
DESY, Notkestraße85
22603 Hamburg, Germany

ABSTRACT

The H1–collaboration is upgrading the backward region of the H1 detector to probe the parton structure of the proton down to very low values of x (10^{-4}). The centerpiece of the upgrade project will be a high resolution spaghetti calorimeter. The new calorimeter is divided in an electromagnetic and a hadronic section. The electromagnetic section consists of scintillating fibers with a diameter of 0.5 mm embedded in a lead matrix. This small radius ensures in combination with a lead to fiber ratio of 2:1 an excellent energy resolution which has been measured to $7.2\ \%/\sqrt{E} \oplus 1.0\ \%$. The electromagnetic calorimeter is designed to measure scattered electrons down to angles of 177°, with respect to the proton direction, giving the opportunity to cover a wide range in Q^2. To measure the hadronic energy flow the second calorimeter section is built with fibers of 1 mm diameter and a lead to fiber ratio of 4:1.

1. Introduction

The electron–proton collider HERA at DESY provides the unique possibility to probe the parton structure of the nucleon up to kinematical regions which have not been accessible to fixed target experiments. Recently new theoretical approaches have stimulated experimental interest in the low–x regime[1].

Events originating from low–x processes are characterized by boosts in the backward (electron) direction, Both the electron and the hadronic final state need to be detected in the backward region. As the H1 detector was designed originally to explore the largest momentum transfers available it was therefore proposed to upgrade the instrumentation in the backward region[2] the centerpiece of which will be a high resolution electromagnetic calorimeter.

The current H1 detector is equipped in the backward region with an electromagnetic lead scintillator sandwich type calorimeter with photodiode readout allowing electron reconstruction up to polar angles of 173° relative to the proton direction. Whereas the new calorimeter is designed to measure electrons up to angles of 177° thus giving access to a wide range of Q^2 for very low values of x.

2. Design and Construction of the Calorimeter

The design of the new calorimeter was guided by the following criteria:

- Access to large electron angles up to 177° to cover a wide range in Q^2,
- Excellent energy resolution, less than 2% at 30 GeV,

- Good time resolution and a high granularity to suppress proton beam background,
- Small radiation length and Molière radius to achieve a compact calorimeter,
- Good electron identification to suppress hadrons from photoproduction background
- Measurement of the hadronic energy flow,
- Operation in a 1.0 Tesla magnetic field.
- Radiation hardness,

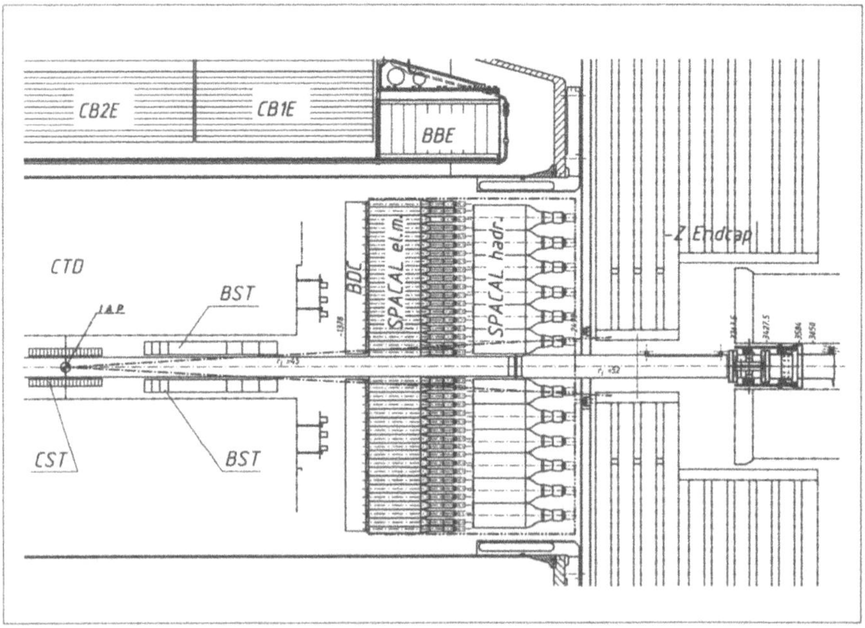

Figure 1: View of the backward scattering region of the H1 detector with the spaghetti calorimeter.

To meet these requirements a spaghetti calorimeter consisting of an electromagnetic and a hadronic section was chosen with a fiber diameter of 0.5 mm and 1.0 mm and a lead to fiber ratio of 2:1 and 4:1, respectively[3]. The detailed design parameters are given in table 1. The overall layout is shown in figure 1.

Table 1: Design parameters of the new calorimeter

Parameter	Electromagnetic Section	Hadronic Section
Pb/Fiber Ratio	2:1	4:1
Fiber Type	Bicron BCF–12	Bicron BCF–12
Fiber Diameter	0.5 mm	1.0 mm
Active Length	25 cm	25 cm
Cell Size	40.5 × 40.5 mm^2	120 × 120 mm^2
Number of Cells	1192	136
Angular Coverage	$\leq$ 177.5°	$\leq$ 178°
Radiation Length	0.91 cm	–
Molière Radius	2.55 cm	–
Interaction Length	25 cm	20 cm
Energy Resolution(elm)	$7\%/\sqrt{E}\oplus1\%$	$13\%/\sqrt{E}$
Angular Resolution	$\leq$ 2 mrad	–
Time Resolution	$\leq$ 1 ns	$\leq$ 1 ns

The basic unit of the electromagnetic section is called a submodule consisting of two cells each 40.5 × 40.5 × 250 mm^3. The fibers are bundeled over a length of 85 mm and held in a bundeling frame. The fiber bundels are glued with black glue over a length of 2 cm to kill the cladding light. It has been observed in the R&D phase that mechanical pressure on fibers at the cell boundaries causes inhomogeneities. Therefore precautions have been taken to avoid high pressure in the bundeling frame which was dimensioned big enough to allow a complete relaxation of all fibers. The submodules were only prepressed with 1.5 tons horizontally and vertically which corresponds to the maximum pressure in the assembled calorimeter due to its own weight. The fiber bundel is coupled via an air gap to a light mixer which in addition concentrates the light on the photocathode of a 1" photomultiplier. The photomultiplier is glued to the light mixer. As the readout needs to be operated in a magnetic field of 1.0 Tesla mesh dynode photomultipliers were chosen [4] (Hamamatsu R–5505). The spectral sensitivity of the bialkali photocathode matches the emission spectrum of blue fibers (Bicron BCF–12) which have been selected for both the electromagnetic and the hadronic section after an intensive R&D program had been carried through[5]. The fibers are mirrored at the front end.

Eight submodules are combined to form a supermodule. Each supermodule has a common housing for light mixers, photomultipliers and bases. It is equipped with a high voltage distribution board and a calibration module which monitors the gain by coupling light of a LED via fibers into the light mixers. The electromagnetic calorimeter consists of 60 regular supermodules plus a combination of four differnt types of edge modules to approximate the round surface of the enclosing cylinder. Figure 2 shows a photo of an assembly 4 supermodules, i.e. 64 cells.

Figure 2: Photo of an array of 4 Supermodules (64 cells).

To avoid channeling effects which have been observed in all SPACAL calorimeters the lead fiber matrix is always orientated in such a way that no main symmetry axis is parallel to a particle trajectory.

The innermost module contains a 2 mm tantalum shield around the beampipe against synchrotron radiation. The fibers closest to this shield are bundeled seperately to act as a veto layer for shower leakage.

The hadronic section is built of single modules which measure 120 × 120 × 250 mm^3. The readout is done by 2" Hamamatsu R–2490-06 photomultipliers.

3. Performance of the Calorimeter

A number of the modules produced were submitted to test beam measurements at DESY and at the CERN SPS and PS. An energy scan was performed at an angle of 5° in the horizontal and 3° in the vertical direction to avoid channeling. Linearity was maintained up to 80 GeV to better than 1%. A common fit for the DESY and CERN data, taken with the 64 cell matrix shown in figure 2, yields a sampling term of $7.2\%/\sqrt{E}$ and a constant term of 1% (figure 3) giving a resolution of well below 2% at 30 GeV. For the DESY measurement 100 MeV have been subtracted in quadrature in order to account for the finite momentum resolution of the beam. Simulations have shown that fiber to fiber variations in light output, which are observed in the order of 5%, contribute with 0.5% to the constant term.

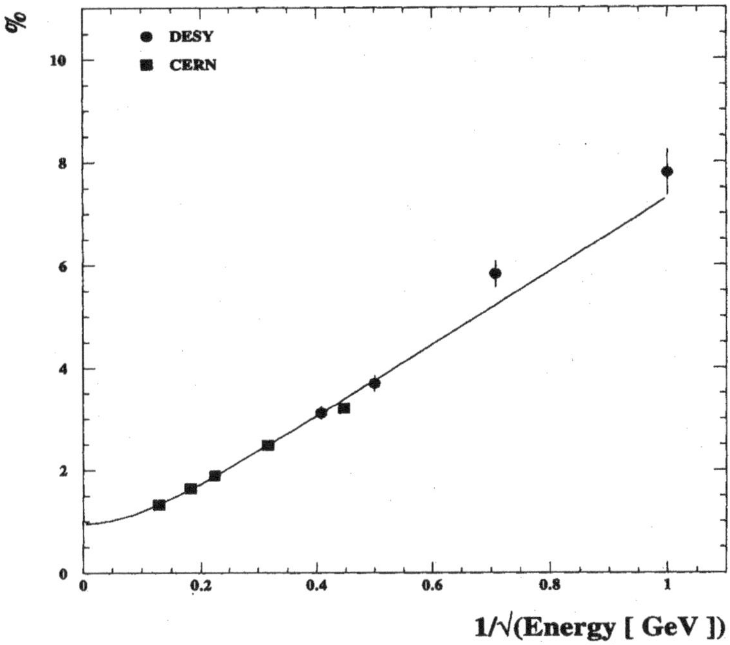

Figure 3: Energy resolution, measured with the modules shown in figure 2.

The light yield has been determined by comparison with a LED to be 2.8 ± 0.3 photo electrons per MeV. For minimum ionizing particles an energy deposition of 400 MeV has been observed. The time resolution was measured to be below 0.5 ns. The calorimeter shows a spatial resolution of 2.8 mm $/\sqrt{E} \oplus 0.4$ mm which translates to 1.9 mrad.

All produced modules were scanned from the front face with a beta source to determine the homogeneity of the supermodules[6]. The modules were scanned with variable step size between 1 and 2 mm in both directions using a ^{90}Sr source with a collimator of 1.5 mm. Within the submodules the response varied only with an RMS of 3 - 4% corresponding to the light yield fluctuations of the individual fibers. The scanning of the final modules revealed a decrease in the response at the cell boundaries of 5% over a width of approximately 2.5 mm. The different orientaions of the lead fiber matrix, to avoid channeling, change the lead fiber ratio at cell boundaries, showing a very localized decrease or increase in response of 10% as observed in the beta-scan. Inhomogeneities are seen in the beam measurements too.

For the hadronic section first test measurements at ITEP give an energy resolution for electrons of $13.3\%/\sqrt{E}\oplus$ 3.6% and for hadrons of $71\%/\sqrt{E}\oplus$ 20% for non leaking hadron showers (1Λ only!). The timing resolution is 0.9 ns and the light yield has been determined to 0.7 photo electrons per MeV. Recently a combined measurement of electromagnetic and hadronic modules has been carried out at CERN. The analysis is under way.

4. Summary

The design criteria for a new calorimeter to upgrade the backward region of the H1 detector have been presented. These goals led to the design and realization of a spaghetti calorimeter. The production for both the electromagnetic and the hadronic section will be completed in November 1994. The calorimeter will be installed in the H1 detector during the shut down 1994/1995.

Test measurements on the final modules show that the design goals for energy and time resolution have been met. Local inhomogeneities at cell boundaries in the order of 5%, measured with a beta scan bench, have been observed. As all modules were scanned carefully we plan a software correction for these inhomogeneities.

We are looking forward to taking the first physics data with our new calorimeter in 1995!

5. Acknowledgements

I wish to thank the backward upgrade group for their permanent support and stimulating interest in this project. Without the outstanding effort of the technical staff of DESY and the collaborating institutes this calorimeter would not be possible. Thanks to the organizers for a very enjoyable and interesting conference in Brookhaven.

6. References

1. J. Bartels and J. Feltesse, *QCD at low x*, Proceedings of the Workshop, Physics at HERA, (DESY, Hamburg, Germany, 1991).
2. H1 Collaboration, *Technical Proposal to upgrade the Backward Region of the H1 Detector*, DESY PRC 92/03.
3. U. Goerlach, *The New Lead Scintillating Fiber Calorimeter for the H1–Detector*, Proceedings of the IV International Conference on Calorimetry in High Energy Physics, (La Biodola, Isola d'Elba, Italy, 1993).
4. R.-D. Appuhn et al., *Performance Tests with a Large Sample of Fine Mesh Photomultipliers for use at 1.2 Tesla*, Proceedings of the IV International Conference on Calorimetry in High Energy Physics, (La Biodola, Isola d'Elba, Italy, 1993).
5. A. Walther and R. Barschke, *Studies of Scintillating Fibres for a Spaghetti Calorimeter to be used for the H1 Backward Upgrade Project*, Proceedings of the Workshop on Scintillating Fiber Detectors, (Notre Dame University, Indiana, 1993).
6. F. Lehner, Diploma Thesis, Hamburg University, unpublished.

THE E864 SCINTILLATING FIBER HADRONIC CALORIMETER

SIMON J. BENNETT
Department of Physics & Astronomy, Wayne State University,
Detroit, MI 48201, USA

for

THE E864 COLLABORATION
Brookhaven National Laboratory, Massachusetts Institute of Technology, Iowa State University,
Penn State University, Purdue University, Wayne State University,
University of Massachusetts, Yale University, University of Bari

ABSTRACT

This contribution details the design and performance of a lead/scintillating-fiber hadronic calorimeter for use in BNL experiment E864. Detailed results are given on the detector performance (linearity, energy resolution, time resolution, and e/π signal ratio) based upon measurements made with a test array of 3×4 towers during the AGS proton running in June '93.

1. Introduction

This paper discusses the design and performance of the lead/scintillating fiber hadronic calorimeter to be used in the AGS experiment E864. E864 is a high sensitivity, high acceptance, search for strangelets (strange multi-baryons), anti-matter and rare composite objects following fixed target collisions of the 11.7 GeV/c per nucleon gold beams at the AGS with a gold target. The experiment utilizes a magnetic spectrometer consisting of three high granularity planes of hodoscopes for time of flight and position information, three planes of X-U-V straw tubes for tracking, and a longitudinally un-segmented calorimeter providing direct energy and time of flight measurements used to form a 'late-energy' trigger essential for the high sensitivity strangelet searches.

The calorimeter is similar in design to the longitudinally un-segmented 'spaghetti' type of the SPACAL collaboration[1,2,3] It uses 1 mm diameter Bicron BCF-12 scintillating fibers spaced 2.12 mm apart within a square lead matrix constructed by laminating the fibers using Bicron BC-600 epoxy between 2% antimonial-lead sheets with rolled semi-circular grooves 1.1 mm in diameter. Both end faces of the tower are machined and the back face of each tower is glued using an optically coupling adhesive to a light guide and thence to a photo-multiplier. The front face is covered with a light absorbing paint. Each individual tower contains 2209 (47×47) scintillating fibers and is 10 cm by 10 cm square and 117 cm or 5.3 nuclear interaction lengths deep. The resultant lead to fiber ratio of a tower is 4.55 to 1 by volume. The entire calorimeter will comprise 754 towers in an array 58 towers wide by 13 towers high and will be completed and installed at Brookhaven National Laboratory by June '95.

Prior to full-scale production, a small 3×4 array of towers was tested during the AGS proton running in June '93 and preliminary results of the linearity, energy resolution, time resolution, and e/π signal ratio will be presented here.

2. Experimental Setup

The 3×4 array of towers was tested using secondary beams produced when the primary proton beam of the AGS struck a production target. Produced particles were momentum analyzed using a magnetic spectrometer and identified using a combination of Čherenkov and time-of-flight counters. The beam line could thus provide a variety of identified particle species including e^-, μ, π^+, π^-, p, $\bar{p}$, and d with momenta ranging from 1 GeV/c to 8 GeV/c.

The test array was mounted on a platform that could move horizontally and vertically with respect to the beam line, with an accuracy of about 1 mm. Further, the detector could be tipped about a horizontal axis, so that particles could be sent into the detector at a chosen angle, θ_z, between the fiber axis and the beam direction. This angle was 54.3 mr for all the data presented here.

3. Experimental Data

Figure 1 shows the results of the mean summed energy response (all 12 towers) for e^-, π^+, π^-, and p for a variety of kinetic energies (incorporating the pion rest mass as appropriate) incident upon the middle of one of the towers just beam-left of the center of the 3×4 array. The data have been calibrated using 2 GeV/c electrons sent into the central region of each tower at an angle of incidence of $\theta_z = 54.3$ mr with respect to the fiber axis. In this way, the electromagnetic showers that developed are nearly completely contained within the tower concerned, and the individual tower calibration constants so obtained can be used to calculate a summed energy response.

In the case of hadrons, a correction has been applied to the summed energy response to compensate for the effects of transverse shower leakage beyond the extent of the 3×4 array, which will not be present in the full calorimeter. This correction has been derived from detailed GEANT simulations[4] of both a 3×4, and a 11×11 array. All physical processes in GEANT were turned on, and shower particles generated using the FLUKA hadronic package were tracked to the lowest kinetic energy allowed (10 keV) in all media. This simulation included such details as treatment of the polystyrene core and PMMA cladding of each individual fiber, saturation effects in the scintillation output of the fibers according to Birk's Law[5], and the double exponential attenuation length of the fiber. The ratio of the simulated calorimeter response for the 11×11 to the 3×4 array was calculated for π^+ and protons for each of the experimental kinetic energies. The mean of these values, 1.16 for π^+ and 1.21 for protons, were used as multiplicative energy leakage correction factors for the data shown in figure 1. The π^- data was corrected using the same factor applied to the π^+ data. This figure shows that the calorimeter displays a reasonably linear energy response across the entire 1–8 GeV range of the measurements for protons, π^+, and

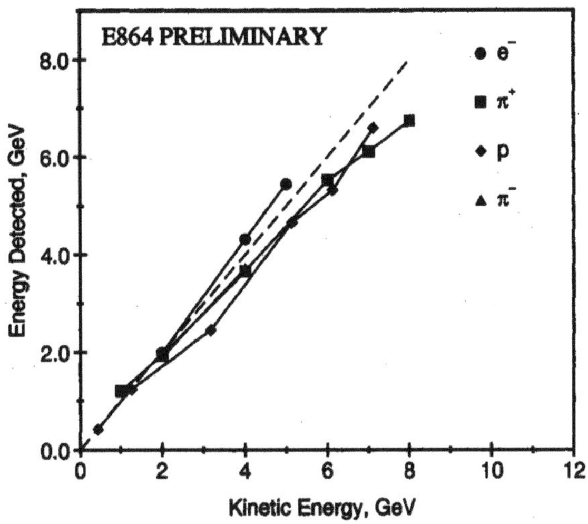

Figure 1: Leakage corrected values for the summed energy response of the 3 × 4 test array for e$^-$, π^+, π^-, and p for a variety of incident kinetic energies

π^-.

Figure 2 shows preliminary results for the e/π ratio extracted from the data shown in figure 1. The indicated errors are based upon estimates of the likely systematic errors in this process. The mean value of e/π is 1.06 over this energy range. Whilst this indicates that the calorimeter is not perfectly compensating, it is in reasonable agreement with that obtained by SPACAL[3].

Figure 3 shows the measured energy resolution of the 3 × 4 array for a variety of particle species and incident energies. The resolution is quoted as k% in $\sigma/E = k/\sqrt{E/\text{GeV}}$. The resolutions have been corrected for transverse leakage effects using the GEANT simulations discussed above. The main effect of this is to reduce the quoted hadronic resolutions because of the increase to the denominator in σ/E. The multiplicative correction factors were 0.889 for π^+/π^- and 0.793 for protons. In the range of interest to E864, energy resolutions are around $20\%/\sqrt{E}$ for electrons and $40\%/\sqrt{E}$ for hadrons.

Figure 4 shows the slewing corrected time resolution, σ, for the tower struck by a variety of incident particles and energies. The time resolution measured for electrons, where the shower is contained within a relatively small volume of detector, is basically determined by the photo-multiplier (Hamamatsu R329), cabling, and electronics, and is around 200 ps. The time resolution for hadrons, around 400 ps, is dominated by fluctuations in the longitudinal deposition of energy within the shower

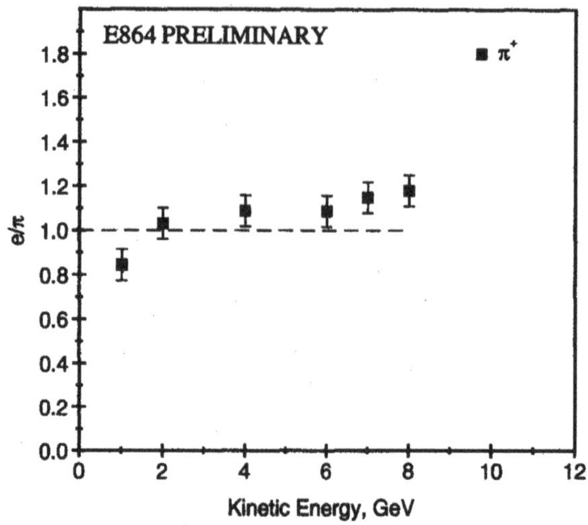

Figure 2: Calculated values for the e/π ratio for the 3×4 test array

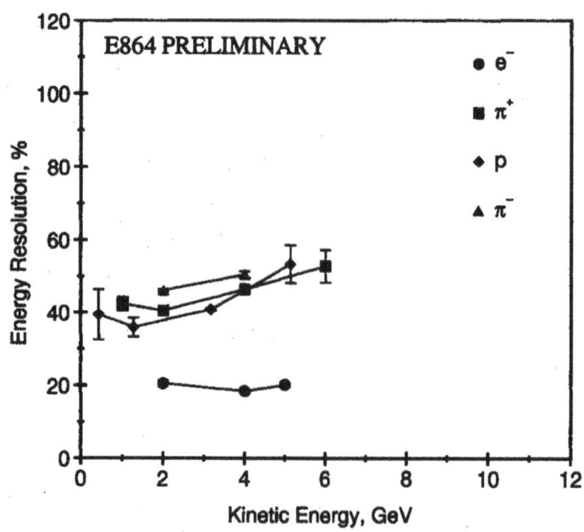

Figure 3: Leakage corrected values for the energy resolution of the 3×4 test array for e^-, π^+, π^-, and p for a variety of incident kinetic energies

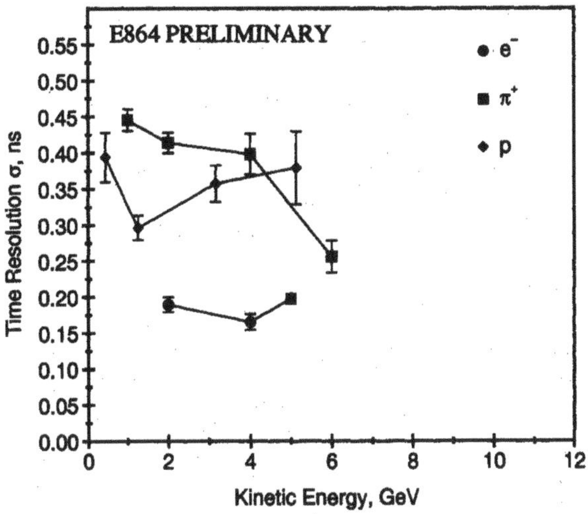

Figure 4: Slewing corrected time resolutions for the central tower in a shower for e^-, π^+, and p for a variety of incident kinetic energies

and the subsequent variation in propagation time for the light to travel down the fiber to the photo-multiplier. Simulations show that at higher energies, an early-time tail develops in the timing distribution from particles that shower late in the module. However, in all cases the time resolution is considerably better than the 500 ps design goal of the calorimeter needed in order for the calorimeter to fulfill its role as a high sensitivity trigger device for high mass objects (such as strangelets) based upon their energy and time of flight.

1. D. Acosta *et al.*, *Nucl. Instr. and Meth.* **A294** (1990) 193.
2. D. Acosta *et al.*, *Nucl. Instr. and Meth.* **A302** (1991) 36
3. D. Acosta *et al.*, *Nucl. Instr. and Meth.* **A308** (1991) 481.
4. T. Cormier *et al.*, in *Calorimetry in High Energy Physics, Proceedings of the Third International Conference*, ed. P. Hale and J. Siegrist (World Scientific, 1993) p. 604.
5. J. B. Birk, *The Theory and Practice of Scintillation Counting* (Macmillan, New York, 1964)

QUARTZ FIBER CALORIMETRY: THE SOLUTION FOR VERY FORWARD CALORIMETERS AT THE LHC.

P. GORODETZKY
CERN, CH 1211, Geneva 23, Switzerland, and
CRN and Université Louis Pasteur, Strasbourg, France
in the name of
Collaboration RD40

ABSTRACT

Quartz fiber calorimetry (QCal) is highly radiation resistant and extremely fast. The special acceptance required for Cerenkov light to be inside the numerical aperture of the fibers makes the visible showers very narrow. This is exactly what is required at small angles for detectors of the future high luminosity collider LHC.

After a brief comparison between the two main fiber geometries (0 and 45°), the last results of the dedicated Research and Development project RD40 are presented.

1. Basics of Quartz Calorimetry

1.1. Basics

When a particle hits a fiber with an angle α between its trajectory and the fiber's axis and if the distance of closest approach between these two straight lines is b, the produced Cerenkov light will stay in the fiber and exit at the fiber's extremity according to the laws of optics. If no perfect (wet) optical contact is applied between this extremity and the photo-detector, then the highly spiralling photons do not exit. Fig. 1. shows the number of photo-

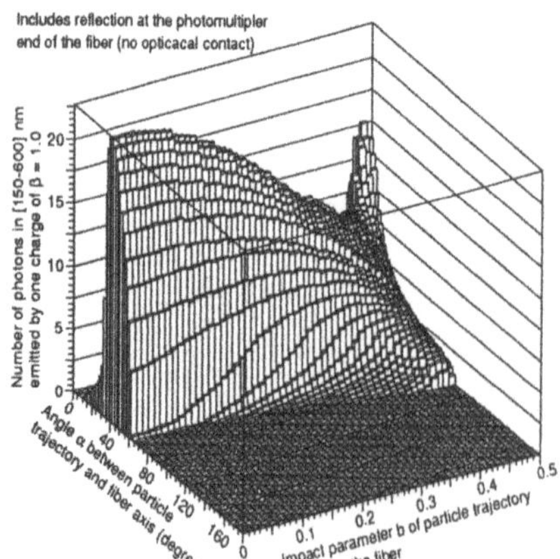

Fig. 1. Number of photo-electrons in a quartz photo-cathode for a 0.37 fiber.

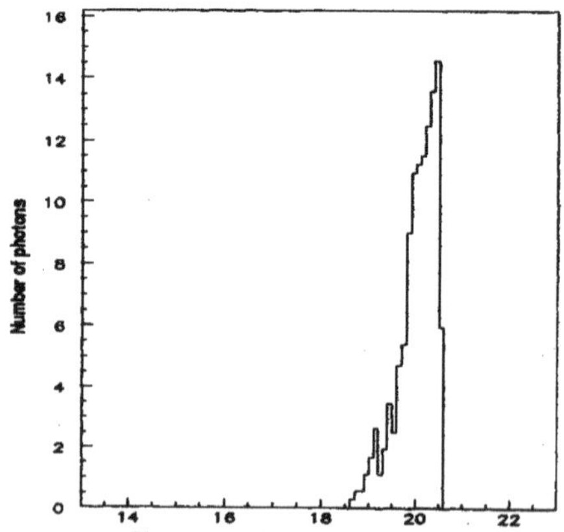

Fig.2. Velocity of photons along fiber axis without optical contact beyween fiber and photo-detector.

electrons created by $\beta = 1.0$ particles crossing a fiber with a numerical aperture of 0.37 (quartz core and doped PMMA cladding). The bulk of the light is in the "fish tail" indicating that only particles at 45° ± 10° with respect to the fibers will create light[1]. This range is narrower for quartz-quartz fibers (numerical aperture of 0.22). The small peak at $\alpha = 0°$ and high b is due to the 5% of the photons that are not in the numerical aperture (this is actually how the numerical aperture is defined) and exit with a large angle $\{\geq \text{Arcsin} \sqrt{(n_1^2 - n_2^2)}\}$. It is possible to get rid of these photons with an angular filter, but that would diminish substantially the total number of photo-electrons, and it is not planned to do so in the future, except for very special applications requiring an extremely good timing. Fig 2. shows the extremely narrow time dispersion (2.5%) of the photon velocities projected along the fiber's axis, making the light pulse slightly larger than one nanosecond (FWHM) for a 4 m long fiber. The speed is largely dominated by the photo-detector as seen in Fig. 3. The rate capability is limited only by the pulse width, for the Cerenkov effect has no dead-time.

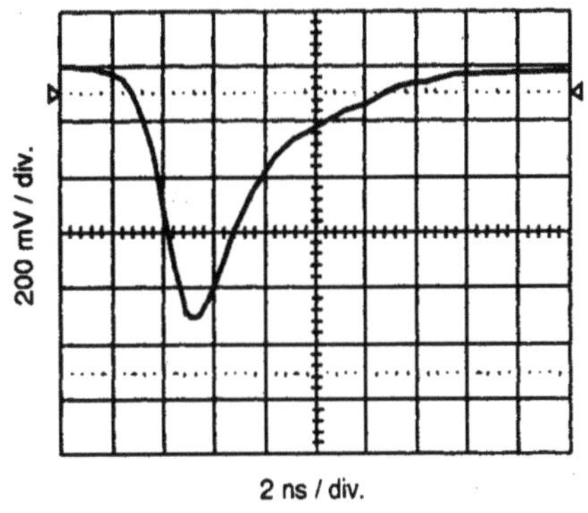

Fig.3. Pulse of a 160 GeV proton on NA50 prototype (Philips XP2262 PMT) after 10 m of cable.

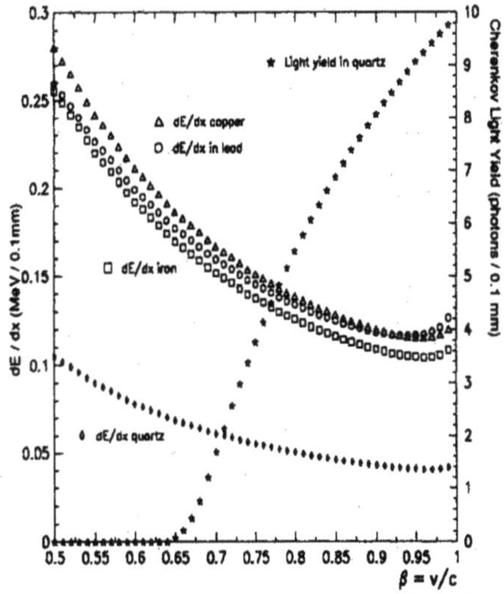

Fig. 4. Right: Cerenkov effect; left: energy loss effect

1.2. Calorimetry

A hadronic shower is composed of EM showers (due to the π^os) linked by charged mesons. Those travel with angles whose distribution is peaked along the incident particle angle (here 0°). Fig. 4. shows the main difference between "classical", ionisation or scintillation based calorimetry, and Cerenkov calorimetry[2]. In classical calorimetry, the signal will be due to the low β electrons which are in isotropic clouds. The charged pions play no role themselves. In Cerenkov calorimetry, only the electrons with a very high β will create light. They are not anymore isotropic (there are twice as many at 0° than at, let's say, 45°).

In Cerenkov calorimetry (QCal) the fibers introduce a narrow angular filter on top of that. There will be more light due to the electrons when the 0° electrons are selected by fibers at 45°. The charged pions with a very high β will travel at small angles and will produce light if the fibers are at 45° and none (or close to none) if the fibers are at 0°. All this is illustrated in Fig.5 and Fig.6. where, in the same calorimetry (fibers with a numerical aperture of 0.37 and a filling ratio of 1%), the responses to electrons and hadrons of various energies are shown for fibers inclined at 0° (or rather 3° to avoid channelling) and 45°. First, the electrons show a perfect linearity, the light being 1.7 times bigger at 45° than 3°. Second, the signal due to the hadrons is 0.8 times the signal due to the electrons at 45°, while it is only 0.6 at 3°. Also, the linearity for pions is better at 45° than 0° and the EM and hadronic parts of the pion showers show why: there is a compensation of these two components at 45°,

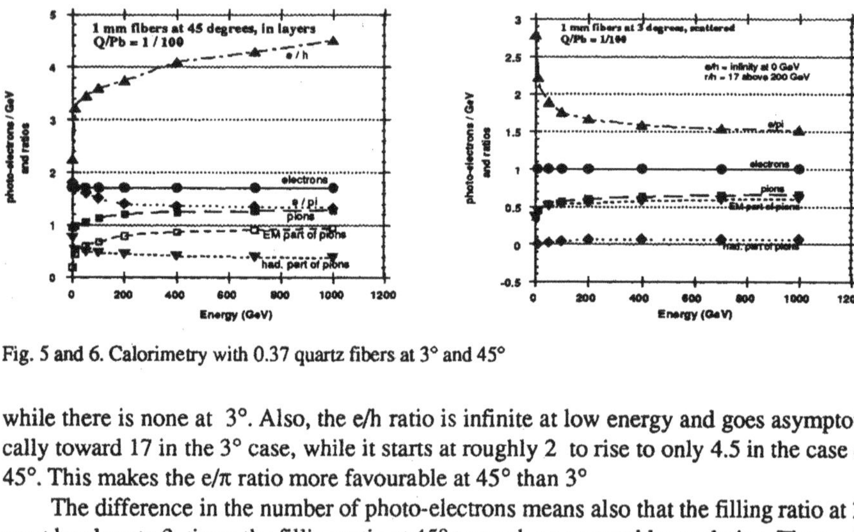

Fig. 5 and 6. Calorimetry with 0.37 quartz fibers at 3° and 45°

while there is none at 3°. Also, the e/h ratio is infinite at low energy and goes asymptotically toward 17 in the 3° case, while it starts at roughly 2 to rise to only 4.5 in the case of 45°. This makes the e/π ratio more favourable at 45° than 3°

The difference in the number of photo-electrons means also that the filling ratio at 3° must be close to 3 times the filling ratio at 45° to reach a comparable resolution. Thus, a 0° fiber calorimeter is at least 3 times more expensive than a 45° one. We will see later that, even then, the constant term is of the order of 14% for 0° while it is always 10% for 45°. The conclusion is that 0° fibers calorimetry should be reserved to very special cases involving for instance projecting towers.

2. RD40 recent results

2.1. Bare fibers measurements

The yield of different fibers was measured at the CERN PS using the set-up shown fig. 7. The fibers to test are cut to a length of 1 m and assembled in a single layer 6 mm wide. A set-up composed of scintillator paddles, scintillating fibers and a drift chamber ensures a good trigger. A Hybrid Photo-Diode[3] catches the Cerenkov light from the fibers. Fig. 8. shows the HPD response to 6 different fibers. The excellent resolution of the HPD

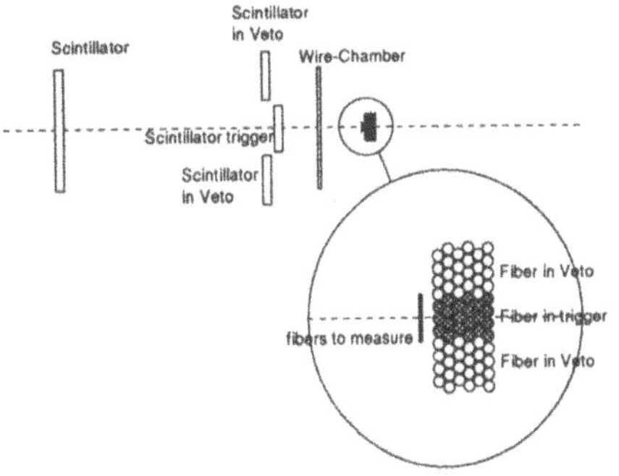

Fig.7. Apparatus used for bare fibers

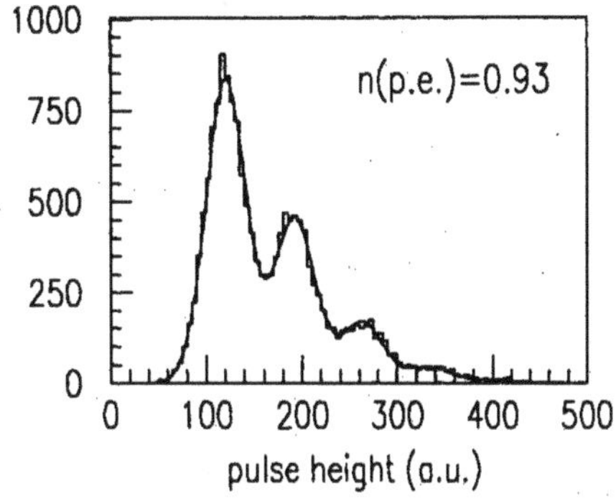

Fig. 8 Typical HPD spectrum showing 0, 1, 2, and 3 photo-electron peaks and the fit.

allows one to separate the 0 photo-electron from the 1, 2, 3,... photo-electron peaks and a fit yields the mean number of photo-electrons with an accuracy of 1%. The angle of the fiber relative to the beam can easily be changed and Fig. 9. shows the angular distribution measured for 0.37 numerical aperture, together with a line which is the projection of Fig. 1 on the angle axis. The agreement shows that Cerenkov light production in fibers is now well understood. Table 1. gives the results for 9 different fibers.

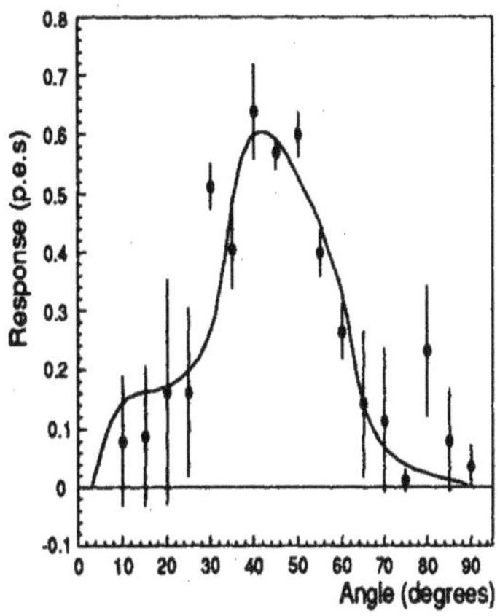

Fig. 9. Angular distribution of a bare fiber.

Table I. Fibers tested with 8 GeV particles at CERN PS.

Manufacturer	Core ϕ [μm]	Cladding ϕ [μm]	Coating ϕ [μm]	NA	ph.e./mm UV+Visible	ph.e./mm Visible
CeramOptec	Q 200	Q(F) 220	Polyimide 245	0.22	1.28 ± 0.31	
CeramOptec	Q 300	Q(F) 318	Polyimide 345	0.22	1.28 ± 0.08	
CeramOptec	Q 300	Q(F) 318	Acrilate 345	0.22	1.25 ± 0.05	0.44 ± 0.07
CeramOptec	Q 600	Q(F) 660	Polyimide 685	0.22	1.39 ± 0.04	
INFOS(*)	Q 300	Q(F) 330	Polyimide 420	0.22	1.48 ± 0.07	0.46 ± 0.07
INFOS (**)	Q 500	Q(F) 530	Polyimide 560	0.22	1.15 ± 0.07	0.49 ± 0.07
CeramOptec (***)	Q 500	HC 540		0.37	0.80 ± 0.03	0.56 ± 0.07
Ensign/Bickford	Q 1000	HC 1030		0.37	0.67 ± 0.02	
CeramOptec (****)	Q 500	Silicone 550	Acrilate 600	0.40	0.98 ± 0.03	0.71 ± 0.08

Table 1. Properties of various fibers.

2.2. Radiation hardness measurement.

The only radiation hardness measurement that has been done up to now is 4 years old. The fibers were irradiated at a 2 MeV electron accelerator along a few centimeters, mimicking the irradiation pattern predicted in a hadronic calorimeter. When looking at the visible wavelength only, quartz-quartz fibers stay unperturbed for more than 2 Gigarad, while quartz-plastic ones have their transparency showing a drop after 700 Mrad due probably to cladding reticulation.

However, a great effort is actually being pursued at LIL, the CERN LEP injector that provides 500 MeV electrons, to measure the radiation damage for different wavelengths. The fibers to irradiate are at 45° with respect to the beam, and the Cerenkov light falls on a grating that illuminates a bank of 32 PIN diodes. Thus, the irradiating particles are also the ones providing the correct test light.

2.3 Electromagnetic calorimeter prototypes.

The cheapest way to check how showers produce light in quartz fibers is to make EM prototypes. The first one built had scattered one mm fibers at 45° and a filling ratio of 1/4 and gave a resolution of $24\%/\sqrt{E} \oplus 3\%$. Since, 3 other prototypes have been built, with fibers at 45° in layers for a filling ratio of 1/8.3. The results (data taken at the CERN PS and SPS) have already been shown[4]. The most interesting result is that the resolution has a zero constant term. This is explained by the fact that, when the fibers are in layers, the response becomes insensitive to the change of aspect ratio of the EM showers when the energy is changed. Also, what was expected to change according to the numerical aperture, was the transverse shower size. This is why each of the layers (5 cm wide) are read in 2 layers of

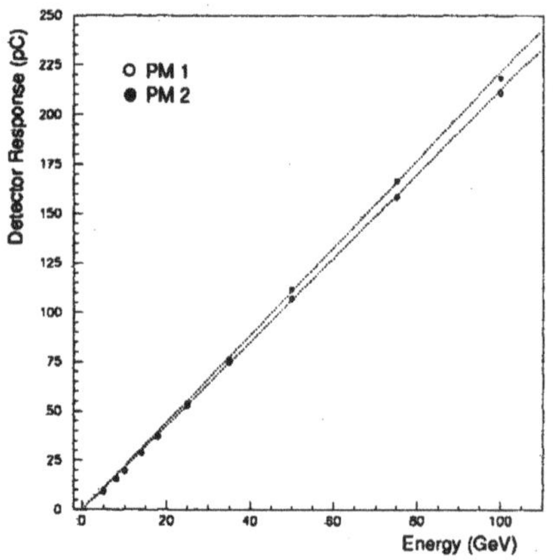

Fig. 10 Linearity of EM prototype.

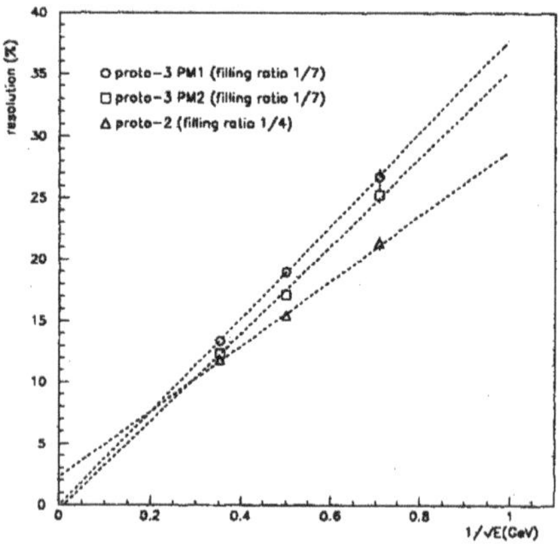

Fig. 11. Resolutions of EM calorimeters with scattered fibers (proto-1) and fibers in layers (proto-3).

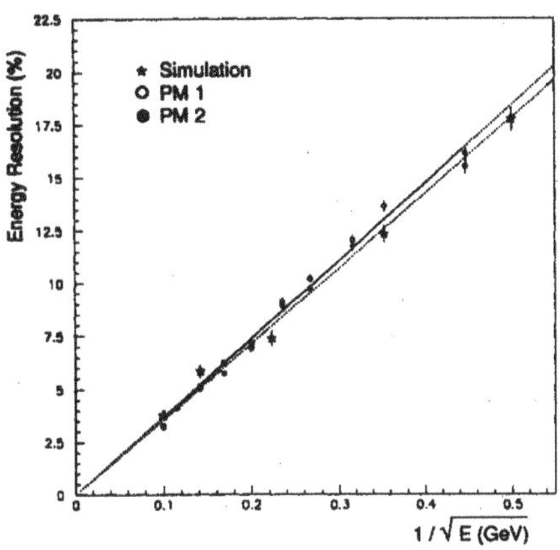

Fig. 12. Resolution of EM prototype with fibers in layers including the high energy (SPS) points.

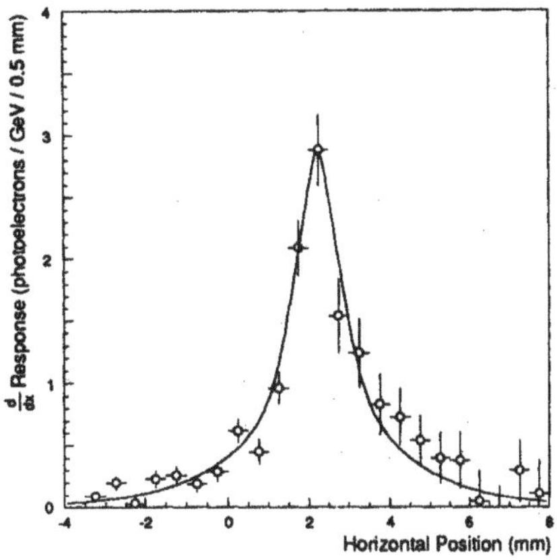

Fig. 13. Scan across the 2 layers side by side.

2.5 cm . One PMT reads all the fibers in the right half, the other the left half. The spectrum for both is perfectly gaussian. Fig. 10,11, 12 and 13 show respectively, the linearity -very good-, the resolution for the two kinds of prototypes (data from CERN PS only), the resolution of the layers prototypes and the shower size. The calorimeter with 0.37 numerical aperture fibers has a containment about 1.5 times greater than the one with 0.22 fibers (90% at respectively 10 and 15 mm). The difference between 8 and 50 GeV is very small. Tables 2. and 3. summarize the EM prototypes results. Minimum ionising particles (muons)

	Proto-3	Proto-4	Proto-R
Fiber manufacturer	CeramOptec	CeramOptec	INFOS
Core ϕ [μm]	500	500	500
Cladding ϕ [μm]	540 (HC)	550 (Silicone)	530 (Q(F))
Coating ϕ [μm]		600 (Acrilate)	560 (Polyimide)
Nominal NA	0.37	0.40	0.22
$V_Q/V_{absorber}$	1/8.3	1/9.2	1/8.6

Table 2. EM prototypes characteristics.

		Proto-3	Proto-4	Proto-R
$\sigma(E)/E$	a	$(35.6 \pm 0.2)\%$	$(41.7 \pm 0.3)\%$	$(42.3 \pm 0.7)\%$
$\left(\frac{\sigma(E)}{E} = \sqrt{\frac{a^2}{E} + b^2}\right)$	b	$(0 \pm 0.4)\%$	$(0.7 \pm 0.6)\%$	$(0 \pm 1.1)\%$
$N_{p.e.}$/mip	(UV+Visible)	9.0 ± 0.6	9.9 ± 0.6	12.6 ± 1.0
GeV/mip		0.57 ± 0.06	0.78 ± 0.08	0.78 ± 0.08
$N_{p.e.}$/GeV (electrons)	(UV+Visible)	15.9 ± 1.6	12.8 ± 1.5	16.3 ± 1.7
pC/GeV		2.113 ± 0.002	1.671 ± 0.001	1.675 ± 0.001
pC/mip		1.20 ± 0.012	1.30 ± 0.13	1.30 ± 0.13
Contributions to the stochastic term:				
Photostatistics		$(25.1 \pm 2.5)\%$	$(28.0 \pm 3.2)\%$	$(24.8 \pm 2.5)\%$
Sampling fluctuations		$(26.6 \pm 6.7)\%$	$(31.1 \pm 8.2)\%$	$(34.3 \pm 5.1)\%$

Table 3. Results of the EM prototypes.

were also shot in the prototypes and the response is totally coherent with the bare fibers measurements. The prototypes were also rotated for 180° around a vertical axis, so that the muon beam entered the calorimeter from the back. The response was 50 times smaller than with the normal operation (normally, no light should then reach the PMTs). This is important when dealing with 400 Hz / cm² beam gas muons at the LHC.

2.4. Hadronic prototypes.

The only prototype with fibers at 45° built up to now is the Zero Degree Calorimeter prototype for NA50. It is only 4 λ deep, and so, its resolution is not really meaningful. However, when tested with Sulphur ions (200 GeV/n), the resolution varies from 10% to 8% according to the density of fibers that are read by PMTs (1/40 to 1/10). It was observed

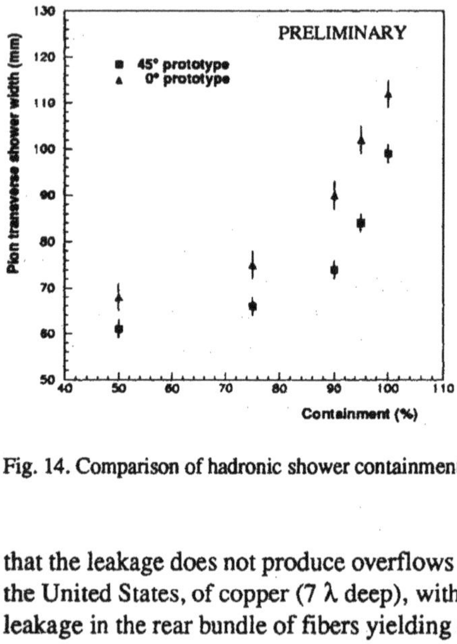

Fig. 14. Comparison of hadronic shower containment.

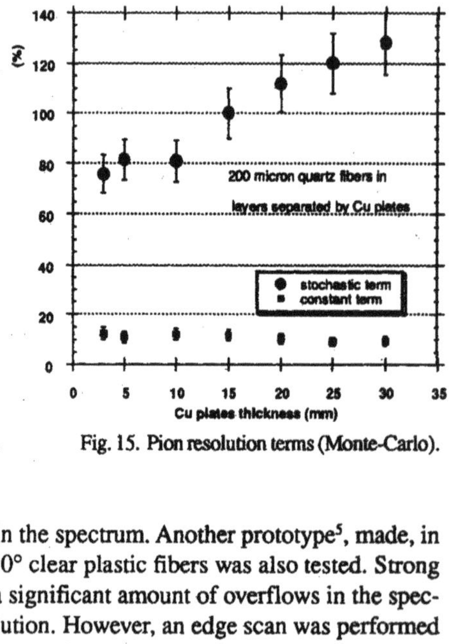

Fig. 15. Pion resolution terms (Monte-Carlo).

that the leakage does not produce overflows in the spectrum. Another prototype[5], made, in the United States, of copper (7 λ deep), with 0° clear plastic fibers was also tested. Strong leakage in the rear bundle of fibers yielding a significant amount of overflows in the spectrum prevents from getting a significant resolution. However, an edge scan was performed on both prototypes with the results given in Fig. 14. Once more, the advantage is for the 45° geometry.

The Monte-Carlo predicts for hadronic calorimeters with 0.2 mm fibers in layers at 45°, separated by copper plates of different thicknesses, resolutions whose stochastic terms and constant terms are given in Fig. 15. The constant term is stable at 10% and the stochastic varies from 80% to 130%. For a 1% filling ratio, the same simulations give for 0° fibers a stochastic term greater than 200% and a constant term of about 15%.

A Zero Degree Calorimeter[6] has been built on this technique for the CERN heavy-ion experiment NA50. RHIC at BNL plans to built 8 of them to measure the spectator neutrons.

3. Conclusion

The radiation resistance of quartz fiber calorimetry is larger than the LHC requirements (200 Mrad / 5 years at $\eta = 5$). The pulses are much shorter (10 ns) than the intercrossings time (25 ns), reducing pile-up to its base level. The rate capability is only limited by the pulse width. Furthermore, the size of the showers (90% in 7 cm for hadrons and 10 mm for EM radiation) makes the observation of particles and jets in the boosted region above $\eta = 2.5$ more easy than in the case of classical calorimetry (90% in 28 cm and 3 cm respectively). Furthermore, the detector is nearly blind to radioactivity (the electrons must have an energy greater than 3 MeV to produce a sizeable signal in a fiber and, due to the angular filter of the fiber's numerical aperture, only 0.5% of them will be seen). Finally, the neutrons cannot produce any signal (the velocities of the recoils are way under the Cerenkov threshold). Such a detector is static (no gases, no liquids), is easy to build, to operate and to move remotely in case one needs to open the main barrel (a very forward calorimeter will become quite radio-active at the LHC luminosity).

4. References

1. B. Grosdidier, *Doctoral Thesis*, CRN 92-45, 10 July 1992.
   D. Lazic, *Doctoral Thesis*, CRN 93-38, 28 May 1993.
   P. Gorodetzky et al., *Radiation Physics and Chemistry*, (Pergamon Press, Oxford, U.K.) **41** (1992) p. 253, eds. R. Clough and K.F. Johnson.
2. K.F. Johnson et al., *Proceedings of the Fourth International Conference on Calorimetry in High Energy Physics*, (World Scientific Pub. Co., Singapore, 1994) pp. 438-442, eds. A. Menzione and A. Scribano.
   P. Gorodetzky et al., *Proceedings of the Fourth International Conference on Calorimetry in High Energy Physics*, (World Scientific Pub. Co., Singapore, 1994) pp. 433-437, eds. A. Menzione and A. Scribano.
   P. Gorodetzky et al., *Proceedings of the Fourth International Conference on Calorimetry in High Energy Physics*, (World Scientific Pub. Co., Singapore, 1994) pp. 462-466, eds. A. Menzione and A. Scribano.
   A. Musso et al., *Proceedings of the Fourth International Conference on Calorimetry in High Energy Physics*, (World Scientific Pub. Co., Singapore, 1994) pp. 457-461, eds. A. Menzione and A. Scribano.
   A. Contin et al., CERN/DRDC/94-4, January 5, 1994.
3. R. DeSalvo et al., NIM **A315** (1992) 375.
   H. Arnaudon et al., NIM **A342** (1994) 558.
4. M. Marino et al., *Proceedings of the 6th Pisa Meeting on Advanced Detectors*, Elba, Italy, May 22-28, 1994.
   V. Gavrilov et al., *Proceedings of the 6th Pisa Meeting on Advanced Detectors*, Elba, Italy, May 22-28, 1994.
   A. Contin et al., *Proceedings of the 6th Pisa Meeting on Advanced Detectors*, Elba, Italy, May 22-28, 1994.
5. D. Winn, Fairfield University, Connecticut, U.S.A.
6. CERN-SPSLC-91-55 and B. Christie et al., RHIC note 13, Brookhaven, May 13, 1994.

BEAM TEST OF LIQUID ARGON TUBE CALORIMETER PROTOTYPE

Alexandre Savine
Department of Physics, University of Arizona
Tucson AZ 85721 USA
M.I.Ferguson, L.Harlow, J.Janesky, P.Loch, M.Qi, J.Rutherfoord, L.Shaver, M.Shupe, J.Steinberg, C.Zeitnitz
Department of Physics, University of Arizona
Tucson AZ 85721 USA
A.Lindenberg
Department of Physics, Columbia University
New York NY 19927 USA
D.Rahm
Physics Department, Brookhaven National Laboratory
Upton LI NY 11973 USA

Abstract A prototype of a liquid argon forward calorimeter suitable for high luminosity hadron collider detectors has been constructed and tested in electron beams at BNL and CERN. The linearity and resolution of the energy response and the position resolution analysis are presented. The constant term in the energy resolution is smaller than 5% and the position resolution is smaller than 0.7 mm..

Extensive studies of the physics requiring a forward calorimeter suggest that the $E_t$miss signatures from SUSY and tagging jets detection provide the most demanding constraints on the measurement capabilities of a forward calorimeter.

Because the forward calorimeters at high energy, high luminosity hadronic colliders are required to satisfy many constraints (high rates, radiation doses, etc.), designers are forced to make many compromises in the course of optimization. For instance technologies suggested for the forward region do not have as good energy resolution as a barrel calorimeter in order to address the other demanding requirements.

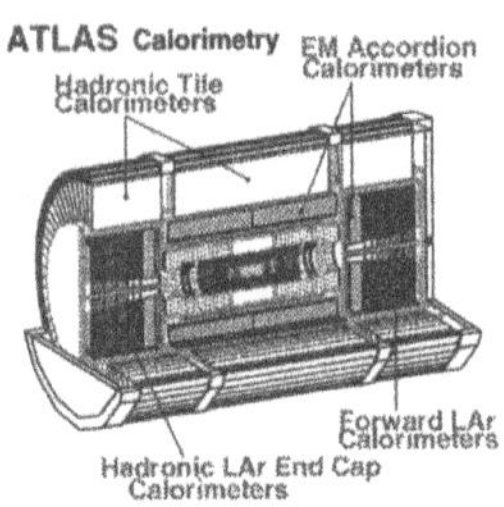

Figure 1

We have developed a calorimeter concept which meets all the demands. 1) The response is fast, of order the bunch crossing time at the LHC. 2) The energy and position resolution are more than adequate to the physics. And 3) the calorimeter is manifestly radiation hard.

This concept was successfully integrated into the GEM detector design[1] and we were particularly pleased with the result. Recently the Liquid Argon Tube calorimeter has became a baseline for the ATLAS detector at the LHC[2] and fits well in the Liquid Argon EndCap calorimetric system (Figure 1). In this report we describe a prototype electromagnetic (EM) module which we constructed. We present results of its performance from test beam runs at Brookhaven (Summer 1993) and CERN (October 1993)[3].

The Basic Design Concept

A novel calorimeter design has been developed into an operating detector prototype by the University of Arizona. The hallmark of this sampling cryogenic noble liquid design is that it employs exceptionally thin active gaps (as small as 100 μm) in order to speed charge collection (as fast as 20 ns) and to avoid the effects of positive ion buildup[4]. These properties, combined with the possibility of using only radiation hard materials in the fabrication, make this design ideal for a high rate and high radiation environment.

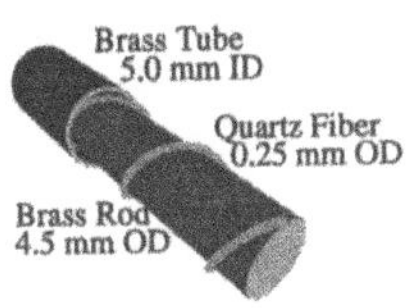

Figure 2 Tube&rod assembly

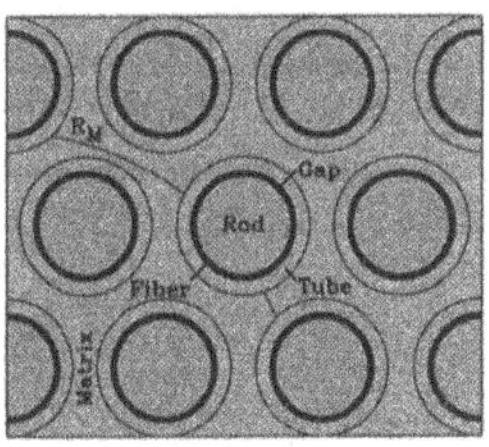

Figure 3
Tube&rod assemblies in absorber matrix

The active gaps of liquid cryogen (liquid argon or liquid krypton) are tubular in shape (cylindrical shells). This geometry allows the use of simple fabrication techniques, while at the same time maintaining precise control of the gap dimension. Clean geometry allows for very fast transfer of collected charge from the electrodes to the electronics.

In the Arizona design, solid rods of metal absorber are inserted in metal tubes with an inner radius larger than the rod radius by an amount equal to the desired gap. Mechanical spacing and electrical standoff in the gap are achieved by winding an insulating thread (quartz fiber having a diameter equal to the gap) around the rod before inserting it into the tube. A typical rod and tube assembly is shown in Figure 2.

The rod and tube assemblies are then inserted into an absorber matrix of the same metal to make up the complete calorimeter assembly, as shown in Figure 3. The parallel tube axes insure that the sampling fraction is constant throughout the calorimeter volume.

This geometry also allows for the possibility to make the calorimeter extremely radiation hard. The tube, rod, and absorber matrix are metal (brass in the EM prototype) and the cryogen is liquid argon. These are not degraded at all by ionizing radiation. The mechanical tolerance of quartz fiber to ionizing radiation is acceptable by several orders of magnitude. These are the only materials near electromagnetic shower maximum. The readout electronics is all located remotely from the forward calorimeter and need not be rad hard.

Mechanical and Electrical Design

We constructed a prototype of a first longitudinal depth segment for a forward calorimeter with design parameters optimized for a testbeam run on electrons at 8 GeV at Brookhaven National Laboratory. To improve the signal to noise measurement at these very low energies we employed a gap of 250 μm, larger than is needed in a typical forward calorimeter application, where the gap might approach 100 μm[5].

The liquid argon active tube gaps have an inner diameter of 4.5 mm, with center-to-center distance on the triangular matrix of 7.5 mm. The active tube gaps are 25.4 cm in length, leading to an active depth of the calorimeter of 14.4 radiation lengths and 1.30 absorption lengths. Absorber matrix, tubes, and rods are all yellow brass.

The absorber matrix is constructed of a stack of round plates with a hexagonal array of drilled holes to allow the insertion of the tube assemblies. A printed circuit board is then attached to one end plate of the module, with the tube pins emerging through holes in the board. Traces on the board create an electrical ganging of the tube assemblies into pairs to accommodate the 96 channels of electronics.

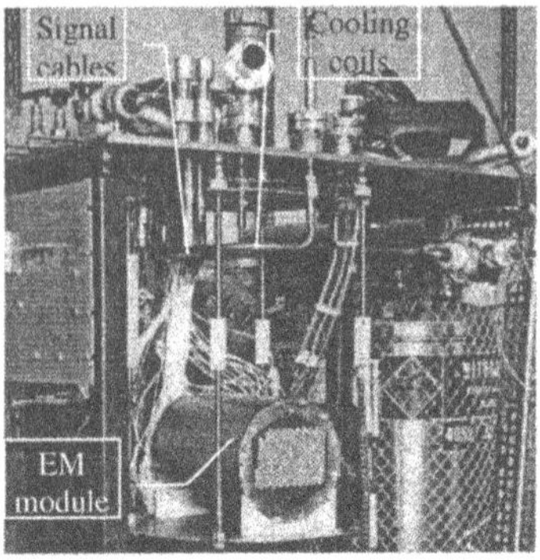

Figure 4 Prototype view

The photograph in Figure 4 is a view of the prototype in its testbeam configuration. Shown are the calorimeter, the dewar top plate and inner support structures, and the cabling inside the dewar. Cryostat cooling (see coils on top) was controlled by an all-pneumatic circuit avoiding the use of electrically operated solenoid valves and their associated electrical noise.

Argon purity is an important issue in liquid noble gas calorimetry due to signal loss through attachment of electrons to electronegative species (oxygen). It was monitored using a β-ionization cell[6]. This technique indicated an oxygen contamination of 0.6-0.7 ppm.

A feature of very thin active gaps is that the drift field of 12 kV/cm in the gap is provided by a 315 volt power supply in this prototype, and would be only 120 volts for 100 μm gaps.

The preamplifier/shaper modules use a common base front-end with a unique feedback arrangement that gives the input a stable and controllable input impedance over a large dynamic range. The shaper is set at 40 ns. The signal peak amplitude is stored in Track-and-Hold modules and digitized by CAMAC based 11-bit ADC's.

The present calibration system pulses all 16 channels simultaneously with the same level, although future designs may allow each channel to be pulsed independently or in combination.

The October 1993 Test at CERN

The first test run of this prototype occurred at Brookhaven in July and August of 1993. Data with 2, 4, 6, and 8 GeV electrons and pions were obtained. Analysis of this data set showed that the prototype was performing as expected. There was no time to manufacture a deeper prototype for the CERN test, and the 14.4 X0 depth chosen for Brookhaven energies leads to the need to make longitudinal leakage corrections at high energies to determine the intrinsic resolution and linearity of the detector.

The CERN test took place during October of 1993 in the H8 beamline. A secondary electron beam was available at 200 GeV/c and tertiary electron beams, at 150, 100, 50, 20, and 10 GeV/c. Other parameters of the test configuration were varied:

1) The calorimeter module could be rotated through small angles about the vertical axis (with $\theta = 0^0$ indicating that the tube axes were parallel to the beam). Data were taken at various angle settings between $\theta = 0.5^0$ and $\theta = 3.4^0$.

2) In addition to cryostat walls of 0.8X0, aluminum blocks could be placed on the platform in front of the cryostat to add 1.1, 2.2, or 3.3 X0 to this in order to simulate the effects of structures which would be present in a full collider detector.

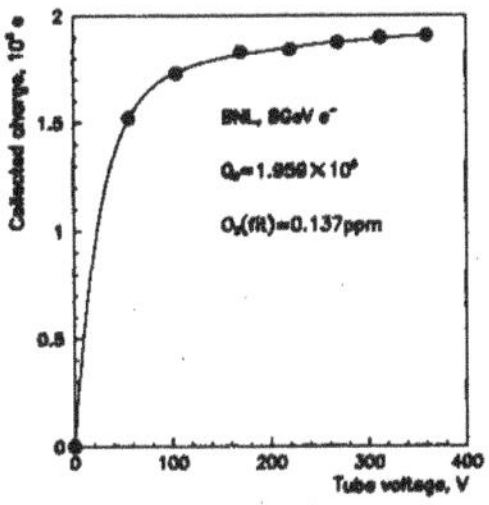

Figure 5. High voltage curve

3) Data sets were taken at various active gap voltages to map out the high voltage curve of the apparatus (Figure 5).

Two scintillation counters S1,S2 were used to generate a level 1 trigger. Surrounding the S1,S2 region transversely was a plane of veto counters insuring that the trigger did not include particles which had interacted in upstream material.

The next element downstream was an XY hodoscope with effective cell size of 0.5mm in the fine central areas (16mm×10mm). Following the cryostat was a lead glass block about 10 X0 deep followed by a bank of leakage counters, L1,2,3. The lead glass measured the leakage energy associated with electron showers in the calorimeter. Following an 80 cm thick concrete block was a scintillation counter to detect muon and pion contamination in the beam. The leakage counters used were also to detect hadronic showers.

The DAQ system was capable of acquiring up to 2400 events per spill, and did not limit the data rate during the CERN test. The data were logged to remote disk via an ethernet interface. In the CERN test, the 'remote' machine was a VAXStation 4000. The diagnostic power in both computers was needed to assure the quality of the data and to develop the run plan dynamically as the test proceeded.

Data Analysis and Corrections.

The analysis sequence for the calorimeter data from each event was fairly straightforward, and involved the following basic steps:

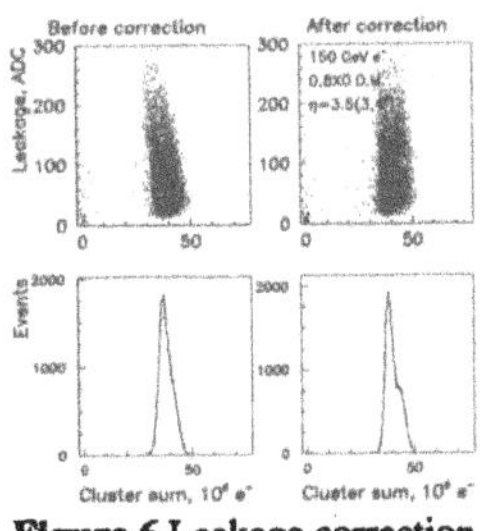

Figure 6 Leakage correction

(1) Subtract the pedestal average for each channel (including Lead Glass Leakage detector) from the raw pulse height in that channel.

(2) Apply the crosstalk correction matrix (of 2 % to 4 % to adjacent channels, from the effect of twisted pair lines driven single-ended instead of differentially).

(3) Multiply by the gain constant determined for each channel through calibration runs, converting the data to units of electrons.

(4) Add in the leakage energy (Figure 6) from the Lead Glass. Results presented were obtained with a fixed calibration constant for the Leakage Lead Glass counter determined from the 200 GeV run.

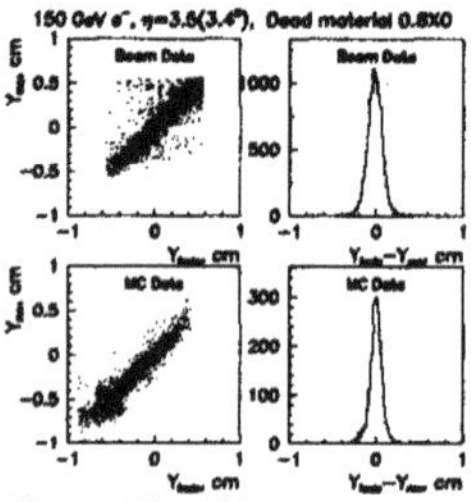
Figure 7 Position reconstruction

(5) Find the x and y position of the shower using a multiparameter likelihood method (to be published). Results of this procedure for 150 GeV, 0.5° run are shown in Figure 7. There are no traditional S-shapes, and position resolution is 3-4 times better than may be observed with the Center of Gravity method. 'Experimental' likelihood tables were also applied to MC simulated data.

(6) The cluster sum with R_{clust}=3.5cm was calculated. Cylindrical clustering with non-linear weighting of channels was employed (to account for non-uniform energy deposit in channel).

$$Aclust = \sum Achan * \sqrt{\frac{Sin}{Schan}}$$

Achan - signal in given channel
Sin - Part of channel covered by cluster
Schan - Total channel area

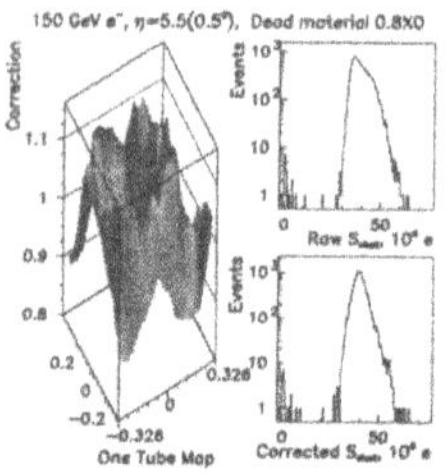
Figure 8 Position correction

(7) Energy was corrected for position dependence, using X and Y from (5) and the correction mapped to one tube 'cell' (Figures 8). This correction is largest when there is no material in front, and becomes smaller compared to other resolution sources as 'realistic' material depths are added.

This sequence was followed for all data sets presented in this analysis. We now give the results.

Observed Noise. At Brookhaven electronic noise of 0.05 GeV/channel (RMS) was observed. It corresponds to 0.23 GeV/cluster and is consistent with preamplifier noise. At CERN the preamplifier gain was lowered by a factor 20 to optimize the dynamic range for 200 GeV beam. As a result, DAQ noise ~2 ADC counts became dominant, and electronic noise (in GeV units) appears to be ~10 times larger.

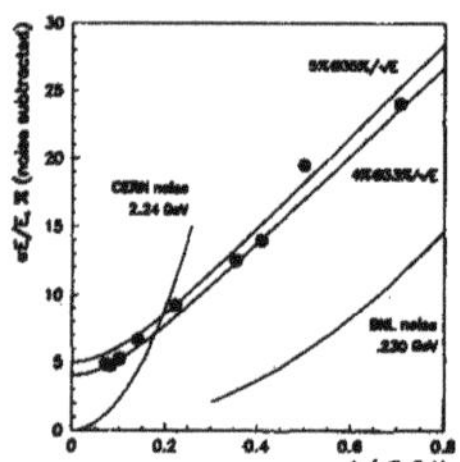
Figure 9 Energy resolution

Energy Resolution Parameters. After all corrections were applied, energy resolution was determined from a 2-Gaussian fit (RMS). The reason for this is that position resolution (~0.7mm) is still larger than the tube gap size, and channeling effects are still present. But after adding 2-3X0 of dead material both Gaussians became almost identical. The Energy resolution for dead material of 1.9X0 in 2+200 GeV range is shown in Figure 9. The stochastic term (34%) may be estimated from low energy data, and the constant term (pessimistically ~5%) from the 150+200 GeV points. The standard parametrization is not suitable for a shallow calorimeter employing leakage corrections. A summary of other CERN runs is shown in Table 1. There is no significant degradation of high energy resolution for a dead material thickness of 0.8-3.0 X0 in front.

Table 1

3.4^{0}	0.8 X0	1.9 X0	3.0 X0	4.1 X0
20 GeV	9.57%	9.24%	12.3%	20.0%
50 GeV	7.12%	6.65%	7.00%	10.0%
100 GeV	5.97%	5.57%	5.50%	6.77%
150 GeV	5.24%	4.92%	4.77%	5.64%
200 GeV	5.58%	5.37%	4.83%	6.15%

Observed Linearity. The difference plot in Figure 10 shows the deviations at most points to be within 1.5%.

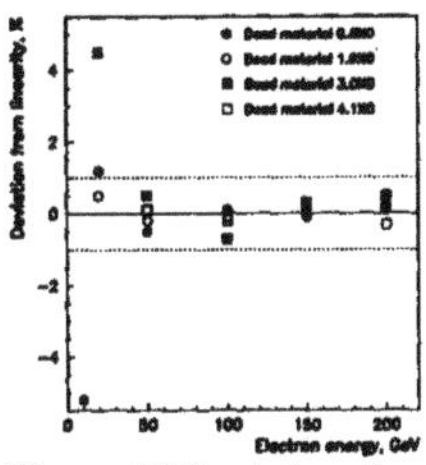

Figure 10 Deviations from linearity

Position Resolution. Very good position resolution (better than 0.7mm) was demonstrated. Table 2 shows the position resolution under various conditions.

Table 2

3.4^0	0.8 X0	1.9 X0	3.0 X0	4.1 X0
20 GeV	1.02mm	1.13mm	1.39mm	1.44mm
50 GeV	0.79mm	0.89mm	0.93mm	1.03mm
100 GeV	0.71mm	0.72mm	0.75mm	0.80mm
150 GeV	0.67mm	0.68mm	0.71mm	0.75mm
200 GeV	0.65mm	0.64mm	0.67mm	0.68mm

Conclusions

Analysis of data from Brookhaven and CERN beam tests of the prototype thin-gap tube liquid argon calorimeter developed at the University of Arizona shows that it is performing as expected, based on analytical predictions and simple simulations. Final results from these tests await the completion of detailed GEANT simulation runs for each data point.

Acknowledgments

We would like to thank J.Ban, K.Bussmann, A.Carroll, D.Dayton, R.Dillon, N.Doble, M.Eklund, K.Elsener, D.Fournier, D.Froidevaux, H.Gordon, R.Helmig, A.Hoffmann, A.Hrisoho, J.Huber, P.Jenni, Y.Kamyshkov, V.Khovansky, D.Lissauer, S.Marino, V.Radeka, S.Rescia, W.Willis, and Z.Zhang for help with these tests and/or fruitful conversations.

[1] B.Barish, W.Willis, et al., 'GEM Technical Design Report', GEM-TN-93-262, SSCL-SR-1219, April (1993)

[2] ATLAS Technical Proposal (to be published)

[3] M.I.Ferguson, et al., 'A Liquid Argon Forward Calorimeter Prototype : Beam Test Results' ATLAS Internal Note, CAL -NO-042, 18 January 1994.

[4] J.Rutherfoord, 'Ion Loading in Liquid Ionization Calorimeters', GEM TN-91-27 (1991)

[5] Tests of tube electrodes with 100 μm gaps are described in W. Don Carlos et al., 'Experimental Data on GEM LAC FCAL Tube Design', GEM TN-92-179 (1992)

[6] G.Blazey, 'Monitoring Liquid Argon purity at D0', International Conference on Calorimetry in High Energy Physics, FNAL, Batavia, IL (1990)

RECENT DEVELOPMENTS IN THE GAS IONIZATION CALORIMETRY WITH PLANAR ELECTRODE GEOMETRY.

S.Denisov, A.Dushkin, N.Fedjakin, Yu.Gilitsky,
V.Kochetkov, V.Mikhailin, Yu.Mikhailov, V.Onuchin,
I.Shein, A.Soldatov, A.Spiridonov, V.Sytnik.
Institute for High Energy Physics
142284, Protvino,Moscow region, Russia

Abstract

The characteristics of two gas ionisation calorimeters with planar electrode geometry and steel and lead absorbers have been studied in hadron and electron beams. The dependencies of calorimeter response and energy resolution on gas pressure , absorber thickness, gate width and delay, hadron energy and thickness of the entrance window have been measured. The contributions of the sampling and intrinsic fluctuations to the energy resolution are considered. The data on electron-to-hadron ratio are discussed. The results on longitudinal profiles and fluctuations of EM showers are presented.

1. Introduction

The first study of a gas ionization calorimeter in hadron beams was performed in 1992 at the 70 GeV IHEP accelerator [1,2]. The absorbers of the calorimeter tested were made of uranium. In 1993-94 we investigated the characteristics of gas ionization calorimeters with steel and lead absorbers. The preliminary results of this study have been reported at the 1993 Calorimetry Conference [3]. The results of a complete data analysis are presented below.

2. Calorimeter Design.

Each calorimeter consists of a stack of ionization chambers interleaved with passive absorbers. Steel and lead absorbers with thickness of 11.8 and $12.9 g/cm^2$ were used[3]. The transverse size of the absorbers is $58 \times 58\ cm^2$. The total thickness of the calorimeters is 7.4λ. Signal electrodes made of 1.5 mm G-10 were placed between the absorbers forming two 4 mm drift gaps. On each side of the electrodes there are 3 vertical signal strips 14 cm wide and 42 cm high.

The stack of absorbers and signal planes is contained in a cylindrical stainless steel vessel designed to operate at up to 40 atm[1,2]. The vessel has a 2 mm steel window for the beam. It was filled with $90\% Ar + 10\% CF_4$ gas mixture. The HV applied to the signal strips was equal to 180 V/atm, which corresponds to a maximum electron drift velocity of 0.12 mm/ns[4]. So the maximum drift time of electrons was 33 ns.

3. Electronics.

Signals from each strip were transported to amplifiers [1,2], using 4 m 50 Ohm cable. The δ-function response of the electronics chain has a 30 ns base width. ADC gate widths of 25, 55, or 145 ns were used in most of the measurements. A part of the measurements were performed with the gate width varied from 19 to 240 ns.

The measured noise level for one channel is $\sigma = 9\ ke$ (55 ns gate) which is in good agreement with the calculated value. The dependence of the noise for all 225 channels on the gate width is shown in fig.1. The value at 55 ns is close to that expected for uncorrelated noise. It means that the contribution of the coherent noise is negligible.

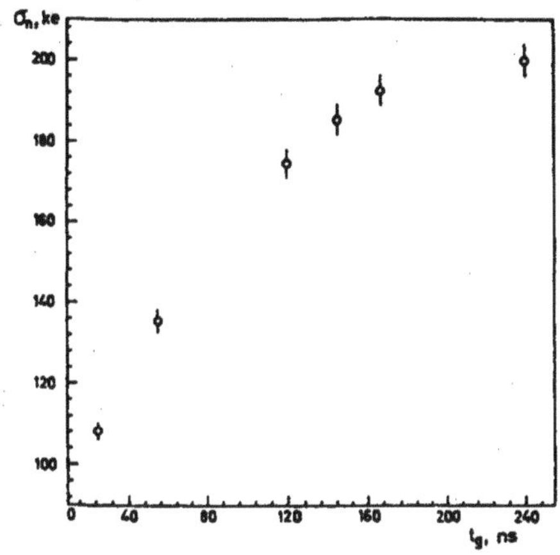

Fig.1. Noise level as function of the ADC gate width.

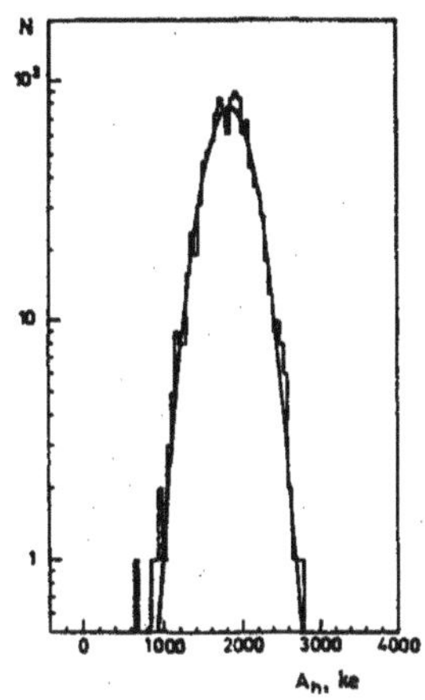

Fig.2. Pulse height spectrum measured with 58 GeV hadron beam.

4. Measurements and pulse height spectra.

Measurements have been performed using $26 \div 58\ GeV/c$ negative hadron beams and a 26.6 GeV/c electron beam produced in the internal targets of the 70 GeV IHEP accelerator. The composition of the hadron beams is π^- (> 96%), K^- and μ^- ($1 \div 2.5\%$), e and $\bar{p}$ (< 1%). The hadron contamination of the electron beam is less than 1%. The typical beam momentum spread was 3%.

The studies consisted in measuring the pulse height distributions at different values of the particle momentum, gas pressure, gate width and gate delay. A typical pulse height distribution is shown in fig.2. This distribution and data presented below were measured at 24 atm gas pressure unless stated otherwise. All the spectra are well fitted by a Gaussian. There is no indication for any tails on the pulse height distributions due to "Texas towers" or other effects.

5. Average pulse heights.

The dependence of the average signal on the gate width t_g and gate delay t_d are presented in figs.3. The following conclusions can be derived from these data:

- 25,55, and 145 gates contain 45,70 and 98 % of total ionization signal respectively,
- maximum pulse height corresponds to $t_d = 0$ ($t_g = 145$ ns), $t_d = 10$ ns ($t_g = 55$ ns) and $t_d = 20$ ns ($t_g = 25$ ns).

Data shown in fig.3 and presented below were measured with $t_d = 0$ ($t_g = 145$ ns) and $t_d = 20$ ns ($t_g = 25, 55$ ns).

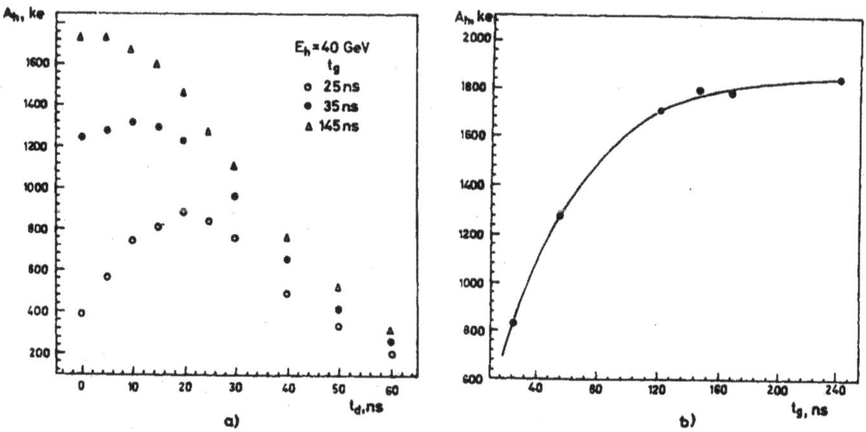

Fig.3. The dependence of the average pulse height on (a) gate delay and (b) gate width measured with (a) steel and (b) lead calorimeters in (a) 40 and (b) 58 GeV hadron beams.

6. Energy resolution for hadrons.

Due to the fine longitudinal granularity of the calorimeters it was possible to study the energy dependence of the energy resolution for the different absorber thickness t by summing the signal not only from all planes, but from odd and even planes separately, from each 3-rd plane and so on. The values of σ/E_h for hadrons obtained by this method are well fitted using the formula [5] (see fig.4):

$$\frac{\sigma}{E_h} = \frac{A}{E_h} \oplus \left(\frac{B}{\sqrt{E_h}} + C \right) , \tag{1}$$

where the parameters A and B represent the equivalent noise energy and the stochastic fluctuations of the shower energy deposited in the working gas. The values of A, B, and C are presented in Table 1. Parameter A was estimated using signal and noise pulse height spectra. The constant term C was assumed to be independent on the absorber thickness t. Typical errors of the parameters are 0.03 for B and 0.004 for C. The stochastic fluctuations are less for the steel calorimeter than for the lead calorimeter; the latter has values of C that are equal to zero within the errors quoted. It is worth mentioning that no corrections were applied to take into account the possible energy leakage in the transverse direction.

Table 1. Parameters in Eq.(1). $t_0 \sim 0.08\lambda$ is the thickness of one absorber. A and B are measured in GeV and $\sqrt{GeV}$, respectively.

Absorber	Gate width,ns	Parameter	Absorber thickness t_0	$2t_0$	$3t_0$	$4t_0$	$5t_0$
steel	25	A	6.0	9.0	11.0	12.9	14.4
		B	0.71	0.82	0.98	1.07	1.22
		C	0.025				
	55	A	5.0	7.6	9.5	11.0	12.3
		B	0.70	0.85	0.94	1.07	1.18
		C	0.045				
	145	A	5.0	7.5	9.2	10.6	11.7
		B	0.77	0.88	1.00	1.18	1.27
		C	0.044				
lead	25	A	7.1	10.2	12.5	14.5	16.3
		B	0.84	1.12	1.32	1.49	1.67
		C	0.001				
	55	A	6.6	9.5	11.7	13.6	15.3
		B	0.82	1.08	1.30	1.42	1.62
		C	0.002				
	145	A	6.1	8.8	10.6	12.3	13.8
		B	0.92	1.16	1.37	1.50	1.70
		C	0.003				

Fig.5 demonstrates that the contribution of hadron shower fluctuations to energy resolution depends weakly on the gas pressure up to 5 atm. The same conclusion was reached for a gas calorimeter with uranium absorbers [2]. For the gas ionization calorimeter operating at high pressure it is important to investigate the dependence of its characteristics on the thickness of the entrance window. This was done by neglecting the signals from the front layers. The results presented in fig.6 show that the passive material in front of the calorimeter has negligible influence on its response and energy resolution up to 0.4 λ.

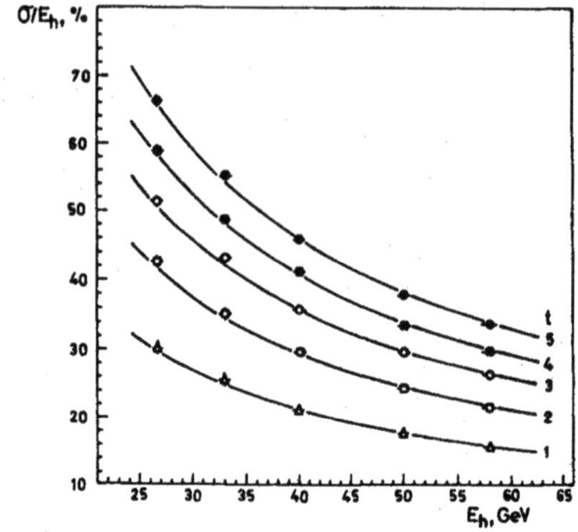

Fig.4. Energy resolution vs hadron energy for lead absorbers ($t_g = 55ns$). Curves represent the Eq.(1). Figures near the curves show the absorber thickness t measured in the units of $t_0 = 0.08\lambda$, where t_0 is the thickness of one absorber.

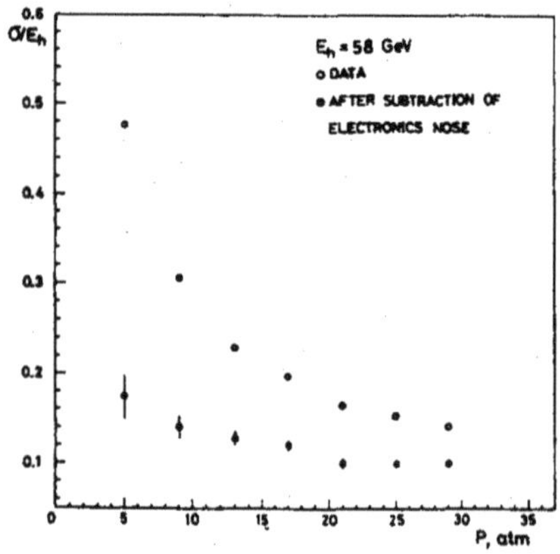

Fig.5. Energy resolution of the lead calorimeter vs gas presure ($t_g = 55ns$.)

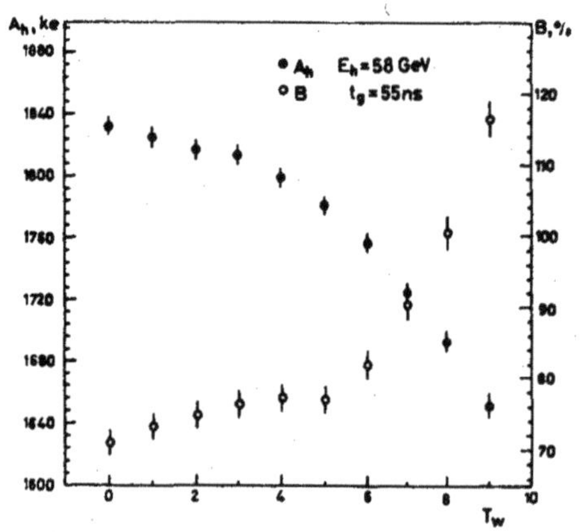

Fig.6. Stochastic parameter and response vs thickness of passive material in front of the steel calorimeter ($t_g = 55ns$). T_w is measured in the units of absorber thickness, t_0.

7. Contribution of the sampling fluctuations to the energy resolution.

The dependence of stochastic parameter B on absorber thickness t is well fitted by either of the formulae (see fig.7):

$$B(t) = b\sqrt{t} \bigoplus c\,, \tag{2}$$

$$B(t) = b'\sqrt{t} + c'\,, \tag{3}$$

Parameters b and c represent the contribution of the sampling and intrinsic fluctuations to the stochastic part of the energy resolution and are shown in fig.7. From fig 7 it follows that for the steel calorimeter the sampling and intrinsic fluctuations give about equal contribution to parameter B while for the calorimeter with lead absorbers the sampling fluctuations are more important. The data on the t-dependence can be used to estimate the absorber thickness required for a calorimeter with a desired energy resolution.

Another way to estimate the contribution of the sampling fluctuations to the energy resolution is a method of interleaved calorimeters[6,7]. Using this method we obtained:

$$b = 0.68 \pm 0.02 \qquad (lead \quad absorbers),$$

$$b = 0.48 \pm 0.02 \qquad (steel \quad absorbers).$$

These results are in good agreement with the b-values derived from the t-dependence (see fig.7).

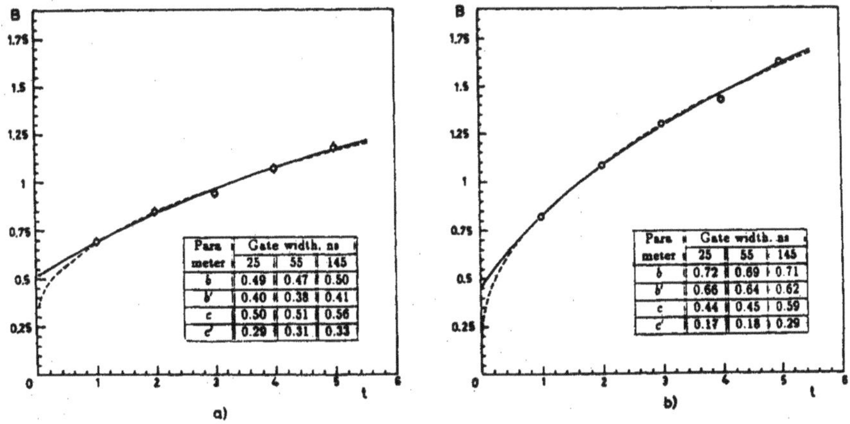

Fig.7. Stochastic parameter B as a function of absorber thickness t for (a) steel and (b) lead calorimeters ($t_g = 55ns$). t is measured in units of absorber thickness,t_0. The solid and dotted lines represent the Eqs.(2) and (3) respectively. Parameters b, b', c, c' are measured in $\sqrt{GeV}$.

8. e/h ratio.

The ratio of calorimeter response to electrons and hadrons (e/h ratio) depends on the gate width and delay (fig.8). This behaviour can be explained by different shapes of hadron and electron signals: an electron signal is shorter than a hadron one. Usually it is important to keep the e/h ratio as close to unity as possible. Fig.8b shows that this can be achieved for the lead calorimeter by using the short gate width ($\sim 25\ ns$) and $t_d \approx 30\ ns$, while at the optimum value of $t_d \approx 20\ ns$ $e/h \approx 1.1$. In the case of the steel calorimeter e/h is higher: 1.15 at $t_d = 30\ ns$ and 1.2 at $t_d = 20\ ns$. It is well known that the constant term in the energy resolution depends on the e/h ratio: the closer e/h is to 1 the smaller is the constant term. So the results on e/h ratio are in qualitative agreement with values of the parameter C in Eq. (1).

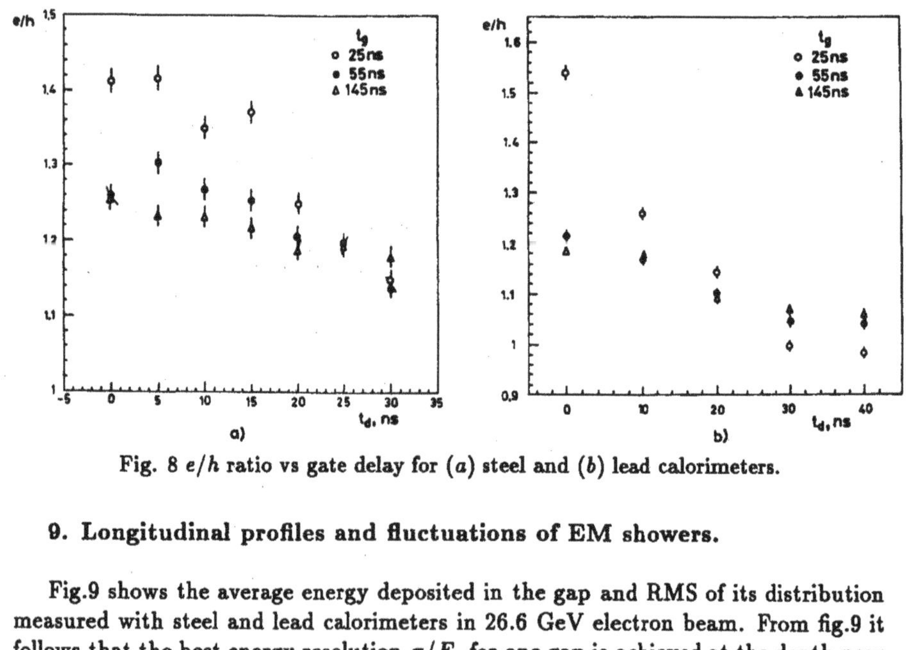

Fig. 8 e/h ratio vs gate delay for (a) steel and (b) lead calorimeters.

9. Longitudinal profiles and fluctuations of EM showers.

Fig.9 shows the average energy deposited in the gap and RMS of its distribution measured with steel and lead calorimeters in 26.6 GeV electron beam. From fig.9 it follows that the best energy resolution σ/E_e for one gap is achieved at the depth near shower maximum and is equal to 0.16 (lead absorbers) and 0.24 (steel absorbers) if the noise is subtracted. These values are in agreement with results obtained by Ts. Amatuni et al.[8,9]

10. Conclusions.

A detailed study of the characteristics of the gas ionization calorimeters with planar electrode geometry shows that they possess good energy, time and position resolutions, high uniformity and stability, simple calibration, and that the e/h ratio can be close to unity. Due to high intrinsic radiation resistance this type of calorimetry is especially promising for high luminosity experiments.

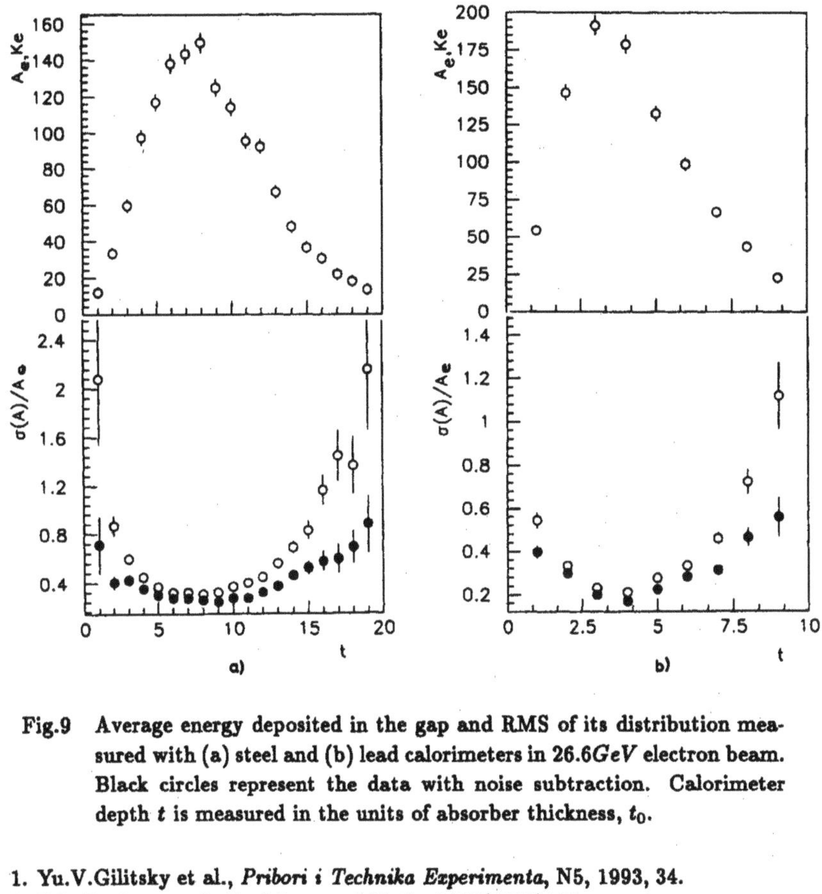

Fig.9 Average energy deposited in the gap and RMS of its distribution measured with (a) steel and (b) lead calorimeters in 26.6GeV electron beam. Black circles represent the data with noise subtraction. Calorimeter depth t is measured in the units of absorber thickness, t_0.

1. Yu.V.Gilitsky et al., *Pribori i Technika Experimenta*, N5, 1993, 34.
2. S.P. Denisov et al.,*Nucl.Instr.and Meth.* A335(1993)106.
3. S.P. Denisov et al.,*Proceedings of the IV Intern. Conf. on Calorimetry*, Elba, 1993.
4. A.Peisert, F. Sauli, *CERN Preprint 84-08*, Geneva, 1984.
5. R. Wigmans , *Nucl. Instr. and Meth.* A259 (1987) 389.
6. C.N.Fabjan et al., *Nucl. Instr. and Meth.* 141(1977)61.
7. G.Drews et al.,*Nucl Instr. and Meth.* A290(1990)335.
8. Ts.A.Amatuni et al.,*Nucl. Instr. and Meth.*, 203(1982) 179.
9. Ts.A.Amatuni et al.,*Nucl. Instr. and Meth.*, 203(1982) 183.

Energy Resolution of a Hadron Calorimeter Made of High-Pressure Gas-Ionization Tubes

S.L. Bagdasarov, P.T. Cox, L. Demortier, N.D. Giokaris, and D.M. Khazins
Experimental Physics Department, The Rockefeller University, New York, NY 10021-6399, USA

M.G. Albrow, D. F. Anderson, S. Cihangir, and J. Zimmerman
Fermi National Accelerator Laboratory, Batavia, IL 60510, USA

G. Fanourakis* and F. Lobkowicz
Department of Physics and Astronomy, University of Rochester, Rochester, NY 14627-0171, USA

H.R. Gustafson
University of Michigan, Ann Arbor, MI 48109-1120, USA

J.A. Budagov and S. Tokar**
Joint Institute for Nuclear Research, Dubna, P.O. Box 79, 101000 Moscow, Russia

M. Morgan
Ability Engineering Technology, Inc., South Holland, IL 60473, USA

Presented by D.M. Khazins.

Abstract

A hadron calorimeter based on high-pressure gas-ionization tubes arranged nearly parallel to the incident particle direction was built and tested in pion and electron beams at the CERN SPS. Data were taken in the energy range of 10 to 100 GeV with the calorimeter tilted to the beam axis at angles varying from 0.9° to 9.1°. Results of the calorimeter response to pions and electrons and its energy resolution are presented. The calorimeter was developed for the forward region of high energy and high luminosity hadron colliders. It proved to be simple and easy to operate. Its parameters meet the needs of forward calorimetry for LHC collider experiments.

* Present address: *University of Crete, Physics Department, 71409 Iraclion Crete, Greece*

** Present address: *Faculty of Mathematics and Physics, Comenius University, Bratislava , 84 215, Slovakia*

The motivation for considering high-pressure noble-gas ionization calorimeters has been discussed in our previous papers [1 - 6]. (For a complete list of references on high-pressure calorimeters see ref. 6.) These calorimeters are radiation hard because they have no gas amplification and are made of radiation hard materials. Our recent measurements [7] demonstrate the ability of such a detector to resist a radiation dose as high as 1 Grad. The high electron mobility in commonly used gas mixtures [5] yields an output signal comparable in width to that of a plastic scintillator counter. The lack of gas amplification is also responsible for the excellent linearity of these detectors. The small signal from conventional gas-ionization chambers is compensated in our case by the high pressure of the working gas and by the high energy of the incident particles in the forward region. At the same time the increased gas density dramatically suppresses the effect of the so-called "Texas Towers" (large signals caused by slow neutrons in one atmosphere gas calorimeters [8]).

Several parallel-plate electromagnetic, EM, [2, 3, 9] and hadron [10] calorimeters have been tested so far. In this paper we describe the beam-test results of the first hadron calorimeter to be built of high-pressure gas-ionization tubes arranged nearly parallel to the incident particle direction. This design was motivated by both safety and signal collection considerations. The features and basic characteristics of the design are discussed elsewhere [6]. The calorimeter was built as a prototype for the high radiation environment of the forward region of high-energy hadron colliders.

A perspective view of the tube calorimeter is shown in fig. 1. It consists of 896 tube ionization chambers (fig. 2) filled with a gas mixture of 95% Ar + 5% CH_4 at a pressure of 100 atm. Details of the calorimeter design as well as the front end electronics and the calibration procedure are described elsewhere [11].

The calorimeter was tested at the CERN - SPS X7B beam line in July 1993. We used negative pion and electron beams in the energy range of 10 to 100 GeV. A scintillator telescope defined a 5 x 5 mm^2 beam at the calorimeter front surface. The calorimeter was installed on a table parallel to the beam. During the measurements we varied the horizontal angle between the calorimeter axis and the beam from 0.9^o to 9.1^o, but the angle in the vertical plane was kept close to 0°. At each angle the horizontal position of the calorimeter was adjusted so that the hadron shower maximum (at about 1 m calorimeter depth) was approximately at the center of the calorimeter cross section. The accuracy of the angle determination was $\pm 0.05^o$, and that of the calorimeter positioning was $\pm$ 2 mm.

An oscilloscope trace of the amplifier output signal for 50 GeV electrons for a calorimeter voltage of 1.2 kV is shown in fig. 3. The signal's full width at this voltage is about 80 ns, and the width of the ADC gate signal (also shown in the figure) was adjusted accordingly.

The calorimeter response to electrons and pions as a function of the beam energy is shown in fig. 4. The data were taken with a calorimeter tilt angle of 9.1^o. Very similar results were obtained at tilt angles

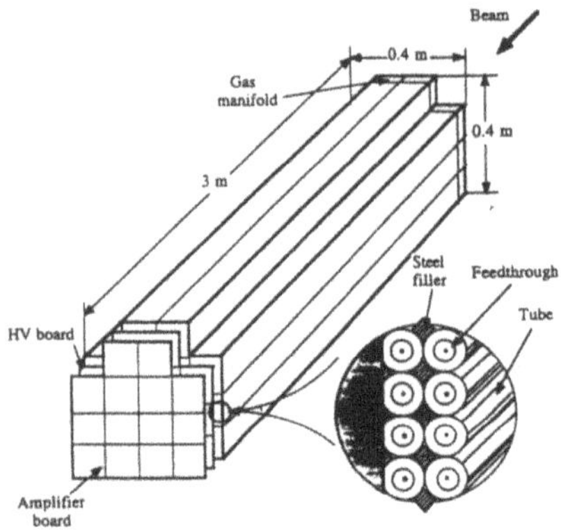

Fig. 1. Perspective view of the tube calorimeter.

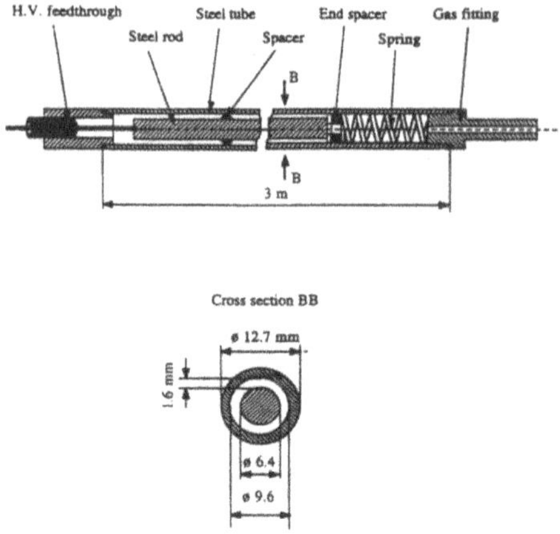

Fig. 2. Schematic of a single tube ionization chamber.

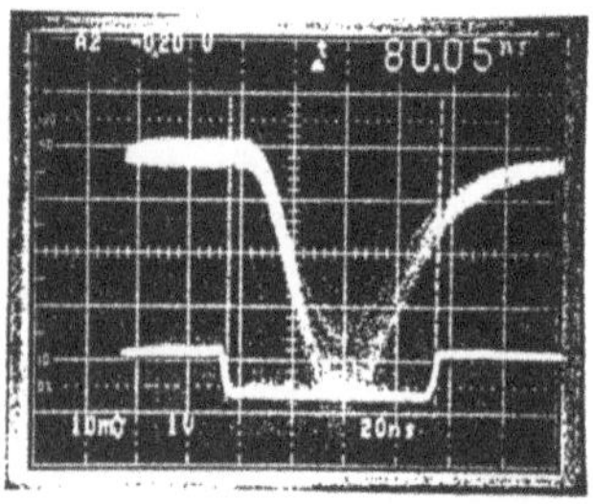

Fig. 3. Oscilloscope trace of the calorimeter output signal for 50 GeV electrons. The lower trace shows the gate signal.

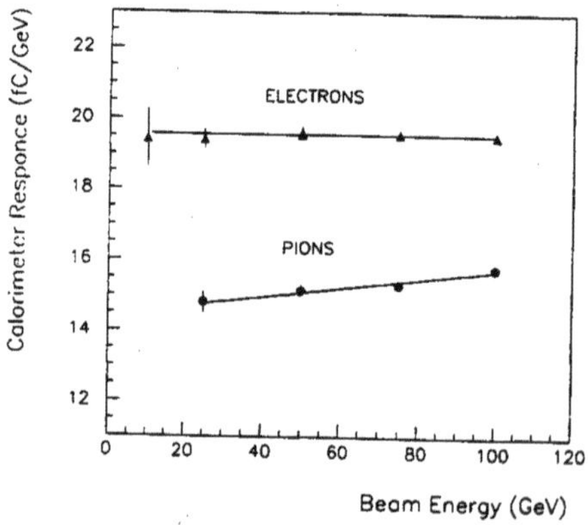

Fig. 4. Collected charge over beam energy for electrons (triangles) and pions (circles), as a function of beam energy.

of 0.9^o and 2.9^o. The average signal amplitudes presented in these plots were corrected for average energy losses in the calorimeter's front plates (gas manifold plus an extra 6 mm lead plate) for the electron data, and for lateral shower leakage for the pion data. The corrections were calculated by Monte Carlo simulations of the calorimeter. The calculated lateral hadron energy leakage shows little dependence on the pion beam energy, it is about 7% for the tilt angles of 0.9^o and 2.9^o, and about 11% for 9.1^o.

The electron data presented in fig. 4 demonstrates a very good linearity. The average charge per deposited energy of 19.6 ± .1 fC/GeV can be compared with the calculated value of 25.1 fC/GeV.The pion data show a small nonlinearity due to the decrease of the e/h ratio with increasing energy .

Fig.5 shows the pulse-height distribution for the 75 GeV pion beam. The spectrum contains two peaks corresponding to pions and muons. The width of the pion peak is the result of intrinsic signal fluctuations, electronic noise, and lateral leakage fluctuations. To obtain the intrinsic calorimeter resolution we quadratically subtracted the electronic noise width and corrected the results for the lateral shower leakage using Monte Carlo simulations. The results are shown in fig. 6 a. Data for three calorimeter tilt angles (0.9^o, 2.9^o and 9.1^o) have been combined because the Monte Carlo simulations predict a negligible dependence of the pion energy resolution on the angle. Furthermore, the experimental errors, especially at low energies, are too large to allow a meaningful fitting procedure for the individual angle data. The result of fitting the combined experimental pion data is:

$$\delta E/E = (70 \pm 12)\% / \sqrt{E} \oplus (7.4 \pm 1.2)\% \qquad (1)$$

The constant term is caused by the inequality of the calorimeter response to electrons and hadrons. A software method of e/h compensation for a well-segmented calorimeter has been developed by the H1 Calorimeter Group [12]. Although our calorimeter has only transverse segmentation, the following analysis demonstrates that even this type of segmentation could be helpful. We proceed from the idea that most of the EM showers, or at least the most energetic ones, are concentrated along the core of the hadron shower. Hence, one can expect a decrease in the signal fluctuations by suppressing the signals from that part of the shower. In fig. 7 the calorimeter energy resolution for 100 GeV pions is plotted versus a suppression coefficient, k, which was applied to the two central modules containing about 70% of the shower energy. The dependence shows a clear minimum at k = 0.6. This coefficient was applied to all the data presented in fig. 6 a, independent of the beam energy and tilt angle, though the optimum value of k is expected to depend slightly on both of these parameters. The results are shown in fig. 6 b. The best fit in this case does not include any constant term and corresponds to:

$$\delta E/E = (64 \pm 6)\% / \sqrt{E} \qquad (2)$$

However, we can not exclude a constant term of ≤ 3%.

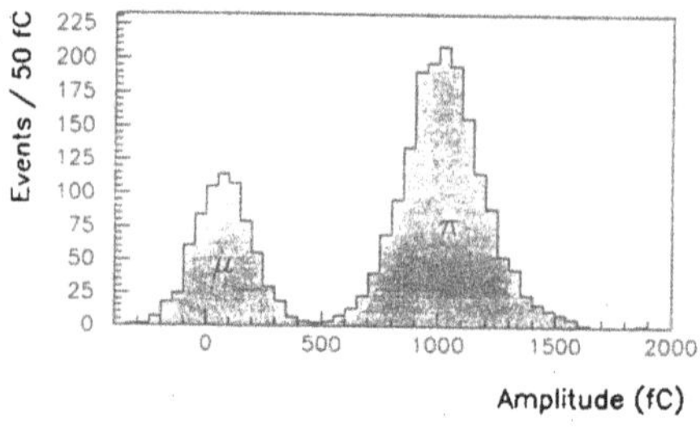

Fig. 5. Pulse-height spectrum for a 75 GeV pion

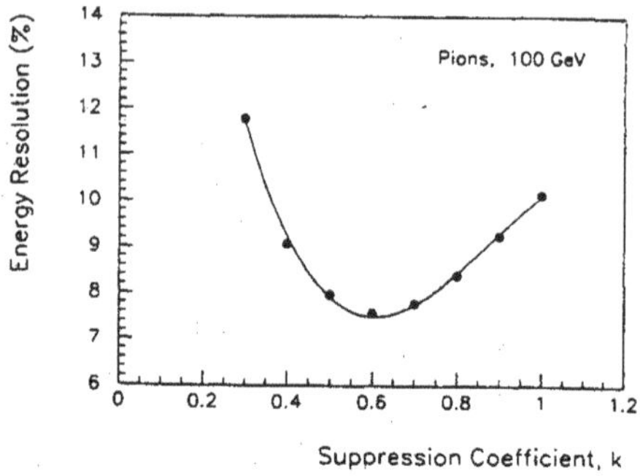

Fig. 7. Pion energy resolution at 100 GeV as a function of the suppression coefficient k, applied to the two central modules. Data are averaged over tilt angle and corrected for electronic noise and lateral shower leakage.

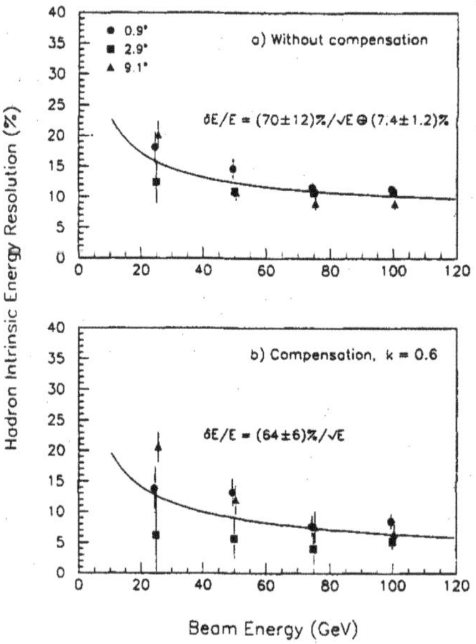

Fig. 6. a) Calorimeter energy resolution for pions, as a function of beam energy. Data are corrected for electronic noise and lateral shower leakage fluctuations. The solid curve is the result of a fit to all points, independently of tilt angle.

b) The same, but after a compensation procedure was applied (see text).

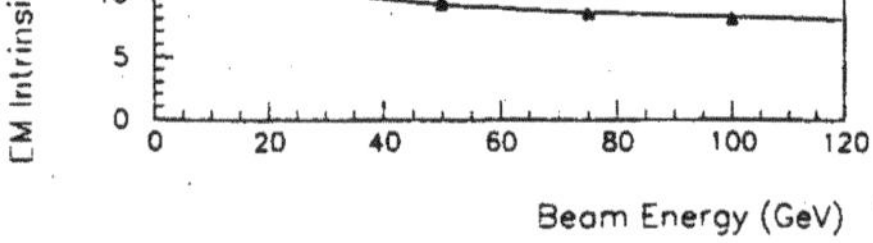

Fig. 8. Energy resolution for electrons as a function of beam energy, for various calorimeter tilt angles. Data are corrected for electronic noise and energy loss in the calorimeter front plates. The solid curves represent fits to the data points.

The electronic noise for 6 modules (10 x 10 cm^2 cross section, each), which contain 80% of the hadron shower, was 4.0 GeV. The intrinsic EM energy resolution of this calorimeter strongly depends on the tilt angle (fig.8).

Summary

The calorimeter proved to be safe and easy to operate, and worked smoothly during the entire beam test. The calorimeter response is linear with energy. The electron collection time for a 95% Ar + 5% CH_4 gas mixture is about 38 ns [5], and we anticipate a 25 ns signal width with proper signal shaping. The intrinsic hadron energy resolution compares well with calorimeters based on scintillator readout. We expect a resolution of about 3 - 4 % at energies above 500 GeV, which are typical in the LHC forward region. The measured electronic noise is negligible at these energies.

Acknowledgements

This work was supported by the U.S. Department of Energy, the National Science Foundation, and the Texas National Research Laboratory Commission. We thank the Fermilab Physics Department for providing support during the construction of the calorimeter. We are grateful to Drs. V. Radeka, S. Hansen and C. Riveta for very productive discussions on the calorimeter electronics design and for their practical help. We are thankful to I. Stauersboll whose excellent welding job provided for smooth construction and testing of the calorimeter. We also thank the CERN beam support personnel, especially R. Dillon and L. Gatignon, for their help during the test.

References

1. N.D. Giokaris et al., Nucl. Instr. and Meth. A291 (1990) 552.
2. D.M. Khazins et al., Nucl. Instr. and Meth. A300 (1991) 281.
3. L. Demortier et al., Nucl. Instr. and Meth. A324 (1993) 77.
4. D.M. Khazins, Nucl. Instr. and Meth. A330 (1993) 316.
5. N.D. Giokaris et al., Nucl. Instr. and Meth. A333 (1993) 364.
6. D.M. Khazins et al., Nucl. Instr. and Meth. A333 (1993) 372.
7. L. Demortier et al., to be submitted to Nucl. Instr. and Meth.
8. S. Cihangir et al., IEEE Trans. on Nuclear Science, vol. 36, No. 1, p. 347 (1989)
9. V.I. Baskakov et al. Nucl. Instr. and Meth. 159 (1979) 83.
10. S. Denisov et al., Nucl. Instr. and Meth. A335 (1993) 106.
11. S.L. Bagdasarov et al., FERMILAB-Pub-94/142, to be published in Nucl. Instr. and Meth.
12. H1 Calorimeter Group, Nucl. Instr. and Meth. A336 (1993) 499.

PIXEL ARRAY DETECTOR FOR FORWARD CALORIMETRY

S. BAGDASAROV, K. GOULIANOS, A. MAGHAKIAN, Q.F. WANG
The Rockefeller University
1230 York Avenue, New York, NY 10021, USA

Presented by Q.F. Wang

ABSTRACT

We present test-beam results obtained with a detector designed to measure the energy and lateral position of both electromagnetic and hadronic showers initiated by high energy particles. The detector is a lead/liquid-scintillator "shish-kebab" type calorimeter read out by wavelength-shifting fibers arranged in a pixel-type towerless geometry suitable for "calorimetric tracking" of particles in the forward region of collider experiments.

1. Introduction

We have designed, constructed and tested a prototype "Pixel Array Detector" (PAD) calorimeter capable of determining the energy and, without using a shower maximum detector (SMD), the position of both EM and HA shower-initiating particles incident at small angles relative to the detector axis. A unique *towerless* geometry, combined with low cost construction, render the PAD suitable for "calorimetric tracking" of particles in the very forward region of collider detectors in moderate luminosity environments like CDF at Fermilab.

The PAD is a "shish-kebab" type calorimeter. It consists of alternating layers of lead plates and liquid scintillator read out by wavelength-shifting (WLS) fibers. The fibers are inserted into an array of aligned holes drilled in the lead plates through the entire depth of the calorimeter and are read out individually or in groups of four by multi-channel photomultiplier tubes (MCPMT's). For particles incident at small angles relative to the fiber direction, the fiber pulse height centroid provides the position of the shower initiating particle. Position determination is obtained not only for electrons and photons but also for hadrons, which upon interaction release on average 1/3 of their energy in the form of π^0's. Thus, a several-interaction-length deep PAD calorimeter can be used to measure the energy and position of both electrons/photons and hadrons. Hadrons may be tagged by a set of WLS fibers which do not sample the first 24 radiation lengths of the detector.

A conceptual design of a PAD calorimeter and Monte Carlo predictions of its expected performance at high energies are described elsewhere[1]. In this paper we report results obtained with the prototype PAD calorimeter tested with 2, 3 and 5 GeV electron beams and an 8 GeV π^- beam at the Brookhaven National Laboratory (BNL). The following sections describe the prototype, the results obtained from the test beam, and our conclusions.

2. The PAD prototype

The PAD prototype is shown schematically in Fig. 1. It consists of 30 parallel lead plates, $15\,cm \times 15\,cm \times 4.8\,mm$ (3/16″) thick, spaced $6.4\,mm$ (1/4″) apart. The plates were laminated with $0.5\,mm$ thick aluminum sheets of 86% reflectivity. Holes of $1\,mm$ dia. drilled through the lead-aluminum plates with a CNC machine in a pattern shown in Fig. 2 were used to accept the $0.83\,mm$ diameter Kuraray Y-11(350)M multiclad WLS fibers. A similar plate made of delrin with holes in the same pattern was used in the front of the calorimeter. After being inserted through the holes, the fibers were glued in the delrin plate, the front surface of the plate was faced and polished, and a reflective aluminum sheet was pressed against the fiber ends to reflect the light back into the fibers. The completed plate/fiber assembly was immersed in mineral-oil based Bicron 517L liquid scintillator. The PAD prototype comprised about 28 radiation lengths or about one interaction length.

The calorimeter was housed in a square steel tube between two aluminum flanges with O-ring grooves for sealing. One of the flanges had a 2.5″ diameter hole through which the fibers were guided to a black delrin "cookie" that served as an optical connector to the MCPMT.

The fibers were grouped into "towers" as shown in Fig. 2. The four fibers in the corners of a square were read by one MCPMT channel, providing an effective tower size of $2\,cm \times 2\,cm$. The fibers in the centers of the squares were read individually. Since the PAD prototype is only one interaction length deep, the single fibers were brought all the way to the front of the calorimeter to test the idea of tagging hadrons using this technique. In the test beam only the shaded towers and encircled single fibers in Fig. 2 were instrumented.

The PAD prototype was read with a Hamamatsu H-5828 MCPMT through a CAMAC based CAEN model C205 32-channel charge integrating ADC controlled by a Macintosh Quadra using National Instruments "Labview" data acquisition software.

The design of the PAD prototype was based on a set of laboratory measurements using a collimated ^{106}Ru (3.54 MeV) β-source directed through a layer of liquid scintillator placed in a plastic cell between two parallel reflective Al sheets of the type used in the PAD. The measured response as a function of the distance of a WLS fiber from the β-beam, shown in Fig. 3, is represented well by the function $L(r) = f(r)\, e^{-r/\alpha}/r$, where f(r) takes into account the finite size of the beam. The fitted values for α are in the range 17–30 mm for scintillator thicknesses 5–8 mm; for the 6 mm scintillator thickness of Fig. 3, $\alpha = 1720\,mm$. From the bench measurements we estimate the *average* response of a 4-fiber-tower to a minimum ionizing particle (MIP) traversing the tower to be ≈ 0.25 photoelectrons (pe's).

The MCPMT was tested and calibrated with a green LED signal illuminating simultaneously all 64 photocathode pixels. We estimate that the cross-tower calibration uncertainty for the instrumented towers is $\pm 6.6\%$.

3. Test beam results

The PAD prototype was exposed to 2, 3 and 5 GeV electrons and to 8 GeV pions at the AGS test beam at BNL. The beam was directed into the central tower of the calorimeter and the energy of the particle was obtained by the "tower-sum" of the energies deposited in the "target tower" and in the adjacent 6+12 towers indicated by the two dashed circles in Fig. 2. The electrons were selected by Cerenkov counters, which however allowed a certain fraction of pions in the data. The apparatus was triggered by a scintillation counter, viewed by two PMT's, whose dimensions transverse to the beam were $\frac{1}{4}\times 2$ inches. A lead collimator confined the electrons to a $\frac{1}{4}'' \times \frac{1}{4}''$ "spot" in the target tower, but the pion "spot" was $\frac{1}{4}'' \times 2''$. The $2''$ dimension covers approximately 3 towers in the vertical direction in the center of the calorimeter as depicted in Fig. 2.

3.1. Electron shower profile, linearity, and energy resolution

For 5 GeV electrons, the target tower contains on average about 18% of the total observed energy, the first circle of towers contains 7% per tower, and the second circle 3% per tower. A LEGO plot of the tower pulse heights is shown in Fig. 4. The tagging fiber pulse heights are on average 2.5 times smaller.

Fig. 5 shows the ADC count distribution for 5 GeV electrons after subtraction of a low energy tail in the raw distribution due to pions. The ratio of the measured energy over the beam momentum, E/p, normalized to that at 5 GeV, was found to be 0.99 ± 0.02 (0.93 ± 0.02) at 3 GeV (2 GeV). The non-linearity at 2 GeV is attributed to lateral electron energy leakage.

A fit to the normalized energy resolution, σ/E, at the three electron beam energies yields $\sigma/E = 23.3\%/\sqrt{E} \oplus 0.4\%$, as compared to $\sigma/E = 14\%/\sqrt{E}$ expected from sampling fluctuations. The broadening of the resolution is attributed to MCPMT channel gain calibration, energy leakage, and photostatistics. The largest contribution comes from limited photostatistics. While Monte Carlo simulations show that at least one photoelectron (pe) per MIP per scintillator layer (1 pe/MIP-layer) is required to render the broadening from photostatistics small relative to that from sampling fluctuations, the PAD yields only $\approx$0.25 pe/MIP-layer. The photoelectron yield could be increased by using 1 mm dia. fibers instead of 0.83 mm, by reducing the effective tower dimensions, which would require employing more fibers, or by reading out the fibers with green-extended-photocathode MCPMT's or with Visible Light Photon Counters (VLPC's have about a factor of 4 to 5 larger photocathode efficiency than MCPMT's).

As a check of the uniformity of response accross a tower, data were collected and analyzed with the 2 GeV electron beam directed into the center of a tower or in between two towers. The response in the latter case was about 2% lower, probably due to the increased lateral energy leakage.

3.2. Position resolution

Position is measured from the "center of gravity" of the shower profile. Fig. 6

shows the measured one-dimensional position distribution obtained for 5 GeV electrons. Since the PAD was triggered with a 1/4" (6.4 mm) wide scintillation counter, the data were fitted with a Monte Carlo simulation in which events were generated with a flat distribution across the face of the counter and the position was smeared with a Gaussian function of width $\sigma(position)$. The best fit to the data, represented by the dashed curve in Fig. 8, was obtained with a Gaussian width $\sigma(position)$=3.2 mm, which is also obtained from the $\sigma(data)$= 3.73 mm of the Gaussian fit to the data shown in Fig. 8 by subtracting in quadrature the contribution of the trigger counter width $\pm 6.4/\sqrt{12}\,mm = \pm 1.8\,mm$. For 3 GeV electrons, we obtained $\sigma(position)$=3.5 mm. For the 8 GeV pions that interact in the calorimeter, the position is determined to an accuracy of $\sigma(position) = 8.2\,mm$, after subtracting the contribution of the trigger counter width (see Fig. 7).

4. Conclusions

A prototype Pixel Array Detector (PAD) calorimeter, designed to measure the energy and position of showers initiated by electrons/photons or hadrons incident at small angles relative to the axis of the calorimeter, was constructed and tested with electron beams of 2, 3 and 5 GeV, and with a pion beam of 8 GeV at BNL. The energy resolution obtained for electrons is $\sigma/E = 23.3\%/\sqrt{E} \oplus 0.4\%$, as compared to $14\%/\sqrt{E}$ expected from sampling fluctuations. The broader than expected resolution is mainly due to limited photostatistics and can be improved, if desirable, with certain modifications to the calorimeter or changes in the readout device. With an effective tower size of $2\,cm \times 2\,cm$, the measured position resolution is σ=3.2 mm(3.5 mm) for 5 GeV(3 GeV) electrons, and 8.2 mm for pions interacting in the calorimeter.

5. Acknowledgements

We thank Vadim Sherman for the skillful machining of the various parts of the calorimeter; Jack Zhao and Iddo Wernick for setting up the data acquisition system and performing the initial MCPMT calibrations and light yield measurements; Anwar Bhatti, Timothy Cox and Philip Melese for their invaluable contributions in the test beam run; the Hamamtsu Coporation for donating to us an experimental MCPMT unit; and the Texas National Research Laboratory Commission for providing initial funding for this project in connection with the Superconducting Super Collider.

6. References

1. S. Bagdasarov and K. Goulianos, *Calorimetric tracking of high energy particles*, Presented by K. Goulianos, Proceedings of the Forth International Conference on Calorimetry in High Energy Physics, La Biodola, Isola d'Elba, Italy, 19-25 September 1993, World Scientific (Aldo Menzione and Angelo Scribano, Eds.) pp.524-529.

Readout and Trigger

Session Chair: Robert Chase (Orsay)

PERFORMANCE OF MONOLITHIC CURRENT-SENSITIVE PREAMPLIFIERS WITH AN ACCORDION LAr CALORIMETER

RD3 Collaboration,
presented by

Daniel V. Camin
Dipartimento di Fisica dell'Universita' and INFN
Via Celoria 16, 20133 Milano, Italy.

ABSTRACT

Monolithic current-sensitive preamplifiers have been used for the first time to read out more than 600 cells of a prototype Accordion LAr calorimeter with integrated preshower at CERN. These preamplifier chips, which have been fabricated using a GaAs MESFET process, improved the noise by a factor close to 2 compared to the noise level obtained with hybrid circuits based either on SiJFETs and/or discrete MESFETs. The noise improved without increasing the power dissipation; this was accomplished by using, at the preamplifier input, MESFETs with large gate width that better match the detector capacitance. The data presented in this paper was recently obtained at the beam test; further data is now being analyzed.

1. Introduction

The program of the RD3 collaboration includes the development and test of different options for the signal readout of the Accordion LAr calorimeter. The scope is to propose for the ATLAS detector, a solution which better fulfills the specifications for a particular detector region. The high radiation levels encountered at the end-cap region, for instance, makes it reasonable to adopt the so-called 0T preamplifier[1] which uses no transistors in the cryogenic liquid. For the barrel calorimeter two solutions have been proposed: one consists of hybrid charge preamplifiers using Si JFETs either in the form of discrete devices or grouped in arrays[2]. The other proposed solution consists of current-sensitive preamplifiers made in a monolithic GaAs MESFET process. This proposal was based on the fact that at cryogenic temperatures GaAs field-effect transistors have very good performance in terms of noise, speed and power dissipation that would make them ideal for fast noble-liquid calorimetry. A discussion of the applications of charge or current preamplifiers for the signal readout follows .

Assuming that the search for a Z' extends up to 5 TeV, the maximum energy deposition in a single cell of the barrel LAr calorimeter of the ATLAS detector will be about 1.5 TeV. In the LAr gap this energy corresponds to a peak detector current of almost 4 mA. The high luminosity expected at LHC makes it necessary to use fast bipolar shaping to limit the contribution of the pile-up noise. The weighting function will be bipolar (CR^2-RC^2) with a peaking time (5% -100%) of about 20 ns (δ response).

The signal of a detector cell will be read out by a low-noise preamplifier connected to the cell's end to reduce the transfer time of the collected charge[3]. Two feedback configurations may in principle be used: a charge-sensitive or a current-sensitive one.

A charge-sensitive configuration produces at the preamplifier output a voltage excursion proportional to the ratio of the charge deposited in the gap after the drift time and the feedback capacitance. For reasons of circuit stability and noise, the feedback capacitance can not be very large, therefore large voltage excursions are expected. A double-differentiator double-integrator shaper follows the charge preamplifier.

A current-sensitive configuration uses instead a resistive feedback making the maximum signal output proportional to the peak value of the triangular detector current. By fixing a sufficiently low-valued resistor the output voltage excursion can be kept limited. For a large detector capacitance, a low-valued resistor at the feedback can be tolerated as the parallel noise it generates will not give a significant contribution to the total noise. For stability reasons a compensating capacitor is required in parallel with the feedback resistor. The time constant of the feedback network can be made small, close to the one used at the shaper: a first differentiator is used at the preamplifier level. A single-differentiator double-integrator shaper follows the preamplifier .

A current-sensitive configuration is, therefore, the most convenient way to handle large detector signals [4]. But, as will be seen in section 2, the preamplifier must have a large gain-bandwidth product (GBW).

The present work reports on the results obtained with monolithic current-sensitive preamplifiers made in a GaAs process which has low noise and very large GBW at cryogenic temperatures[5].

In section 2, we recall the constrains determined by a fast current-sensitive feedback configuration. In section 3 the characteristics of the chip fabricated are summarized. In section 4 we discuss on the evaluation of a production run. In section 5 we report the results recently obtained at a beam test and finally in section 6 results on radiation damage are presented.

2. The monolithic current-sensitive preamplifer.

The circuit was realized using a dominant-pole amplifier fed back with a parallel combination of a resistor and a capacitor, as shown in Fig 1. In that figure C_D and C_A represent the detector and FET input capacitances respectively.

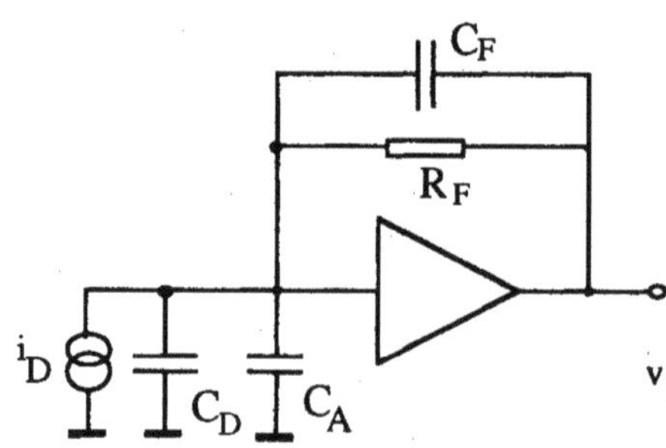

Fig 1: The dominant pole amplifier fed back with a parallel combination of a resistor and a capacitor. The time constant R_FC_F relative to the shaper time constant defines either a charge or a current-sensitive configuration.

The transfer function v_o/i_D of that circuit (providing that $1/\tau_i$ is higher than $4/\tau_F$) is

$$\frac{v_o(s)}{i_D(s)} = \frac{R_F}{(1+s\tau_F)(1+s\tau_i)} \tag{1}$$

where $\tau_F = R_F C_F$ is the feedback time constant, $\tau_i = C_T/2\pi f_T C_F$ is the time constant determined by the GBW f_T of the dominant-pole amplifier, and $C_T = C_D + C_A + C_F$ is the total capacitance seen at the preamplifier input.

The transfer function can be rewritten as:

$$\frac{v_o(s)}{i_D(s)} = \frac{1}{sC_F}\,\frac{s\tau_F}{(1+s\tau_F)}\,\frac{1}{(1+s\tau_i)} \tag{2}$$

which corresponds to the transfer function of a charge-sensitive configuration followed by a differentiator with time constant τ_F. This differentiator, located at the preamplifier, is the first CR network of the overall CR^2-RC^2 bipolar shaping.

For the present case $\tau_F = 18$ ns : a value very close to 12 ns used at the shaper was adopted[6]. To reduce the noise contribution of the shaper following the preamplifier, a large noise gain is required. If it is fixed to a reasonable value $C_T/C_F \sim 65$, then the GBW must be larger that 3 GHz. This value is easily obtained at LAr temperature with GaAs MESFET technology.

3. Characteristics of the chip fabricated

The monolithic preamplifier uses an input MESFET[7] with gate dimensions (L x W) = 3 x 24000 μm^2. Two channels have been accomodated in a die of 2.5 x 1.5 mm^2 mounted in a 28-pin ceramic package[6]. The circuit has internal input protection diodes and the frequency compensation can be selected by setting DC control voltages. The feedback network is connected externally to the chip, making it possible to use different combinations of resistors and capacitors to better adapt the different rapidity regions and longitudinal compartments of the final detector. In addition, the external feedback network allows for non-linear response if it would be required. The following measured parameters, valid in the temperature range 77 K to 150 K, define the main characteristics of the chips:

- Low frequency voltage gain: 900
- Main pole at: 6 MHz
- White series noise: 0.23 nV/√Hz
- Input capacitance: 140 pF.
- Output voltage excursion on 100 Ω load: 4 V @ Vcc= 9 V; 2 V @ Vcc= 7 V
- Power dissipation: 54 mW @ Vcc = 7 V; Vco = 5 V; Vee = -3.5 V.

The series noise can be reduced to 0.19 nV/√Hz but at the expense of higher power dissipation, 150 mW. This can be done by simply increasing the value of the separate voltage supply Vco used to bias the input transistor. Still, we consider the adopted values indicated above a good compromise of noise/power dissipation .

To evaluate the noise performance at different detector capacitances and shaping times, noise measurements in a charge-sensitive configuration have been performed at 77 K. The results are indicated in Fig. 2. From the measured data a white series noise density of 0.23 nV/√Hz was extracted. This noise level, obtained for the best chips, was verified by direct measurements of the preamplifier series noise spectral density.

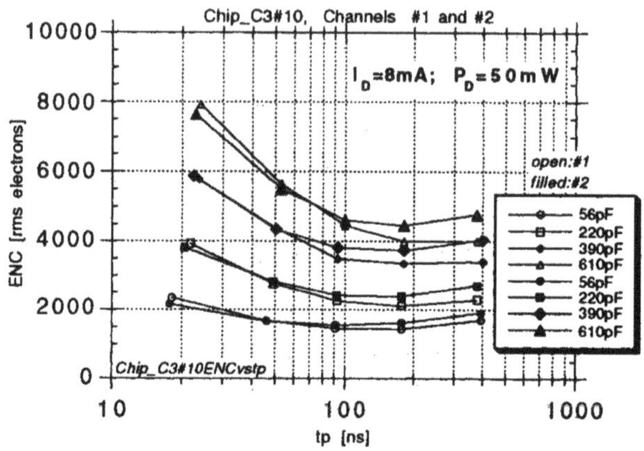

Fig 2: ENC as a function of peaking time for different detector capacitances of both channels of the monolithic preamplifier chip fed back in a charge-sensitive configuration.

To evaluate the effect of the temperature on the noise, measurements of ENC at different peaking times for a fixed detector capacitance have been performed (Fig 3). In that figure it can be observed that for 20 ns < tp(δ) < 200 ns the noise is virtually constant from 87 K to 160 K.

Regarding dynamic range, we point out that in the current-sensitive configuration with R_F = 1500 Ω, this preamplifier in its present version (still not optimized for large dynamic range), is able to read detector signals up to 3 mA with a +/- 1 % integral nonlinearity (Fig 4).

A detailed description of the chip is given in ref.[6].

4. Evaluation of a production run.

A number of chips exceeding the present needs of RD3 beam tests have been manufactured to have statistics of a production foundry run. Four wafers have been processed although not exclusively for the preamplifier chips, but sharing the area with other projects. A total of 1536 devices (3072 channels) have been manufactured.

Two screen tests have been performed on 892 devices randomly distributed in three wafers. A functional test was done at room temperature with a yield higher than 77 %. Later-on, dynamic and noise performance were verified at cold, for a small sample (160 units, i.e. 320 channels) of all chips manufactured.

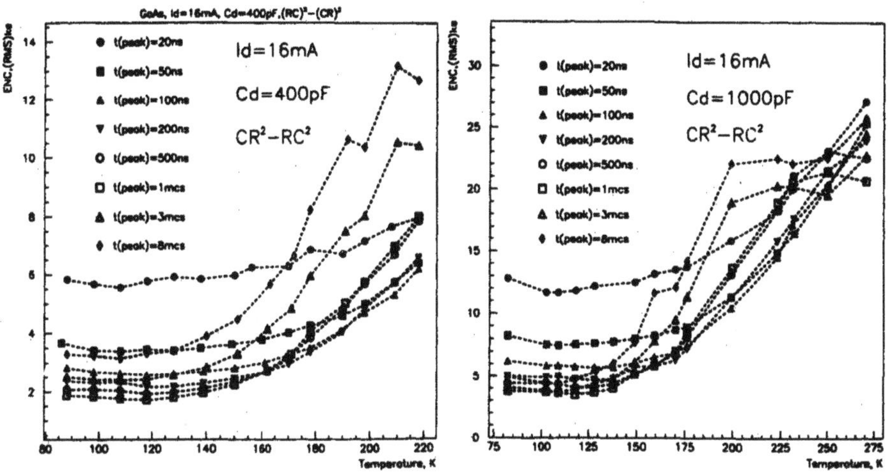

Fig 3: ENC as function of temperature for a) C_D = 400 pF and b) C_D = 1000 pF, at different peaking times (bipolar shaping). These measurements have been done at I_D = 16 mA.

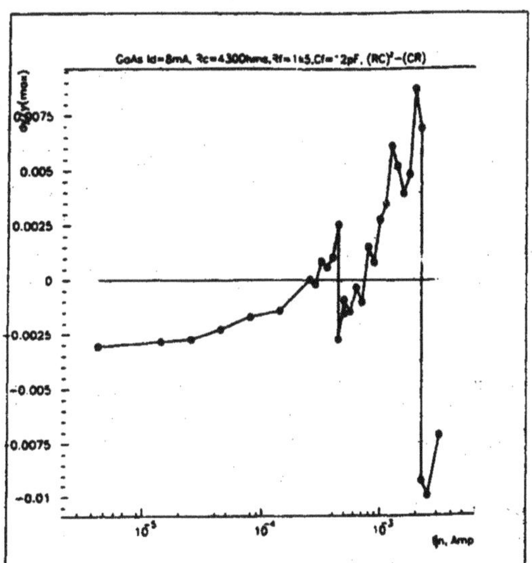

Fig 4: Residuals of output voltage linear fit for input current up to 3 mA.

In Fig 5 we show the ENC distribution at 77 K for tp(δ) = 20 ns (bipolar shaping) after cutting 47 more noisy and 14 of more slower chips out of 160 units that passed the room temperature test. The plot therefore shows the noise distribution of 62 % of the chips that passed the wafer test or, equivalently, 48 % of the chips fabricated randomly located in the three wafers. The average value is somewhat higher than the value expected from a series noise of 0.23 nV/√Hz. A contribution of noise coming from the second stage was identified as the cause of this excess noise. In Fig. 6, the peaking time distribution of the sample is shown.

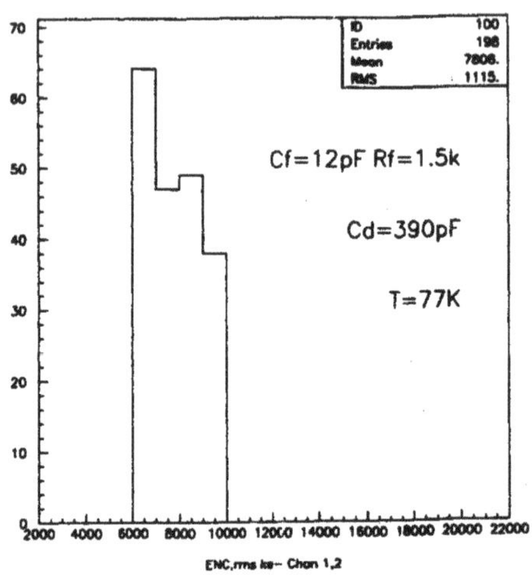

Fig. 5: Noise distribution of a 99 chips sample (198 channels). An CR-RC² filter was used.

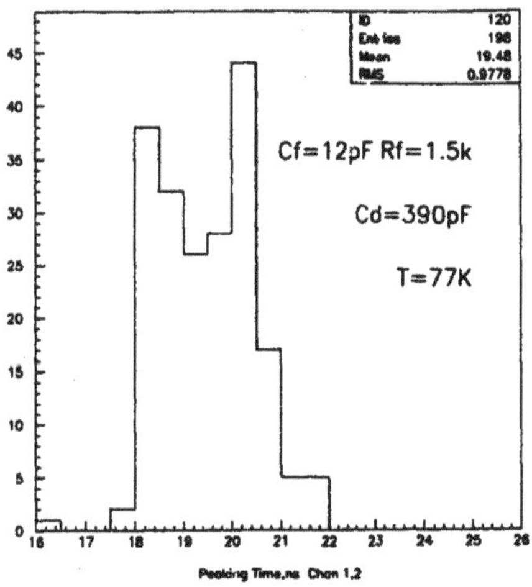

Fig 6: Peaking time after CR-RC² shaping of the same sample as Fig 5.

Most of the chips with very high noise were also slow. We plan to determine the correlation between parameters of the DC test and high noise or slow speed. Future screen tests will be more selective to filter those bad channels at the wafer test level.

In addition, it is expected to obtain lower noise contribution from the second stage in a new chip now being designed.

5. Results obtained at a beam test.

More than 600 channels of the prototype Accordion calorimeter and integrated preshower have been readout with GaAs monolithic current-sensitive preamplifiers. Pedestal fluctuations have been recorded in the region read out with these chips. The results, for the different detector compartments, are shown in Fig. 7. In the same figure, pedestal fluctuations of hybrid circuits are also indicated for comparison.

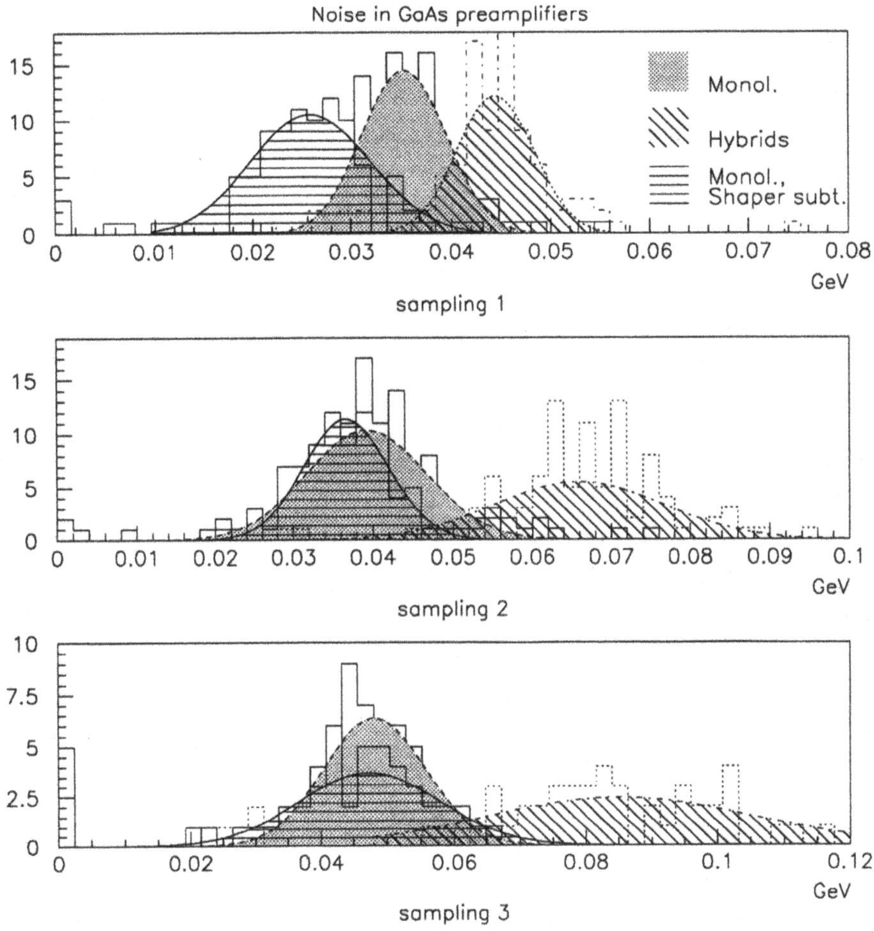

Fig 7 Pedestal fluctuations in the different calorimeter samplings, read out with monolithic current-sensitive GaAs preamplifiers and with charge-sensitive hybrid circuits. *Monol* and *Monol.Shaper*: see note 1.

It can be noted that in sampling 1, where the detector capacitance is smaller than in samplings 2 and 3 , the shaper/ADC readout system contributed to the total noise by the same amount (26 MeV) than the preamplifier itself.

The monolithic preamplifiers have also been used to read out the integrated (UV) preshower. The results have been reported in a separate paper[8].

Note 1: *Monol.* means noise measured by the system reading the monolithic preamplifiers. *Monol.Shaper substraction* distribution is obtained after quadratically substracting the pedestal fluctuation read out for every channel with the preamplifiers off. The approximate capacitance values of sampling 1, 2 and 3, are 400 pF, 600 pF and 900 pF respectively.

6. Radiation damage.

We have evaluated the radiation damage at cold of monolithic preamplifiers fabricated with the same process used to make the present chips [9,10]. Those chips use a small input transistor to match a preshower detector[11]. The results have shown a 15 % noise increase after a neutron fluence of 4.5 x 10^{14} n/cm^2 (1 MeV equivalent) and 10 % noise increase after 0.5 MRad total dose. That indicated fluence and total dose values correspond, in the barrel region, to 10 years of LHC operation at a luminosity of 10^{34} [12].

Just recently, we have performed an irradiation test at cold of the actual monolithic current-sensitive preamplifiers reported in this paper. Details of the irradiation tests follows.

Three chips containing four preamplifiers and a single FET have been installed in a cryostat in which the temperature could be varied from 77 K to 300 K [10]. Two channels, fed-back in a charge-sensitive configuration, had at their input a capacitance of 390 pF and 1000 pF respectively, simulating the detector. The circuits have been biased at the nominal conditions.

The cryostat was put in front of a ^{60}Co gamma source and has been irradiated at a rate of 100 KRad/ day. The Compton edge of the ^{60}Co source is at about 1 MeV, well above the threshold of 0.4 MeV below which electrons do not create displacement damage and vacancies in the lattice in GaAs[13,14]. After 500 KRad a Lorenztian component is observed in the series noise spectrum.

Measurements of ENC after shaping with an CR^2-RC^2 filter, have been taken after each step of 100 KRad. The noise increased by about 10 %, Fig 8. Negligible change was observed in the peaking time after the shaper and in the DC biasing voltages or currents.

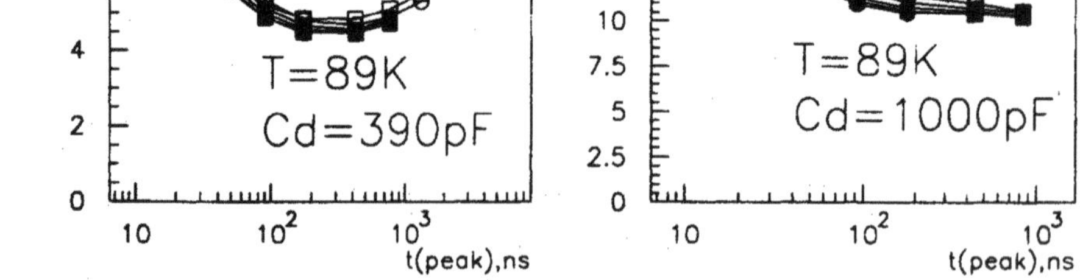

Fig 8: ENC as a function of peaking time beforeirradiation and after 5 steps of 100 KRad at 89 K for two different detector capacitances, C_D = 390 pF and C_D = 1000 pF.

7. Summary and conclusions.

Monolithic current-sensitive preamplifiers fabricated in a GaAs ion-implanted process, have been used to read out the signal of a prototype Accordion calorimeter and integrated preshower.

The evaluation of a production run have shown that 48 % of the chips fabricated have low noise and fast speed at LN temperature. Noise and speed change by a negligible amount in the temperature interval 77 K-160 K.

A noise reduction of about 50 % on the average, compared to hybrid preamplifiers also used with the detector, has been accomplished. The noise improvement was obtained with a power dissipation in the monolithic preamplifier equal or even lower than that of the hybrid circuits.

Radiation damage tests have shown that the preamplifier chips, designed for the barrel LAr calorimeter, are radiation hard to the levels expected after 10 years of LHC operation.

8. References

1. R.L.Chase, C. de la Taille, S.Rescia and N.Seguin *Nucl. Instr. and Meth.* **A330** (1993)pp. 228-242.
2. V.Radeka, S.Rescia and D.Sephani, "Performance and Reliability of BNL Cryogenic Preamplifiers for Noble Liquid Calorimetry", this Conference.
3. V.Radeka and S.Rescia, *Nucl. Instr. and Meth.* **A265** (1988)p. 228
4. C. de la Taille, RD3 Internal note 45 (1993).
5. D.V.Camin, G.Pessina and E.Previtali, *IEEE Trans. on Nucl. Sci.*,**40** (1993)pp.759-763.
6. D.V.Camin, N.Fedyakin and G.Pessina, "Monolithic Current-Sensitive Preamplifier for the Accordion LAr Calorimeter", Proc of the 6th Pisa Meeting on Advanced Detectors, 22-28 May, 1994. to be pub. in *Nucl. Instr. and Meth.*
7. D.V.Camin, G.Pessina and E.Previtali, *IEEE Trans. on Nucl. Sci.*, **41**(1994)pp.1260-1266.
8. RD3 Coll. presented by I.Winzerter-Seez, "Performance of an Accordion Calorimeter Equipped with Integrated UV Strips", this Conference.
9. RD3 Coll, proc of the IV Intnl. Conf. on Calorimetry in High Energy Physics, Ed; A.Menzione and A.Scribano, World Scientific (1994) pp.618-626.
10. M.Citterio, V.Radeka and S.Rescia, "Radiation Effects on Front-End Electronics for Noble Liquid Calorimetry", this Conference.
11. D.V.Camin, N.Fedyakin and G.Pessina "Monolithic GaAs Readout for the LAr Preshower Detector" proc of the IV Intnl. Conf. on Calorimetry in High Energy Physics, Ed; A.Menzione and A.Scribano, World Scientific (1994)
12. G.Battistoni, A.Ferrari and P. Sala, "Background Calculations for the ATLAS Detector and Hall" ATLAS (1994).
13. G.P.Summers , E.A.Burke,P.Shapiro et al. "Damage correlations in semiconductors exposed to gamma, electron and proton radiations". IEEE Transactions on Nucl.Sci.Vol.40, No6, (1993) pp.1372-1379.
14. V.A.J. van Lint, T.M.Flanagan,R.E.Leadon ,J.A.Naber and V.C.Rogers. "Mechanisms of Radiation Effects in Electronic Materials.Volume1." NY, John Wiley & Sons, 1980. p.343.

A FAST MONOLITHIC SHAPER FOR THE ATLAS E.M. CALORIMETER

R.L. Chase, C. de La Taille, J.P. Richer, N. Seguin-Moreau
Laboratoire de l'accélérateur Linéaire, IN2P3-CNRS et Université Paris-Sud, 91405 Orsay Cedex, France

Presented by C. de La Taille

ABSTRACT

A monolithic CR RC^2 shaper has been realized using BiCMOS technology. It provides a high gain and a low gain output to accomodate the 16 bit dynamic range of the ATLAS electromagnetic calorimeter. The peaking time for a step input can be tuned with a 4 bit logic between 25 and 40 ns. The noise referred to the input is $e_n = 1.5\ \mathrm{nV}/\sqrt{\mathrm{Hz}}$ with 100mW power dissipation.

1. Introduction

The shaper, located between the preamplifier and the readout chain, optimizes the signal to noise ratio which is the quadratic sum of the electronic noise (decreasing with slower shaping) and pileup noise (increasing with slower shaping). For the ATLAS electromagnetic calorimeter, the high levels of pileup at 10^{34} luminosity require a peaking time (5-100%) for the calorimeter signal around 40 ns (corresponding to $t_p(\delta) \approx 20$ ns)[1]. At lower luminosity, an equivalent slower shaping is achieved by digital filtering with multiple samples[2]. A good peaking time uniformity would be an advantage to reduce the cost and complexity of the digital readout.

The shaper also provides additional amplification to bring the preamp noise into the mV range rendering the noise of the readout chain negligible. However, the very large dynamic range (16 bits) anticipated in ATLAS makes this last feature unachievable. One solution is to compress the dynamic range into 10-11 bits, which matches the detector intrinsic resolution[3]. Another option is to cut the 16 bits in two times 13 bits, with two different outputs in a ratio between 8 and 16. These 13 bits match well the dynamic range of switched capacitor arrays and ADCs that could be used for the readout[4]. Due to the large number of channels (200,000) and their favoured location in a calorimeter crack, the realization of an ASIC was highly desirable.

We will now describe the requirements, design and experimental results obtained with a 4 channel monolithic circuit using AMS 1.2μ BiCMOS technology.

2. Shaper characteristics

2.1. Overview

Two different preamplifiers are foreseen for the ATLAS EM calorimeter[1,4] :

- for the barrel calorimeter, current preamplifiers using Si or GaAs FETs would be located close to the electrodes, operating at LAr temperature[5,6]. They deliver positive output pulses, with a triangular shape similar to the detector current, but of opposite polarity. These signals are sent outside the cryostat through 50 Ω cables, terminated at both ends.

- for the End-cap calorimeter, due to the high radiation environment, cable-coupled current preamplifiers[7] (ØT) would be located outside the cryostat, at the end of a 50 Ω cable. They terminate the cable in its characteristic impedance, but are non inverting, giving a negative pulse to the shaper located nearby.

Thus, the requirements for the shaper are very different : in the first case, it must accomodate a positive signal and show a constant 50 Ω input impedance over the whole dynamic range, whereas in the second case, it deals with a negative pulse, with no restriction on the input impedance. However, in both cases, the signal has the same shape, leading to a similar architecture. The present availability of only NPN transistors in multiproject fabrication led us to start with the negative signal of the ØT.

The architecture chosen consists of one differentiation (CR) and two integrations RC, usually referred to as CR RC^2. The differentiation minimizes the pileup contribution due to the long decay (~400 ns) of the signal from the calorimeter[1,8]. It also delivers a bipolar signal, which allows AC coupling in the following stages without baseline shifts, as the signal area is zero. The integrations filter out the high frequencies to reduce the electronic noise. It is not necessary here to have many integrations, which usually slightly improve the signal to noise ratio, as this is achieved digitally further down in the readout[2]. This allows minimizing the number of stages and decreases the power consumption.

2.2. Temporal waveform

We give here a few theoretical expressions of the waveform at the shaper output. The transfer function of a CR RC^2 shaper is

$$H_{sh}(s) = \frac{\tau s}{(1+\tau s)^3}$$

in which τ is the shaper time constant (RC product).

By taking the inverse Laplace transform, we can calculate the response to any input signal. For example, the response to a step is :

$$h_{step}(t) = \frac{x^2}{2} e^{-x}$$

in which x=t/τ.

The signal reaches its peak at $t_{max}(step) = 2\,\tau$ and has an amplitude $h_{max}(step) = 0.270$. For an impulse, the peak is $h_{max}(\delta) = 0.2306/\tau$ at $t_{max}(\delta) = 0.59\,\tau$.

At the fast shaping considered here, it is not possible to neglect the rise-time of the preamp τ_{pa}, which modifies the overall response. In the case of the barrel with inverting current preamps, $\tau_{pa} = R_f C_f$, in which R_f and C_f are the feedback resistor and capacitor. In the ØT case, $\tau_{pa} = R_0 C_d$, in which R_0 is the cable characteristic impedance and C_d the detector capacitance. Detailed analytic expressions of signal and noise, including the preamp, can be found in ref.[1]. A quantity often used to calculate the noise with little sensitivity to the preamp rise-time is the peaking time, between 5% and the peak, of the overall chain[1]. For the shaper alone, it gives : $t_p(\delta) = 0.574\,\tau$ and $t_p(step) = 1.82\,\tau$.

2.3. Shaper noise contribution

The shaper noise e_{sh}^2 can be referred to the preamp input by dividing by the preamp transfer function $H_{pa}(s) = \pm R_f/(1+\tau_{pa}s)$. This gives two terms : one which looks like series noise and one like parallel noise.

$$i_n^2 = \frac{e_{sh}^2}{R_f^2} \qquad e_n^2 = \frac{\tau_{pa}^2}{R_f^2 C_d^2} e_{sh}^2$$

When $\tau_{pa} \ll \tau$, the preamp is close to an ideal current preamp and the shaper noise behaves like a parallel noise. When $\tau_{pa} \gg \tau$, the preamp is a charge preamp and the shaper noise looks like series noise, referred to the input with the classical noise gain C_t/C_f. The plots are shown in ref.[1].

3. Shaper design

3.1. Overview

The requirements for the shaper are the following :

i) Input impedance : it should be precisely 50 Ω in order to terminate the cable from the preamplifiers. Any change over the whole dynamic range will result in non-linearity as part of the signal gets reflected. For the ØT, the input impedance is unimportant, but as we aim at a single design, we also impose a constant input impedance (although possibly higher).

ii) Shaping time : as shown in ref.[1], the optimum peaking time at 10^{34} luminosity is around $t_p(\delta)$=20 ns, which corresponds to τ=18 ns in the shaper, assuming τ_{pa}=15 ns. A uniformity better than 1 ns would eliminate the need for channel to channel timing correction.

iii) Noise : the very low noise achieved by the preamps[6] ($\sim 0.3\ \mathrm{nV}/\sqrt{\mathrm{Hz}}$) combined with their limited noise gain due to the large dynamic range lead to an upper limit of e_{sh}= $2\ \mathrm{nV}/\sqrt{\mathrm{Hz}}$ for negligible contribution.

iv) Power dissipation : the foreseen location of the shapers in the crack between barrel and end-cap calorimeters requires a high level of compactness, with typically 8 shapers per chip to avoid expensive ceramic packages, and this requires $P_d < 100$ mW/channel.

v) Linearity : the choice of having two linear outputs with high and low gain instead of dynamic compression[3] argues for very good linearity (better than 1%) in order to take the full benefit of linear systems.

vi) Output : typically a 4 V swing in a 1 kΩ load impedance, is compatible with the following stage located nearby.

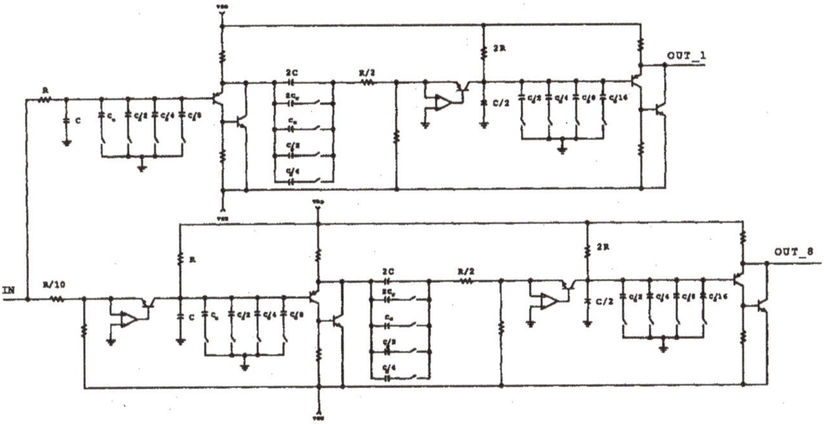

Fig. 1. Block diagramm of the shaper

The architecture retained consists of two sections in parallel : high gain and low gain, with similar design, except for the input stage which also provides the cable termination. In order to accomodate to the large dispersion in resistor values (±20%) and capacitor values (±10%) of the process, an adjustment has been included with extra capacitors and switches in parallel with those that determine the shaping time. 4 bits, common to the whole chip, allow a tuning of ±40% around the central value, with a resolution of 5% corresponding to 1 ns. The global schematic is shown in Fig. 1.

3.2. High gain input stage

The high gain input stage (cf. Fig. 2) provides a first amplification in order to make the noise from following stages negligible, a first integration (RC) and a constant 50 Ω input impedance.

A common base configuration (Q3) has been chosen to make an active termination

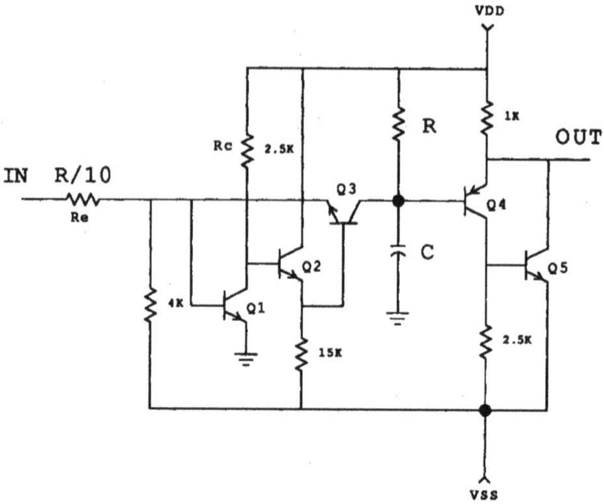

Fig. 2. Schematic diagramm of the high gain input stage

as it minimizes noise and exhibits good overload behaviour. In order to improve the linearity, Q3 has been inserted in the feedback loop of a rudimentary amplifier (Q1-Q2), which decreases Q3 input impedance ($\sim 1/g_{m3}$) by the open loop gain $g_{m1}R_c$. Thus, the input impedance is :

$$R_{in} = R_e + \frac{1}{g_{m1}R_c g_{m3}}$$

which is very close to the value of R_e, set to 50 Ω.

The transfer function of this stage is simply given by :

$$\frac{V_{out}}{V_{in}} = \frac{R}{R_e}\frac{1}{1 + sRC}$$

in which the resistors ratio determine the gain with a good precision. The pole $RC = \tau$ yields the first integration. The signal is buffered to the second stage with a White follower (Q4-Q5) described in the next paragraph.

The noise is dominated by the series noise due to R_e and Q1 :

$$e_{sh}^2 = 4kT(R_e + R_{bb'} + \frac{1}{2g_{m1}})$$

in which $R_{bb'}$ is the base spreading resistance of Q1.

The noise is minimized by using a large transistor for Q1 to reduce $R_{bb'}$ and operating it at a relatively large collector current, around 3 mA.

3.3. Low gain input stage

The low gain input stage (cf. Fig. 3) only provides the first integration (RC). The noise performance here is non critical and no amplification is needed, so that a simple RC network followed by a buffer is adequate.

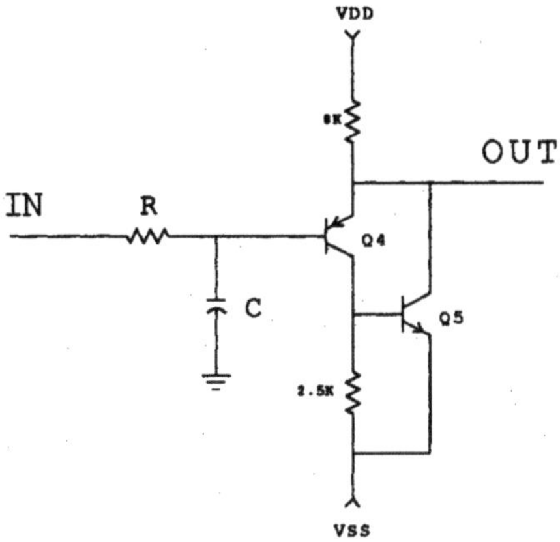

Fig. 3. Schematic diagramm of the low gain input stage

The buffer is built around a White follower (Q4-Q5) as a low output impedance is needed here to insure good linearity. This overcomes the weaknesses of lateral PNP transistors such as Q4. The transistor pair forms a closed loop configuration, in which local feedback is used to decrease the output impedance to a fraction of an Ohm. However, the stability of such a configuration needs careful attention, especially when driven from a low impedance or with capacitive loading.

3.4. Second stage

The second stage is similar in both sections : it provides the differentiation, the second integration and some gain. The same common base configuration is used (cf. Fig. 2), although operated at lower current to reduce the power dissipation. A gain of 4 is taken here in order to recover the signal attenuation (0.27) due to the differentiation, whereas the second integration is given by the pole on the collector. Again, a White follower is used as the output stage.

4. Experimental results

The first prototype described hereafter was designed slighltly differently from the requirements listed above, mainly for historical reasons. It is a little faster, the time constant being set to τ = 13 ns, with R = 1 kΩ and C = 12 pF + 1 pF from the transistors. The high gain was set to 10 using ten resistors R in parallel for R_e , which led to an input impedance of 100 Ω. There were 4 bi-channels per chip.

i) DC parameters : single test transistors have been included on the chip, a NPN an a lateral PNP. Their measured characteristics have been found in good agreement with the Spice parameters used for the simulations. The current gain for the NPN was β_{NPN} = 80 at I_c = 1 mA, whereas for the lateral PNP, β_{PNP} was only 10 at I_c = 0.2 mA (typical operating points). Concerning passive components, the resistor R and capacitor C determining the shaper time constant were implanted separately. R was measured as 950 Ω and C gave 12.5 pF.

ii) Dynamic measurements : the step response of the shaper (Fig.4) gives a temporal waveform close to that predicted by simulation. For the unity gain section, simulation indicated a gain of 0.91 and a peaking time (5-100%) for a step input of 24 ns (all bits OFF). It differs from the ideal expression of par. 2.2 because of second order poles. Measurements gave a gain of 0.84 and a peaking time of 26 ns. The slight discrepency in peaking time is probably due to some extra capacitance and explains the smaller gain. The high gain section gave a gain of 7.2 with the same peaking time.

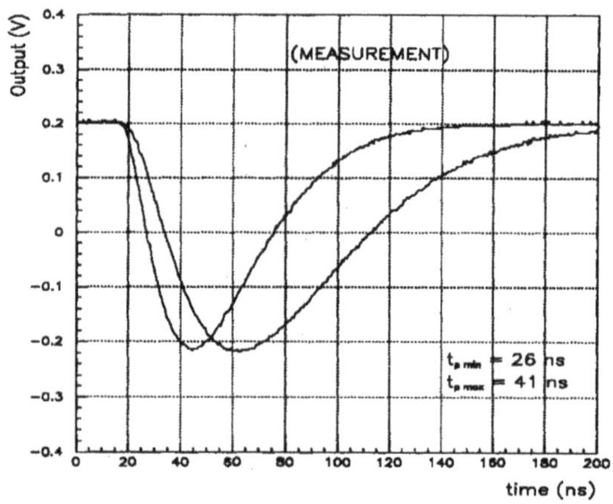

Fig. 4. Measurement of the shaper time response for a step input, with minimum and maximum adjustment

iii) Peaking time adjustment : the peaking time (5-100%) has been measured on 8 chips with 4 bi-channels[1]. The *rms* of the dispersion amounted to 0.95 ns, which confirms our assumption of good uniformity over a same chip, and even over a wafer. The peaking time adjustment with the 4 bits is shown in Fig. 4. With all bits OFF, the peaking time was measured $t_p(step)$ = 26 ns, whereas all bits ON gave 42 ns. Thus, one bit gives a variation of 1 ns, as required.

iv) Linearity : the linearity has been measured on both high gain and low gain sections, with the peaking time set at a median value ($t_p(step) = 32$ ns) to take into account a possible effect of the switches. The residuals to a best line fit indicate an integral non-linearity [data - fit)/max] better than ±0.3% for both gains[1]. The "relative" non-linearity [(data - fit)/data], which is actually much more demanding, is also satisfactory (in the percent range) as shown in Fig. 5. This confirms the stability of the input impedance over the dynamic range.

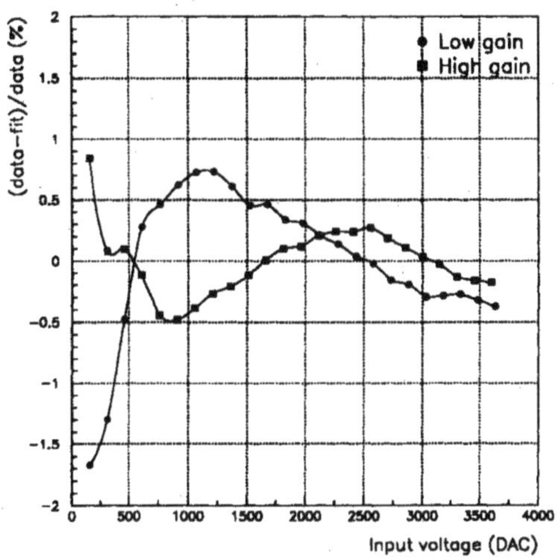

Fig. 5. "Relative" non linearity of the shaper

v) Crosstalk : unfortunately, it has not been possible to measure the crosstalk between adjacent channels, because of a high frequency oscillation which occurs when several channels are powered simultaneously.

vi) Noise : the noise spectral density has been measured[1] on the high gain output, using Fourier transform on a digital oscilloscope. Referring it to the input gives $e_{sh} = 1.75$ nV/$\sqrt{\text{Hz}}$, from which R_e contributes $\sqrt{4kTR_e} = 1.3$ nV/$\sqrt{\text{Hz}}$. Thus, the shaper alone exhibits a noise density of 1.2 nV/$\sqrt{\text{Hz}}$, corresponding to 85 Ω noise resistance; it would give 1.5 nV/$\sqrt{\text{Hz}}$ with 50 Ω input impedance.

5. Conclusion

Experimental measurements on the first prototype of a monolithic NPN shaper for ATLAS have given good results concerning the speed, the peaking time adjustment and the noise. A high frequency oscillation problem will almost certainly be resolved in the next iteration by the addition of a few "stopper" resistors in the White followers.

6. References

1. R.L. Chase et al. "A fast monolithic shaper for the ATLAS EM calorimeter" ATLAS internal note LARG-No-010
2. W.E. Cleland, E.G. Stern "Signal processing considerations for liquid ionization calorimeters in high rate environment". NIM A338 (1994) 467-497
3. FERMI collaboration : Status Report CERN/DRDC/93-21
4. ATLAS Technical Proposal CERN/LHCC/94
5. V. Radeka et al. "Performance and reliability of Si cryogenic preamplifiers for noble liquid calorimetry", this Conference
6. D.V. Camin et al. "Performance of monolithic current-sensitive preamplifiers with an accordion LAr calorimeter", this Conference.
7. R. L. Chase et al. "Experimental results on cable-coupled preamplifiers (ØT)" NIM A343 (1994) 598-605.
8. V. Radeka, S. Rescia "Speed and noise limits in liquid ionization calorimetry" NIM A325 (1988) 228-242

Liquid Ionization Calorimetry with Time-Sampled Signals*

O. Benary[9], W. Cleland[7], H. Cunitz[5], I. Ferguson[2] A. Gordeev[6], H. Gordon[3], E. Kistenev[3], P. Kroon[3], M. Leltchouk[5], D. Lissauer[3], H. Ma[3], D. Makowiecki[3], A. Maslennikov[4], S. McCorkle[3], D. Onoprienko[6], A. Onuchin[4], Y. Oren[9], V. Panin[4], J.A. Parsons[5], V. Radeka[3], D. Rahm[3], L. Rogers[3], S. Rescia[3], J. Rutherfoord[2], M. Seman[5], W. Sippach[5], M. Smith[3], J. Sondericker III[3], U. Sonnadara[7], R. Steiner[1], D. Stephani[3], E. Stern[5,7], I. Stumer[3], H. Takai[3], H. Themann[8], Y. Tikhonov[4], W.J. Willis[3,5]

1) Adelphi University, 2) University of Arizona, 3) BNL, 4) Budker Institute for Nuclear Physics, 5) Columbia University, 6) ORNL, 7) University of Pittsburgh, 8) SUNY at Stony Brook, 9) Tel-Aviv University.

Abstract

We have carried out a test using time-sampled signals from a liquid krypton electromagnetic calorimeter with lead absorber plates in the accordion geometry. The data were collected by sampling the shaped waveforms at 18 ns intervals. The samples are used to reconstruct the signal amplitude and its time origin. We study the amplitude resolution for a 5×5 tower analog sum and the timing resolution for signals from individual calorimeter channels. The results of the amplitude resolution measurements using time-sampled data are essentially identical to those found in the same calorimeter using the conventional method of sampling the signal at its peak, in which a value of $6.72\%/\sqrt{E}$ for the amplitude resolution was found. We find that the precision in the reconstructed time origin of the signal is excellent, and the expected $1/E$ dependence of the time resolution has been observed. For a single section with energy deposit E, we find that the resolution can be written in the form $\sqrt{(c/E)^2 + \sigma_{\rm cal}^2}$, with a value of c of 0.38 GeV·ns for the front section and 0.70 GeV·ns for the back section. The energy-independent resolution term for our system is found to be $\sigma_{\rm cal} \approx 150$ ps. When the timing measurements from the front and back sections are combined in a weighted mean, the time resolution of calorimeter is found to be approximately 0.16 ns at 15 GeV/c.

*This research was supported in part by the National Science Foundation, the U.S. Department of Energy under contract DE-AC02-76CH 00016, and TNRLC under grants RGFY92-109 and RGFY93-208.

1 Introduction

The use of calorimetry at accelerators with a beam structure in which the bunch spacing is shorter than the duration of the shaped signal poses special problems for data processing. Trigger latency requires that the signals from the detector be stored for a few microseconds, and large channel counts preclude the use of cable delays. It is therefore convenient, during the trigger latency, to temporarily store samples (either digital or analog) in synchronism with the bunch structure of the machine. Once a trigger decision has been made, the samples can be used to reconstruct the important parameters of the waveform.

It is therefore useful to study the properties of time-sampled calorimetry data, since the information may be useful in the design of data acquisition systems for collider experiments currently being contemplated. For this reason, during a beam test of a liquid krypton calorimeter at the Brookhaven AGS[1, 2], we devoted a short period of beam time to the collection of data in which the waveform was sampled in 18 ns intervals.

The calorimeter was located in a negative beam, which could provide electrons over the energy range 5-20 GeV. The arrival time t_0 of the beam particle was determined with two small scintillation counters in the beam to a precision of ± 0.13 ns. The preamplifiers are located in the liquid, close to the electrodes, and the signals are sent through a set of intermediate amplifiers into the counting house. The signals are received by a set of variable gain shaping amplifiers, which permit equalization of the gains to an accuracy of 0.1%. Analog sums of an area consisting of 5×5 towers are formed (front and back sections are separate sums) and also sampled at 18 ns intervals. The shaping time for the individual channels has a δ-response peaking time of 40 ns, whereas the shapers for the analog sums have a shaping time of 50 ns.

The sampling of the signals is carried out in a switched capacitor array, and the samples are then digitized in a 12-bit ADC. The system used was an analog pipeline[3] designed for the ZEUS calorimeter readout. Each channel of the pipeline was run at 13.9 MHz, but one calorimeter readout channel was sampled by four interleaved pipeline channels to achieve an effective sampling rate of 55.6 MHz. The dynamic range of the readout was approximately 1500:1.

Data were collected beam over a period of about four hours. Three runs consisting of 100K events each were taken at beam momenta of 5, 10, and 15 GeV/c, and one run of 20K events was taken at 20 GeV/c. An example of the digitized waveforms is shown in Fig. 1. In any given event, samples are taken every 18 ns, as indicated by the crosses in the figure. However, since the sampling clock and the beam signal are incoherent, the position of the sampling pattern is random with respect to the origin of the waveform. Therefore the overlaying of many samples, when plotted relative to the arrival time of the beam, creates a continuous image of the waveform, which is shown by the dots in the figure. Note that there is a region of about 200 ns digitized ahead of the waveform, which is used in this analysis to establish the noise level.

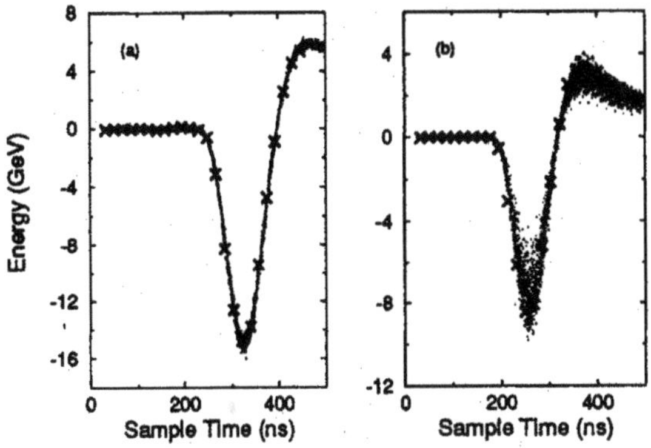

Figure 1: Data taken by the sampling system, summed over many waveforms. For any particular waveform, a set of samples (indicated by $\times$) are taken at 18 ns intervals. Because the beam enters the apparatus at random times with respect to the sampling clock, summing over many samples gives a complete picture of the waveform. The signals are the sum of the front and back sections of the (a) analog sum of 5×5 towers, (b) individual tower centered on the beam line.

2 Description of the method

The treatment of time-sampled data using the method of optimal filtering is discussed in detail in Reference 4. We give here a brief outline of the method. We assume that the shape of the signal is a known function $g(t)$ and that the samples $S_1, S_2, ..., S_n$ are taken at times $t_1, t_2, ...t_n$. The unknown parameters are the amplitude of the signal A and its time origin τ. The sample S_i taken at time t_i is then $S_i = Ag(t_i - \tau) \approx A[g(t_i) - \tau g'(t_i)]$. The amplitude A and timing parameter τ can be found from the samples S_i through the relations

$$A = \sum_{i=1}^{n} a_i S_i \quad \text{and} \quad \tau = \frac{1}{A}\sum_{i=1}^{n} b_i S_i, \tag{1}$$

where the coefficients a_i and b_i are determined from the optimal filtering formalism.

The variances of A and τ are given by

$$\sigma_A^2 = \sum_{ij} a_i a_j \langle n_i n_j \rangle = \sum_{ij} a_i a_j R_{ij} \quad \text{and} \quad \sigma_\tau^2 = \frac{1}{A^2}\sum_{ij} b_i b_j R_{ij}, \tag{2}$$

in which R_{ij} is the autocorrelation function evaluated at time $t_i - t_j$. We solve for the a_i and b_i, minimizing the expressions for σ_A^2 and $A^2\sigma_\tau^2$, under the constraints $\sum a_i g_i$=1, $\sum a_i g_i' = 0$, $\sum b_i g_i$=0, and $\sum b_i g_i'$ =-1. Using the method of LaGrange multipliers to impose these constraints, we find, in vector notation

$$\vec{a} = \lambda \mathbf{R}^{-1}\vec{g} + \kappa \mathbf{R}^{-1}\vec{g}', \qquad \vec{b} = \mu \mathbf{R}^{-1}\vec{g} + \nu \mathbf{R}^{-1}\vec{g}' \tag{3}$$

in which λ, ν, μ, κ, and Δ are quantities which depend only on **R** and the elements of the line vectors $\vec{g}$ and $\vec{g}'$, which are the values of the signal waveform $g(t)$ and its derivative $g'(t)$, evaluated at the times of the samples. Thus, with a knowledge of the impulse response function $h(t)$ of the pre-filter, the autocorrelation function, and the current waveform $i(t)$, the coefficients a_i and b_i are completely defined.

In this analysis, two different waveforms are used to study the amplitude and timing reconstruction. For the amplitude analysis, we use the analog sums of a matrix of 5×5 calorimeter towers, centered on the tower hit by the beam. The total energy signal is defined to be the sum of the two reconstructed amplitudes. Because of small timing differences between different channels making up this sum[2], this signal is not suitable for the timing reconstruction. Instead we use the signals from the individual channels in the center of the matrix. Signals from the front and back sections are reconstructed to give two independent time measurements for each event.

The electronic pre-filter consists of one stage of integration with time constant τ_A at the preamplifier input followed by a shaping amplifier with two stages of differentiation and four stages of integration, with time constant τ_F. However, the nominal form of the δ-response function does not provide an adequately accurate description of the signal waveform. Thus we found it necessary to parameterize the integration stage with a fourth-order polynomial in frequency space, whose coefficients were determined by a least squares fitting procedure. Using this technique, we were able to achieve fits to the waveform whose mean deviations from the samples was within a factor of two of the thermal noise.

3 Results

3.1 Amplitude resolution

In order to obtain the amplitude resolution we perform a Gaussian fit to the central part of the total energy distribution, after correction for showers which start upstream of the calorimeter. The values found at each energy are in agreement with those found from the T/H data[1]. To obtain the contribution of this width coming from the calorimeter stochastic term, we subtract in quadrature the width of the pedestal distribution and the fractional width of the beam momentum spectrum (0.5%). The resulting values, plotted against $1/\sqrt{E}$ are fit to a line constrained to pass through the origin. From this fit we obtain the result $\sigma_E/E = 6.7\%/\sqrt{E}$, assuming a constant term of zero, in excellent agreement with the value obtained using the T/H method. In addition, we have tested the linearity of our data and find it to be better than 0.2%. We find that the noise level achieved by combining the time-sampled data is lower than that of a single sample by about a factor of two, although in this case the noise does not dominate the resolution measurement. This reduction in noise can be understood by a lengthening of the effective measurement time through the use of multiple samples[4].

3.2 Timing resolution

We study the time resolution of the front and back sections of the calorimeter independently. We form the quantities $\delta_f = t_f - t_0$ and $\delta_b = t_b - t_0$, in which t_0 is the start time. From Eq. 2 we see that the noise contribution to the timing resolution varies inversely with the amplitude. The constant of proportionality depends on the optimal filter coefficients and the series noise autocorrelation function for the front and back sections. In addition to this stochastic term in the timing resolution, we expect an energy independent term $\sigma_{\rm cal}$ arising from timing jitter in the sampling electronics, timing fluctuations arising from variations in the shower starting position, and fluctuations in the parameters introduced by the fitting procedure. The largest of these is the analog pipeline cell-to-cell sampling time jitter, which has been measured to be approximately 0.15 ns.

The experimental variances of δ_f and δ_f have been fit to an equation of the form $\sigma^2 = c^2/E^2 + \beta^2$, and the values for c_f, c_b, and β are shown in Table 1. Our values for the intercept β and the start time resolution σ_0 yield a value for $\sigma_{\rm cal}$ of 0.15±0.01 ns.

Table 1: Fitted values of the noise parameters for the individual calorimeter sections.

Parameter	Expected value	Fitted value
c_f (GeV·ns)	0.40	0.38±0.01
c_b (GeV·ns)	0.57	0.69±0.01
β(ns)	$\sqrt{(0.13)^2 + \sigma_{\rm cal}^2}$	0.20±0.01

If one considers using the calorimeter as a timing device, from which one wishes to extract the maximum timing precision, we use a weighted mean time, where the weights are given by the inverse of the variance of the timing measurement. Since the energy dependence of the variance is known, this can be calculated from the parameters given in Table 1 on an event-by-event basis. To measure the resolution of the weighted mean T, we measure the width of the distribution of the quantity $\Delta_T = T - t_0$. We determine σ_T from this width, subtracting σ_0 in quadrature. The resulting values for σ_T are 0.27, 0.17, and 0.16 ns for beam momenta of 5, 10, and 15 GeV/c respectively.

We have utilized the present data to study experimentally the change in precision as the number of samples used to reconstruct the waveform is decreased. For the amplitude resolution, we have found a weak dependence of the number of samples on the resolution, since the main contribution to the resolution comes from sampling fluctuations rather than electronic noise, which is the only component of the resolution that is reduced with multiple samples. For the timing resolution, there is a stronger dependence on the sampling pattern, since the timing resolution is dominated by the noise. We have found that eliminating every other sample (*i.e.*, going from 7 to 3 or 4 samples) has almost no effect on the resolution, but reducing the number of samples to 2 results in a worsening of the resolution by 30-35%.

4 Discussion

In this paper we have shown how the technique of optimal filtering can be used to extract amplitude and timing information from the time-sampled signal of a liquid ionization calorimeter. The results of the amplitude resolution measurements are essentially identical to those found in the same calorimeter using the more traditional method of sampling the signal at its peak. The timing measurements show that the timing precision of the calorimeter is excellent. Using the weighted mean of the time measurement from the front and the back sections of the calorimeter, we have obtained a timing resolution of 0.16 ns at a beam energy of 15 GeV.

One of the interesting applications for the method of multiple sampling is for the recording of calorimetry data at high luminosity colliders. For this type of operation, the required timing accuracy (for bunch crossing identification) is only a few ns. For the front section of an EM calorimeter, our results indicate that a time resolution of 4 ns can be obtained for energy depositions in a single calorimeter channel of as low as 100 MeV, with a sampling fraction of 0.325. However, it must be remembered that for operation at a high luminosity collider, pileup noise will also contribute to a degradation of the resolution. The optimal filtering formalism maximizes the resolution for both the amplitude and timing for the combination of all noise in the system. The fact that one of these sources of noise (pileup) is time-dependent due to changing machine luminosity strengthens the case for recording the data through multiple samples.

A more complete account of this work is contained in Reference 5.

References

[1] O. Benary, et al., Nucl. Inst. and Methods **A344**, 363 (1994).

[2] O. Benary, et al., Nucl. Inst. and Methods **A332**, 78 (1993).

[3] J. Möschen, et al., Nucl. Instr. and Meth. **A288** (1990) 180.

[4] W. E. Cleland and E. G. Stern, Nucl. Inst. and Methods **A338**, 467 (1994).

[5] O. Benary, et al., Nucl. Inst. and Methods **A344**, 363 (1994).

THE PSALM PROJECT

J. Ardéléan, E. Augé, D. Breton, R.L. Chase, G. Martin, E. Plaige
Laboratoire de l'Accélérateur Linéaire
F-91405 Orsay Cedex, France

E. Delagnes
CEN Saclay / DAPNIA
F-91191 Gif sur Yvette Cedex, France

presented by Etienne Augé

ABSTRACT

An Analog Linear Memory with Pedestal Subtraction (PSALM) is being developed jointly by LAL-Orsay and CEN-Saclay for the readout of the electromagnetic calorimeter of the ATLAS experiment at the future LHC collider. Preliminary results from a second prototype, received in august 1994, are presented. The dynamic range achieved on a test bench is 12 bits.

In 1993, LAL Orsay and CEN Saclay started a joint development for an analog memory that might be part of the readout electronics of the ATLAS electromagnetic calorimeter. Such analog memories were already in use on the ZEUS experiment at DESY [1], and similar developments were under way in the University of Alberta (in cooperation with LBL) [2] and in Nevis labs [3].

We have used an original design for all components in our memory chip, in particular for the operational amplifiers and analog buffers, and an original layout. We have also used a different process from our competitors [4]. However, the main originality in our project is the systematic subtraction, when reading a data from the memory, of the parasitic charges added to the signal during the write and read operations. This pedestal depends in general on the position used in the memory. Its subtraction should increase the dynamic range of the memory, in particular when no further memory cell-dependant corrections can be applied.

1. Architecture of the calorimeter readout

In the analog approach, the signal from each calorimeter cell is sampled permanently at the LHC bunch crossing frequency (40MHz) and each sample is stored in an analog memory during the latency of the level 1 trigger (2 μs). In case of a positive trigger decision, a number of samples, centered on the corresponding bunch crossing, are read back from the analog memory and digitized. At the maximum trigger rate (100 kHz), we may digitize up to 5 samples per event. We may use even more samples at lower trigger rate.

After digitization, the ADC words are calibrated. Then, the several samples from each calorimeter cell are combined to give a single estimate of the energy in the cell, of the time of the incoming particle, and of the signal quality. The result is made available in a convenient form for the level 2 trigger and for the data acquisition system.

The main advantage of this solution is that only interesting data are digitized. This may be achieved with a limited number of commercially available ADCs, at a limited cost. Because of the important R&D effort, analog memories just fitting the requirements already exist in the form of prototypes, and even better chips will be available at low cost (less than 10 SF/channel) in the next few years.

2. Requirements on the analog memory

2.1. Dual port memory

To limit the dead time, the sampling of the calorimeter signals and their storage into the analog memory should not stop when data are being read back from the memory.

2.2. Thirteen bit dynamic range

The dynamic range of the electronics for each calorimeter cell should be in excess of 16 bits, and should not degrade the calorimeter resolution. To cope with this difficult requirement, we send the signal from a single pre-amplifier (associated to a single calorimeter cell) into two shapers, each with a dynamic range of the order of 10,000, and with a gain ratio of the order of 12. We associate to each shaper channel an analog memory channel, with a 13 bit dynamic range.

For example, if we use this dual gain system associated to a 13 bit ADC with an rms precision equal to half the lsb to measure energies between -100 GeV and +2 TeV in a cell, it will introduce an rms noise of 24 MeV, to be added in quadrature to the pre-amplifier noise. This is adequate.

3. Principle of an analog memory

A "write address" is given to the memory chip by an external controller each 25 ns. It is distributed on a bus inside the memory chip. Switches allow to sample the input signal and store the resulting voltage in the capacitor corresponding to the given write address. To retrieve a sample, the controller sends its position to the memory chip. Switches allow to place the corresponding capacitor in the feed-back loop of the read amplifier.

4. Limitations of the dynamic range

The limitation of the dynamic range does not come from true noise, but from two systematic effects :

- cross-talk from the digital lines to the input signal. With simultaneous write and read operations, the cross-talk may occur either when sampling the signal or when reading it back. If the capacitors were always accessed in the same sequence, then the cross-talk affecting a given capacitor would always be the same, and this effect would be corrected by using an effective pedestal, different in each capacitor. However, this does not hold in the case of random accesses, and this effect should be considered as a noise.

- parasitic charges injected by the switches. The switches to write and read are not exactly identical from capacitor to capacitor. As a result, the parasitic charge they inject may be different, and the pedestal value then depends on the capacitor. Using the same average pedestal for all 128 capacitors associated to a given input channel results in measurement errors which should also be considered as noise. The PSALM has been designed to get rid of this effect.

5. Description of the PSALM

One PSALM channel is shown on figure 1. A signal is sampled in a capacitor C_P by switch 1. Switch 2 is then opened to isolate the capacitor. To read, one uses switches 3, 4 and 13 *. To perform the pedestal subtraction, a signal just read is temporarily stored in capacitor C_C using switches 7 and 8. Then C_P is cleared using switches 1 and 5, and a second read operation is performed, the result being temporarily stored in capacitor C_S using switch 6. Switches 9,11 and 10 allow to transfer the difference between the two readings to the output amplifier.

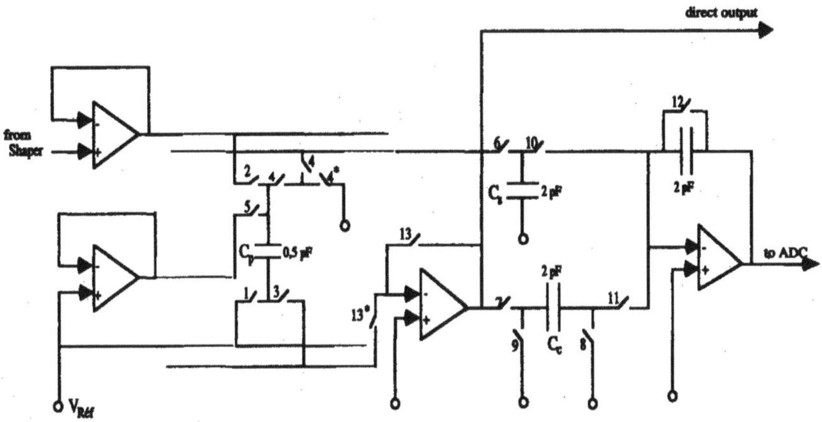

Figure 1: *Schematic view of one PSALM channel (only one of the 128 capacitors is shown)*

After a first prototype received in march 1994, we have received a second version in august 1994. This chip has 4 channels, each with 128 capacitors.

6. Description of the test bench

For the first measurements, we have developed a test bench split in two boards, linked by a short cable :

- The analog board carries a PSALM chip, a 12 bit 10 MHz ADC associated to a Fifo. Multiplexers allow to send either an external signal or the reference voltage to any of the four inputs of the PSALM. Multiplexers also allow to read either the pedestal subtracted output, the direct output, or the reference voltage, and to perform any subtraction between two of these signals externally to the PSALM chip. The signal may be amplified by a factor 10 at the input of the ADC for a better measurement of pedestals and noise.

- The digital board sits in a VME crate. It provides write addresses (at 40 MHz) and read addresses to the PSALM, but does not allow at the moment simultaneous write and read operations. It also provides a VME interface with a MacIntosh computer, where data are collected from the Fifo and analysed.

7. Preliminary results

The power dissipation of the PSALM chip is 0.5 W, mainly due to the 40 MHz clock. This dissipation should be substantially lower for the next versions. The power dissipation on analog voltages is small (4.5 mW per input channel).

7.1. Dynamic range

We have first measured the system linearity, using DC levels. With a non-linearity below 0.2%, the dynamic range is in excess of 3.3 V. It should be increased by 1 Volt changing the NMOS switches 2, 5 and 4 to CMOS ones. Measurement using the output of present RD 3 shapers are being performed.

The cell to cell dispersion of the pedestal for a given channel, measured without any pedestal subtraction is 0.85 mV rms. We obtain the same number when measuring the difference between two adjacent channels, which shows that only a fraction of this dispersion is coherent over the channels.

If we clear each capacitor before reading it, still using no pedestal subtraction, the cell to cell dispersion goes down to 0.64 mV rms. This contribution should be completely suppressed by the pedestal subtraction, leaving a cell to cell dispersion as small as 0.6 mV. However, when we really use the pedestal subtraction, we obtain a dispersion of 0.8 mV. We hope to obtain better result whith a better tuning of the present chip.

7.2. Noise level

At the moment, we measure an rms noise of 0.8 mV per capacitor. However, we have characterised the op-amplifier which is used in the chip and measured a noise level less than 0.1 mV rms, for a 1 GHz bandwidth. We therefore anticipate a noise level per capacitor of 0.5 mV rms. A better design of switch 13 (reset of the read amplifier) should allow to go down to 0.35 mV.

The noise level is still 0.8 mV rms per sample when measuring the difference between two adjacent channels. Part of this noise is therefore coherent over the channels. This contribution comes from a coupling between the the PSALM chip and the analog board. We experience that the simple analog board on the test bench needs a very careful design in order to fully benefit from the PSALM performances.

8. Conclusions

Using only one pedestal per channel, our system reaches a dynamic range of 2750 (11.5 bits). With one pedestal per capacitor, we obtain 12 bits.

With a better design of the analog board, we anticipate a dynamic range of 3500 with only one pedestal per channel, and 6600 (12.7 bits) with one pedestal per capacitor.

The next version of the chip will have a dynamic range in excess of 4 Volts, with an rms noise of 0.35 mV. It will carry 16 channels and is expected in Orsay in february 95.

A new digital board for our test bench will allow measurements on the present PSALM chip with simultaneous write and read operations in december 1994.

We prepare a system to be used in the readout of the prototype accordeon calorimeter in a test beam at CERN in 1995.

In the end, our present experience tells us that the design of a board carrying analog memories (such as the future front-end board for the ATLAS calorimeter) is especially delicate.

References

1. J. Möschen et al. *Nucl. Inst. and Meth.* **A288** (1990) 180.
2. S.A. Kleinfelder, M. Levi, O. Milgrome *Nucl. Phys. B* **23A** (1991) 382.
3. A. Gara, J.A. Parsons, W. Sippach *Proceedings of the second International Conference on Electronics for Future Colliders* LeCroy Corp. (1992) 61.
4. 1.2 μm CMOS process from Austrian Micro-Systems, Schloβ Permstätten, A-8141 Unterpermstätten Austria

DESIGN AND PERFORMANCE OF AN ANALOG MEMORY BASED CALORIMETER READOUT

J.A. PARSONS

and

H. CUNITZ, A. GARA, B. SIPPACH, L. ZHANG

Nevis Laboratories, Columbia University, P.O.Box 137, Irvington, NY, USA

ABSTRACT

We have designed a calorimeter readout architecture, based upon a custom switched-capacitor analog memory IC (SCA), and have proposed it for the readout of the ATLAS calorimeter. In the design of the SCA, careful attention has been paid to minimizing systematic effects due to the simultaneous Read and Write operations which are required in order handle the Level 1 trigger rate of 100 kHz with minimum deadtime. Bench tests of the latest SCA prototype have shown a dynamic range in excess of 13 bits. A VME-based readout board has been designed in order to allow instrumentation of a LKr electromagnetic calorimeter prototype during a testbeam run at CERN during August 1994. Details of the architecture will be presented, as well as online results of the testbeam run, which include a sub-nanosecond timing resolution.

A key element of the ATLAS detector, being designed for operation at the LHC at CERN, is a high precision liquid ionization electromagnetic calorimeter. The system requirements which must be met by the electronics readout of the ATLAS calorimeter include:

- a dynamic range in excess of 16 bits,
- excellent energy resolution,
- precise time resolution,
- an electronics contribution to the constant term in the energy resolution of less than 0.2%,
- deadtimeless operation at a maximum Level 1 Accept rate of 100 kHz,
- formation of analog sums of course granularity for formation of the Level 1 trigger,
- storage of the calorimeter signals during the Level 1 latency of 2 μs,
- the transfer of fully digitized and processed data to Level 2, with up to 5 samples per channel,
- low cost and low power for a system of about 200k channels.

Preamplified signals will be driven to crates of readout electronics which will be mounted either inside the ATLAS detector or nearby outside. There, the signals are split into two gain scales and shaped before being sampled at the bunch crossing frequency of 40 MHz. The samples are stored in analog form using a switched-capacitor analog memory (SCA). Upon receipt of a Level 1 trigger, the relevant samples are buffered in the SCA before being read through the on-chip multiplexor for digitization by a commercial 12-bit ADC. The digitized samples are buffered in FIFO's, and then processed by DSP's (one for every 8 channels) before transfer to Level 2. The DSP's process multiple samples for each signal in order to extract the deposited energy as well as the time of deposition. Each DSP has access to a RAM in which are stored the calibration constants it needs.

Details of the readout architecture, including studies of the multiple levels of buffering, discussion of the multiple sampling technique utilized, and a description of the formidable front-end processing provided, may be found in reference [1]. Here we will concentrate on the design and measured performance of prototypes of the SCA and readout.

SCA Design and Performance

To achieve deadtimeless operation with an SCA requires a design which incorporates the ability to perform simultaneous Read and Write operations. The layout for such a SCA is shown schematically in figure 1. Each of the m pipeline cells comprises a sampling capacitor (C1 through Cm) and four switches. Two of the switches (W1 and W2) connect the input voltage across the capacitor during a Write operation, while the other two (R1 and R2) connect the capacitor across the output op-amp for a Read operation. The n channels are multiplexed and driven off-chip to the ADC by closing switches M1 through Mn in the appropriate sequence. After a sample is read, the Rst switches are closed to reset the op-amps as well as any parasitic charge on the Read bus.

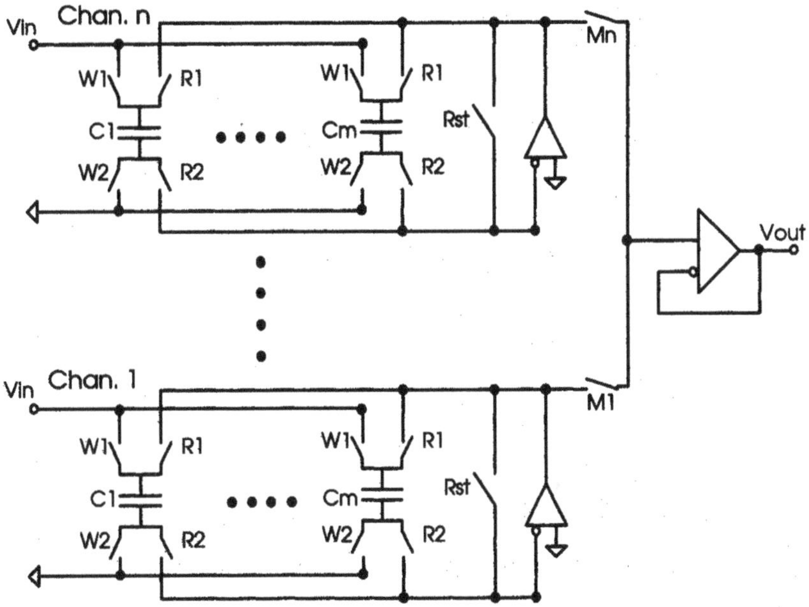

Figure 1. Schematic diagram showing the SCA architecture.

The Read and Write switches are controlled independently by separate buses (not shown on the figure). Thus, while one sample is being read out, it is possible to continue to sample the input voltage and write the samples to other cells in the same pipeline. In this scheme, buffering of the data for Level 2 is achieved by transferring the addresses of the relevant samples to a separate list of cells awaiting digitization. The chip is designed for random access to any cell, with the read and write addresses generated by off-chip logic and decoded on the SCA chip.

The dynamic range of SCA's is limited mainly by systematic noise introduced by clock feedthrough (CFT) during the opening/closing of the cell switches during Write and Read operations. This problem is, of course, only exacerbated by the requirement to perform Write and Read operations at the same time. Furthermore, the need to add up large numbers of channels of the calorimeter to form jet and global sums implies that very small effects in individual channels can lead to significant systematic errors if channel-by-channel correlations are present. We have taken special care in the design of the SCA in order to minimize these effects, through a careful design of the cell switch architecture and control scheme (for details, see reference [1]).

We have developed our SCA design through a sequence of 6 prototypes using a 1.2 μm CMOS process. During this systematic development, we have investigated several methods for minimizing the systematic CFT noise. The latest prototype consists of a 6 channel X 128 cell analog pipeline. The architecture is fully differential, with two pipelines used per channel, and an on-chip differential output multiplexor. This prototype also incorporates all of the on-chip logic required for the addressing/control scheme we have developed. This logic includes differential PECL level shifters for receiving all control lines, an accumulator for the Write address, and a shift register for the Read address. Some of the parameters measured in bench tests of this prototype are summarized in Table 1.

Parameter	Value
Input Signal Voltage Range	3.5 V
Pedestal Noise	350 μV (RMS)
Dynamic Range	≈ 13 bits
Cell-to-Cell Pedestal Variation	700 μV (RMS)
Cell-to-Cell Gain Variation	< 0.03%
DC Nonlinearity	< 0.05%
Signal Droop	< 1 mV/10 ms
Time Jitter	< 10 ps
Input Time Constant	≈ 1 ns
Crosstalk	< 0.01% (RMS)
Induced Noise due to Simultaneous Read/Write	<0.1 mV (RMS)

Table 1. Measured parameters of the SCA prototype.

Prototype VME Readout Board

In order to allow testing in a beam environment, we designed and produced a 12-channel VME-based readout board incorporating two SCA chips. A block diagram of this readout board is shown in figure 2 below.

The data from the SCA's are digitized by 14-bit 5 MHz commercial ADC's and then buffered in FIFO memories which are accessible over the VME bus. The addressing signals for the SCA's, the Convert signals to the ADC's, and the Write clocks to the FIFO's are all provided by an on-board FPGA, which is booted upon power-up from an EPROM.

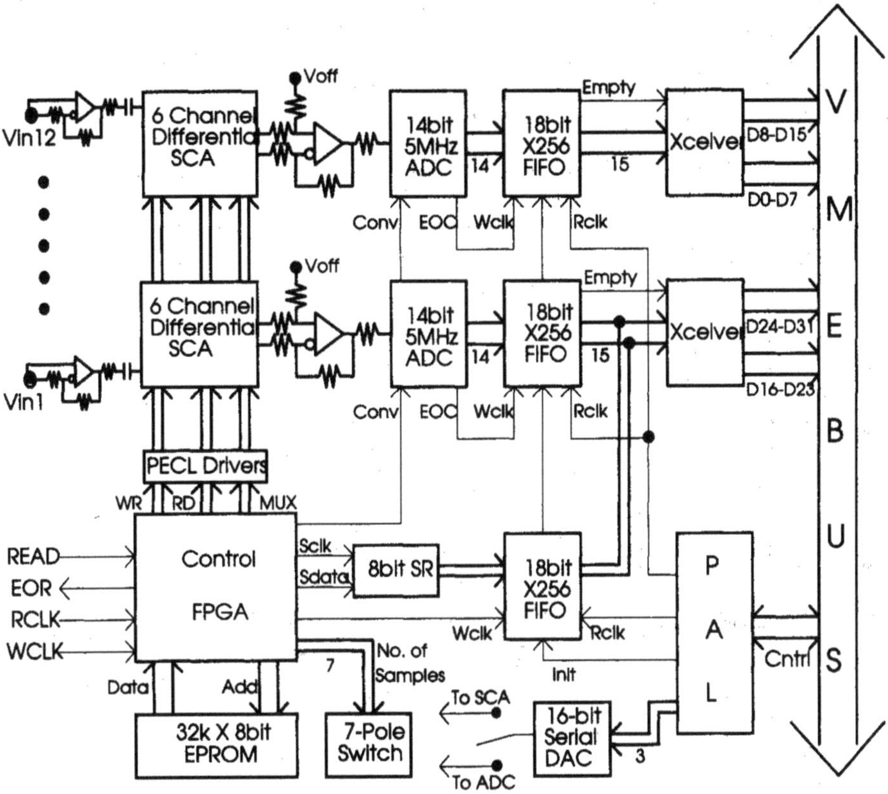

Fig. 2. Block diagram of the 12-channel VME readout board.

As a calibration tool, an on-board 16-bit serial DAC, programmable via VME, provides a precise DC voltage which can be switched either into the inputs of the SCA's or directly to the ADC inputs. We have used this DAC to measure the DC performance of the boards. The quantization noise, when measured with the DAC switched directly to the ADC, is 0.8 ADC counts (RMS). This increases to typically 3.5 counts (RMS) when the DAC level is written into, and subsequently read from, the SCA's, implying a dynamic range of the readout board slightly in excess of 12 bits. Given the 2.5 V input range of the 14-bit ADC, the noise corresponds to about 530 μV. The DC nonlinearity is less than 0.1% over the full range of the ADC.

Testbeam Studies

During a two-week testbeam run at CERN during August 1994, we used these electronics to measure the response of a LKr electromagnetic calorimeter prototype to electron and pion beams of momenta between 20 and 200 GeV/c. (For details of the calorimeter prototype and other features of the testbeam, see the talk by D. Lissauer in these proceedings.)

To minimize cost, only 36 channels of SCA electronics were used; with these we instrumented a total energy analog sum, separate analog sums for the three longitudinal depths of the calorimeter, and blocks of individual channels in each of the three depths near the center of the beam spot. The sampling by the SCA's was controlled by a free-running 40 MHz clock; the phase of the arrival time of particles relative to the SCA clock was measured with a TDC with 50 ps bin size. Figure 3 shows the samples taken of the total energy sum as measured in response to a 50 GeV/c electron. The LKr signal has been shaped with a bipolar shaping function which has a 40 ns risetime in response to a delta function.

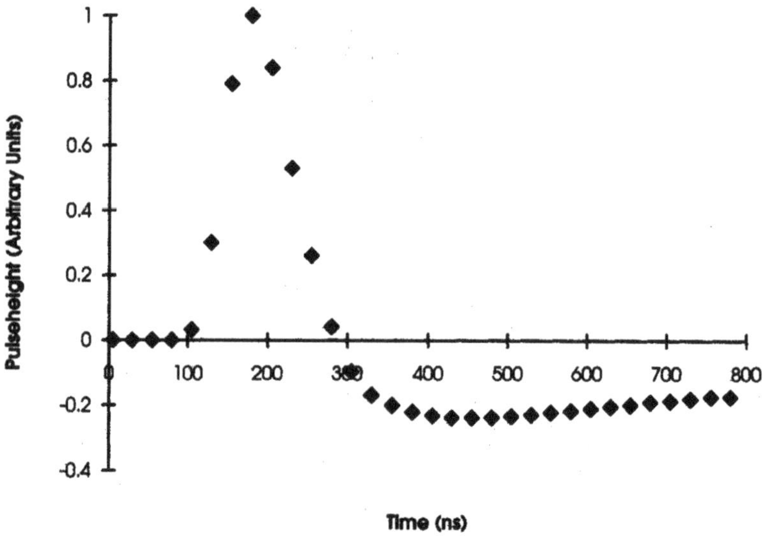

Fig. 3. Response to a 50 GeV/c electron. Successive samples are separated by 25 ns.

The analog sums will be used to study the energy resolution, while the individual channels yield the timing resolution. A comparison of the energy measurement can be made with results of a Track-and-Hold system which was used in parallel to read out all channels of the calorimeter. Other studies to be made include:

- measurement of the energy resolution versus shaping time (data were taken with both 40 ns and 20 ns shaping applied to the total energy sum),
- investigation of events with multiple interactions or "pile-up",
- studies of extracting the final calibration constants from the pulser calibration data, a

key issue for the final system of 200k channels,

- optimal filtering studies (eg. energy/timing resolution versus the number of samples used, the sampling pattern, etc.), similar to those we performed in a previous study (for more information, see reference 2 and the talk by W. Cleland in these proceedings).

Detailed analyses of the data are currently underway. The high quality of the data is evidenced by results obtained online during data taking. For example, a crude timing estimate was calculated from 5 samples of the total energy sum; comparison with the time as measured with a TDC revealed a timing resolution of order 600 ps for 50 GeV/c electrons, which still includes smearing effects due to any differences in cable lengths, etc. for the individual channels making up the total energy analog sum.

Summary

We have proposed, for the ATLAS calorimeter, a readout architecture based upon a custom switched-capacitor analog memory we have developed through a sequence of prototypes. The SCA has a dynamic range of about 13 bits, and has been carefully designed to minimize systematic noise due to simultaneous Read and Write operations.

To permit testing in a beam environment, we have designed and fabricated a 12-channel VME-based readout board incorporating two SCA chips followed by 14-bit 5 MHz commercial ADC's. These boards were used to instrument a LKr calorimeter prototype during a testbeam at CERN during August 1994. While detailed analyses of the data are underway, preliminary results, including sub-nanosecond timing resolution, are very encouraging.

Acknowledgements

It is a pleasure to acknowledge our colleagues from Brookhaven National Laboratory, the University of Milan, the University of Pittsburgh, the University of Rochester, and the University of Washington, without whose diligent efforts the testbeam would not have been possible.

This work has been supported in part by the W. M. Keck Foundation, the National Science Foundation, and the Department of Energy. The excellent technical work of the Nevis electronics shop is gratefully acknowledged.

References

1. A. Gara, J.A. Parsons, W. Sippach, Proc. of the 2nd International Conference on Electronics for Future Colliders, LeCroy Corp. (1992) 61;
H. Cunitz, et al., "A High Rate, Precision Calorimerty Readout", To appear in Proc. of the 4th International Conference on Electronics for Future Colliders, LeCroy Corp. (1994).

2. O. Benary, et al., BNL Preprint 60279 (1994); submitted to Nucl. Instr. and Meth.

CURRENT-SAMPLING WAVEFORM ACQUISITION FOR DIGITAL OSCILLOSCOPE AND DETECTOR APPLICATIONS

WALTER O. LeCROY, JR.
LeCroy Corporation
Chestnut Ridge, NY 10977

ABSTRACT

The technique of current sampling arose from the need for gated, charge-integrating ADCs in high energy physics. In these circuits, the gate completely encloses the signal pulse, and speed is not a primary requirement. A single acquisition at a time is performed. These circuits have been adapted to sample continuous currents continuously, and now form the basis for digital oscilloscopes sampling at real-time speeds of 5 Gigasamples/sec and more. Simplified, lower speed versions may be suitable in cost and performance for detector waveform acquisition.

1. The Charge-Integrating ADC of High-Energy Physics

In most particle physics detectors, the primary information of interest is carried in the form of charge. It is the area of the pulse, and not its peak amplitude or shape, that is of interest. The conventional charge-integrating ADC (Fig. 1) commonly used for these signals gates the signal pulse current onto a capacitor by means of a bipolar transistor differential stage, and then discharges the capacitor with a constant rundown current. The result depends only on the charge in the input signal and the size of the rundown current, and not on the size of the integrating capacitor or the voltage developed on it.

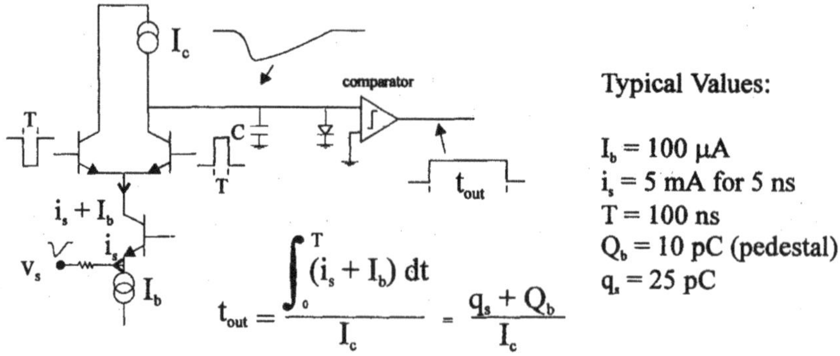

Figure 1. The HEP gated charge-integrating ADC

These circuits utilize gate widths of a few tens to a few hundreds of nanoseconds. They are intended to completely enclose the signal pulse in order to acquire all its charge. Their inputs respond only to negative-going pulses, because standing current is made small in order to minimize pedestal in the output.

2. Current-Sampling for Waveform Capture

To be useful in recording transient waveforms, the current-sampling circuit of the gated-integrator ADC must be modified to sample repetitively at high speed, and to handle bipolar signals. Since the gate must now sample the signal itself, rather than simply enclose it, the gate width is no longer a secondary factor; its stability and speed become critically important.

The waveform-capture current-sampler uses the same bipolar differential stage as its predecessor, but now both output collectors are used to achieve continuous sampling. Successive samples appear on alternate collectors, as shown in Fig. 2. (Their subsequent fate is discussed later.) In addition, a standing current somewhat larger than half the signal swing allows for bipolar signals.

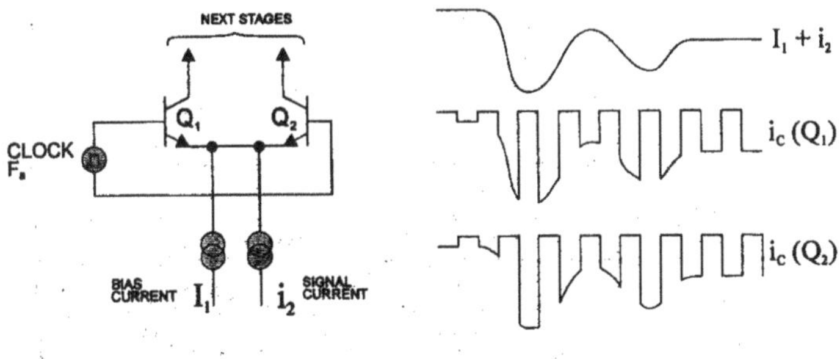

Figure 2. Basic current-sampler

For a sampling rate of 1 GS, each sample duration is 1 ns and the clock rate applied to the sampler is 500 MHz. The size of the sample is proportional to the sum of the instantaneous signal amplitude and the standing current, and to the gate width. So jitter sources in the sampler drive must be thoroughly suppressed; to calculate the jitter level required for this technique is to abandon it.

The frequency response of a current-sampler is of sin(x)/x form as shown in Fig. 3. The frequency roll-off for the current sampler of the Model 9360 digital oscilloscope, which utilizes a 600 ps sample duration, is shown in Fig. 4, in comparison with more familiar 6 dB per octave and 12 dB per octave curves having the same 3 dB point.

3. The Current-Sampler System

The overall operation of the current-sampler system is illustrated conceptually in Fig. 5 as a two-pole rotating switch. The input signal current is steered onto successive capacitors, of which there is a limited number (624 in the Model 9360 sampler). Preceding the write operation is a read/clear operation, which clears each storage capacitor of its previous charge just before it is rewritten with a new sample.

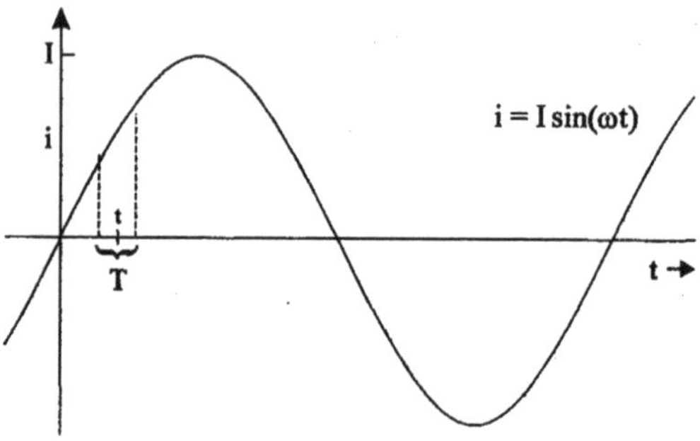

Sample (integrating) time $\equiv T$

$$\text{Output charge} \equiv q_o = \int_{t1-T/2}^{t+T/2} I\sin(\omega t)\,dt$$

$$= \frac{2I}{\omega}\sin(\omega t)\sin\left(\omega\frac{T}{2}\right)$$

$$\text{Gain} \equiv \frac{\text{output}}{\text{input}} = \frac{\frac{2I}{\omega}\sin(\omega t)\sin\left(\omega\frac{T}{2}\right)}{T\,I\sin(\omega t)}$$

$$= \frac{\sin\left(\omega\frac{T}{2}\right)}{\omega\frac{T}{2}}$$

Figure 3. Frequency dependence of current-sampler gain

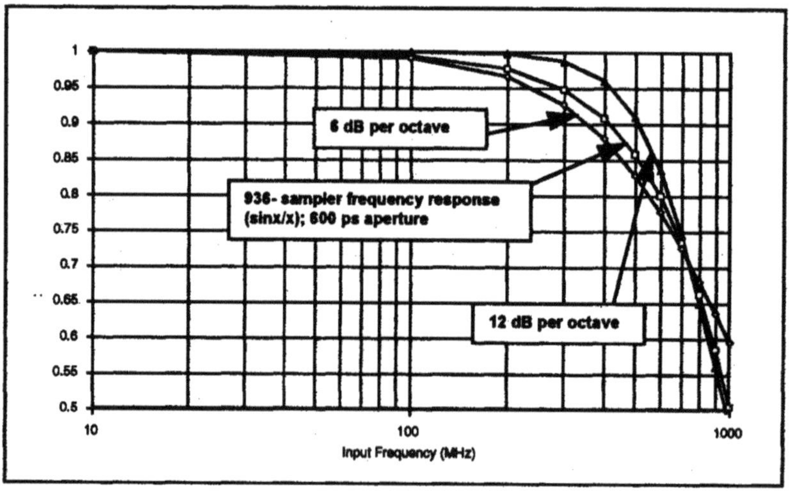

Figure 4. Frequency response of a 600 ps sampler compared with conventional roll-offs

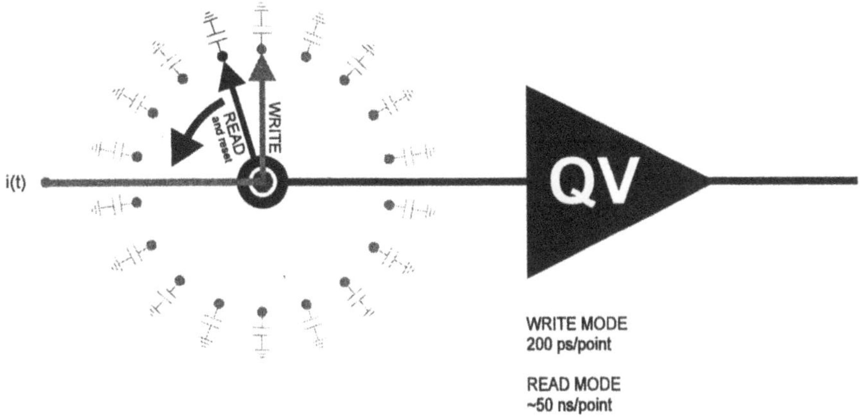

Figure 5. Easy conceptual view of current-sampler system: rotary switch wipers rotate at high speed for write (dwell time = sample gate width), low speed for read.

The system runs continuously, so that at all times the capacitors contain a record of the last full cycle of samples. This time interval is equal to slightly less than the sampling width times the number of storage capacitors, the difference being due to the clearing operation and other housekeeping. A "trigger" stops the writing, and the system switches to read mode, in which the clearing process continues as in write mode but at a slower rate. In the read mode each discharge current, representing the last sample on each capacitor, is not discarded but is integrated in a charge to voltage converter, whose output is digitized by a flash ADC. The reading process is identical to the clearing process, except for speed.

In the actual circuit, a series of stacked common-emitter current switches routes the charge packets (current samples) from the input sampler stage to the storage capacitors (Fig 6). In each level, only a single transistor is "on" at a time, defining the route of any signal injected into the emitters. For any charge packet appearing at a collector of the input sampler (Figs. 2 and 6), one and one only path is open, to a specific storage capacitor. The two identical trees, one for each collector of the sampler, function alternately. While one side is conducting, the other changes state to point to another storage capacitor. As seen by the charge packet, these routing paths are thus static, and consist simply of a series of grounded-base transistors.

It is a strength of the system is that all sampling thus occurs at a single point, in contrast to techniques that obtain their speed through highly parallel structures of basically slow or low repetition rate circuits. Design challenges, however, include:

The need to stack successive steering stages requires many supply voltages and drive signal levels. The clearing of the storage sites involves clamping a capacitor to a voltage through a diode (not a resistor), resulting in a "time constant" that is essentially infinite, since the resistance of the diode increases as the current through it decreases. And, base-emitter leakage currents are not very important for usual integrated circuit applications, and so are normally not well specified or controlled. These difficulties must be dealt with, but none are fatal.

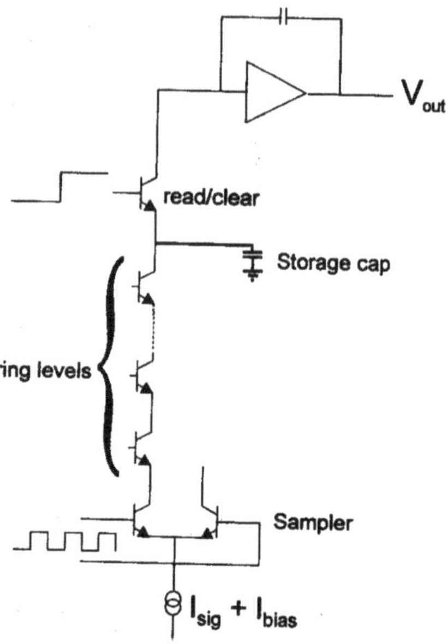

Figure 6. Current-sampler write and read paths

4. A Waveform-Capture Chip for Detector Applications

Digital oscilloscope requirements demand sampling circuits that stress accurate waveform capture at the highest possible speeds. Usual detector applications don't demand multi-gigasample speeds, and relaxing the speed requirement greatly simplifies both the sampler itself and its support circuitry. This leads to the thought that the technique may be usefully applied for detector waveform capture. Because it is intrinsically fast, even a reduced speed implementation would offer sampling speeds higher than directly necessary for most detector signals. This excess sampling speed can be translated directly into increased resolution through either averaging or more sophisticated digital processing, which can be valuable.Fig. 7 illustrates a single-chip waveform capture system, including current-sampling front end, ADC, and memory. Sampling speed is shown as 2 ns and record length as 500 point, but these numbers are flexible. The possibility of "on-the-fly" hardware correction prior to digitization is included, but might not be necessary. Two or more memory pages are foreseen, allowing simultaneous write and read.

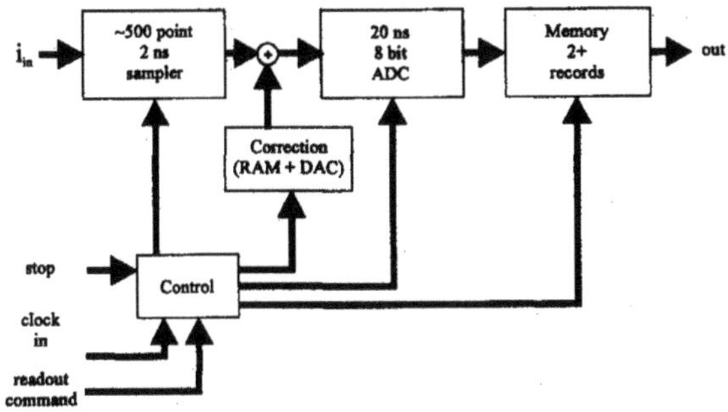

Fig. 7. HEP waveform recorder chip with fast current-sampling input

Summary chip characteristics are:

- Integrating, current-sensing input
- 8-bit resolution per sample, with higher resolution possible for slow signals
- 1-3 ns integrating (sampling) time for each sample
- About 10 microseconds digitizing time (deadtime)
- Low-frequency time reference - no high speed clock to distribute
- Ping-pong memory - no deadtime on read
- Power consumption less than one watt
- Cost less than $100 per channel

Such a chip would provide, in a single channel, the possibility of good energy resolution, shape discrimination, pile-up rejection, and time information, with considerable flexibility in choosing the trade-offs among these for a specific application by choice of the processing of the stored information.

5. Summary

Current-sampling is the sampling technique of choice in applications where very high speed is required and limited record length is acceptable. In addition, for slower signals, the speed of the technique can be traded for added resolution by simple digital signal processing, which extends its applicability. This flexibility can make the same circuit useful over a range of signal speeds and resolutions.

6. Acknowledgements

I would like to thank Brian Cake, Brian Yamrone, Bob Talambiras, of LeCroy Corporation NY, and Martin Miller, of LeCroy SA, for their essential contributions to the successful development and application of this technology.

A HIGH SPEED DIGITIZING PHOTOMULTIPLIER TUBE BASE FOR THE KTEV CsI CALORIMETER

JULIANA WHITMORE
Fermi National Accelerator Laboratory
Batavia, IL 60510, USA

ABSTRACT

A circuit has been designed to digitize PMT signals over an 18-bit dynamic range with 8-bits of resolution.[1,2,3] The crucial element of the circuit is the custom charge integrating and encoding (QIE) ASIC. This chip is designed to operate at rates up to 53 MHz, and, in conjunction with an 8-bit FADC, generates 12-bit floating point output. Bench tests of a 17-bit version of the digital base demonstrated excellent noise performance, linearity and pedestal and gain stability. Twenty-five channels of digitizing PMT bases have been built and used for readout of a CsI array in a test beam at CERN. Performance of these devices in a beam environment is discussed.

1. Introduction

A high speed digitizing base has been developed for readout of the pure CsI calorimeter in KTeV*, a Fermilab fixed target experiment. The major focus of the KTeV experiment is a precision measurement of the direct CP violating parameter ϵ'/ϵ.[4] To achieve this goal, the CsI calorimeter must have an energy resolution of 1% for 15 GeV photons and be extremely linear over a wide dynamic range. The digitizing circuit must be compatable with these stringent requirements. A detailed description of the KTeV calorimeter can be found in reference 5.

The required performance of the KTeV calorimeter has serious implications on the CsI crystals, photomultiplier tubes, and the readout electronics. Linearity specifications limit the peak current in the phototube to 30 mA, setting the high end of the dynamic range at 64 GeV. In addition, longitudinal crystal response needs to be monitored. Muons transverse to the crystal axis deposit 16 MeV of energy. A 5% measurement on these events necessitates 0.5 MeV resolution at the low end of the dynamic range. The result is a requirement of 17-bits to cover KTeV's full operating range. The ADC is allowed to contribute no more than 0.2% to the energy resolution between 1 GeV and 64 GeV. Digitization error from an 8-bit ADC meets this specification and is much less than the 0.5-1% CsI intrinsic resolution. In addition, the energy resolution is sensitive to pile-up problems†and to accidental energy deposits in the calorimeter. A detailed time profile will allow monitoring of these effects. The digital base circuit features the low input noise, wide dynamic range, and high frequency operation required for the KTeV calorimeter.

*A collaboration of Chicago, Colorado, Elmhurst, Fermilab, Osaka, Rice, Rutgers, UCLA, UCSD, Virginia and Wisconsin.

†The tail of a CsI pulse has a 25 ns fast and a 1 μs slow time component.

2. Overview of Digital Base Operation

The three fundamental components of the digital PMT base (DPMT) are the charge integrator and encoder (QIE) custom designed ASIC, a commercial FADC and a custom buffer/cable driver chip. Figure 1 shows a schematic of the primary functions of the digital base. The QIE receives signals from a PMT, splits the current

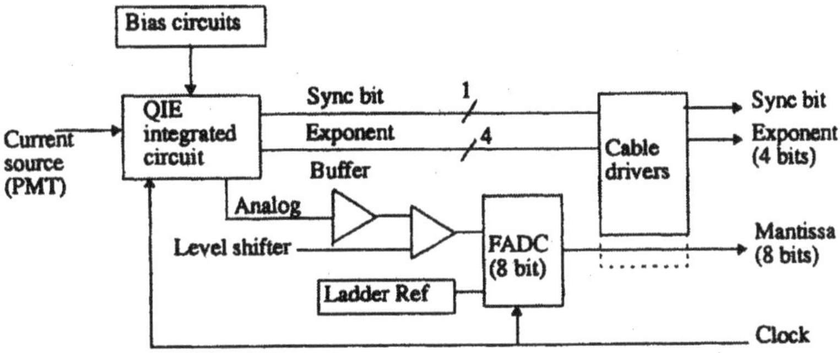

Figure 1: The KTeV multi-range digital PMT base block diagram.

into binary weighted ranges and integrates the divided signal on capacitors for each range. The QIE selects the integrated signal in the appropriate range of interest and presents the analog signal for digitization by the 8-bit FADC. Simultaneously, the QIE generates 4 bits of exponent data, identifying the range of the input signal. The 12-bit floating point data is buffered (to eliminate phase differences between mantissa and exponent data) and then piped to a FIFO using a custom cable driver circuit.

In order to better understand the performance of the QIE, a detailed description of the operating principles is useful. The QIE performs 4 distinct functions. It first adds the PMT signal current to a bias current, divides the total current into 10 binary weighted ranges (I,I/2,...,I/512), and simultaneously integrates the charge on 10 1pF capacitors. The capacitor voltages are then applied to a bank of comparators having a common reference voltage. The comparators are latched, and the latched output is encoded into a 4-bit Gray code number which represents the exponent.[‡] Because of the binary weighting, only one capacitor voltage will be within the required -1V to -2V range. The voltage from this capacitor is output and mapped into the input range of the FADC. The capacitors are then reset. Fig. 2 depicts the device behavior over 4 clock cycles. A bank of four capacitors in each range is necessary for deadtimeless operation of the circuit. Each capacitor performs one operation per clock cycle with capacitors switching in round-robin fashion. Fig. 3 shows a schematic of the DPMT

[‡]The exponent lines from the QIE are differential, reducing pickup on the analog section of the device.

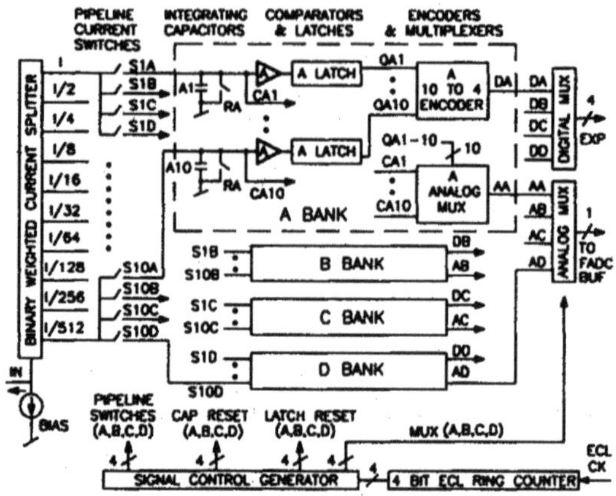

Figure 2: A schematic of the pipelined operations of the QIE ASIC.

analog and digital response versus input charge. The QIE generates a synchronization bit every fourth clock cycle to label the first capacitor in the pipeline.

The DPMT base incorporates both the PMT high voltage divider and the digital readout section. The circuit board dimensions are 0.94" x 9.75", with approximately 2.75" of the length designated for the HV divider. It is designed to fit directly behind a PMT in the dense array of phototubes in the KTeV calorimeter. The PMT anode signal is sent to the DPMT via a 6" coaxial cable. A 50 pin low profile header connector transmits low voltage power to and digital data from the board.

Because of space constraints, the QIE, FADC and buffer/driver chip are wire-bonded directly to the DPMT boards. Thermal vias located on the die pads conduct heat to the back of the card. A 1/32" copper plate is attached to the back of the board as a heat sink. The power dissipation for the digital part of the base is 3.5 Watts; the HV divider adds an additional 0.5 Watt.

3. Bench Tests of the DPMT

Measurements have been made using a 9-range ($I/2 \rightarrow I/512$) QIE device and an 8-bit Analog Devices AD9002 FADC. Bench tests have been performed using a 16-bit DAC programmable charge pulser. To contain the pulse within a single time sample, the pulse was shaped to have a 15 ns tail and the DPMT was clocked at 10 MHz.[§] To simulate a PMT, the pulser output was fed through a current buffer to the AC coupled input of the DPMT.

[§]This device has been run at 53 MHz. However, parasitic capacitance differences between ranges result in slightly different pulse shapes in each range. This is not a problem if the charge is deposited in a single time sample, but it effectively leads to charge loss for a pulse which stradles multiple integration periods and switches ranges. Newer versions of QIE have addressed this problem.

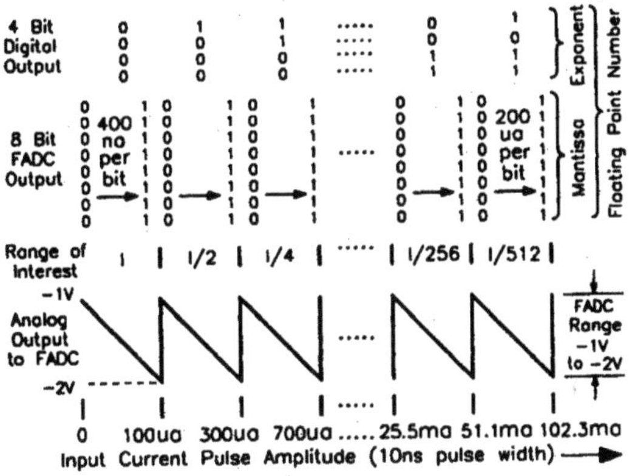

Figure 3: The analog and digital reponse of the QIE ASIC versus input charge. Also depicted is the digitization of the analog signal by an 8-bit FADC.

Tests of noise performance, capacitor gain and pedestal variation, and constants stability were made. DPMT output response for a single time sample versus input charge is shown in Fig. 4. The LSB for the DPMT in the I/2 range was 8 fC per count. The charge limit of the pulser system restricts the ranges probed to the lower six (I/2→I/64). Pedestal noise of 3-4 counts rms was primarily due to the pulser circuit. Similar measurements using a PMT input showed a pedestal noise of 0.3 count rms (2.4 fC). The four capacitors in a range had gain variations of 0.1-0.3% and mean pedestal variations of ± 2 counts. Differences in FADC switch points arise from the 20% variation in range comparator voltages within the QIE.

To study the stability of the correction constants, gains and offsets were derived and applied to data taken 1 week later. Fig. 5 shows the DPMT data after linearization using the "stale" constants. Also shown is the percent difference between corrected data and linear fit. The constants appear to be quite stable over a period of 1 week. The figures are nearly identical to ones made with current constants. The average mean residual is 0.2-0.3%. The remaining structure is common to all ranges (probably due to the FADC) and could be correctable. The resolution versus input charge is plotted in Fig. 6. The dominant contribution to the resolution at the low end is noise from the pulser circuit. The high end resolution is 0.1-0.3%, comparable to the resolution expected from digitization.

4. Performance of the DPMT in a Beam Environment

A test-array consisting of 25 CsI crystals has been instrumented using the DPMT cards described in Sec. 3 and tested in the X1 beam line at CERN. The PMT anode signals were DC coupled to the DPMTs, which were operated at 10 MHz. The 17-bit

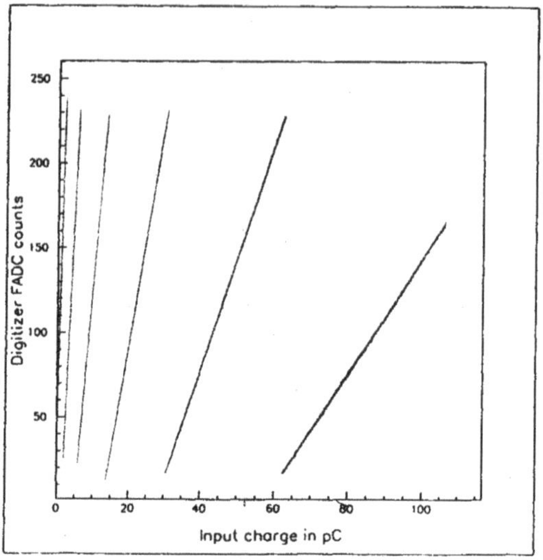

Figure 4: DPMT output in FADC counts versus input charge.

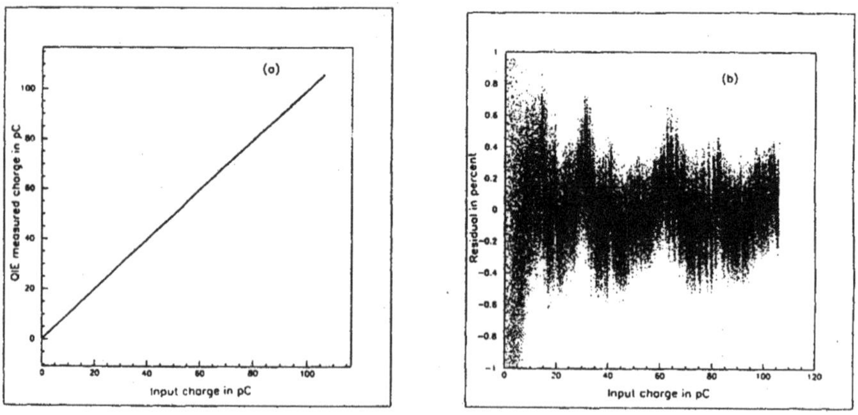

Figure 5: (a) Linearized DPMT output in pC versus input charge, (b) Percent residual from a straight line versus input charge.

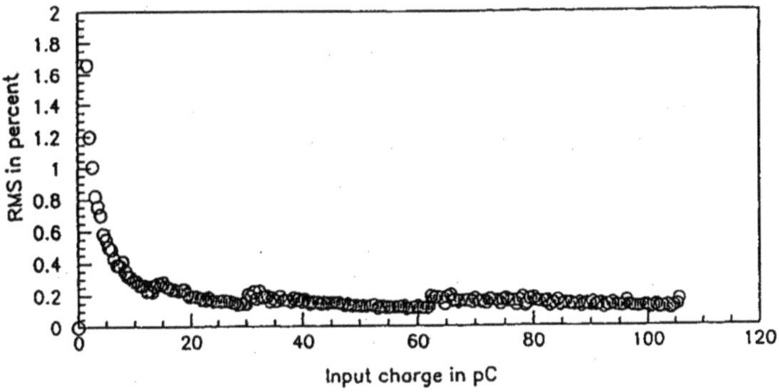

Figure 6: DPMT resolution in percent versus input charge. Breaks in the smooth function arise from switching from a higher sensitivity range to a lower sensitivity range.

dynamic range of the DPMT was tuned to have a lowest range (I/2) LSB of 13 fC, corresponding to 1 MeV, and a full scale energy of 102 GeV. A KTeV prototype laser system[7] was used to calibrate the DPMTs *in situ* over the full dynamic range. A detailed description of this mini-KTeV calorimeter and its performance in the test beam can be found in reference 6.

Beam data covered most of the dynamic range of the digitizer. Typical DPMT pedestal noise for a channel was less than 1 count rms. Pedestals summed over all 25 channels had a 5 count (5 MeV) rms. Data included 60 GeV electrons, 40 GeV pions and transverse muon data which was taken with the calorimeter rotated by 90^0 with respect to the beam. Transverse muons deposit 30 MeV of energy in the 5 cm crystals which were used in the array. The dynamic range of the digitizers is illustrated in Fig. 7. CsI calorimeter E/p resolutions of better than 1% were achieved for electrons between 5→60 GeV energies.

5. Conclusions

A 17-bit dynamic range digitizer circuit has been built and tested. Bench tests of the device show very low noise floors (2.4 fC), good linearity, and stable gains and pedestals over a period of 1 week. A 25 channel CsI array has been successfully instrumented with DPMTs and tested in a beam environment with excellent resolution.

6. Acknowledgements

The author would like to thank the Fermilab Research Division Electrical Engineering Department for their development of the digitizer circuit and the Fermilab Physics Department Electronics Group for their work on the clock modules, power

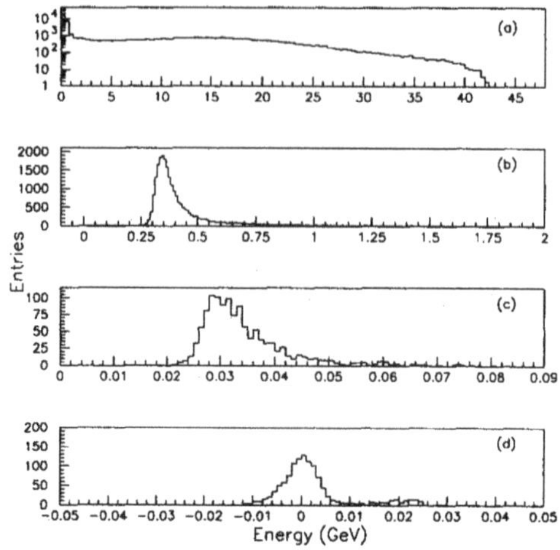

Figure 7: Energy distributions derived from summing 25 channels for data using one 100 ns time sample from a) 40 GeV pions, b) same spectrum for the pion MIP peak, c) 30 MeV transverse muons, and d) pedestals.

sequencers, ribs and CAMAC FIFOs.

7. References

1. R. J. Yarema *et al.*, *A Pipelined Multiranging Integrator and Encoder ASIC for Fast Digitization of Photomultiplier Tube Signals*, Second Annual Conference on Electronics for Future Colliders, (Chestnut Ridge, New York, 1992).
2. R .J. Yarema *et al.*, *A Fast, Wide Range Charge Integrator and Encoder ASIC for Photomultiplier Tubes*, IEEE Trans. on Nuclear Scence, **Vol. 4**, (August 1993), pp.750-752.
3. R. J. Yarema *et al.*, *A High Speed, Wide Dynamic Range Digitizer Circuit for Photmultiplier Tubes*, presented at the 6th Pisa Meeting on Advanced Detectors, (22-28 May, 1994), La Biodola, Isola d'Elba, Italy. Also to be published in NIM.
4. K. Arisaka *et al.*, *KTeV Design Report, Fermilab Report*, **FN-580**, (January 22, 1992).
5. R. Ray, *The KTeV Pure CsI Calorimeter*, these proceedings.
6. R. Kessler, *KTeV CSI Test Beam Results at CERN*, these proceedings.
7. H. Nguyen, *The KTeV PMT Laser Monitoring System*, these proceedings.

ANALOG OPTICAL LINKS IN LIQUID ARGON

Sylvain Tisserant
Centre de Physique des Particules de Marseille (CPPM)
IN2P3-CNRS et Université d'Aix Marseille II
Faculté des Sciences de Luminy, 13288 Marseille, France

ABSTRACT

High granularity liquid argon preshower detectors raise a serious problem for the feedthroughs. Analog optical links may be suited to this application with limited dynamic range. A solution with LED and multimode fibers is studied. Results on dynamic range, linearity and radiation resistance are presented.

1. Introduction

1.1. Initial Motivation

The study, we presente here, was motivated by the read out of the preshower proposed for the Atlas experiment[1]. In the Letter of Intent[2] about 200000 channels were forseen. It would be almost impossible to bring out such a large number of signals through the cryostat using classical copper links and feedthroughs. However due to the reduce dymanic range of the response for a preshower device, a more compact solution may be considered using optical fibers. We report here on the status of a faisability study carried out in the framework of the RD3 program. We will recall first the physics and technical requirements for the read out electronics of a preshower detector in a LHC experiment. We will then describe the components of an optical link. In the following sections we will report on the results obtained on test benches in laboratory and preliminary data from beam tests.

1.2. Requirements

The dynamic range of the preshower detector is mainly driven by two of its main functions. First, the photon identification, which needs the high granularity, requests an electronics noise smaller than about 0.3 mip per channel. Second, the energy correction, to compensate for the fluctuations of the energy loss in the material in front of the calorimeter, yields to 1000 mips for the upper value. The linearity should be better than $\pm 2\%$ on this dymanic range. The bandwidth of the link has to be 50 MHz to cope with the signal characteristics. The emitting components have to work in liquid argon at $87^{\circ}K$. This implies a low power dissipation (≤ 20 mW per channel) and a full reliability over 10 years. Especially this emitting part has to resist to a high radiation level : roughly 10^{14} neutrons/cm^2 and 1 MRad integrated over 10 years at high luminosity.

2. Faisability Study

2.1. Optical Link Description

Different solutions can be forseen to build analog optical links. Some are the purpose of dedicated R&D programs. For complementarity, we chose to study the faisability of a link based on Light Emitting Diodes. Other types of light emitters, like laser diodes or multi quantum well diodes, will be studied in a near future. One purpose of this study was to replace a copper link between the cold preamplifier output and the warm electronics (shaping and digitization) input. As the already existing preamplifier acts as a current to voltage converter while the LED light yield is proportional to its input current, a driver was used in between the preamplifier and the LED. This solution is provisional, if optical links are chosen to read out the Atlas preshower the LED driver function would be integrated within the preamplifier.

The fiber has to be radiation hard. Good results were obtained during previous irradiation tests using step index fibers with a pure silica core, a fluorine-doped silica cladding and a polymide primary coating. The polyimide coating allows the fiber to keep a good mechanical and optical behaviour in the cold. For these fibers two diameters are commercially available : 125 and 250 μm. In order to have the bigger coupled optical power and so to make our life easier during this faisability study we started with large diameter fibers : typically 200/220/245 μm for the core, cladding and coating respectively. One goal of this study was to install a few links to read-out a part of the preshower prototype in a beam test. For this purpose a compact (up to 128 fibers in a 4 mm diameter hole) and cheap feedthrough was made easily using an epoxy resin. Tests were performed at the liquid nitrogen temperature without any leakage nor light attenuation. This confirms that optical links, if well suited to our specification, could easily solve the feedthrough problem.

For the receiver, outside the cryostat, two solutions were available : PIN diodes or Avalanche Photo Diodes. An attractive feature of the APDs is their intrinsic gain. However the stability of this gain is difficult to monitor. For this reason we chose to use PIN diodes as light converters. In order to compensate for the light losses, mostly the coupling effiencies, an amplification is needed. Then the output of each PIN diode is connected to a transimpedance amplifier. Another role of this last device is to match the characteristics of the optical link to the following already existing electronics.

PIN diodes are held to be very linear and stable. So the most critical item in such an analog optical link is the light emitter. The following subsection is devoted to its choice and to its characteristics.

2.2. LED Characteristics

The light emitter acts as an electrical to optical power converter. For the current design of the used preamplifier its maximum input signal is 2.5 V over a 50 Ω impedance. We chose to start with the more powerfull LED type : Gallium Arsenide (AlGaAs) LED which emits a 850 nm wawelength light. Several diodes from different suppliers were tested at room and liquid nitrogen temperatures. Unless otherwise specified, the results shown in the following text correspond to the HFE4050 diode of Honeywell, with which the most part of our systematic measurements was performed. However some other LEDs from other manufacturers fulfil the requirements. For example we also used the 1A362 diode from Abb Haffo during beam tests.

Static and dynamic behaviours, like current-voltage static characteristic, current to optical power conversion, etc..., were studied and compared. For example the figure 1 shows the optical power coupled to the fiber at room temperature. The black squares correspond to the static measurement, while the white ones correspond to the dynamic data. The figure exhibits a good agreement between the two sets of results. This allows us to use static measurements, easier to handle, to sort LEDs. The comparison of the characteristics of a sample of a few tens of LEDs from the same production batch showed a 10% dispersion of the light yield (about 0.01 mW/mA in liquid nitrogen). Such a dispersion can be monitored with the calibration system of the read out electronics chain. The figure 2 was obtained from the static measurement of the coupled optical power in liquid nitrogen (77°K). It shows the residual non linearity of the emitted power with respect to the best straight line fit. From this plot we can determine the polarization point of the diode : 10 mA. This input current corresponds to a bias voltage of 1.6 V and yields to a power dissipation of about 16 mW. The dynamic signals are added with respect to this polarization point. From the figure 2 we see the excursion available for dynamic signal : from 0 up to 50 mA, with a linearity well better than ±1%.

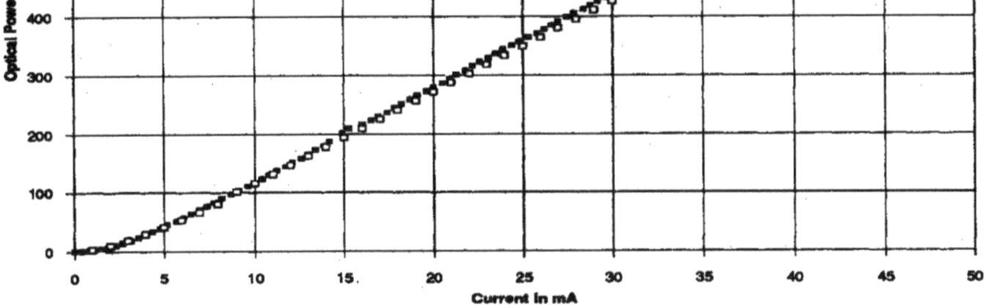

Fig. 1. Static and dynamic coupled power at room temperature for an HFE4050 Honewell diode

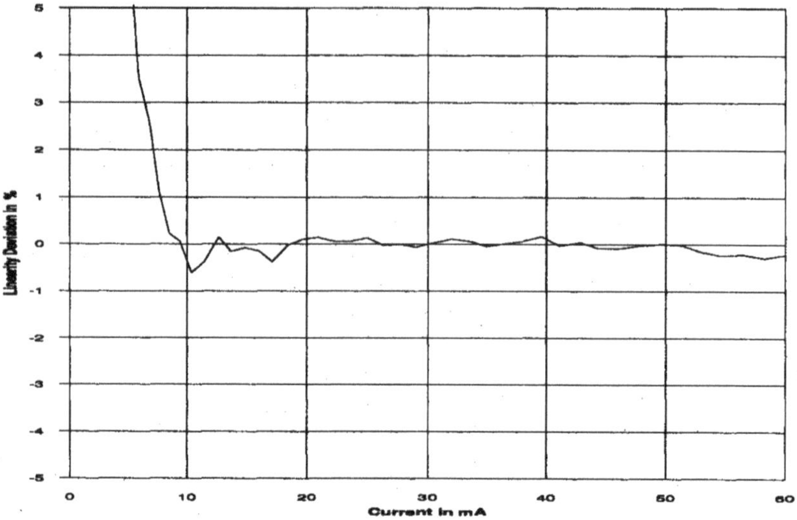

Fig. 2. Linearity deviation for an HFE4050 Honeywell diode at 77°K

Concerning the LED radiation hardness, neutron irradiation tests were performed at Grenoble showing, for two types of LED (HFE4050 from Honeywell and HBFR1404 from Abb Haffo), a 10% light yield loss for a 1 MeV equivalent neutron integrated fluence of 10^{14} cm^{-2}.

2.3. Optical Link Characteristics

Starting from these very encouraging results we set up a few complete optical links, including LED with its driver, multimode fiber, PIN diode and transimpedance amplifier. For this faisability demonstration only on the shelf available components were used. The characteristics of such a full optical link were measured in laboratory at room and liquid nitrogen temperatures. The figure 3 shows the non linearity of its response (defined as V_{out} versus V_{in}). The experimental points are distributed between ±2%, mainly due to the measurement accuracy, over a range running from a few mV up to 1 V. A saturation effect can be observed for larger values, it is induced by the transimpedance amplifier. This amplifier has also a limited bandwidth of about 30 MHz. The bandwidth of the optical link without this transimpedance amplifier is 50 MHz, reaching the specifications.

In order to compare the noise contribution of each optical link component with respect to the input noise, we need to take into account the different gain or attenuation factors of the link. The output noise can be written as :

$$N = G_t \sqrt{N_t^2 + (A\,G_d)^2 (N_d^2 + N_i^2)} \qquad (1)$$

with :

G_d : transconductance driver gain ($G_d \approx 20\ mS$);

A : optical link attenuation factor : takes into account conversion efficiencies at the emitter and receiver level, light losses in the connections and ligth attenuation in the fiber ($A \approx 1/200$);

G_t : transimpedance amplifier gain ($G_t \approx 4000\ \Omega$);

N_i : detector noise, it includes also the preamplifier noise ($N_i \approx 300\ \mu V$);

N_d : equivalent input noise of the driver ($N_d \approx 100\ \mu V$);

N_t : equivalent input noise of the transimpedance amplifier ($N_t \approx 75\ nA$).

The noise induced by the optical link has to be negligeable with respect to the detector noise N_i, that means :

$$N_d \leq \frac{1}{3} N_i \quad and \quad N_t \leq \frac{1}{3} A\, G_d\, N_i \tag{2}$$

to keep the respective contributions smaller than 5% each. Using the typical values, indicated in brackets, we see the driver noise is small enough, but the transimpedance amplifier noise is dominant.

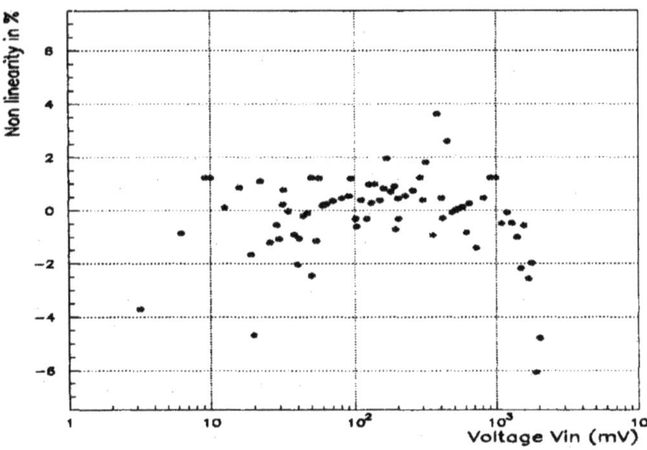

Fig. 3. Nonlinearity over the dynamic range of the optical link

2.4. Conclusions and Near Future

The results presented above show that the performances of this first analog optical link are close to the requirements. The limiting element of the chain is identified as the transimpedance amplifier. Developments of a new one suited to our needs are going on. In parallel 32 links were used to read out the preshower prototype in beam test in September and October at CERN. Preliminary results, using the calibration system, confirm our laboratory measurements.

The faisability studied may be considered as achieved. We are already working toward a solution suited to the Atlas experiment. This includes developpments, in close contact with industry, on the integration of the cold components (LED and driver) in a radiation hard technology. Multi-fiber ($\sim$ 128 fibers) compact connectors will also be tested with fiber ribbons.

3. References

1. J. Collot, Construction and test of a fine-grained liquid argon preshower prototype, these proceedings.
2. Atlas, Letter of Intent for a General-Purpose pp Experiment at the Large Hadron Collider at CERN, CERN/LHCC/92-4.

THE LEVEL-1 CALORIMETER TRIGGER FOR THE CMS DETECTOR AT LHC *

S. Dasu[†] T. Gorski, J. Lackey, D. Panescu, W. H. Smith, W. Temple
Department of Physics
University of Wisconsin
Madison WI 53706

ABSTRACT

The level-1 calorimeter trigger system for the CMS detector at LHC collider provides a deadtimeless pipelined trigger that implements simple algorithms which satisfy the physics efficiency and data rate requirements. We plan to run the trigger system at 160 MHz in order to provide a compact system by quadrupling the use of most components and data paths within an LHC crossing of 40 MHz. Many logic elements in the design are implemented in custom integrated circuits with much attention paid to the trigger data optimization and the latency in processing. Through extensive simulation of the level-1 calorimeter trigger, we have determined the background rates and efficiencies for various triggers such as those for electrons, photons, jets and neutrinos.

1. Introduction

We present here a conceptual design of a level-1 calorimeter trigger system for the CMS detector [1] which is being built for operation at the LHC collider at CERN. For the nominal LHC design luminosity of $10^{34}\text{cm}^{-2}\text{s}^{-1}$, an average of 25 events occur at the beam crossing frequency of 40 MHz. This input rate of $\approx 10^9$ interactions every second must be reduced by a factor of at least 10^7 to 100 Hz, the maximum rate that can be archived by the on-line computer farm. CMS has chosen to reduce this rate in two steps. The first level stores all data for 3 μs, after which no more than a 100 kHz rate of the stored events is forwarded to the higher level triggers. This must be done for all channels without dead time. A second level of trigger is provided by a subset of the on-line processor farm, which passes a fraction of these events for further filtering by the remainder of the on-line farm. During the 3 μs of the level 1 trigger processing time, trigger decisions must be developed that discard a large fraction of the data while retaining the small portion coming from interactions of interest. The large physical size of the detector and the short decision time present a series of technical and system problems.

2. Level 1 Calorimeter Trigger Design

The level 1 calorimeter trigger system consists of regional and global processors.

*Research funded by U.S. Department of Energy contract DE-AC02-76ER00881 and Texas National Laboratory Research Commission grant RGFY93-205.
†Speaker

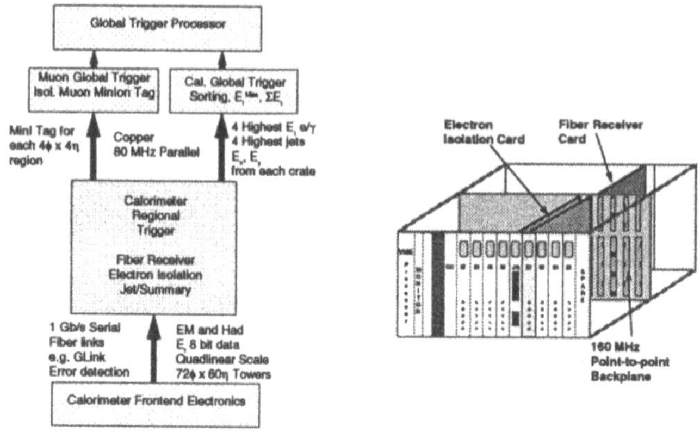

Figure 1: CMS Level 1 Calorimeter Trigger Overview and details of one crate.

The regional system processes the electromagnetic and hadronic trigger tower sums from the calorimeter front end electronics and delivers regional information on electrons, photons, jets and partial energy sums to the global calorimeter level 1 trigger system. The regional system begins after the data from the front end electronics is received on optical fibers and translated to signals on copper and ends with cables that transmit the results to the calorimeter global level 1 trigger system.

After establishing requirements for the performance of the trigger system [2] which fulfill the physics goals of the CMS experiment, we have produced a conceptual design of the level-1 calorimeter regional trigger system. General considerations that have been emphasized in this conceptual design include access to components, requirements of power, space and cooling, information for diagnostics, efficiency and performance monitoring, traffic and timing on backplane, feasibility of I/O connections, and interface to DAQ and clock/control. The design is implemented by using off-the-shelf technology whenever feasible. ASICs are used only where fully justified. The logic design maximizes flexibility and programmability using memory lookup tables and programmable gate arrays.

A conceptual block diagram of the CMS Level 1 Calorimeter trigger system with detail of one calorimeter regional trigger processor crate is shown in Figure 1. The input data to the regional trigger processor consists of EM and Had transverse energies arranged in $72\phi \times 60\eta$ trigger towers of size $0.087\phi \times 0.087\eta$ each. Each regional processor crate contains 8 Receiver cards, 8 Electron Isolation cards, a Jet/Summary card, a crate processor card and a sytem monitor card. It is designed to fully process 256 trigger towers. The rear section of the Fiber Receiver Card shown in Figure 2

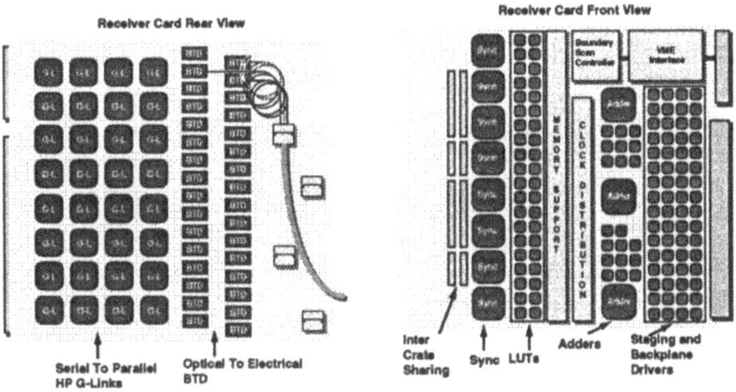

Figure 2: Fiber Receiver Card front and rear views.

converts serial data, received from thirty two optical fibers (each containing eight bit data from two towers and five error detection bits) to parallel data and transmits through the board to the front section of the card (See Figure 2). A custom ASIC synchronizes the incoming data to the trigger system 25 ns clock and aligns it to the correct crossing number. The fiber receiver does error detection, using error codes transmitted with the data, and logs the errors for subsequent readout. The 8-bit nonlinear scale data is decoded with memory lookups and fed out at 160 MHz. $4\phi \times 4\eta$ jet sums are processed on board using the adder ASIC discussed below. The Jet/Summary card shown in Figure 3 receives two $4\phi \times 4\eta$ jet sums from each of the eight Fiber Receiver cards and computes total E_t, E_x and E_y for the 256-tower region within the crate. A place for τ trigger and luminosity monitoring circuitry is reserved on this card. The Electron Isolation card shown in Figure 3 receives data at 160 MHz in a staged fashion from two neighboring Fiber Receiver cards and performs the electron identification algorithm in a custom ASIC. The results are sorted in another custom ASIC and the top four candidates within a 32-tower region are transferred to the Jet/Summary card. The Jet/Summary card does a further sort to output the top four electron, jet and tau candidates in addition to the calculation of total E_t, E_x and E_y information in the 256-tower region.

3. Adder ASIC

As a test of the technology proposed here, we are completing an integrated circuit that adds 8 11-bit numbers (in addition to overflow bits, totalling 13 bits) in 6 ns. This

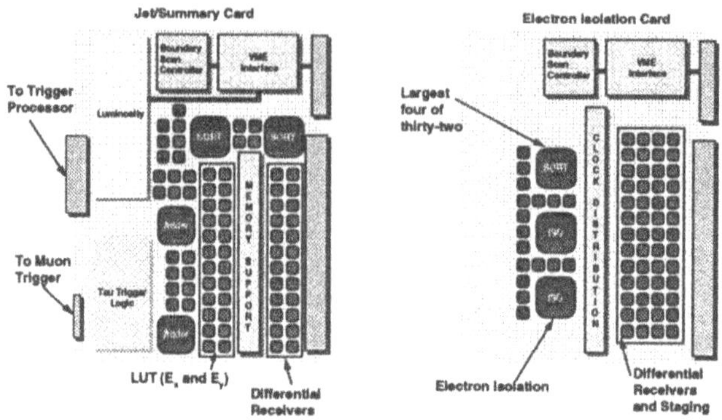

Figure 3: Jet/Summary and Electron Isolation Cards.

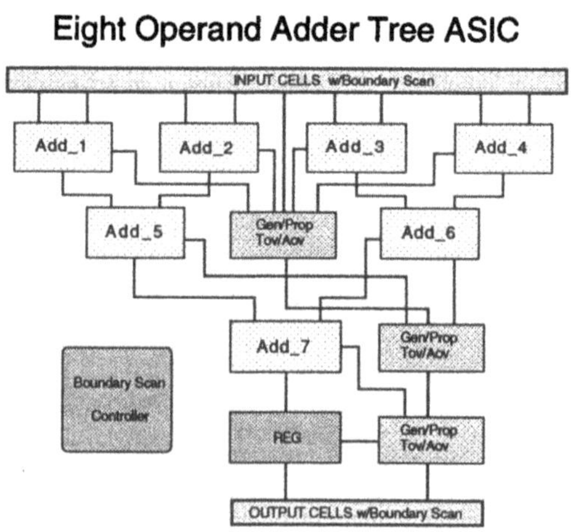

Figure 4: Design of GaAs 8 x 11 bit Adder ASIC.

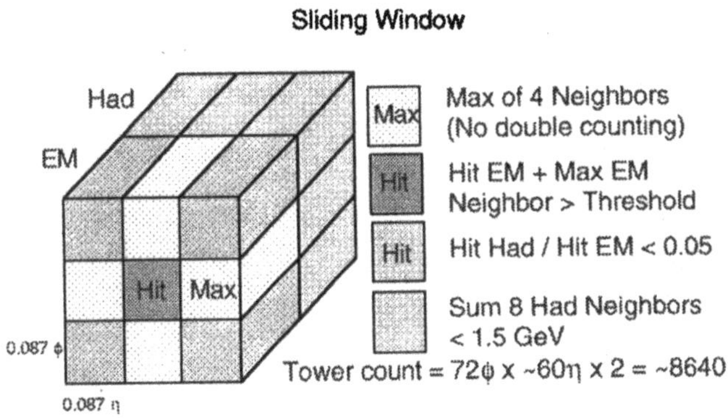

Figure 5: CMS Level 1 Electron Trigger Algorithm.

device is being developed in cooperation with Vitesse corporation in GaAs technology. The design is shown in Figure 4. The high speed GaAs technology was chosen to provide an opportunity to run sections of the trigger logic at four times the LHC clock frequency and for its ECL I/O capability. This can result in a considerable reduction of circuitry in the trigger data path due to a single circuit performing the job of 4 circuits and elimination of ECL/TTL translators.

4. Calorimeter Trigger Simulation

We have used a Monte Carlo program to simulate several hundred thousand QCD jet and other events in the CMS detector using a simplified geometry with parameterized detector response. The high resolution EM calorimeter transverse energy data was summed in regions of $0.087\phi \times 0.087\eta$ to provide trigger data. Both EM and HAD data were further reduced in resolution and dynamic range to account for 8 bit quadlinear scale transmission through the fiber optics system. Figure 5 shows our preferred proposal for an EM trigger algorithm. The electron trigger algorithm considered all EM transverse energy sums of adjacent trigger tower pairs over a threshold. The algorithm involves two separate cuts on the longitudinal and transverse isolation of the EM energy deposit. The first cut involves hit tower HAD to EM energy ratio (H/E < 0.05). A second cut requires transverse isolation, i.e. a cut on the sum of HAD transverse energies in the nearest eight towers surrounding the hit tower (ΣH < 1.5 GeV). In order to reduce the number of bits of information exchanged between electronics cards the dynamic range of neighboring tower HAD information

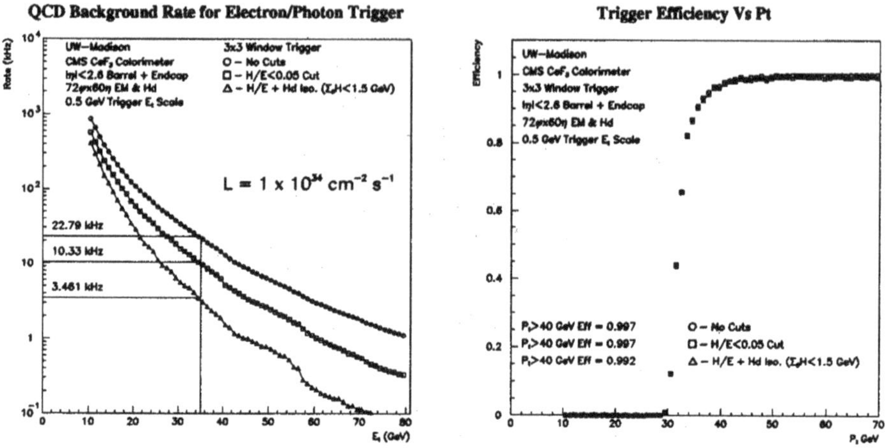

Figure 6: Single electromagnetic cluster trigger background rate for CMS is plotted versus the two tower sum transverse energy deposited with various cuts. The circular points are without any cuts, the square points are after H/E cut and the triangular points are after isolation cut. (b) Efficiency of triggering on electrons is plotted versus p_T of the electron for various algorithms.

is limited to 2 bits. Overflows of both the 8 bit scale for EM and central HAD towers, and 2 bit scale for neighboring HAD towers are treated as maxima. The jet trigger algorithm involves sums of transverse energy over $4\phi \times 4\eta$ towers. An overlapping sums of transverse energies in $8\phi \times 8\eta$ regions is also considered. Missing E_t and single hadron triggers also make use of the $4\phi \times 4\eta$ transverse energies.

We have simulated the above algorithm, including the details of bit resolution and dynamic range, as designed in the hardware and obtained the electron/photon trigger QCD background rate and efficiency for detecting single electrons with minimum bias background. The rates from QCD two jet production background and the efficiency for electrons are shown after various cuts in the algorithm in Figure 6. We have also studied jet trigger rates and efficiencies (see Figure 7) and found that transverse energy sums in $0.4\eta \times 0.4\phi$ region or overlapping regions of $0.8\eta \times 0.8\phi$ provide sufficient performance. The rate and efficiency performance of both electron/photon and jet triggers is within the physics requirements of the CMS detector.

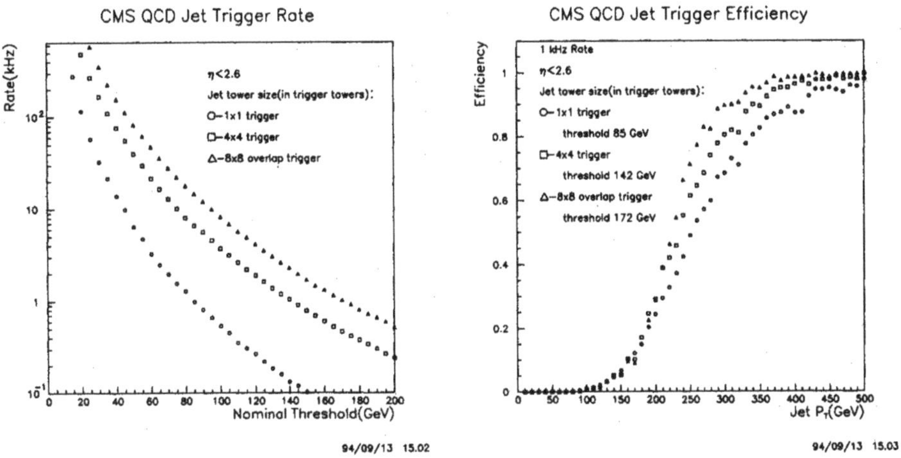

Figure 7: (a) Jet trigger rates using various tower region sums are plotted versus the E_t deposited in the calorimeter region. The rate includes only barrel and endcap calorimeters, i.e. $|\eta| < 2.6$. (b) Efficiency of jet trigger for various tower region sums is plotted versus the jet p_T.

5. References

1. CMS, The Compact Muon Solenoid, Letter of Intent, CERN/LHCC 92-3, 1 October 1992; Status Report and Milestones, CERN/LHCC 93-48, 15 October 1993.
2. S. Dasu et al., *CMS Level-1 Calorimeter Trigger*, Proc. of the Int. Conf. on Computing in HEP, San Fransisco, 1994.

Optical Readout, Monitoring, Radiation Damage

Session Chair: C. Wuest (LLL)

Report on Tests of Avalanche Photodiode Readout for Fast Scintillators

Charles Findeisen, Universität Zürich
Thomas Flügel, Dieter Renker, PSI
Eckart Lorenz, Max-Planck-Institut für Physik und Astrophysik, München
Nicolay Root, Boris Schwartz, Institut of Nuclear Physics, Novosibirsk

Introduction

Future calorimeters for LHC, for a B-factory or for nuclear medical instruments need readout devices, which are cheap because of the large number of elements, fast, compact, insensitive to magnetic fields, which can stand high radiation levels and which produce only little noise because fast scintillators like CeF_3, pure CsI, BaF_2 or $PbWO_4$ have only little light output.

Silicon Photodiodes (PD) are candidates for photosensors in strong magnetic fields, but there are still minor problems with noise because of the leakage currents and a large parallel capacity and there is the big problem with the so-called nuclear counter effect. For low yield scintillators the stray particles, leaking out at the rear end of the crystals, can produce a signal as large as the scintillator signal itself.

New state of the art avalanche photodiodes (APD) can overcome the problems of PDs . Both, the noise and the nuclear counter effect are in first order reduced by the gain factor (typically 50). Only the very thin (1-2 micrometer) photosensitive layer of silicon contributes fully to the signal from stray particles, while the rest of ≈30 micrometer in and behind the amplification region only contributes with reduced amplification . This means, that in a PIN diode 20 000 photons are needed to produce a signal as big as the signal from one single minimum ionizing particle (80 electron-hole pairs per micrometer in a 200 micrometer wafer, quantum efficiency 80%), while in an APD this number is only 1700 (assuming a gain of 50).

The small sensitive area of most of the APDs (10-20 mm^2) is not necessarily a drawback, since the small noise contribution, which can be lower than an equivalent of 30 photoelectrons [1] , allows to use only a fraction of the scintillation light. With the help of fluorescent flux concentrators, thin sheets of lucite doped with a dye with large Stokes shift which are located at the rear end of a crystal and which are read out by APDs at the side, it is possible to collect 25 % of the total light produced and this, because of the high quantum efficiency, brings the performance of APD readout close to that of photomultipliers (and reduces the nuclear counter effect even more) [2]. But already now there are APDs on the market with areas greater than 200 mm^2 [3].

Compared to APDs from earlier times, the new ones [4] are very stable against radiation [1], careless operation [5] and ambient influences like temperature changes.

Test of the Nuclear Counter Effect

The potential of APDs has been demonstrated in a 80 GeV test beam (H2 at CERN). A $PbWO_4$ crystal of 18 cm length was read out by a PIN-PD and alternatively by an APD. While the response of the PD readout showed a long tail at the higher energy end of the spectrum caused by rear end leakage, the response of the APD readout had a Gaussian shape without a visible tail [6].
For a direct comparison of the nuclear counter effect a PIN-PD (S3590-01)and an APD (S5345 SPL, 85 pF) were mounted in a beam of 84 MeV protons (πM1 at PSI) and the response was measured under identical conditions. The energy loss of a proton in silicon at this energy is four times higher than the loss of a minimum ionizing particle giving 320 electron-hole pairs per micron and therefore 64000 pairs are created in the 200 micron wafer of the PIN-PD. The APD was operated at a gain of 24. Calculating from the relative position of the APD response (Fig. 1a and 1b) and taking the average gain into account a thickness of 30 micron was determined, a value which was as well derived by comparing the capacities. The suppression of the nuclear counter effect in the APD relative to PIN-PD is given by the ratio of the peak positions in figures 1a and 1b divided by the gain. Even at the relative low gain of 24 the suppression factor is 13.

Fig. 1a: Response of a PIN-PD to 84 MeV protons. Pedestals (channel 160) are not subtracted.

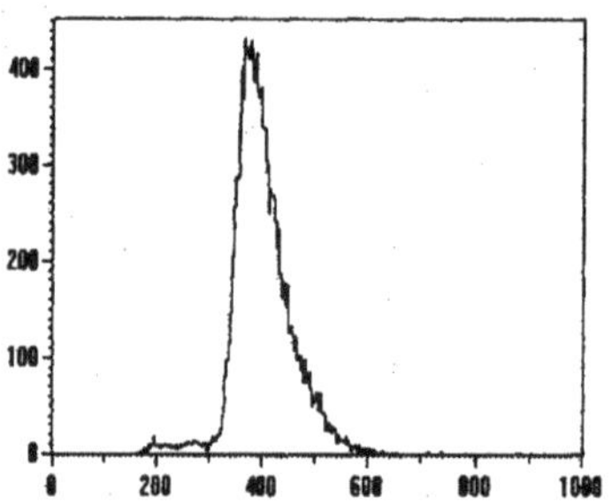

Fig. 1b: Response of an APD to 84 MeV protons. Pedestals not subtracted. The peak at channel 180 is caused by the noise due to the higher dark current in the APD.

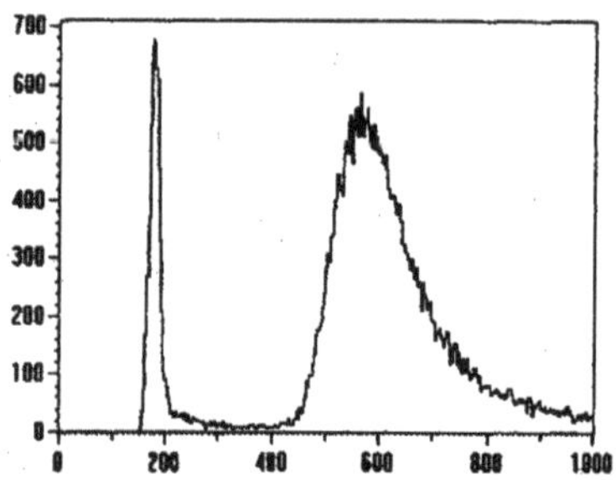

Radiation Hardness

An APD type S5345 of the high capacity (320 pF) type and a high quality PIN photodiode S1790 (both from Hamamatsu) were exposed to a Co^{60} source of 70 kRad/hour. Both devices were held under operation tension throughout the exposure. Fig. 2 shows the increase of the dark current after 0.25, 2 and 8 Mrad. In both devices the dark current increased substantially. While the PD showed a serious degradation in resolution (tested by the widening of the signal from direct conversion of the 60 keV γ line from an Am^{241} source in the silicon), no degradation was observed in the APD operated at a gain of 50. The increase of the dark current in the APD is mostly caused by a higher surface leakage while the bulk current in the photosensitive volume in front of the avalanche zone was less affected. This was confirmed by the line resolution for the low energy γ lines (13.9, 17.6 and 21.0 keV) from Am^{241} converted in the photosensitive volume which were not significantly broadened.

Fig. 2:

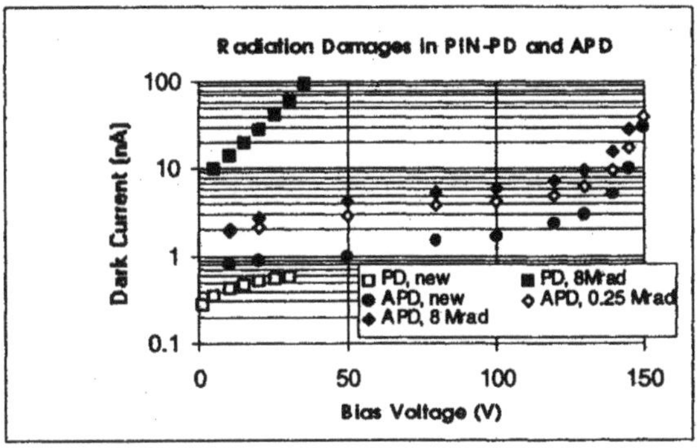

For one minute an APD S5344 (120 pF) was exposed to a 72 MeV proton beam giving a dose of 200 kRad in one shot. Again the diode was under bias voltage. The dark current went up by a factor of 500 immediately after the irradiation, but came down to a factor of 170 after two days. In order to speed up the healing the diode was annealed for one day at 250 °C. The dark current then was 30 times higher than it was before the irradiation.

Temperature Dependence

The temperature dependence of the APD gain was studied in the range of -20 to +50 °C. In Fig. 3 the gain of a low capacity (85 pF) type APD is plotted as a function of bias voltage for different temperatures. It was possible to derive a rule (Fig. 4) which is valid in the temperature range from -20 to +20 °C :

$$1/G * dG/dT = -(0.025 \pm 0.003) * G \quad \%/^{\circ}C \qquad ; \qquad G = \text{gain}$$

At a typical gain of 50 this makes −1.25 %/°C, which is close to the temperature dependence of the light output of most scintillating crystals (e.g. $PbWO_4$).

The gain of the high capacity APD was measured too as a function of temperature. At a gain of 50 the dependence is −0.85 %/°C in the range of 0 to 50 °C.

Fig.3:

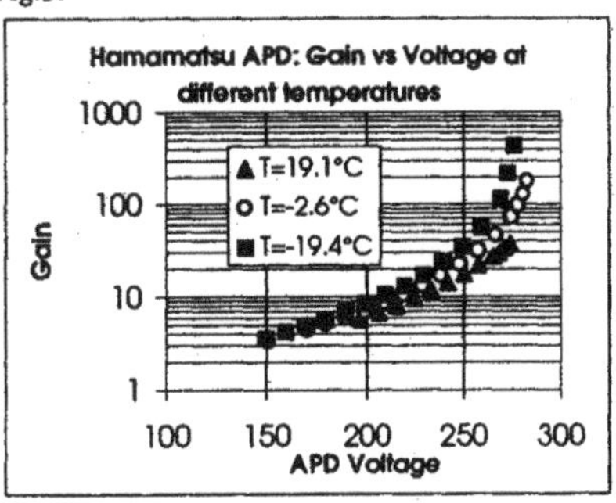

Fig. 4:

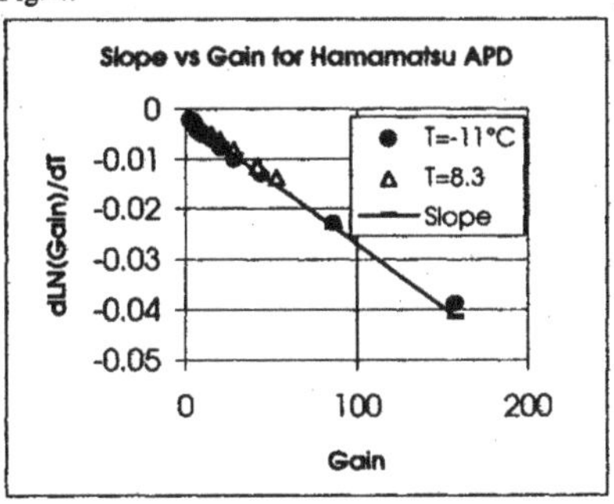

An avalanche photodiode produced by EG&G (CD2192, 5x5 mm^2)was tested in parallel. In contrast to the Hamamatsu diode, which is made by epitaxial growth, the EG&G diode is made by ion implantation. Therefore its capacity (25 pF compared to 85 pF) and the dark current are smaller. But the temperature dependence of the gain (Fig. 5) and the dark current is stronger and it was not possible to derive a corresponding rule.

Fig. 5:

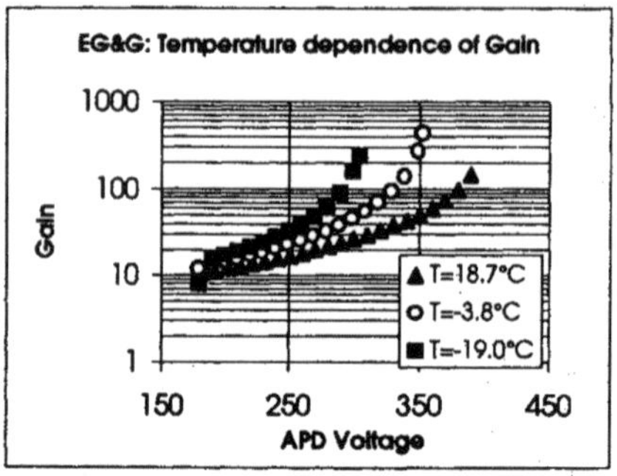

Sensitivity to Magnetic Fields

As expected there is no gain variation in magnetic fields. The stability of the gain - it was set to 50 - was measured in fields up to 5 T parallel and vertical to the electric field in the diode by determining the direct conversion of low energetic γ lines from an Am^{241} source in the photosensitive volume of the APD (Fig.6a and 6b). The same result was obtained at lower bias voltage (gain 25). The small variations of less than one percent are mostly caused by changes of the temperature during the long ramping time of the magnet, which was not controlled in these measurements.

Fig. 6a:

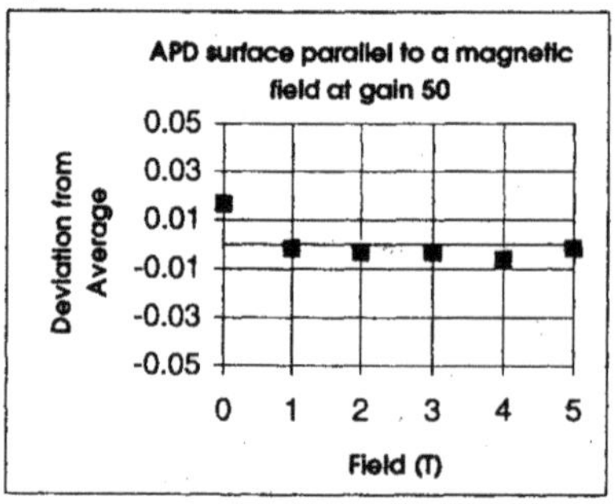

Fig. 6b:

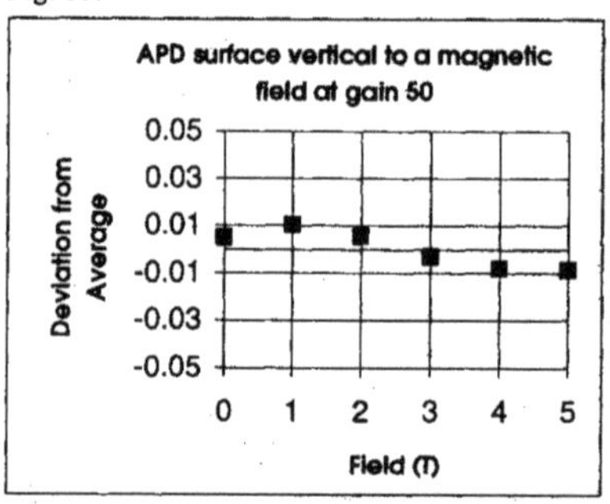

Conclusion

It should be mentioned that the use of large area APDs is still in its infancy compared to the well established PD readout and significant progress can be expected in the near future. In summary APDs present a large improvement over PIN PDs in the readout of low level, fast signals. In many applications they can even replace photomultipliers.

References

[1] Fast readout of plastic and crystal scintillators by avalanche photodiodes, E. Lorenz et al., Nucl. Instr. Meth. A344 (1994) 64

[2] E. Lorenz et al., Nuc. Inst. Meth. A249 (1986) 235

[3] Advanced Photonix, Inc., Camarillo, CA 93012, USA

[4] we used for our tests the APDs S5344 and S5345 (the type with 320 pF capacity as well as the new low capacity type with 85 pF) and for comparison a PIN-PD S3590-01, all from Hamamatsu Photonics K.K., Ichino-cho, Japan

[5] Test of a fast, low noise readout of pure CsI crystals by avalanche photodiodes, E. Lorenz et al., to be published

[6] Ch. Seez, Workshop on Electromagnetic Calorimetry in CMS, Annecy, 13.-14. January 1994

The DEP Hybrid Photomultiplier Tube

L.Boskma (*), R. Glazenborg, R. Schomaker

Delft Electronische Producten (DEP), PO Box 60, 9300 AB Roden, Holland
fax 31-5908-13510

(*) corresponding author

Abstract
The concept of the DEP Hybrid Photomultiplier Tubes (scientifically better known as Hybrid PhotoDiode) is explained and an overview of the key performance data is given.

1 Introduction

Although the concept of a Hybrid vacuum tube with photocathode and diode is twenty years old, the idea has been revived for physics 6 - 8 years ago by Dr. R DeSalvo et al. (CERN). The combination of high energy resolution with total insensitivity for magnetic field should make these devices the ideal read-out for e.g. calorimeters. Several development projects have been completed successfully and today a range of products is commercially available.

Many people and groups have joined the HPD collaboration over the last four years. Presently the majority of the work is done by CERN and INFN (evaluation, application), Canberra Semiconductors (Diode development and manufacturing) and DEP (tube development and manufacturing).

2 Principle

The HPMT is a vacuum tube with a photocathode coupled to a silicon PIN diode. Photoelectrons, created in the photocathode, are accelerated towards the PIN diode with several keV and create a track of electron/hole pairs near the surface of the silicon chip. As the diode is reversely biased, electrons and holes are collected on the electrodes.
An incoming photon that creates a photoelectron will produce a charge pulse on the electrodes.

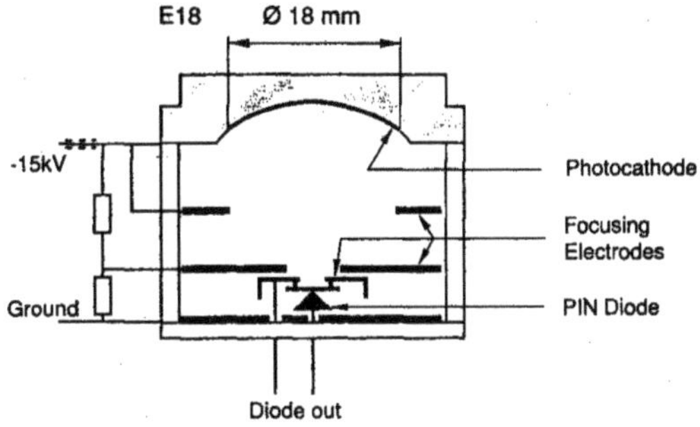

Fig. 1
Electrostatically focused
Hybrid Photomultiplier Tube

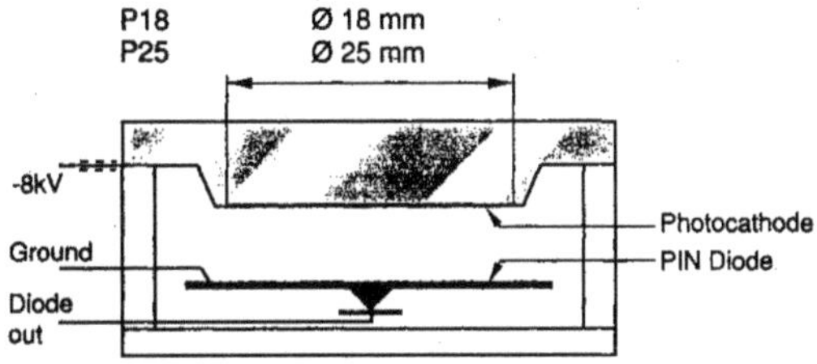

Fig. 2
Proximity focused
Hybrid Photomultiplier Tube

Figures 1 and 2 show the two basic concepts, which differ in electron focusing. Electrons will leave a point on the photocathode with an energy of a few eV oriented in random directions. Focusing is necessary to collect all these electrons on about the same spot on the diode plane. Of course this is especially relevant in pixellated versions.
Fig 1 shows the electro-statically focused version. In this concept a large photocathode can be combined with a small diode, thus giving the fastest timing, and a high photo-electron resolution (refs. 1).
Fig 2 shows the proximity focused version, which is total insensitive for strong magnetic fields (refs. 2 and 3). Ref. 4 gives a recent overview of the concepts.

3 Gain

The gain is linear with the high voltage following the formula
G = (HV-V(th)) / 3.6, where "HV" is the applied high voltage, "V(th)" the threshold voltage and "3.6" the energy, expressed in eV, necessary to create an electron/hole pair. For an acceleration voltage of 15 kV and a threshold of 2 kV this will give a typical gain number of 3500 electrons per photo-electron (Fig 3). Presently the proximity focused versions operate at a lower voltage. A version that operates at a higher voltage, combined with a lower threshold, is just coming into production.

It should be noted that the gain stability of such a device is much better compared to conventional Photomultipliers. This is because the gain is linear rather than exponential with the high voltage, and the gain is virtually independent of the environmental temperature.

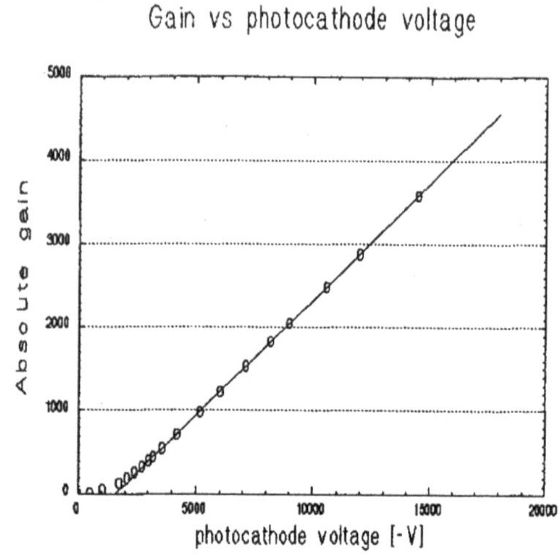

Fig. 3 Gain versus Acceleration voltage

The dynamic range has been measured on proximity focused devices. Fig 4 shows both the output pulse height and the output pulse integrated charge as a function of the # of photoelectrons per pulse. Eight orders of magnitude are achievable linearly. The jump is due to a change in amplifier.

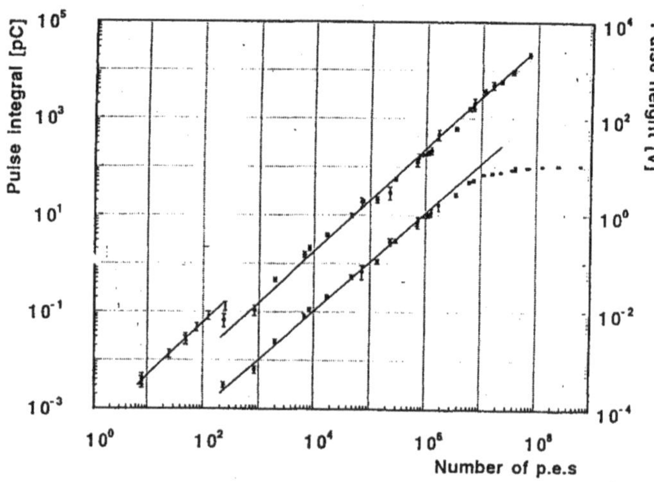

Fig 4 Dynamic range

4 Photocathodes

Highly efficient thin S20 and bialkali photocathodes have been developed for these devices to make the ideal read-out for e.g. scintillators. Fig 5 gives the spectral response (Quantum efficiency versus wavelength) of some examples of available photocathodes. Especially developed for the HPMT are a low darkcount bialkali photocathode and a high QE (>30%) S20 version for UV. Special windows can easily be implemented at low cost due to the photocathode transfer technique.

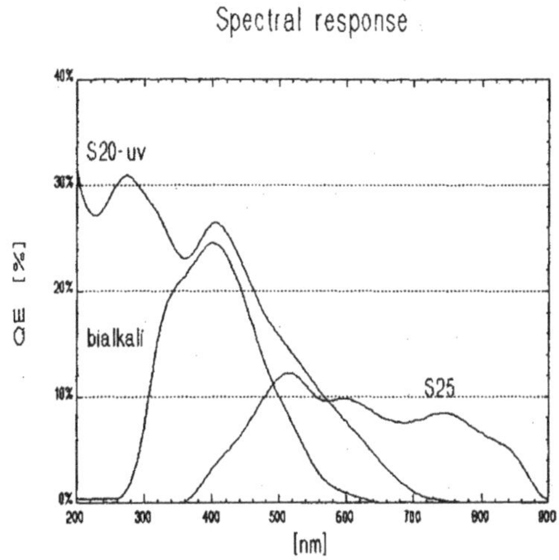

Fig. 5 Typical HPMT photocathodes

5 Noise

Three noise sources are being discussed. The first one is the variation in photoelectron gain. The second source is the thermal electron emission from the photocathode (darkcounts), and the third one is a collection of second order sources like the silicon read-out noise, cosmics and small discharges.

5.1 Gain variation and Photoelectron resolution

Because of the absence of electron multiplication stages, the counting statistics of an HPMT are much better than in regular PMT's. In fact since the relative photon to photon gain variations are extremely low, a peaked pulse height distribution is obtained with separate peaks for separate photo-electron numbers. Fig 6 shows a photoelectron response distribution in an electrostatically focused version, as measured by CERN-LAA (refs 5).

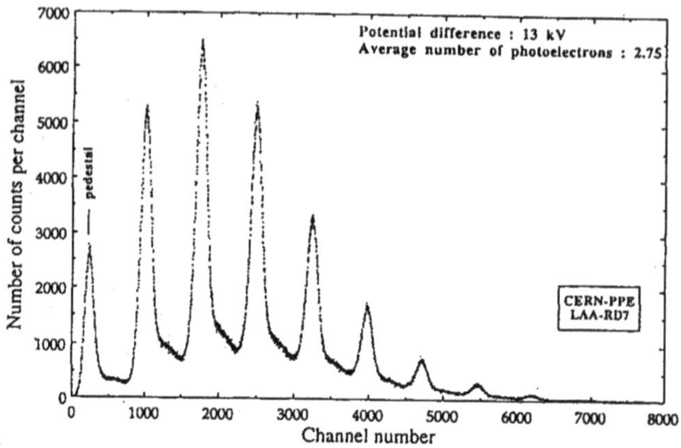

Fig 6
Photo-electron resolution of the electrostatically focused HPMT

This noise (gain variation), expressed as equivalent photoelectrons, is depending on the electron gain and on the chip capacity. A high acceleration voltage, combined with a low capacity (e.g. small chip), will of course give the best results, as is shown in fig. 6. The chip capacitance will contribute to the electronics noise of the preamplifier. For the worst case, an 8 kV proximity focused version, with a large 200 pF capacity chip and an optimized amplifier, a noise equivalent to 4 photoelectrons is achieved. Lower capacity chips, combined with a 20 kV tube design, bring this down to about 1 or 2 photoelectrons. This is important, as the # of photoelectrons in e.g. a Cherenkov counter, may be only a few.

5.2 Darkcounts

The thermal electron emission of the photocathode is a noise source. The signal from such an electron cannot be distinguished from a single photo-electron. The countrate is depending on the type of photocathode and on the temperature. At room temperature an S20 photocathode will give a few hundreds of counts per second per cm^2, for a bialkali this is at least an order of magnitude lower. Cooling decreases these figures with a factor of two for every 7 degr Celsius.

5.3 Additional sources

The sources described in par 5.1 and 5.2 are by far the dominant noise sources.
The noise from the silicon itself is in the order of tens of electrons. As the signal is a few thousands electrons, this source is negligible. Cosmics may add spurious signals. These are comparable with the effect in silicon strip detectors. The interaction with the photocathode is zero due to the low thickness, but in theory cosmic rays may also create Cherenkov light in the photocathode input window. Incidentally small electrical discharges in the tubes may give spurious peaks.

6 Magnetic field behavior

The gain of proximity focused versions has been measured as a function of the magnetic field. All components of these tubes were non-magnetic. Fig 7 shows the normalized gain versus the magnetic field up to 1.6 Tesla, for several angles between the tube axis and the magnetic field. The modulation occurs because the spiraling electron trajectories lead to variable impinging angles on the chip and thus to variable effective thicknesses of the threshold layer. (See for calculations ref. 3) PMT's, including fine mesh versions, are always sensitive for magnetic fields, especially if tilted, because their gain mechanism (bouncing over dynode structures) is heavily influenced by the magnetic field bending. Recent tests with HPMT's have proven that operation up to 2.5 Tesla is still without loss of gain. No breakdown is foreseen in stronger magnetic fields.

Fig 7
Gain versus
magnetic field

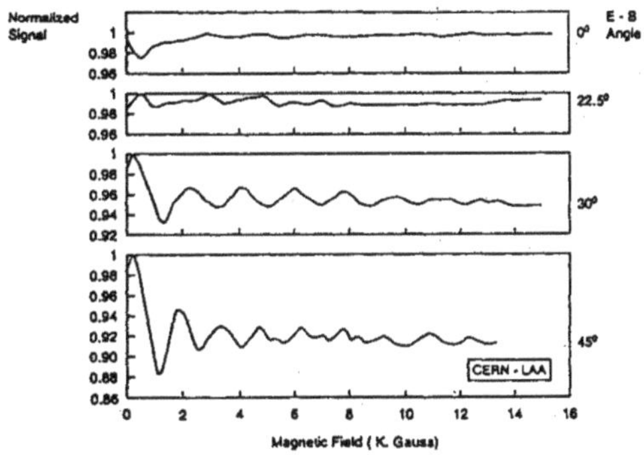

7 Timing properties

Small chips combined with a tube design with small electron transit time differences, gives the shortest rise and fall times. Fig 8 shows the output pulse for an electrostatically focused tube.
Work is done on prototypes for several large projects at CERN and INFN Italy, which require both proximity focusing and fast timing. Therefore a lot of effort has been put in combining the large chips necessary for proximity focusing (ϕ 25 mm) with fast timing properties. Timing requirements concern both the rise- and fall times. Transit time spread is not an issue in this type of tube. As a result of improvements in diode design, tube design and photocathode processing an extremely small pulse delay variation across the tube of only 20 ps (!) has been achieved. Rise and fall times were both in the order of 4 ns. The time resolution as a function of the # of photons per pulse is given in fig. 9. See for details ref. 6 (June 1994). This makes these devices quite suitable for time of flight detectors and for fast triggers.

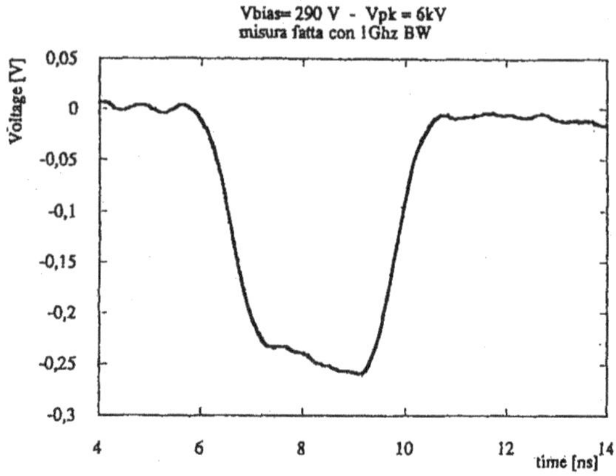

Fig 8 Pulse of Electrostat. focused HPMT (ϕ 18 mm cathode, ϕ 2 mm chip)

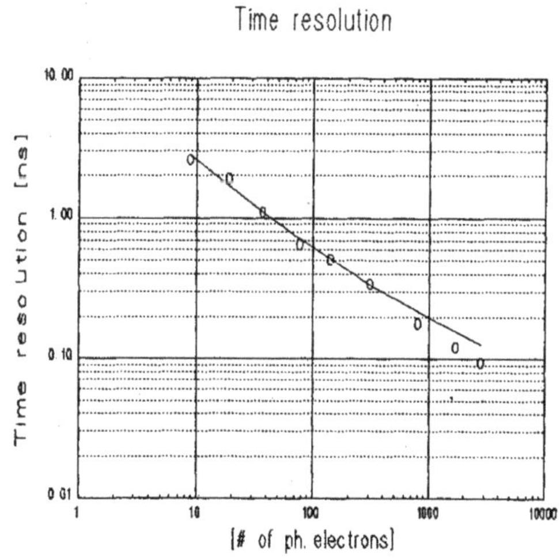

Fig 9
Time resolution versus # of photons

8 Pixellation

The collaboration over the last two years has resulted into the development of 40 mm devices with 19 hexagonal pixels. The electrostatically focused version has been tested. The main properties, as explained for the single pixel devices were maintained, while the timing was faster than expected. The pixel edge response spread was as small as 40 µm, so the crosstalk is negligible (ref. 3).
Also a 40 mm proximity focused version has been built which is presently still under test.
Fig 10 gives a view on the diode of the proximity focused version, mounted in the tube envelope. The insert shows a photo of the E-version.
25 mm proximity focused versions with 7 pixels are in preparation for mass production, prototypes are under evaluation.

Fig 10
40 mm HPMT with
19 pixels,
anode and tube

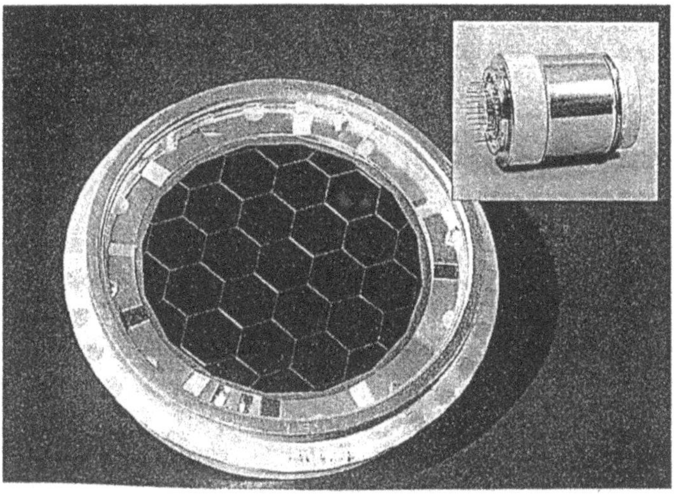

9 Commercial status

Several hundreds of tubes are produced and the devices have been proven to be a reliable and flexible tool. 18 mm electrostatically focused and 25 mm proximity focused, single pixel versions are commercially available and can be bought in larger series right now. New brochures are available, describing the tube families with the main properties. In a few months 18 mm single pixel proximity focused versions, and 25 mm proximity focused versions with 7 pixels, will be available in larger series too.

10 Future plans

Our primary emphasis is still on further improvement of the timing properties, and on further reduction of the equivalent noise for proximity focused versions. A second issue is further improvement of the so-called packing fraction, the ratio of active diameter versus outer diameter. Furthermore we are working on the integration of side electronics, like preamps, power supplies, and possibly shapers and discriminators.
Versions with avalanche diodes are not as promising as initially thought, because they suffer some severe disadvantages for physics applications (ref. 7). However, to stay informed, we do some experiments with these units. Special tubes for gamma camera's, and tubes for astronomical applications, are in study. In a larger collaboration the development of a thousand pixel version is considered, however funding is not in place yet.

Acknowledgments

The development of this technology and these devices is the direct result of a long term fruitful collaboration between CERN, INFN, Canberra and DEP. We at DEP are grateful to our partners for the endless stream of ideas, their substantial funding and the fruitful working relationship. Almost all the tube evaluation, as presented here, is done by CERN.

References

1 R. DeSalvo et al, NIM A315 (1992), 375-384
L.K. van Geest et al, NIM A310 (1991), 261-266
2 S. Basa et al, NIM A330 (1993), 93-99
3 H. Arnoudon et al, NIM A342 (1994), 558-570
4 R.J. Schomaker "Performance of the HPMT", Proc. Opto '94, Paris
5 C. D'Ambrosio et al, NIM A332 (1993), 134-141
C. D'Ambrosio et al, NIM A345 (1994), 279-283
6 P. Benetti et al. "Speed Properties of Large Surface Proximity focused Hybrid Photodiodes", CERN-LAA/HC/94-15 (1994)
7 G. Anzivino et al "Failure modes of large surface Avalanche .." CERN-LAA/HC/94-4 (1993)

THE KTEV PMT LASER MONITORING SYSTEM

Hogan H. Nguyen
Fermi National Accelerator Laboratory
Batavia, IL 60510, USA

ABSTRACT

The KTeV CsI electromagnetic calorimeter will be monitored *in situ* by a light pulser system. Pulsed light from a liquid dye scintillator, itself pumped by a Nd:YAG laser, will be injected into each PMT of the calorimeter. The system is designed to linearize the response of the combined PMT and readout electronics, as well monitor the gains of the PMT's.

1. Introduction

The KTeV (Kaons at the TeVatron) collaboration*is currently building a high-precision, pure CsI electromagnetic calorimeter[1] with sophisticated digital readout.[2] One of the goals of the calorimeter is linearity in response, i.e. $E_{true} = E_{meas}^{1-\alpha}$, where $\alpha < 0.5\%$. The purpose of the monitoring system is to measure *in situ* the combined PMT and readout electronics nonlinearities over the entire dynamic range, as well as rate-related gain drifts and other short-term drifts.

A prototype system has been installed at Fermilab. In addition, another prototype system was used during a CERN test beam run of a 25 channel system.[3] The results from these systems, concentrating on the stability and linearity measurements, are discussed in section 3.

The monitoring system will consist of light pulses, emitted from 4 liquid dye scintillators, injected into the calorimeter via 3100 quartz optical fibers. The liquid dye scintillators will be pumped by a frequency tripled Nd:YAG laser running at 5 Hz pulsed-mode. Hence, a light pulse (from the dye reemission) will be fanned-out in common to 775 channels. A liquid dye scintillator has been chosen such that its reemission matches in pulse shape to that of light from CsI, ensuring that the nonlinearities measured will be valid for real data.

The light pulses from the scintillators will be monitored independently by PIN photodiodes. These diodes have been measured to have good linearity and gain stability. A quartz filter wheel will control the light intensity common to all 4 scintillators. The intensity is such that each PMT channel will span an equivalent of 10^6 photoelectrons. Good pulse-to-pulse stability is achieved by keeping the location of the emission point (in the dye) stable in space. Hence, the measurement error from each laser pulse is limited predominantly by PMT photostatistics.

2. Design

*Chicago, Colorado, Elmhurst, Fermilab, Osaka, Rice, Rutgers, UCLA, UCSD, Virginia, and Wisconsin.

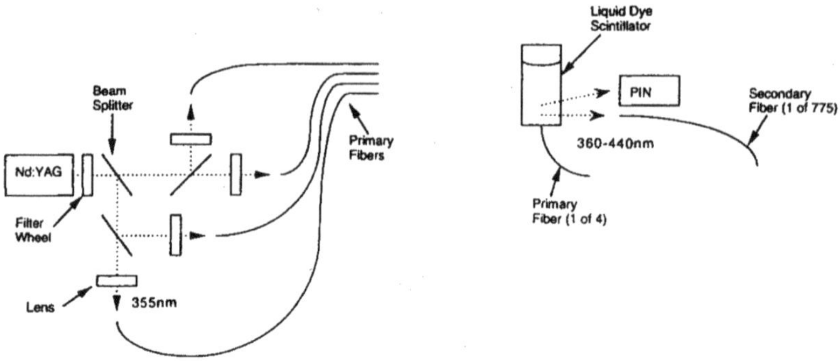

Figure 1: Conceptual Layout.

The conceptual layout is shown in figure 1. A 355 nm pulsed beam from a Continuum Surelite I f-3x Nd:YAG laser is split equally 4 ways and focused into 4 quartz *primary* optical fibers approximately 10 meters in length. 1 mm core CeramOptec Silica/Silica-Clad fibers will be used in order to handle high peak laser powers. Before the split, the beam traverses through a variable attenuation filter wheel. The wheel controls the light intensity common to all 4 fibers and allows one to 'scan' the light level over the required dynamic range. The focussing optics are 50 mm focal length lenses used to ensure efficient transfer of energy into each fiber.† The laser will run at 5 Hz pulsed mode at an energy $\approx$ 5 mJ/pulse.

Each of the 4 primary fibers terminate in a liquid dye scintillator of atleast 50 ml volume contained in a glass vessel spherical in shape. The termination is such that the fiber face will be mechanically locked and butted against the glass vessel wall. The liquid dye scintillator is Sodium Salicylate in Ethanol[4] with a concentration of 7 grams/liter. Prior to exposure to laser light, the scintillator will be bubbled with N_2 and then hermetically sealed in its vessel. The reemitted light spectrum has an average of 400 nm, a width of 40 nm, and a natural lifetime of 25.2 nsec. With this concentration, the 355 nm laser beam will be absorbed in a length of $\approx$ 1 mm. The exit angles of the rays is bounded by 15° due to the fiber critical transmission angle. Therefore the emission spot is a cone volume of 15° and hypotenus of 1 mm.

Viewing each dye will be 775 *secondary* 5 meter length, 600 μm core silica/plastic-clad-silica Spectran optical fibers. These secondary fibers will have pointing geometry and will be located $\approx$ 6 cm from the emission spot. Each of the 775 secondary fibers terminate at the back of a CsI crystal. Hence, light enters the PMT's via reflection from the internal surfaces present in the crystals. Since the reflection efficiencies and tube gains will vary channel-to-channel, the PMT charge outputs will also vary channel-to-channel. Equalizing the PMT charges can be achieved by individual adjustments of the distance between secondary fibers and the emission spot. The PMT's

†For high energy laser beams, it is important to focus the beams entirely into and illuminating 90% of the fiber core. Energy entering the cladding and jacket causes fiber damage.

also have UG11 filters which transmit light between 300-400 nm in order to block out the slow, long wavelength component of the pure CsI emission. Hence, only the dye spectrum between 360-400 nm is viewed by the PMT's.

The heart of the monitoring system are Hamamatsu R1722-02 PIN photodiodes which monitor the brightness the emission spots on a pulse-by-pulse basis. These unity gain devices will be operated with a reverse bias of -10 volts, and with a 1 kΩ series resistance to the output. These precautions serve to limit the output PIN current to $\leq$ 5 mA for a charge of 250 pC, which is important for maintaining good linearity. UG11 filters will also be used for spectral matching with the PMT's.

A crucial consideration is the readout electronics of the PIN diodes. No preamplifiers will be used. Since the PMT electronics readout will be a multi-ranging, 17-bit dynamic range device, one needs a device with comparable or better dynamic range. Although a final device has not been chosen, one that has given promising preliminary results is a 20-bit Burr-Brown DDC101 ADC operating at 1 msec gate.

3. Results and Discussions

In this section, results will be presented from a prototype setup at Fermilab, concentrating on PIN diode linearity and long term stability. Results will also be shown for another prototype system taken to CERN as part of a beam test, concentrating on how it was used for calibration.

For the Fermilab prototype, figure 2a shows the response of 2 PIN photodiodes (PIN1 and PIN2), each digitized by a LeCroy 2249W ADC, when they both view a common emission spot from a liquid dye scintillator. The light level has been varied over the range 50 to 1000 counts. The line-fit residual would show linearity better than 0.1%. However, an independent measurement of the linearity using a 2.5-to-1 technique[5] indicates a linearity of better than 0.5% for peak current < 3 mA (see figure 2b).

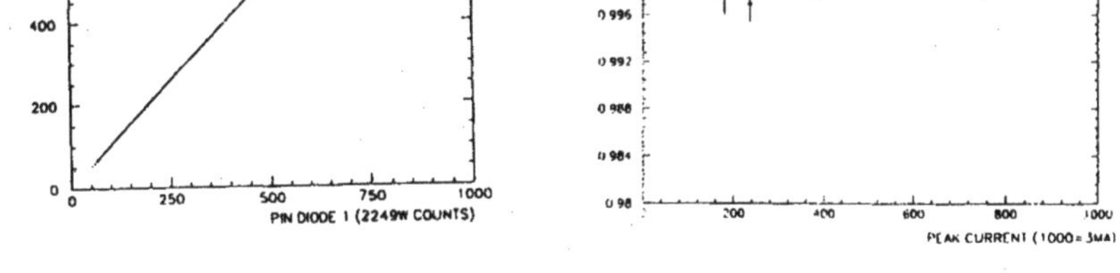

Figure 2: (a) Response of PIN2 vs. PIN1 and (b), a 2.5-to-1 measurement of the PIN nonlinearity.

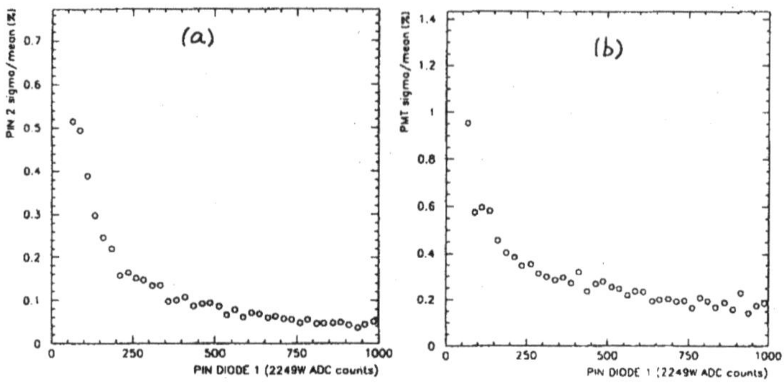

Figure 3: (a) rms(PIN2)/<PIN2> vs. PIN1 and (b), rms(PMT)/<PMT> vs. PIN1.

The accuracy from each laser pulse is reflected in the rms width of the curve in figure 2a. The smaller the percentage rms width, the more accurate the measurement. Figure 3a shows rms(PIN2)/<PIN2> as a function of PIN1. Hence, each laser pulse has a measurement error of 0.05% at 1000 PIN1 counts. The measurement error per pulse degrades at low light levels due to ADC noise and finite bin width. The photostatistical errors are negligibly small throughout the range since even at 10 2249W counts, the PIN diode measures 10^7 electron-hole pairs.

For photomultiplier tubes, the measurement errors have contribution from photostatistics. Figure 3b shows rms(PMT)/<PMT> as a function of PIN1, indicating a measurement error of 0.15% per pulse at 1000 PIN1 counts. Here, the PMT was a Hamamatsu R5364 operating at gain $\approx$ 5000 and also digitized with a LeCroy 2249W.

3.1. Stability

The stability of the system is gauged by how well two PIN diodes track other with time, i.e. for a stable system, the ratio of their signals is constant with time. An important limitation of the system dye damage due to the laser. Laser damage can be summarized as having the following effects:[6] (1) an overall decrease in the dye light output, and (2) a shift in the transmission wavelength cutoff of the solvent. The damage is expected to be worse for higher laser energy. Effect (1) is not a serious limitation. However, when the spectrum from the dye emission overlaps that of the transmission cutoff, effect (2) would cause an apparent spectral shift in the output of the dye. If the monitoring device and PMT's differ in spectral response, a shift in the dye spectrum would cause an apparent change in gain.

For the measurement of stability, two of the PIN diodes (PIN1 and PIN3) viewed the entire dye spectrum, while the other (PIN2) viewed only the spectrum between 360-400 nm. This allows sensitivity to spectral shifts in the dye. The liquid dye scintillator itself received 1 mJ/pulse. The running consisted of 30 minutes of 5 Hz running, followed by 15 minutes rest in order to observe recoveries. This cycle

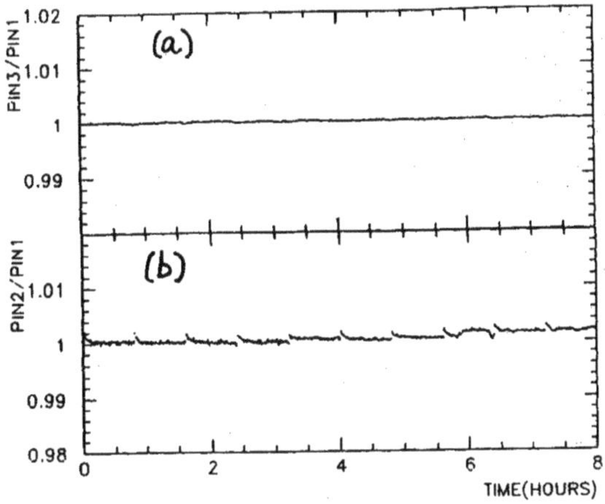

Figure 4: (a) Average PIN3/PIN1 vs. time and (b), average PIN2/PIN1 vs. time.

is repeated 10 times in one 8 hour running period. Figure 4a shows the average PIN3/PIN1 ratio as a function of time for 8 hours of running. Figure 4b shows the average PIN2/PIN1 ratio for the same running period. Based on these results, the system is stable to 0.1% over 8 hours. However, a small spectral shift is observed at the beginning of each 30 minute running period.

3.2. Operation of Device at Test Beam

At the CERN test beam, another prototype monitoring system was used for calibrating the combined PMT and prototype KTeV digitizer. While the preliminary testbeam results using this procedure have been given,[3] final details on the calibration will be published at a later time. The discussion that follows describes only the calibration procedure.

The prototype KTeV digitizer developed at Fermilab has 9 different sensitivity scales (or ranges), each twice as sensitive as its neighboring range. The device contains internal comparators which determine the range of validity for a given signal size. During the test beam, this prototype KTeV digitizer was operated such that the top of range 9 corresponds to 510 pC. Finally, an 8-bit FADC digitizes $(Q_{signal} + Q_{bias})/2^{n-1}$, where n is the range of validity, and Q_{bias} is a constant bias charge added to the signal. Ignoring nonlinearities, calibrating this device consisted of determining the sensitivity (slope) and comparator thresholds (intercepts) for each of the 9 ranges. A more complete description of the KTeV digitizer is given in the reference.[2]

Figure 5a shows the response of the 8-bit FADC versus PIN diode (digitized by a LeCroy 2249W) for a given laser scan. The lines show the FADC outputs as the range of validity switches between 6 through 9. Calibration of this device consisted of fitting for the slopes and intercepts of each of the 9 ranges. Hence, for a given FADC value and range, one then converts to a linear scale given by a PIN diode.

Since the 2249W has only 11 bits dynamic range, one lacks the ability to fully cover all 9 ranges in a single laser scan. To cover the more sensitive ranges, one

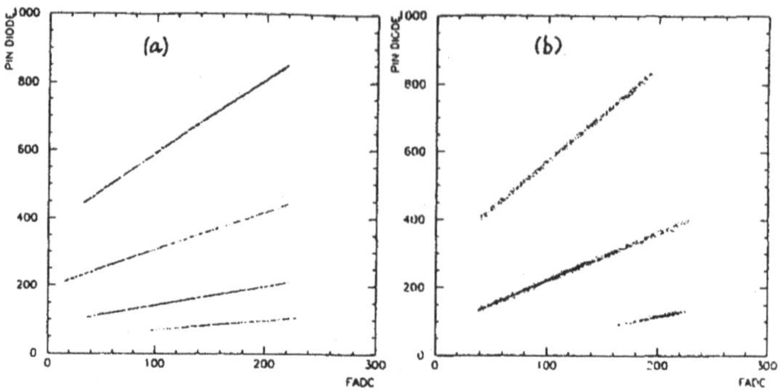

Figure 5: (a) PIN vs. FADC for ranges 6 through 9 and (b), PIN vs. FADC for ranges 2 to 4.

attenuates the light into the PMT's (using neutral density filters) *while* keeping the light into the PIN diode unattenuated. Figure 5b shows the ability of this technique to map the linearity of the lowest ranges of the KTeV digitizer. Finally, the data from scans of different filter values are merged together, thereby covering the full dynamic range of the KTeV device.

4. References

The author would like to thank co-workers Kamran Vaziri, Anatoly Ronzhin, and Yuri Pischalnikov for their contributions. Special acknowledgements is given to Anna Pla-Dalmau of the Fermilab Research Division for the sharing of her expertise and help in preparing scintillators.

1. R. Ray, *The KTeV Pure CsI Calorimeter*, these proceedings.
2. J. Whitmore, *A High Speed Digitizing Photomultiplier Tube Base for the KTeV CsI Calorimeter*, these proceedings.
3. R. Kessler, *KTeV CsI Test Beam Results at CERN*, these proceedings.
4. I. Berlman, *Handbook of Fluorescence Spectra of Aromatic Molecules*, (2nd edition, Academic Press, New York and London, 1971), p. 166.
5. *Photomultiplier Handbook, Theory, Design, and Applications*, (Burle Industries), p. 49.
6. C. Zorn, *Radiation Physics and Chemistry* Volume 41 Numbers 1/2 (1993) p. 37.

A HIGH STABILITY LIGHT EMITTING DIODE SYSTEM FOR CALIBRATION AND MONITORING OF THE LEAD GLASS ELECTROMAGNETIC CALORIMETER IN THE NOMAD EXPERIMENT AT CERN

Cern[1] – *Cosenza*[2] – *Firenze*[3] – *Moscow*[4] – *Padova*[5] – *Pavia*[6] – *Pisa*[7] Collaboration *

Presented by Dario Autiero
INFN, Sezione di Pisa and Dipartimento di Fisica dell'Universita' di Pisa,
Via Vecchia Livornese 582/A, 56100 Pisa, Italy

ABSTRACT

The Nomad (WA96) experiment at CERN is searching for $\nu_\mu \leftrightarrow \nu_\tau$ oscillations. A favoured channel for this search is the electronic decay of τ. The electromagnetic calorimeter plays then a crucial role in the Nomad detector and a precise knowledge of its energy scale is very important. The calorimeter, operating in magnetic field, is made of an array of 875 lead glass blocks and the Cerenkov light produced in the glasses is detected by photomultipliers able to work in magnetic field (tetrodes). We developed a calibration and monitoring system based on blue light emitting diodes and high stability current pulse generators. The system has to fulfill two major tasks: correct for the presence of magnetic field the calibration constants obtained at an electron test beam with no field and to monitor long term stability of the system. We aim to keep the calibration constants of the calorimeter within 1% during the full duration of the experiment.

1. Calibration of the Nomad electromagnetic calorimeter

The e.m. calorimeter is made of 875 lead glass blocks arranged in an array 25x35 † The calorimeter absolute calibration was done with 10 GeV electrons at the X5 SPS beam at CERN in october-november 1993. The blocks were calibrated in subsamples, in absence of magnetic field. The calibration and monitoring system was designed to fulfill two major tasks:

a) Translation in magnetic field of the calibration done with electrons at zero field;

b) Monitoring of the calibration constants from the calibration to the end of the experiment (more than 2 years), in order to exploit the excellent energy resolution of the lead glass counters;

with the following consequences and requirements:

*(1): G.Stefanini, R.Grabit. (2): L.La Rotonda, M.Valdata-Nappi. (3): G.Conforto, E.Iacopini, A.Marchionni, F.Martelli, M.Veltri. (4): S.Gninenko, A.Kovzelev, S.Volkov. (5): M.Baldo-Ceolin, F.Bobisut, A.Cavestro, D.Gibin, A.Guglielmi, M.Laveder, M.Mezzetto, A.Sconza, L.Visentin. (6): G.Ambrosini, P.Cattaneo, C.Conta, R.Ferrari, M.Fraternali, A.Lanza, M.Livan, D.Orestano, F.Pastore, E.Pennacchio, R.Petti, G.Polesello, A.Rimoldi, V.Vercesi. (7): D.Autiero, A.Cardini, V.Cavasinni, T.DelPrete, A.DeSanto, V.Flaminio, G.Renzoni, C.Roda.

†See Mauro Mezzetto's talk at this conference

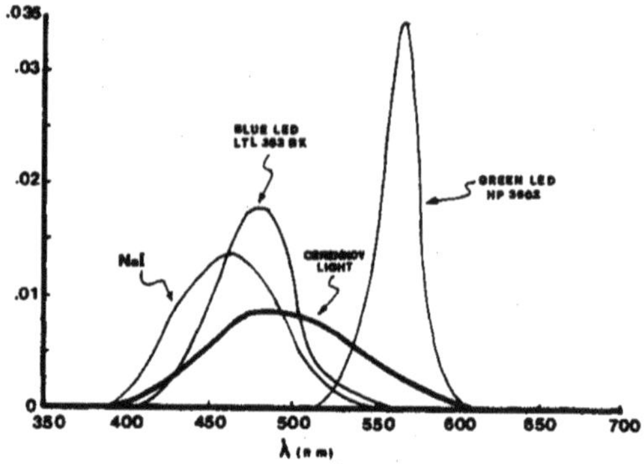

Fig. 1. Spectra from different light sources as detected by the photocathode

In order to have final calorimeter configuration completely flexible and independent of the geometry of blocks at test beam calibration, we chose to have individual light sources on each block. The alternative solution: a unique light source (laser or flasher), distributed to blocks through an optical fibre network, would have been much less flexible in mounting and dismounting the calorimeter. The effects of magnetic field have to be the same for the light generated in electron showers and for calibration light sources. Moreover the calibration system has to be intrinsically very stable to preserve the calibration from the test beam to the final installation and to guarantee no variations during the experiment. The monitoring of calibration constants requires the spectrum of the calibration light sources to be as close as possible to the spectrum of Cerenkov light detected by the tetrodes. The monitoring system has to keep under control any contribution to the stability of the final signal: ageing of paper used to wrap the glass and of the glue used to fix the tetrode to the glass, change of optical properties of the glass, change of the gain of the electronic chain (preamplifier,shaper,ADC). The gain of the electronic chain is also monitored with the help of reference charge pulses injected in the test input of the preamplifiers. The most critical component of the chain is the tetrode: in this case possible changes due to ageing or malfunctioning, involving photocathode quantum efficiency, gain and behaviour in magnetic field, have to be precisely monitored.

Given the severe requirements, the final choice was to use blue LEDs as individual light sources on each block. In Figure 1 a comparison between different light sources is shown. The spectra are presented as detected by the photocathode; the calculation includes the spectrum at the source, the glass transmissivity and the tetrode quantum efficiency. In contrast with green LEDs, the blue LEDs appear to match very well the spectrum of the Cerenkov light produced by an electromagnetic shower in the block. Blue LEDs are built by a new technology (silicon carbide) and are slower and less efficient than the green ones. Their characteristics required the use of dedicated current drivers. We used blue LEDs LTL 353BK [‡] They have a peak wavelength of 470 nm, a luminous intensity of 35 mcd at a forward current of 10 mA, and a forward

[‡]Taiwan Liteon Electronic Co. LTD, Taipei

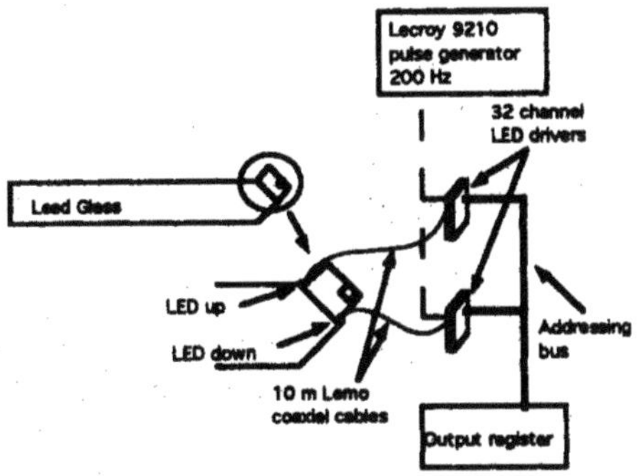

Fig. 2. Calibration system for a lead glass block

voltage of 3V at a forward current of 20 mA.

2. The LED drivers

New custom LEDs drivers were developed within a joint project with the CERN ECP division. The drivers are high stability current pulse generators especially designed for the use with blue LEDs. We set the drivers working point to generate current pulses with an amplitude of 26 mA and a duration of 220 ns. The LEDs mounted on the calorimeter were preselected in order to have a maximum difference in light yield of a factor two. The current pulses corresponding to the chosen working point give an equivalent shower energy of about 17 GeV. The stability of the driver current pulses is crucial for the stability of the emitted light (1ns variation over 220 or 1% in amplitude can change by 3% the light yield). The current pulses duration is accurately defined by using a unique external precise reference pulse distributed to all drivers. This reference pulse is generated by a Lecroy 9210 pulse generator in conjunction with a 9211 250 Mhz variable edge module. This configuration provides an accuracy of 35 ps on the duration of the pulses and several facilities like self compensation of temperature effects and self calibration of the pulser. Figure 2 shows the calibration system of one lead glass block. On each glass we mounted two blue LEDs facing the glass surface in two corners of the face where also the tetrode is glued. The LEDs are held in black plastic cases firmly fixed on the glass surface and connected to the led drivers via 10 m long coaxial cables. The two LEDs mounted on the same glass are called up and down and they provide redundancy to perform useful crosschecks. The LED drivers specifications are:

- 32 channels VME boards (VME WES V-430), current pulse generators;
- Pulse duration and frequency defined by an external timing signal ranging from $> 50ns$ to $< 1ms$;
- Pulse amplitude defined by an internal reference level adjustable from 0 to 50 mA with a precision of 10^{-4} ;

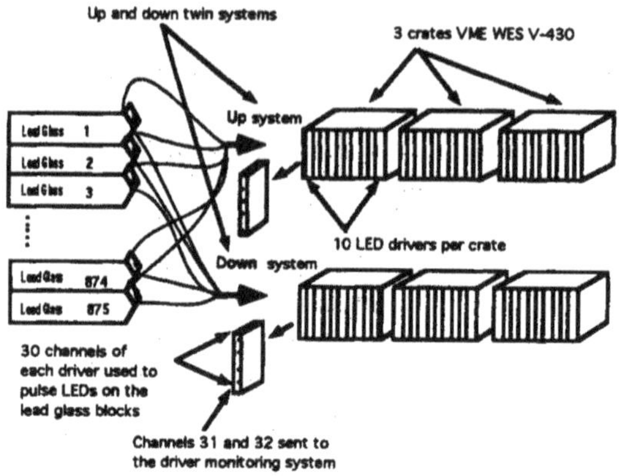

Fig. 3. Calibration system for a lead glass block

- Output for monitoring the current reference level;
- Possibility to enable whatever configuration of channels in a driver and whatever configuration of drivers;

The drivers have also some special features to optimize their stability:

a) Absorbed power is constant in time to limit rate effects and load dependance. In fact during the duration of the current pulse (220 ns) the current is absorbed by the LEDs mounted on calorimeter while for the remaining time the current is absorbed by 32 LEDs internal to the driver.

b) The drivers are temperature self stabilized. During laboratory tests no appreciable variations were seen in the signals while heating the drivers up to 60 C.

The drivers provide output current pulses which are identical to the reference pulses they receive as input; the amplitude is determined by an internal reference level common to the 32 channels. Once a timing pulse is received all the channels enabled via the VME bus produce the corresponding current pulse. The calibration system makes use of a large redundancy in order to perform crosschecks to distinguish a real variation, to be accounted in the calibration constants, from a possible malfunctioning. As shown in Figure 3 there are 2 LEDs on each block fed by two independent 'twin' drivers. In case of a real variation of the calibration constant the two LEDs have to show a common behaviour. Moreover since the duration of the pulses is defined by the timing source common to all the drivers and the amplitude is defined by a reference level common to all the 32 channels on each driver, and continuosly monitored, it is easy to disantangle possible mulfunctionings which have to be common to groups of channels. In addition, on each driver, only 30 channels are used to feed the LEDs mounted on the calorimeter while the two (again redundancy) last channels are used to feed LEDs mounted on an independent monitoring system which allows at the same time the monitoring of the LED drivers and a measurement of LEDs stability, done on a significative subsample in a completely independent way.

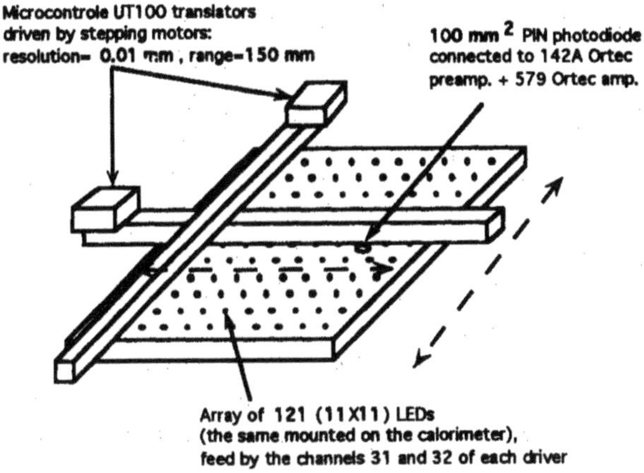

Fig. 4. XY photodiode monitoring system

These last two channels of the 60 drivers used in the calibration system feed 120 LEDs arranged in an array 11x11 scanned with only one PIN photodiode, moved by an XY motorized system with a positioning precision of 0.01mm, and connected to a different electronic chain. The use of an unique readout channel cancels possible dependance on the readout devices and allows relative comparisons between the monitored channels. The setup is shown in Figure 4.

3. System performances and conclusions

The calibration performed with no field has to be transferred when the calorimeter is in presence of the magnetic field using the LEDs system. This procedure corrects for the loss of the tetrode gain in **B** field (about 20%). This operation is possible only if the behaviour of the tetrodes in magnetic field affects in the same way the detection of the light produced in electromagnetic showers and of the light emitted by the LEDs. The effect of the magnetic field on the tetrode is mainly to make blind a fraction of the surface of the photocathode. This geometrical inefficiency can be explained by the motion of the electrons emitted by the photocathode in a strong electromagnetic static field. Since electromagnetic showers are diffuse light sources while LEDs are pointlike sources and the optical paths can be very different one can expect differences in the losses in magnetic field for the light emitted by these two classes of sources. To minimize these differences we decided to use diffusive LEDs and to mount them at the end of the blocks, near the tetrodes, to double the optical paths. As shown in Figure 5 the ratio between the loss in magnetic field measured with the electron showers and the one measured with LEDs was accurately checked at test beam. Even if the ratio is not 1 the spread of the distribution of the ratios is a measure of the precision of the correction. Using LEDs up we found a ratio of

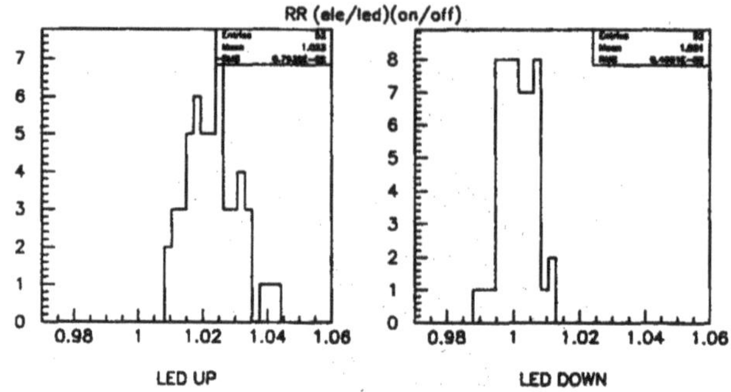

Fig. 5. Ratio Bon/Boff LED/shower

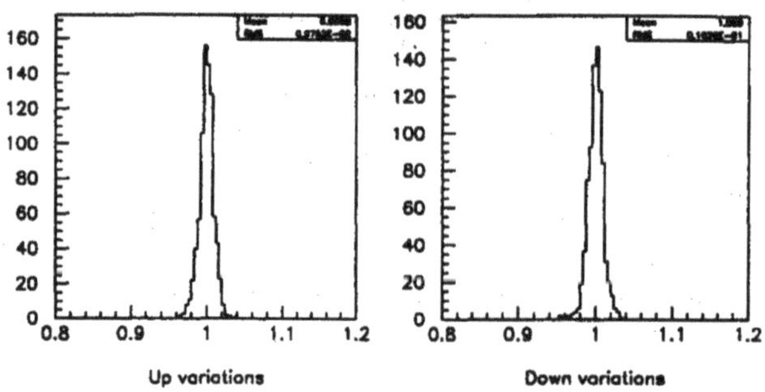

Fig. 6. LEDs stability over a period of 150 days

1.023 with a spread of 0.8%. Using LED down we measured a ratio of 1.001 with a spread of 0.5%. The difference in the ratios can be explained with the difference in the optical paths for LEDs up and down.

In the laboratory tests and during 150 days of 1994 data taking we checked carefully the intrinsic stability of our calibration system and we found that it was dominated by temperature effects on the LEDs themselves. The temperature dependance is relatively small (on average -0.3%/C) and changes from LED to LED but it is easy to control and correct for it. Temperature was then measured and controlled during test beam calibration and in the experiment and all the LEDs were characterized measuring their temperature dependance from calibration data taken at various temperatures. After correction for the temperature variations, no real variations to be accounted in the calibration constants were seen in four months of life of the experiment, with the exception of a few pathological readout channels with electronic problems (Figure 6). The system is adequate for long term monitoring of the calibration constants: over a period of 150 days the stability was within 1%. The LED calibration system allows conversion of Boff to Bon calibration constants to an accuracy of about 0.5% (using LED down).

Radiation Effects On Front-End Electronics For Noble Liquid Calorimetry*

Mauro Citterio, Sergio Rescia and Veljko Radeka
Brookhaven National Laboratory, Upton, NY 11973 USA

Abstract

Front-end electronics for liquid ionization chamber calorimetry at hadron collider experiments may be exposed to substantial levels of ionizing radiation and neutron fluences in a cryogenic environment. Measurements of devices built with rad-hard technologies have shown that devices able to operate in these conditions exist.

Si-JFETs and GaAs MESFET devices have been irradiated and tested at a stable cryogenic temperature up to doses of 55 Mrad of ionizing radiation and up to neutron fluences of $4x10^{14}$ n/cm^2. Radiation effects on DC characteristics and on noise will be presented.

I. Introduction

Fast calorimetry for hadron collider experiments will have their front-end electronics mounted on the electrodes operating in a cryogenic environment (90K for a liquid Argon calorimeter or 120K for a liquid Krypton) [1].

Simulations of radiation levels show that at this experimental location the electronics might be exposed to level of ionizing radiation of 10 kGy (1 Mrad) and neutron fluences up to $2x10^{14}$ n/cm^2 during ten years of operation at a luminosity of $2x10^{34}$ cm^{-2}s^{-1} [2, 3].

To evaluate the radiation effects in this regime of operation, special cryostats which allow cryogenic irradiation and DC and AC testing of active devices have been built.

Some commercially available Si-JFET and GaAs MESFETs have been irradiated under bias and monitored over several months at a steady cryogenic temperature. Several devices from each technology have been selected for identical characteristics and subdivided in two groups. A first group has been exposed to gamma rays using a Co^{60} source at a temperature of 90K. A second set of devices has been irradiated with high energy neutrons at a temperature of 98K using a spallation source, protons against a Li target.

At the end of the tests, a total dose up to 55 Mrad and a neutron fluence up to $4x10^{14}$ n/cm^2 have been achieved.

The goal of this research has been not only the assessment of the radiation hardness of Si-JFET and GaAs MESFETs for cryogenic operations but also the achievement of a better understanding of the physical mechanism involved with radiation induced effects in these active devices.

Radiation effects on noise power density (measured up to frequencies of 10-30 MHz), equivalent noise charge (ENC) and DC characteristics of the different technologies will be presented.

* This research was supported by the U.S. Department of Energy: Contract No. DE-AC02-76CH00016.

II. Irradiation Tests

The ionizing radiation environment in which the electronics will operate in the experiments has been reproduced using a 20 KCi Co^{60} gamma ray source at Brookhaven National Laboratory. The photons emitted by a Co^{60} source (with energy of 1.1 and 1.3 MeV) represent a reasonable approximation of the spectrum of the albedo photons within a typical calorimeter configuration in which the photon spectrum decreases rapidly with energy and has a characteristic energy of ~ 1 MeV [4].

The large neutron fluences have been simulated using a fast neutron irradiation facility at the 5.5 MV CN Van de Graaff accelerator, University of Massachusetts, Lowell [5]. This irradiation facility uses a $^7Li(p,n)^7Be$ reaction to produce neutrons in the 0 - 2 MeV energy interval ($<E>=1$ MeV), close to the neutron energy spectrum expected in the experimental conditions [4] and contains minimal portions of thermal and near thermal neutrons. The physical characteristics of the source are summarized in Table 1.

Reaction type	$^7Li(p,n)^7Be$
Proton beam energy	4 MeV
Proton beam current	~ 25 μA
Neutron fluence	~ 10^{14} n/cm^2 per day
Neutron energy spectra	0 - 2 MeV $<E> = 1.1$ MeV
Slow neutrons	(E<10KeV) < 0.05 %
γ-ray contamination	~ 10 Krad/day
Neutron Dosimetry	- ^7Be target activity - Ni-foil activation as described by ASTM E264-87

Table 1. Neutron Irradiation Facility Characteristics.

This neutron source was selected because it is practically free from any gamma ray background so that neutron effects alone could be studied.

Both the irradiation facilities allowed to carry out the irradiation studies at cryogenic temperatures, however they did not give the possibility to measure on-line the characteristics of the devices under tests. Transient effects during radiation were not investigated.

III. Gamma ray induced effects

A. Silicon JFETs

Silicon JFETs are remarkably radiation hard with respect to γ-rays. DC characteristics variations are hardly measurable up to 55 Mrad. Even the gate current, that had been generally observed to increase with radiation at room temperature, remained constant at 90K.

The only sensitive parameter to ionizing radiation is noise.

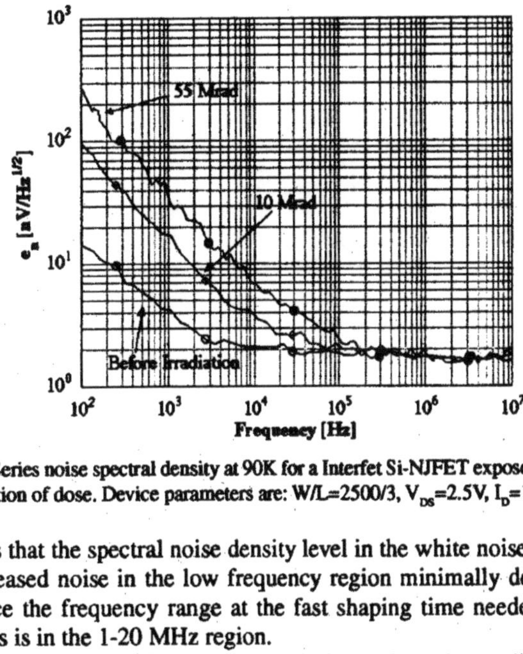

Fig. 1 Series noise spectral density at 90K for a Interfet Si-NJFET exposed to Co^{60} γ-rays as a function of dose. Device parameters are: W/L=2500/3, V_{DS}=2.5V, I_D=1mA.

Fig. 1 shows that the spectral noise density level in the white noise region is unaffected by γ-rays. Increased noise in the low frequency region minimally degrades the signal to noise ratio since the frequency range at the fast shaping time needed for calorimetry at hadron colliders is in the 1-20 MHz region.

The noise increase at low frequencies could be attributed to radiation induced single point defects in the device channel which act as generation-recombination centers for the majority carriers. The drain current fluctuations due to these centers can be represented as a sum of lorentzian terms in the series noise power density e_n [5,6]:

$$\frac{\overline{de_n^2}}{df} = \sum_{i=1}^{N} \frac{a_{Li}}{1+\omega^2 \tau_{Li}^2} \qquad (1)$$

Each term is constant up to the characteristic frequency $f_{Li}=1/2\pi\tau_{Li}$ and then rolls off as $1/f^2$; the physical nature of the defect is described by the two constants a_{Li} and τ_{Li}.

A measurement of the series noise spectral density versus temperature after a total dose of 55 Mrad of ionizing radiation has been performed. A typical result of this measurement at the temperatures of 90K and 120K is shown in Fig. 2.

In silicon JFETs the lorentzian noise components before irradiation are negligible at frequencies above a few hundred Hz, at any temperature.

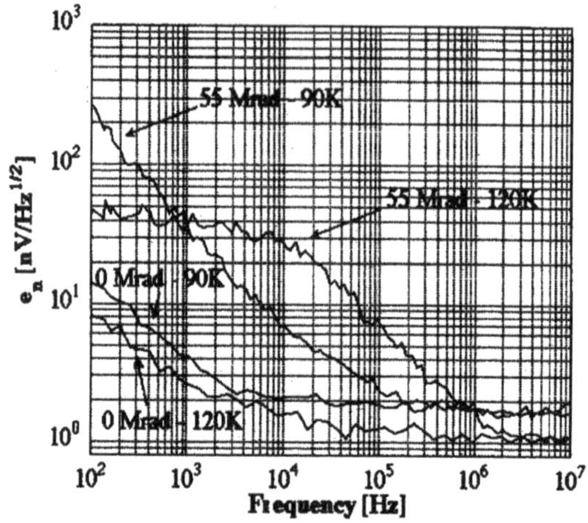

Fig. 2 Series noise spectral density at 90K and 120K for an Interfet Si-NJFET exposed to Co^{60} γ-rays before and after irradiation. Device parameters are: W/L=2500/3, V_{DS}=2.5V, I_D=1mA.

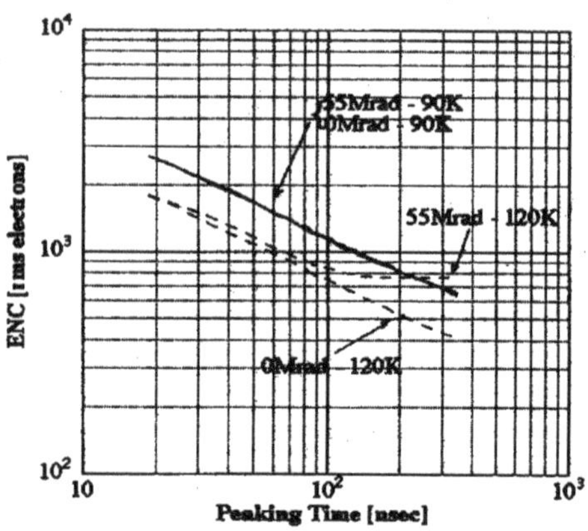

Fig. 3 Computed Equivalent Noise charge at 90K and 120K for an Interfet Si-NJFET exposed to Co^{60} γ-rays before and after irradiation. Device parameters are: W/L=2500/3, V_{DS}=2.5V, I_D=1mA, C_{iss}~10pF. For the calculation: C_D=33pF, no parallel noise contribution.

After gamma rays irradiation these noise contributions are enhanced and become the dominant component over a broader frequency range especially as the temperature increases. In the 90K to 120K temperature range the increase in noise can be attributed to one radiation generated lorentzian component with characteristic frequency lower than 100 Hz at 90K and of ~10kHz at 120K.

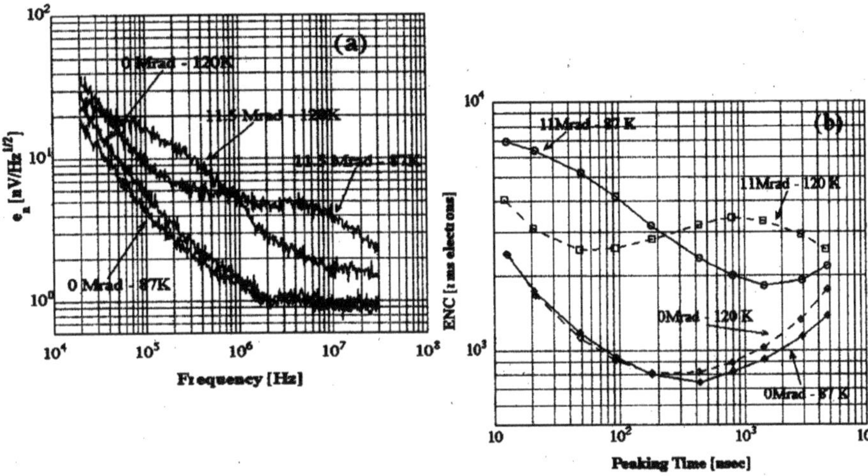

Fig. 4 Series noise spectral density (a) and equivalent noise charge (b) at 87K and at 120K for the input device of a GaAs preamplifier (Triquint process) exposed to Co^{60} γ-rays before and after irradiation. The bias current in the device is 2mA approximately.

At short processing times the noise degradation is still negligible even at 120K. A computed ENC in the case of a bipolar shaping, obtained from the series noise spectral density data of fig. 2 is shown in Fig. 3.

B. GaAs MESFETs

No degradation has been observed in the DC characteristics of GaAs devices up to the total dose of 55 Mrad.

GaAs MESFETs experience a larger increase in noise than Si-JFETs after Co^{60} exposure. Spectral noise density data show that the increase in noise is not limited to the low frequency portion of the spectra.

The radiation induced noise can be described even for GaAs MESFETs by lorentzian terms as in equation (1). The analysis of the noise versus temperature after irradiation shows that at least two lorentzian contributions, with characteristic frequencies in the region of interest, degrade the noise performances at liquid Argon and liquid Krypton temperatures (see Fig. 4a).

This hypothesis is confirmed by the analysis of ENC measurements, shown in Fig. 4b, in which the strong temperature dependency of this induced noise is shown.

The absolute noise variation as a function of dose varies with the manufacturing process and even with different runs of the same process. The radiation induced noise increase varies from a best case of 10-20% (NEC, Triquint) to a worst case of 200% (Triquint) at 90K at 40 nsec peaking time after a total dose of 2Mrad.

C. Annealing

No annealing has been observed at cryogenic temperatures over a period of several months from the conclusion of the gamma ray irradiation for the three different technologies. No recovery in the properties has been observed even after several thermal cycles between 90K and 350K.

III. Neutron Induced Effects

A. Silicon JFETs

Silicon JFETs are affected by neutrons. The lattice damage in the crystal structure generated by the neutrons affects the DC characteristics of JFETs: the pinch-off voltage as well as I_{DSS} are reduced due to the generation of acceptor-like state in the channel and the deactivation of donor atoms. The trasconductance decreases too, by about 25% at constant current at $4x10^{14}$ n/cm^2 as shown in Fig. 5.

The variation of noise as a function of the fluence is shown in fig. 6. The noise increases in the low frequency region of the spectrum and the reduction in the trasconductance causes an increase in the white noise region. This degradation in the noise translates in a 30-35% increase in ENC at 40 nsec shaping time.

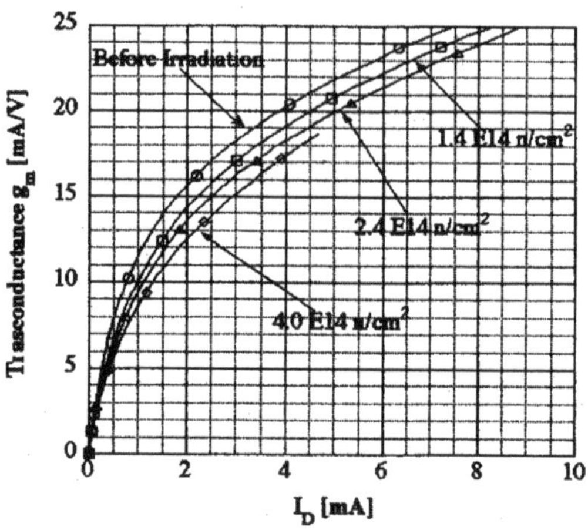

Fig. 5 Trasconductance for a Si-NJFET (Interfet) as a function of neutron fluence. The device parameters are: W/L=3350/3, V_{DS}=2.5V.

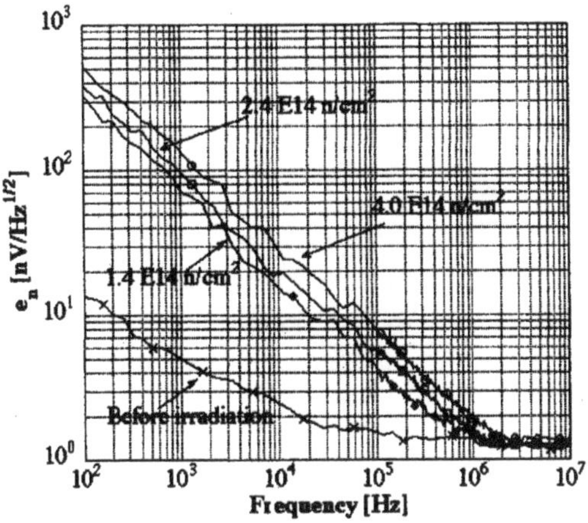

Fig. 6 Series noise spectral density at 98K for a Si-NJFET (Interfet) as a function of neutron fluence. More measurements were performed at high frequency to obtain a better resolution. The device parameters are: W/L=3350/3, V_{DS}=2.5V, I_D=1mA.

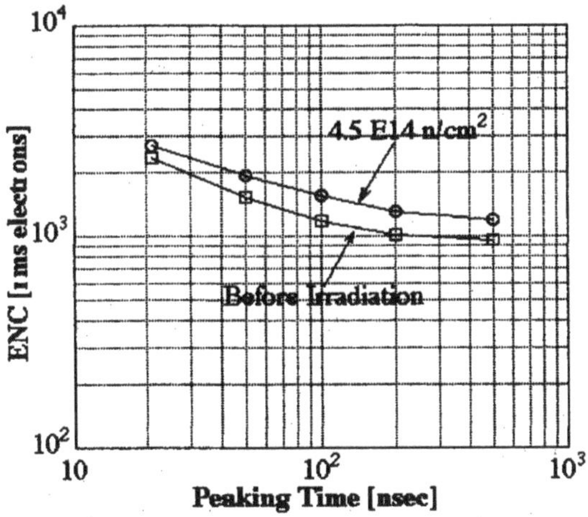

Fig. 7 Equivalent Noise Charge for GaAs preamplifier at 87K before and after the exposure to a total fluence of 4.5 E10[14] n/cm[2].

B. GaAs MESFETs

GaAs Mesfets are less affected by neutron irradiation in comparison to Si-JFETs. The geometrical position of our sample with respect to the highly collimated neutron source allowed a total neutron fluence of 1×10^{14} n/cm^2. No significant variations have been observed up to this level of fluence, both in noise and in DC characteristics.

In other tests performed at SARA, Grenoble [7], GaAs preamplifiers have been irradiated up to neutron fluence of to $4.5x10^{14}$ n/cm^2 at 87K [8]. The preamplifiers were able to withstand the test and have shown a noise increase of about 20% at 90K at 40 nsec peaking time as shown in Fig. 7.

C. Annealing

No annealing has been observed at cryogenic temperatures over a period of few months from the conclusion of the neutron exposure for the three different technologies.

No thermal cycling has been performed yet.

IV. Conclusions

The device noise has been identified as the parameter most sensitive to radiation for all the technologies under study.

No variations in the DC properties have been observed for both technologies against ionizing radiation. Si-JFETs devices show a DC characteristic variation at very high neutron fluences.

Both Si-JFETs and GaAs devices have been proved able to operate in the experimental conditions foreseen for fast calorimetry at hadron colliders. Si-JFETs are less sensitive to ionizing radiation than GaAs MESFETs. GaAs Mesfets are less sensitive to neutron irradiation than Si-JFETs.

V. References

[1] V. Radeka, S. Rescia, "Speed and Noise Limits in Ionization Chamber Calorimeters", NIM A265 (1988), pp 228-242

[2] ATLAS: Letter of Intent. CERN/LHCC/92-4, LHCC/I2

[3] GEM: Technical Design Report, GEM-TN-93-262, SSCL-SR-1219, April 30, 1993

[4] D.E. Groom, Editor, "Radiation Levels in the SSC Interaction Regions", SSC-SR-1033

[5] G.H.R. Kegel, "Fast Neutron Generation with a Type CN Van De Graaff Accelerator", NIM B40-41 (1989), pp 1165-1168

[5] K. Kandiah, M.O. Deighton, F.B. Whiting, "A Physical Model for Random Telegraph Signal Currents in Semiconductor Devices", J.Appl.Phys. 66 (2), 15 July 1989

[6] K. Kandiah, "Improvements in Low Frequency Noise of MOSFETs For Front End Amplifiers", NIM A288 (1990), pp 150-156

[7] J. Collot et al., "SARA, a Neutron Irradiation Station Dedicated to Large Hadron Collider Detector Design", Proc. of the IV International Conference on Calorimetry in High Energy Physics, Sept. 19-25, 1993, La Biodola, Isola d'Elba, Italy

[8] D. V. Camin, "Radiation Resistance of Front-End Electronics for Liquid Argon Calorimetry at LHC", Proceedings of the IV International Conference on Calorimetry in High Energy Physics, September 19-25, 1993, La Biodola, Isola d'Elba, Italy Calorimetry in High Energy Physics, September 19-25, 1993, La Biodola, Isola d'Elba, Italy

SCINTILLATORS AND FIBERS FOR CALORIMETRY

V. HAGOPIAN, M. BERTOLDI, S. JONES, M. RABALAIS
J. THOMASTON, G. TOWNSEND and H. WHITAKER
Department of Physics, Florida State University
Tallahassee, FL. 32306-3106, U.S.A.

ABSTRACT

Scintillators with Wave Length Shifting (WLS) fibers have been employed for the active region of calorimeters in high energy physics experiments. The WLS fibers allow more hermetic modules with the the photodetectors far from the scintillators. The scintillators and fibers do damage in a high radiation environment and have fewer photoelectrons than directly coupled scintillators. In addition the radiation hard fluors in WLS fibers have decay times of the order of 10 nsec.

Large calorimeters often employ scintillators and WLS fibers for the active region as the detectors can be build hermetic, the signals carried out by means of a 1 mm diameter fibers and the overall cost is lower than other similar performance technologies. The detectors can be made compensating (i.e. similar response to electrons and hadrons) and have good resolution. Several years ago, the scintillators were very radiation soft and the scintillator-fiber combination produced too few photoelectron for good energy resolution. Typical number of photoelectrons were about 1 to 2 using an ordinary bi-alkali photomultiplier tube. An extensive research effort was started to increase the photoelectrons, measure the radiation damage, response time and uniformity. Various commercial samples were tested, as well as, custom designed scintillators and fibers. In general the longer the scintillation wavelength, the more radiation hard, as the plastic base shows less opacity for longer wavelengths. This article gives results of three commercially available blue scintillator with green WLS fiber, and one special manufactured green scintillator and orange fiber.

Prior reported work[1] has shown that a rectangular scintillator with a embeded WLS fiber meets all of the calorimeter requirements, except in the very high radiation areas (more than 10 Mrad). Unfortunately the fastest WLS fiber so far is radiation soft and the search continues. The number of photoelectrons produced by a relativistic particle traveling across a 4 mm thick scintillator-fiber system is now more than 10 times the original numbers, where most of the gain is in capturing more photons from the scintillator, which also improved the uniformity across a rectangular plate. The best tile fiber combination is a blue scintillator with green WLS fiber.

A summary of the best rectangular tile fiber system follows. The system uses a WLS fiber with one end sputtered with aluminum and the fiber completely embeded in the scintillator. The other end of the WLS fiber that comes out is fused to a clear long fiber. Clear fibers are much more radiation hard and have much longer attenuation length. The tile fiber preparations are as follows:

1. Tile edges are diamond polished and coated white with an enamel based paint, such as Bicron 620.

2. The fiber groove is machined with a dental drill bit at high speed and as close to the edge as possible, typically one mm from the edge. The corner radii should not be smaller than 2.5 cm for a 1 mm fiber. The radius can be smaller for smaller diameter fibers.

3. The photon yield is proportional to the fiber diameter to the power 1.2.

4. The fiber scintillates a small amount and the coupling efficiency of the blue to green photons is lower than 10%, so the depth of the fiber can be adjusted to give uniform response across the scintillator fiber junction. For a 4 mm thick scintillator, the optimum depth for the fiber is 1.6 mm.

5. Fibers are diamond cut at the both ends and aluminum sputtered at the end that is embeded in the scintillator.

6. The scintillator fiber system is placed between two TYVEK 1055D paper.

7. Use of double clad fiber increases the light yield by about 50%.

8. Using a green extended cathode photomultiplier tube increases the light yield from the K27 fluor green WLS fiber by another 40%.

The number of photoelectrons were determined using a cosmic muon telescope where the muons were required to pass through 5 cm of lead to insure minimum ionizing relativistic velocities. The test for uniformity were performed using both X-ray excitation, and Ru^{106} source with a trigger that required the electron to pass through the tile. The two methods agreed within measurement errors. The average uniformity across a tile is better than 2%. Except for the corners of the tile, the rest of the tile uniformity is better than 1%. The decay times were measured using both cosmic muons and electrons from a Ru^{106} source with the same results.

The radiation damage was measured by irradiating the scintillators and the fibers using a 3 MeV electron accelerator. Both scintillators and fibers recover somewhat with time and the results quoted are after recovery of a month or so. The table below is a summary of three years of effort.

For the tests the tile size used were 10 cm $\times$ 10 cm $\times$ 0.4 cm; fiber: 1 mm in diameter. Fibers: Y11-200 and Orange are double clad, the rest single clad. Fluor of Y11 and BCF91A is K27, fluor of BCF92 and BCF20 is G2; K27 peak at 495 nm; G2 peak at 506 nm. Fiber is 45 cm long outside the tile. Fusing a 3 meter clear fiber reduces light yield by about 50%. SCSNxx, 3HF, Yxx and Orange fiber are made by Kuraray. BCxx are made by Bicron[2]. All quoted numbers have roughly 10% error. Light yield reduction due to radiation damage is fitted to the curve $\exp[-D/D_o]$ where D is the radiation dose and D_o is the fitted constant.

The table lists the various characteristics of the scintillators and WLS fibers. In general polyvinyl toluene scintillators such as NE110 or BC408 are more radiation soft but have light yields comparable to the BC499-52 scintillator. The fluor of BCF92 (called G2) is very radiation hard in liquid mode, but radiation soft after polymerization. The polymerization process has a major effect on radiation hardness. The paint, TYVEK paper and aluminum sputtering are radiation hard to at least 30 Mrad. The light yield has been measured by three photomultiplier tubes. R329 and R580-17 are made by Hamamatsu Corporation where R329 has an ordinary Bi-alkali cathode while R580-17 has the green extended photocathode. 8852 is a red sensitive phototube made by Burle Corporation.

Table 1. *Scintillator With WLS Fiber Performance*

Scintillator	*Fiber Fluor*	*# Photons*	*D_o Mrad*	*Decay nsec*	*Photoelectrons R329 Bi-alk*	*R580-17 Green ext*	*8852 Red*
SCSN38	Y11-200	115	4.9	10.3	15	21	8
SCSN38	BCF91A	82	5.0	10.3	11	15	6
SCSN38	BCF92	71	3.6	5.4	6	9	5
SCSN38	BCF20	71	4.1	5.4	6	9	5
SCSN81	Y11-200	92	6.5	11.1	12	17	6
SCSN81	BCF91A	66	6.6	11.1	8	12	5
SCSN81	BCF92	48	4.3	5.9	5	7	3
BC499-52	Y11-200	157	9.6	8.5	19	27	11
BC499-52	BCF91A	115	9.8	8.5	15	21	8
BC499-52	BCF92	95	5.7	4.1	8	12	7
3HF	Orange	30	8.4		-	1	2

In summary, the blue scintillator BC499-52 with the green double clad WLS fiber Y11-200 has the required characteristics for the active medium of a calorimeter. Faster and more radiation hard scintillators and fibers are still needed for the harsher environment. We would like to acknowledge Drs. A. Byon-Wagner, D. Green and J. Freeman of Fermilab and Dr. K. Johnson of Florida State University for many discussions and encouragement. We would like to specially thank Dr. Chuck Hurlbut of Bicron for many discussions and for sending us samples. This work was supported in part by the US Department of Energy, National Science Foundation REU program.

References:

1. V. Hagopian, et al, Proceedings of the III International Conference on Calorimetry in High Energy Physics, Corpus Christy, Tx. (1992), P. 575, Eds. P. Hale, J. Seigrist, World Scientific Press. (1993). V. Hagopian, et. al., Proceedings of the IV International Conference on Calorimetry in High Energy Physics, Elba, Italy (1993), P. 655, Eds. A. Menzione, A. Scribano, World Scientific Press (1994).
2. BICRON Corp. 12345 Kinsman Rd. Newbury, OH. 44065, USA; Kuraray Corp, Shuwa Higashi Taesu Bldg., 2 Chome, Chuo-Ku, Tokyo, 104, Japan.

SCINTILLATORS AND WAVELENGTH SHIFTING FIBERS IN MAGNETIC FIELDS OF UP TO 20 TESLA

V. HAGOPIAN, M. BERTOLDI, S. JONES,
J. THOMASTON, G. TOWNSEND and H. WHITAKER
Department of Physics, Florida State University
Tallahassee, FL. 32306-3106, U.S.A.

ABSTRACT

The blue scintillators SCSN81, SCSN38, BC499-52, BC404A, NE110 and the green wavelength shifting fibers BCF91A and Y11 have been placed in magnetic fields of up to 20 tesla. The fibers and scintillators were excited by Co^{60}, UV and cosmic muons. Excitations with a UV light source did not increase the light yield with increasing magnetic fields. When excited by Co^{60} gamma rays the light yield of the scintillators increased by 6% to 8% in 2 tesla magnetic field. From 2 to 20 tesla the light yields stayed about the same.

Many calorimeters in High Energy Physics use scintillators and fibers as the active medium for the detection of charged particles. It has also been known for some time that the light yield increases with magnetic field[1]. Most of the previous measurements were performed in magnets that had maximum fields of about 2 tesla. Modern calorimeters, using superconducting coils are capable of higher fields. One such detector is the Compact Muon Solenoid (CMS) that is proposed to be built for LHC at CERN. This detector is designed to have a field of 4 tesla. Since the reason of the light yield increase is not known, scintillators have been measured in higher fields. Normally in calorimeters the excitation of the scintillators is by means of charged particles. Since high energy charged particles are not easily available, the test was performed using sources that are not effected by magnetic field. In this measurement a Co^{60} source was used to excite the scintillators. In addition a short wavelength UV light was also used to excite the fluors in the scintillators and fibers. Finally a low field conventional magnet at 0.3 tesla was used to measure light yield increase using relativistic cosmic muons.

The warm coil magnet at the National High Magnetic Field Laboratory has a 51 mm bore and capable of 20 tesla field. For this test five types of scintillators and two types of wavelength shifting (WLS) fibers were used: SCSN38 and SCSN81 made by Kuraray, BC404A and BC499-52 by BICRON, NE110 by Nuclear Enterprise and the WLS fibers BCF91A made by Bicron and Y11 by Kuraray[2]. Both BCF91A and Y11 use the same fluor K27. In addition the same type scintillators and fibers were also measured in a small magnet with a maximum field strength of 0.8 tesla. In the small magnet measurements, the Co^{60} source was placed both outside and inside the magnetic field and the results were the same. The light yield increase in both magnets were the same between 0.1 and 0.8 tesla. The zero current field of the 20 tesla magnet was unknown and probably about 10 gauss.

In the small magnet, the scintillators were directly coupled to phototubes by means of UV transmitting light guides. In the 20 tesla magnet, the scintillators were coupled to the WLS fibers and read with a green sensitive Hamamatsu phototube R580-17. The length of the fibers were 4 meters with about 3.5 meters outside of the magnet. The scintillators were cut in a circle of 46 mm diameter and a 2 mm deep groove was machined near the edge. Each scintillator edge was painted white with BICRON BC620 paint that resulted in an increase of light yield of about 50%. The assembly included an 8 micro curie Co^{60} source that was sandwiched between two scintillators. An Intel 486 based computer with an ADC card read the current of the phototube (typically about 300 namp) about 30,000 times. The scintillators and fibers were also excited by a short wavelength UV light source in both the 0.8 tesla and 20 tesla magnets.

The largest source of error in these type of measurements is temperature stability of the photocathode. For blue light the stability is not very critical as the photocathode quantum efficiency is fairly stable. But for green wavelengths, such as the K27 fluor which peaks at 495 nm, the quantum efficiency of about 18% is very temperature sensitive. The best procedure is to perform the measurements quickly before the temperature has a chance to change. Figure 1 shows the light yield increase due to magnetic fields of up to 14.5 tesla. Even though the measurements were performed up to 20 tesla of magnetic fields, the shielding of the phototubes were not perfect and a 2% drop in light yield was observed that was due to the fringe field. The point to point errors are usually less than 0.5%, while the overall errors are less than 1%. Both SCSN38 and SCSN81 are polystyrene and show the same light yield increase. Repeated measurements gave very consistent results. In general terms, when excited by a Co^{60} source the light yield increase occured in the first 2 tesla or so and then saturated above that value. The curves on the figure is to guide the eye and are not fits. Several thicknesses of scintillators were used and the light yield increase was independent of thickness.

To measure the light yield increase due to the fluors, the excitation was performed by short wavelength (about 300 nm) UV light and transported by a 4 meter long quartz fiber. The quartz fiber illuminated the blue scintillator, which was coupled to the green WLS fiber. The lower portion of figure 1 shows that the light yield increase due to fluor excitation as a function of magnetic field is essentially consistent with zero ± 1% up to 20 tesla of magnetic field.

Finally as a check on the Co^{60} excitation, the small magnet aperture was opened to 6 cm and a flat scintillator was placed in the magnetic field of 0.3 tesla. The cosmic muons used for this measurement were required to pass through 5 cm of lead to ensure relativistic velocities. For calibration, another scintillator outside the magnet was also connected to the same phototube but triggered with a second cosmic muon telescope. At 0.3 tesla the light yield increase for BC404A was 3.7% by cosmic muons and 3.6% by Co^{60} source.

The tested scintillators and fiber survive well in high magnetic fields of up to 20 tesla. The light yield increases between 3% and 5% at 0.5 tesla and between 6% to 8% at 2 tesla. At higher fields the light yield essentially saturates and stays

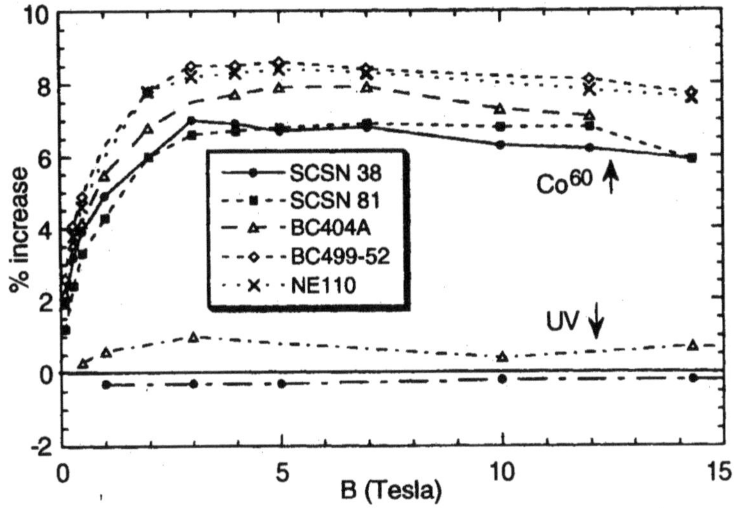

Figure 1. Relative light yield change of scintillators in magnetic fields with excitation by a) Co^{60}, upper curves b) short wavelength UV, lower curves.

constant up to highest measured field of 20 tesla. Similar measurements were performed at Fermilab in fields of up to 2.5 tesla [3]. The Fermilab and our measurements agree within the 0.5% measurement error. The light yield increase in calorimeters could very well be different as the many low energy particles could easily curl thereby increasing the path length. A test will be performed at CERN during the summer 1995. We would like to thank Prof. J. Crow, Director and Dr. S. Hannahs liason physicist of the National High Magnetic Field Laboratory (NHMFL), for the help in using the high field magnets. We also acknowledge Drs. D. Green and A. Ronzhin of Fermilab for sharing their results with us. We also acknowledge the Florida State University physicists Prof. H. Wahl, Dr. J. H. Goldman, Dr. K. Johnson and Mr. M. Rabalais for helping us at various times. The research of this project was funded partially by US Department of Energy and National Science Foundation REU program.

References:

1. S. Bertolucci, et. al., *Nuc. Instr. & Methods*, **A254**, (1987) 561; J.P. Cumalat, et. al., *ibid.*, **A293**, (1990) 606; D. Blomker, et. al., *ibid*, **A311**, (1992) 505.
2. BICRON 12345 Kinsman Rd, Newbury, Oh. 44065, USA; Kuraray Shuwa Higashi Yaesu Bldg., 2-Chome, Chuo-Ku, Tokyo,104, Japan; N.E. Tech., Bankhead Medway, Sighthill Indust Est, Edinburgh, Britain.
3. Drs. D. Green and A. Ronzhin, Fermilab, private communications.

Conference Program

Sunday, September	25th Brookhaven Center

15:00	Registration	

Monday, September 26th Berkner Hall	Antonio Ereditato (Chair), INFN - Naples Overviews

07:30	*Breakfast*	
09:00	Howard Gordon, BNL	Welcoming Remarks
09:15	Jacques Colas, LAPP - Annecy	Advances in Liquid Ionization Calorimetry
10:00	Harvey Newman, California Inst. of Technology	Crystal Calorimeters in Particle Physics
11:00	Coffee Break	
11:15	Marvin Weber, LBL	Scintillator Materials for Calorimetry

Monday, September 26th Berkner Hall	Bruno Mansoulie (Chair), Saclay Experience from Large Calorimeter Systems

12:00	Dean Schamberger, SUNY-Stony Brook	The D0 Calorimeter
12:45	*Lunch*	
14:00	Alexei Babaev, DESY	Performance of the H1 LAr Calorimeter
14:45	James Crittenden, University of Bonn	The Performance and Understanding of the ZEUS Calorimeter
15:30	*Coffee Break*	
16:00	Paolo Franzini, Univ. di Roma-"La Sapienza", Dipart Di Fisica	The KLOE Electromagnetic Calorimeter
17:00	*Buses Leave for Dinner on the Bay Mist Cruising the Great South Bay*	
21:15	*Buses Start Returning Back to BNL*	

Tuesday, September 27th BNL Hall		Craig Woody (Chair) EM Crystals
07:30	*Breakfast*	
09:00	Paul Lecoq, CERN	The Choice of a Scintillator for an Homogeneous Calorimeter in the CMS Experiment at LHC
09:30	Pierre LeComte, CERN	Test Beam Results of a Cerium Fluoride Crystals Matrix
09:50	Alexander Singovsky, IHEP - Serpukhov	Lead Tungstate Crystals and Their Application for Calorimetry at LHC
10:30	*Coffee Break*	
11:00	Steven Kettel, BNL	A Highly Efficient CsI (undoped) Endcap Veto System for BNL E787
11:20	Ronald Ray, Fermilab	The KTeV Pure CsI Calorimeter
11:40	Richard Kessler, Univ. of Chicago	CsI Test Array Results
12:00	Gerald Eigen, Cal. Inst. of Tech.	The Electromagnetic CsI (T1) Crystal Calorimeter BABAR Detector at PEP-II
12:20	*Lunch*	
13:00	**Instrumentation Tour (Sign up)**	

Tuesday, September 27th Berkner Hall		Konstantin Goulianos (Chair), Rockefeller U. Hadronic Calorimetry
14:00	Marzio Nessi, CERN	ATLAS Tile Calorimeter
14:20	Antonio Amorim, LIP/CFNUL, Lisbon	ATLAS Tilecal Testbeam Results
14:40	Jerome Schwindling, CE-Saclay	Performance of a Liquid Argon Accordion Hadronic Prototype
15:00	Qifeng Shen, Purdue University	Status of the Hadron Calorimeter in the CDF Plug Update
15:20	*Coffee Break*	
15:40	Yoshihisa Iga, DESY	Simulation of the ZEUS Calorimeter
16:00	Johannes Knapp, Enrico Fermi	A Calorimeter for Cosmic Ray Hadrons up to 10 TeV
16:20	Virgil Barnes, Purdue University	Construction and Calibration of an SDC Hadron Calorimeter Wedge
16:40	Paolo Vitulo, Univ. of Pavia and INFN	A Neural Network Algorithm to Reconstruct Neutral Showers
17:30	*Cocktails*	
18:30	*Dinner*	
20:00	*Concert*	"*The* VANGUARD" *Chamber Players*

Wednesday, September 28th		Craig Wuest (Chair),LLL
Berkner Hall		HAD/Optical Readouts, Monitoring/Radiation Damage
07:30	*Breakfast*	
08:45	Dieter Renker, PSI	Report on Tests of Avalanche Photodiode Readout for Fast Scintillators
09:10	Lieuwe Boskma, DEP/Scientific	The DEP Hybrid Photomultiplier Tube
09:30	Hogan Nguyen, Fermilab	The KTEV Laser Monitoring System
09:50	Dario Autiero, INFN-Pisa	High Stability L.E.D. System for Calibration Monitoring of the Lead Glass EM Calorimeter of WA96 Experiment at CERN
10:10	*Coffee Break*	
10:30	Carl Zorn, CEBAF	Review of Recent Results on Radiation Damage to Plastic Scintillator Based Detectors
10:50	Dimitri Seliverstov, St. Petersburg NPI	New Scintillators Based on CeF_3 Doped with Different Elements
11:10	Mauro Citterio, BNL	Radiation Effects on Front-End Electronics for Noble Liquid Calorimetry
11:30	Vasken Hagopian, Florida State University	Scintillators and Fibers for the Active Region in Calorimetry

Wednesday, September 28th	Excursion to Montauk Point
12:00	*Buses leave for Excursion to Montauk Point*
13:00	*Lunch/Wine Tour at Duck Walk Vineyards*
15:00	***1st Bus** leaves for Montauk Point Lighthouse*
17:00	***1st Bus** leaves for Gosman's Dock*
18:00	*Dinner at Gosman's Dock, Lake Montauk*

Thursday, September 29th Berkner Hall		Bernard Aubert (Chair),LAPP EM Liquid and Others
07:30	*Breakfast*	
09:00	Laurent Serin, LAL - Orsay	First Results from an Electromagnetic EndCap "Accordion" Prototype
09:20	David Lissauer, BNL	Precision Electromagnetic Calorimetry with Liquid Krypton
09:50	Mauro Savrie, INFN & U. of Ferrara	The Liquid Krypton Calorimeter of the NA48 Experiment at CERN for a Precision Measurement of ε'/ε in CP Violating $K^{\circ}\rightarrow 2\pi$ Decays
10:10	Fulvio Mauri, INFN - Pavia	ICARUS: A Second Generation Proton Decay and Neutrino Detector at Gran Sasso
10:30	*Coffee Break*	
11:00	Scott Teige, Indiana University	The Brookhaven National Laboratory E852 Lead Glass Calorimeter System
11:20	Daiva Mikunas, ANL	Test of a ZEUS BCAL Module Including Presampler and Shower Max Detector in a Low Energy Beam
11:40	Mauro Mezzetto, INFN - Padova	Test Beam Results of the WA96 Electromagnetic Calorimeter
12:00	*Lunch*	
13:00	**RHIC Collider Tour (Sign up)**	

Thursday, September 29th Berkner Hall		Robert Chase (Chair), Orsay Readout & Trigger Electronics
14:00	Daniel Camin, INFN - Milan	Monolithic Front-End Electronics for the Accordion LAr Calorimeter
14:20	Sergio Rescia, BNL	Monolithic Silicon Preamplifiers for Liquid Ionization
14:40	Christophe de La Taille, LAL - Orsay	Integrated Shapers
15:00	Wilfred Cleland, U. of Pittsburgh	Liquid Ionization Calorimetry with Time-Sampled Signals
15:20	Etienne Auge, LAL - Orsay	Analog Pipelines
15:40	*Coffee Break*	
16:00	John Parsons, Columbia U.	Design/Performance of an Analog-Memory Based Readout
16:20	Walter LeCroy, LeCroy Corp.	Current-Sampling Waveform Acquisition for Digital Oscilloscope and Detector Applications
16:40	Juliana Whitmore, Fermilab	A High Speed Digitizing Photomultiplier Tube Base
17:00	Sylvain Tisserant, Centre de Physique des Particules de Marseille	Analog Optical Links
17:20	Sridhara Dasu, U. of Wisconsin	Level-1 Trigger System for CMS Detector at Large Hadron Collider
18:00	*Cocktails*	
19:00	*Dinner at Berkner Hall*	
20:30	*E. J. Wagner, Crime Historian*	*Murder in the Laboratory*

Friday, September 30th Berkner Hall		Sebastian White (Chair), BNL EM Scintillator
07:30	*Breakfast*	
09:00	Ludwik Dobrzynski, Ecole Polytechnique LPNHE	Shashlik Type Calorimetry for LHC
09:25	E. Guschin, INR, Moscow	Multi-bundle Shashlik Towers
09:40	William Brower, Univ. of California - San Diego	Test of a High Resolution Electromagnetic Calorimeter of Wavelength Shifting Fiber Readout
10:00	David Underwood, ANL	Test of a Small Prototype STAR Calorimeter
10:15	Gary Westfall, Michigan State University	The STAR Electromagnetic Calorimeter
10:30	*Coffee Break*	
10:45	Edouard Kistenev, BNL	PbSc Electromagnetic Calorimeter for the PHENIX Experiment at RHIC (Results of the Beam Test)
11:10	Will Brooks, CEBAF	An Electromagnetic Shower Calorimeter with Stereo Fiber-Optic Readout for a Large Acceptance Spectrometer at CEBAF
11:30	Antonio Capone, U. di Roma - La Sapienza	Chorus Collaboration on the Calorimeter
11:50	Guido De Zorzi, U. di Roma - La Sapienza, INFN	Performance of a Scintillating Fibers Semiprojective Electromagnetic Calorimeter
12:30	*Lunch*	
13:15	**RHIC Test Facility Tour (Sign up)**	

Friday, September 30th		Richard Wigmans (Chair), Texas Tech University
Berkner Hall		HAD/Forward
14:30	Fabio Garufi, INFN - Napoli	Test Beam Results from a Very Forward Liquid Scintillator Calorimeter for the LHC
14:50	Gerhard Mueller, DESY	A Spaghetti Calorimeter for the H1-Detector
15:10	Simon Bennett, Wayne State University	The E864 Scintillating Fiber Hadronic Calorimeter
15:30	Philippe Gorodetzky, CERN	CMS Silica Fiber Calorimetry. The Solution for Very Forward Calorimeters
15:55	*Coffee Break*	
16:20	Alexander Savine, U. of Arizona	Results from Electron Testbeam Experiments for a Liquid Argon/Brass Tube Calorimeter
16:40	Sergei Denisov, IHEP - Protvino	Recent Results in Gas Calorimetry at IHEP
17:00	David Khazins, Rockefeller U.	Beam Test of a Hadron Calorimeter Made of High Pressure Gas-Ionization Tubes
17:20	Nikos Giokaris, Rockefeller U.	Recent Results on High Pressure Gas Calorimetry
18:00	*Barbeque at*	

Saturday, October 1st		Giorgio Apollinari (Chair), Rockefeller University
Berkner Hall		EM Preshower
07:30	*Breakfast*	
09:00	Johann Collot, ISN - Grenoble	A Fine Grained Liquid Argon Preshower Detector
09:20	Isabelle Wingerter-Seez, LAPP/Annecy	Performance of an Accordion Calorimeter Equipped with Integrated UV Strips
09:40	Jon Eisenberg, U. of Washington	Beam Test of a Noble Liquid "Preshower" Detector for Small π^0 Identification
10:00	*Coffee Break*	
10:30	Vincent Hedberg, U. of Lund	The New DELPHI Small Angle Calorimeter
10:50	Qinfang Wang, Rockefeller University	PAD: Pixel Array Detector for Forward Calorimetry
11:10	Jean Ernwein, CE - Saclay	Shower Maximum Detector for SDC Calorimetry
12:00	*Lunch*	

LIST OF PARTICIPANTS

Amorim, A.	LIP/CNFUL – Lisbon, Portugal
Apollinari, G.	Fermilab, USA
Aubert, B. L.	LAPP, France
Auge, E.	LAL – Orsay, France
Autiero, D.	INFN – Pisa, Italy
Babaev, A.	DESY, Germany
Barnes, V. E.	Purdue University, USA
Bennett, S. J.	Wayne State University, USA
Blankman, A.	LeCroy Corporation, USA
Boskma, L.	DEP/Scientific, Netherlands
Brooks, W.	CEBAF, USA
Brower, W. S.	University of California – San Diego, USA
Camin, D. V.	University of Milano/INFN, Italy
Capone, A.	University di Roma – La Sapienza, Italy
Chase, R. L.	LAL – Orsay, France
Chevalley, J.-L.	CERN, Switzerland
Citterio, M.	BNL, USA
Cleland, W. E.	University of Pittsburgh, USA
Colas, J.	LAPP – Annecy, France
Collot, J.	Institut des Sciences Nuclearires – Grenoble, France
Cormier, T. M.	Wayne State University, USA
Costa, G.	INFN – Milano, Italy
Crittenden, J. A.	University of Bonn, Germany
Dasu, S. R.	University of Wisconsin, USA
Davison, P. W.	Thorn EMI Electron Tubes, USA
de La Taille, C.	LAL – Orsay, France
De Zorzi, G.	University di Roma – La Sapienza, INFN, Italy
Denisov, S.	IHEP – Protvino, Russia
Dobrzynski, L.	Ecole Polytechnique, France
Eek, L.-O.	Royal Institute of Technology, Sweden
Eigen, G.	California Institute of Technology, USA
Eisenberg, J. K.	University of Washington, USA
Elias, J. E.	Fermilab, USA
Ereditato, A.	CERN, Switzerland
Ernwein, J.	CE – Saclay, France

Evangelist, L. A.	ELECTRON Tubes Inc., USA
Franzini, P.	Laboratori Nazionali di Frascati, Italy
Garufi, F.	INFN – Sezione di Napoli, Italy
Gedcke, D.	EG & G Nuclear Instruments, USA
Giokaris, N. D.	Fermilab, USA
Glebov, V.	BNL, USA
Gorbics, M. S.	LeCroy Corporation, USA
Gordon, H. A.	BNL, USA
Gorodetzky, P.	CERN, Switzerland
Goulianos, D.	Rockefeller University, USA
Guschin, E.	INR – Moscow,
Hagopian, V.	Florida State University, USA
Hallman, T. J.	LBL – Berkeley, USA
Hanson, T.	NE Technology Limited, Scotland
Hedberg, V.	CERN, Switzerland
Hergert, E. D.	Hamamatsu, USA
Hrisoho, A.	LAL – Orsay, France
Hsiung, Y. B.	Fermilab, USA
Iga, Y.	DESY, Germany
Kessler, R. S.	The University of Chicago, USA
Kettell, S. H.	BNL, USA
Khazins, D.	Rockefeller University, USA
Kistenev, E. P.	BNL, USA
Knapp, J.	Enrico Fermi Institute, USA
Komatsubara, T. K.	BNL, USA
Korbel, V. M.	DESY, Germany
Kotcher, J.	BNL, USA
Kusner, M.	BICRON, USA
Lanni, F. M.	INFN – Milano, Italy
Lecomte, P. M.	CERN, Switzerland
Lecoq, P.	CERN, Switzerland
Lecroy, W.	LeCroy Corporation, USA
Lee-Franzini, J.	Laboratory Nazionali Di Frascati, Italy
Lefebvre, M.	LAL – Orsay, France
Lewis, D.	BICRON, USA
Lissauer, D. A.	BNL, USA
Ma, H.	BNL, USA
Majka, R.	Yale University, USA
Makowiecki, D. S.	BNL, USA

Mansoulie, B. P.	CE – Saclay, France
Mauri, F.	INFN – Pavia, Italy
McDonough, J.	EG & G Nuclear Instruments, USA
McAlpine, R.	THORN EMI Electron Tubes, England
Mezzetto, M.	INFN – Padova, Italy
Mikunas, D. V.	Argonne National Lab., USA
Miller, L.	Phillips Scientific, USA
Mueller, G.	DESY, Germany
Nakamura, M.	Hamamatsu, USA
Nessi, M.	CERN, Switzerland
Newman, H.	California Institute of Technology, USA
Nguyen, H. H.	Fermilab, USA
Oxford, R. P.	NE Technology, Inc., USA
Parsons, J.	Columbia University, USA
Phillips, T.	Phillips Scientific, USA
Radeka, V.	BNL, USA
Ray, R. E.	Fermilab, USA
Renker, D.	Paul-Scherrer-Institut, Switzerland
Rescia, S.	BNL, USA
Samsonov, V.	St. Petersburg NPI, Russia
Savine, A.	University of Arizona, USA
Savrie, M.	INFN – Universita Di Ferrara, Italy
Schamberger, D. R.	SUNY at Stony Brook, USA
Schiller, N. H.	Hamamatsu Corporation, USA
Schwindling, J.	CE – Saclay, France
Seemungal, W. A.	Philips Photonics, USA
Seliverstov, D.	St. Petersburg NPI, Russia
Serin, L.	LAL – Orsay, France
Shaevitz, M.	NEVIS Laboratories, USA
Shah, G. G.	NE Technology, Inc., USA
Shen, Q.	Fermilab, USA
Singovski, A.	IHEP – Serpukhov, Russia
Skuja, A.	University of Maryland, USA
Stumer, I.	BNL, USA
Swirnow, W. M.	LeCroy Corporation, USA
Takai, H.	BNL, USA
Teige, S. W.	Indiana University, USA
Tisserant, S.	Centre de Physique des Particules de Marseille, France
Underwood, D. G.	Argonne National Lab., USA

Vitulo, P.	University of Pavia and INFN, Italy
Wang, Q.	Rockefeller University, USA
Weber, M. J.	Lawrence Berkeley Laboratory, USA
Westfall, G. D.	Lawrence Berkeley Laboratory, USA
White, S. N.	BNL, USA
Whitmore, J.	Fermilab, USA
Wigmans, R.	Texas Tech University, USA
Wingerter-Seez, I.	LAPP – Annecy, France
Woody, C.	BNL, USA
Wuest, C. R.	LLNL, USA
Yost, B.	LeCroy Corporation, USA
Ziemer, E. A.	Ziemer & Associates, Inc., USA
Zorn, C. J.	CEBAF, USA

AUTHOR INDEX

Errata

Proceedings of the Fifth International Conference on Calorimetry in High Energy Physics

p. vi, line 2 from bottom; Howard J. Gordon should read as Howard A. Gordon

Please note:
The following last two pages of the article "Pixel Array Detector for Forward Calorimetry", pp. 394–397 were left out at the time of printing.

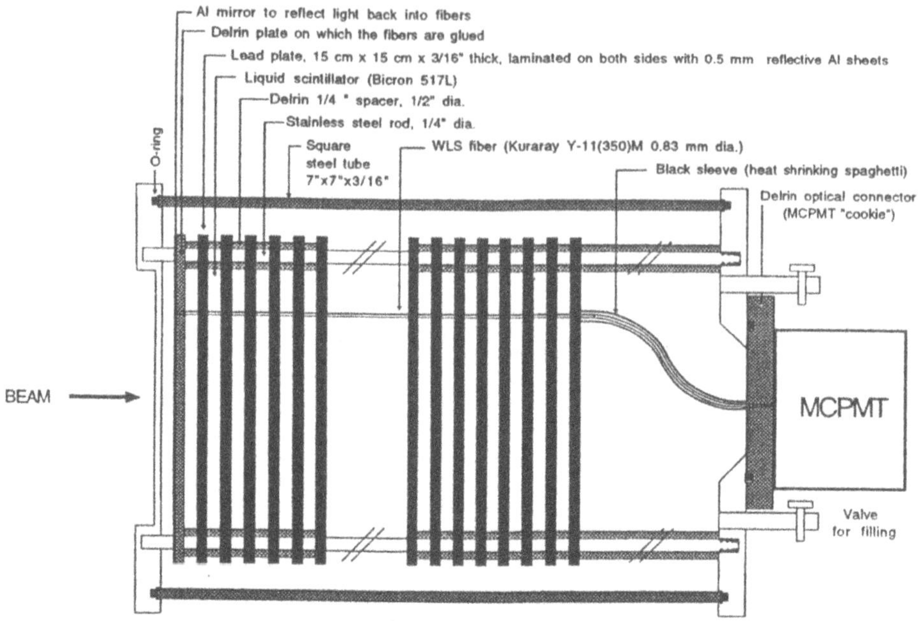

FIG. 1: Schematic side view of the PAD prototype

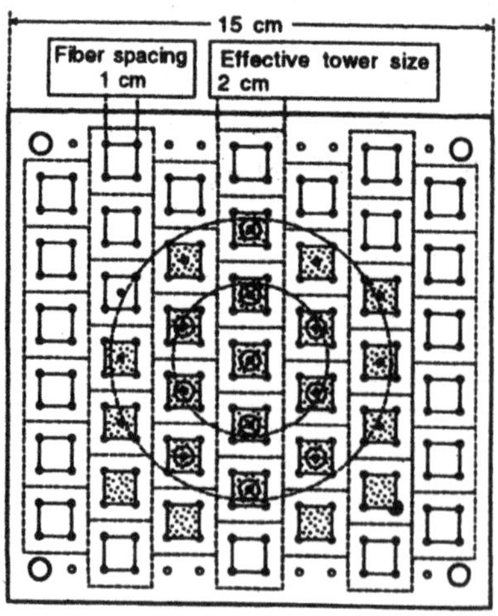

Figure 2: Front view of PAD prototype.

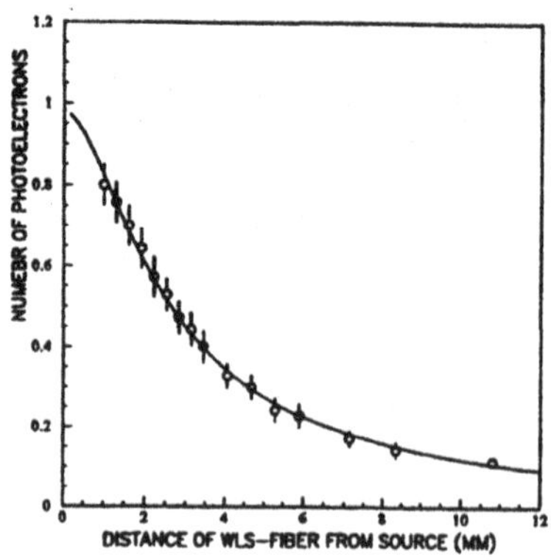

Figure 3: Bench measurement light yield results.

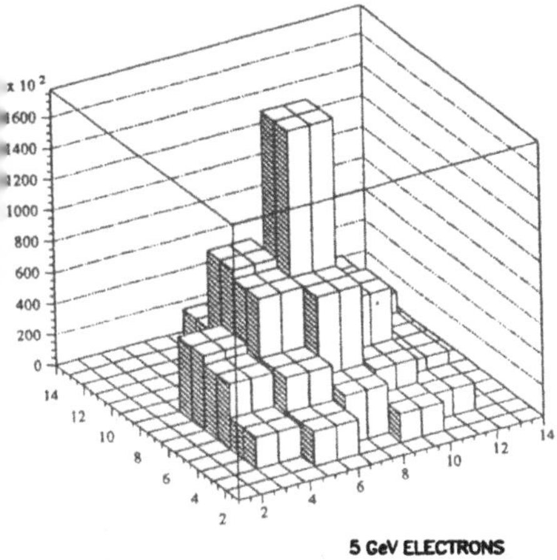

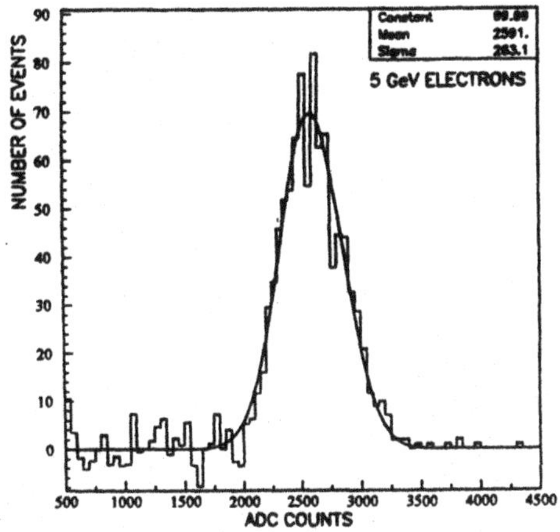

Figure 4: *left* LEGO plot of tower ADC count for 5 GeV electrons.

Figure 5: *right* Tower-sum ADC count distribution for 5 GeV electrons.

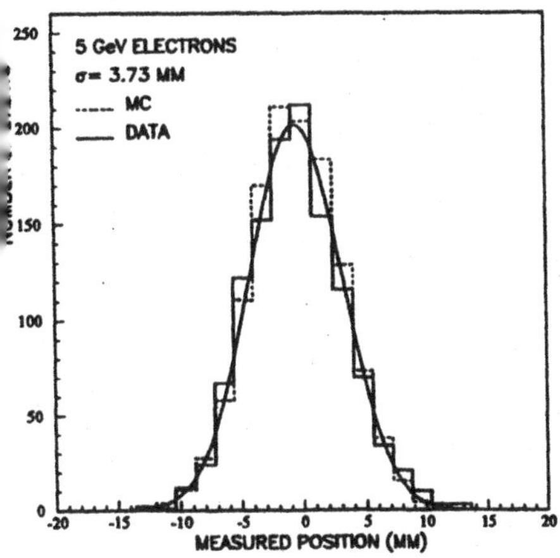

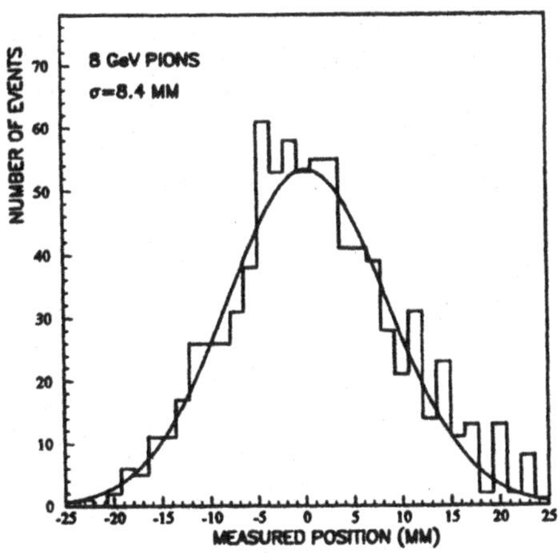

Figure 6: *left* Number of events versus position (in one dimension) measured from the centroid of the energy deposited in the calorimeter for 5 GeV electrons triggered with a 6.4 mm wide scintillation counter.

Figure 7: *right* Number of events versus position (in one dimension) measured from the centroid of the energy deposited in the calorimeter for 8 GeV π^- triggered with a 6.4 mm wide scintillation counter.

www.ingramcontent.com/pod-product-compliance
Lightning Source LLC
LaVergne TN
LVHW051924120926
847902LV00009B/69